APPLICATIONS OF MATHEMATICAL HEAT TRANSFER AND FLUID FLOW MODELS IN ENGINEERING AND MEDICINE

Wiley-ASME Press Series List

APPLICATIONS OF MATHEMATICAL HEAT TRANSFER AND FLUID FLOW MODELS IN ENGINEERING AND MEDICINE

Abram S. Dorfman

University of Michigan, USA

This Work is a co-publication between ASME Press and John Wiley & Sons, Ltd.

Registered office
John Wiley & Sons Ltd, The Atrium, Southern Gate, Chichester, West Sussex, PO19 8SQ, United Kingdom

For details of our global editorial offices, for customer services and for information about how to apply for permission to reuse the copyright material in this book please see our website at www.wiley.com.

Library of Congress Cataloging-in-Publication Data

Names: Dorfman, A. Sh. (Abram Shlemovich), author.
Title: Applications of mathematical heat transfer and fluid flow models in
    engineering and medicine / Abram S. Dorfman.
Description: Chichester, UK ; Hoboken, NJ : John Wiley & Sons, 2017. |
    Series: Wiley-asme press series | Includes bibliographical references and
    index.
Identifiers: LCCN 2016033003 (print) | LCCN 2016049187 (ebook) | ISBN
    9781119320562 (cloth) | ISBN 9781119320739 (pdf) | ISBN 9781119320746
    (epub)
Subjects: LCSH: Heat–Transmission–Mathematical models. | Fluid
    mechanics–Mathematical models.
Classification: LCC QC320 .D59 2017 (print) | LCC QC320 (ebook) | DDC
    621.402/2–dc23
LC record available at https://lccn.loc.gov/2016033003

A catalogue record for this book is available from the British Library.

Set in 10/12pt, TimesLTStd by SPi Global, Chennai, India.

Printed and bound in Malaysia by Vivar Printing Sdn Bhd

10 9 8 7 6 5 4 3 2 1

In memory of my favorite co-author,
who passed away too soon,
Professor of University of Toledo,
Ella Fridman.

Contents

Part II   APPLICATIONS IN FLUID FLOW

Part III   FOUNDATIONS OF FLUID FLOW AND HEAT TRANSFER

Series Preface

The *Wiley-ASME Press Series in Mechanical Engineering* brings together two established leaders in mechanical engineering publishing to deliver high-quality, peer-reviewed books covering topics of current interest to engineers and researchers worldwide. The series publishes across the breadth of mechanical engineering, comprising research, design and development, and manufacturing. It includes monographs, references, and course texts. Prospective topics include emerging and advanced technologies in engineering design, computer-aided design, energy conversion and resources, heat transfer, manufacturing and processing, systems and devices, renewable energy, robotics, and biotechnology.

Applications of Mathematical Heat Transfer and Fluid Flow Models in Engineering and Medicine

No problem can be solved from the same level of consciousness that created it.

Albert Einstein

Preface

This textbook for advanced graduate and post-graduate courses presents the applications of the modern heat transfer and fluid flow mathematical models in engineering, biology, and medicine. By writing this work, the author continues the introduction of brand-new efficient methods in fluid flow and heat transfer that have been developed and widely used during the last 50 years after computers became common. While his previous two monographs presented these contemporary methods on an academic level in heat transfer only [119] or in both areas heat transfer and fluid flow on the preliminary level [121], this manual introduces the modern approaches in studying corresponding mathematical models—a core of each research means determining its efficiency and applicability. Two types of new mathematical models are considered: the conjugate models in heat transfer and in fluid flow and models of direct numerical simulation of turbulence. The current situation of applications of these models is presented in two parts: applications of conjugate heat transfer in engineering (Part I) and applications of conjugate fluid flow (peristaltic flow) in medicine and biology and applications in engineering of direct numerical simulation (Part II). These parts contain theory, analysis of mathematical models, and methods of problem solution introduced via 134 detailed and 231 shortly reviewed examples selected from a list of 448 cited original papers adopted from 152 scientific general mathematic, computing, and different specific oriented journals, 42 proceedings, reports, theses, and 37 books. This list of 448 references comprehends the whole period of the new methods existing from the 1950s to present time, including more than 100 results published during the last 5 years among which more than half of the studies were issued in 2014 to 2016.

The term conjugate, or coupled problem, was coined in the 1960s to designate the heat transfer strict investigation that requires matching temperature fields of bodies flowing around or inside the fluids. Later on, it became clear that these terms and procedures are important to many other natural and technology processes, consisting of interactions between elements and/or substances. In particular, the peristaltic flow is an inherent conjugate phenomenon because such flow occurs due to interaction between the elastic channel walls and the fluid inside channel. The conjugate formulation reflects the basic features of a studied phenomenon. Due to that, the models of this type are reliable and significantly improve the correctness and physical understanding of the results. Conjugate methods constitute a powerful tool for solving contemporary problems, substituting the previous approximate approaches. At the same time, it is important to know when the common more simple approaches may be used with comparable exactness, instead of more complex conjugate procedures. The textbook answers this question, as well as significant other questions governing the applications of conjugate methods.

The other group of new methods considered in this text is based on direct numerical solution of exact unsteady (without averaging) Navier-Stokes equations. Because the unsteady Navier-Stokes equations describe the complete space- and time-dependent field of turbulent flow, the results of direct numerical simulation are considered as an experimental data gained computationally. Such results provide highly accurate instantaneous turbulence characteristics giving further insight into physics of turbulence, opening new possibilities, fresh ideas and improving applications.

The discussion goes along with 239 exercises and 136 comments. Whereas the former allowed the reader to improve his or her skills and experience, the latter are used to clear up specific terms and to note some instructive historical facts. The majority of exercises are used by the author to divide the derivation of particular expressions or formulae with a reader. To realize such an offer, the way of solution and the result are given in the text. However, the mathematical procedure is left for the reader as an exercise. Such a type of exercise gives a person a choice to be satisfied only by results, or use the suggested drill to improve their own expertise. For convenience, it is pointed out in the text when each exercise should be performed, and to find a specific one, the reader may apply the contents where the locations of exercises are indicated. Also, for the reader orientation, the more sophisticated exercises and examples (and hence, corresponding publications) are marked by an asterisk (*).

As mentioned above, comments provide significant information required for understanding. Such valuable subjects as, for example, special means, like tridiagonal matrix algorithm (TDMA), or alternating direction implicit method (ADI), or scientific terms, such as order of the value of magnitude, function singularities, tensor or factor of nonisothermicity, are explained via comments incorporated in the text. Meaningful historical notes are also introduced through comments at the relevant manual points. Thus, after discussing the benefits of the boundary layer theory, it is noticed that boundary-layer methods was not utilized for the first 25 years until Prandtl's lecture at the Royal Aeronautical Society meeting in 1927. The other examples of historical notes given by comments are explanations of the name BBO of differential equation, the Saffman slip boundary condition, the Paul Erlich role in monoclonal antibodies, and the Smagorinsky contribution in the direct numerical simulation of turbulence.

In view of the intended audience, special attention is given to the balance between strictness and comprehensibility of the writing. Such a compromise is realized using a strict formulation of the problems on one side and the detailed explanation of definitions, special terms and procedures on the other. For example, it is justified that both problems—heat transfer of flow past a body and peristaltic flow in a flexible channel—are similar, and both are inherently conjugate. At the same time, it is explained in detail why a nonlinear model of peristaltic flow differs in essence from a linear heat transfer pattern.

In contrast to exciting college courses on heat transfer presenting basically simple empirical approaches based on the heat transfer coefficients, the conjugate methods are grounded on contemporary fluid flow and heat transfer models. Therefore, to help the reader to understand the conjugate principles and procedures, the third part of this textbook offers fundamental laws of laminar and turbulent fluid flows and applied mathematic methods frequently used in engineering (Part III). Setting subsidiary chapters behind the body text, it is assumed that the reader takes a relatively small part of the information required to understand only a specific thesis or topic, rather than studying the whole subject in advance. In addition, the references given in the text in the form: Chap. (Chapter), S. (Section), Exam. (Example), Exer. (Exercise), and Com. (Comment) help the reader to find directly the desired portion of knowledge. Such a book structure permits the reader to get explanation step by step during studying. At the same time, an experienced person may read the text ignoring those citations.

As a whole, the textbook is written so as to be usable to senior and post-graduate students and engineers with the prerequisites of calculus, fluid mechanics, and heat transfer college engineering courses.

The textbook begins with Part I presenting applications in heat transfer, which starts with an introduction containing two pieces. The first writing, "When and why conjugate procedure is essential" explains in detail where the term conjugate came from, what it means, and in which cases conjugation procedure is important. The second piece entitled "A core of conjugation" presents the qualitative analysis of a simple problem of heat transfer from a plate heated from one end. This assay clarifies a physical meaning of the conjugation principle by showing the contrasting distributions of the heat transfer characteristics on the interface in two flow directions, from heated and unheated ends.

This part consists of four chapters, incorporating the theory of conjugate heat transfer based on universal functions (Chapter 1) and three chapters of applications: universal function applications (Chapter 2), conjugate problem applications in flows around bodies and inside channels (Chapter 3), and special application of conjugate heat transfer models in industrial and technological processes (Chapter 4).

The first chapter begins from the formulation of conjugate heat transfer problems specifying two sets of equations, the initial and boundary conditions governing the conjugate problem for a body and fluid. Each equation, such as Navier-Stokes or Laplace equation, is followed by references to chapter or section from the third part, presenting appropriate explanation. The initial conditions, the three kinds of common boundary conditions, and the Dirichlet and Neumann problem formulation for elliptic differential equations are considered in detail. The conjugate conditions on fluid/body interface (fourth kind boundary condition) and specific methods for performing the conjugate procedure are discussed also.

The next section introduces the universal functions that are widely used in this text. It explains what universal functions are, and shows that these types of functions are proper and convenient tools for nonisothermal and conjugate heat transfer analyzing. Two forms of universal functions, integral and differential, are employed.

First, the special form of Duhamel's integral containing influence function is derived which, in fact, presents a universal function for the heat transfer on a plate with arbitrary temperature and zero pressure gradient flow.

Then, the equivalent differential universal function, in the form of a series of temperature head derivatives, is obtained by the consecutive differentiation of the Duhamel's integral. The calculation data for the series coefficients and for the exponents of influence function in Duhamel's integral conclude the determination of the universal functions for laminar flows.

Because the universal functions are valid in the same form for other regimes and conditions, the remaining part of the first chapter specifies only the series coefficients and the appropriate exponents of the influence function for differential and integral forms of universal functions. These results are obtained for the following cases: turbulent flow, compressible zero pressure gradient flow, power-law Non-Newtonian fluids, moving through surrounding continuous sheet, plate with unsteady arbitrary temperature distribution, flow past axisymmetric body, inverse universal functions for arbitrary heat flux distribution, and functions for recovery factor.

Chapter 2 provides applications of universal functions. General properties of conjugate heat transfer are investigated, considering the conjugate problem as a case of heat transfer from arbitrary nonisothermal surface. The results are obtained analyzing universal functions and are supplemented with relevant examples. It is found that: (i) the second term of series with the first derivative in differential universal function basically determines the effect of the temperature head gradient because the first coefficient of series is from 3 to 10 times larger than the second one, whereas the others are negligible small, (ii) because the first coefficient is positive, the increasing temperature heads (positive derivative) leads to greater and the decreasing temperature heads (negative derivative) results in lesser heat transfer coefficients than that for an isothermal surface, (iii) strikingly large effects, resulting in zero heat transfer if the negative derivative is large or the surface is long enough, (iv) the positive and negative pressure gradients respectively decrease and increase the heat transfer coefficient of nonisothermal surface, (v) the higher the Prandtl number is, the less the effect of nonisothermicity in turbulent flow is, and the higher the Prandtl and Reynolds numbers, the less the effect of nonisothermicity in turbulent flow is, (vi) the effect of nonisotermicity caused by variable time temperature is greater than that of variable space temperature, and (vii) the Biot number specifies the degree of problem conjugation and shows that in both limiting cases, $Bi \to \infty$ and $Bi \to 0$ conjugate problem decays, so that the greatest effect of conjugation occurs at comparable body/fluid resistances at $Bi \approx 1$.

The second part of Chapter 2 describes some inherent characteristics and phenomena for conjugate heat transfer, indicating that:

- The differential universal function builds up the general convective boundary condition testifying that a series with only the first term constructs the boundary condition of the third kind, taking into account only isothermal effect, whereas retaining of the followed terms increases the accuracy of boundary conditions, accounting for the effect of the first and higher temperature head derivatives.
- Because the second term with the first derivative basically defines the effect of nonisothermicity, the calculation of its value gives an estimation of error caused by a boundary condition of the third kind telling us whether the conjugate solution is required or the simple common approach is acceptable.

- Using a general boundary condition allows reducing the conjugate problem to equivalent conduction problem for the body only.
- There exists a gradient analogy, which means that the temperature head gradient affects the heat transfer coefficient, as the free stream velocity gradient influences the friction coefficient.
- In the case of decreasing temperature head, the heat flux inversion might occur when the heat flux vanishes—a phenomenon analogous to separation of boundary layer in flows with adverse pressure gradient.

Chapter 3 presents results of conjugate heat transfer investigations in flows around bodies (external flows) and inside the channels and tubes (internal flows) in general, without specifying a concrete application in any device or process. The examples reviewed in this chapter differ by methods of problem solution, form of objects, boundary conditions, flow regimes, and state of flow (initial or developed). These results present effects of different parameters and conditions on conjugate heat transfer intensity in external and internal flows in general, without reference to particular application. The specific practical applications of conjugate heat transfer are discussed in the next chapter.

Special attention is given to conjugate heat transfer in flows past the thin plate, which are considered first. Due to the relative simplicity of this type of problem, we used the universal functions to create effective methods and obtain significant results, which include: (i) investigation of the temperature singularities on the solid/fluid interface, (ii) creation of the charts for simple conjugate problems solution, (iii) consideration of examples to help a reader to possess the charts usage, and (vi) computation of the inequalities for quasi-steady approach validation.

The other part of this chapter contains 15 reviewed and 27 indicated as other works of original studies of conjugate heat transfer in external and internal flows. Here, as well as in the following Chapters 4 and 6, the original studies are presented describing problem formulation, a mathematical model as the system of equations, ideas of solutions, and the basic results, but without exercises, which would be difficult for beginners.

The following examples are reviewed:

- Past plate and bodies: laminar flow past finite rectangular slab, flush sources on an infinite slab, free convection on vertical and horizontal thin plates, elliptic cylinder in laminar flow, translating fluid sphere, radiating plate with internal source, and radiating thin plate in compressible flow.
- Inside channels and tubes: fully developed laminar flow in a pipe heated by uniform heat flux, turbulent flow in parallel plats duct at periodical inlet temperature, fully developed flow in thick-walled channel with moving wall, hydrodynamically and thermally developed flow in a thick-walled pipe, laminar flow in the entrance of plane duct, flow in a channel of finite length, unsteady heat transfer in a duct with laminar flow, and transient heat transfer in a pipe with constant surface temperature.

Chapter 4 contains specific conjugate heat transfer applications in different industrial areas and technology processes. Thirty-one original papers are reviewed in four sections considering heat exchangers and finned surfaces (12 examples), thermal treatment and cooling systems (9), simulation of industrial (3) and technological (7) processes. Chapter 4 begins with conjugate solution of the classical problem of overall heat transfer coefficient of two flows separated by a thin plate, which is usually considered as a model of heat exchanger. Six conjugate solutions of this problem using different methods are analyzed, showing how much the conjugate strict results differ from data obtained by simple common approach. The following solutions are considered: two solutions of concurrent and countercurrent laminar and turbulent flows, solutions for two quiescent and two flowing fluids separated by vertical plate, and vertical thin-walled pipe with forced inside and natural outside flows.

The conjugate results for overall heat transfer coefficient obtained for a thin plate are compared with exact two-dimensional conjugate solution to understand where assumption of thin plate is applicable, and how otherwise such results should be corrected. It is found that conjugate results for thin plate are practically accurate, except a small area close to the leading edge, where two-dimensional effects are important and should be taken into account. The next three examples present more reliable heat exchanger models: two conjugate models using double pipes and a special model for the microchannel exchanger. The two last samples of this section consider models of finned surfaces.

In the second section of Chapter 4, the thermal treatment of moving continuous materials is analyzed in the first three examples, and conjugate heat transfer in different cooling systems is studied in the other six examples. The heat transfer in electronic packages is discussed in the first two examples, the results for cooling turbine blades and vanes are presented in the next two examples, and the last two samples analyzed the protection of systems in reentry rocket, and in a nuclear reactor at emergency loss of coolant. The next section gives three examples of simulation of the processes in industrial equipment. Because of complexity, there are relatively few publications of this type. The three models considered here simulate processes in twin-screw extruder, optical fiber coating, and continuous wires casting.

The last section of Chapter 4 presents heat transfer investigations in seven technological processes. The first three examples examine heat and mass transfer in multiphase flows using models of such complicated processes; wetted-wall absorber, concrete production, and Czochralski crystal growing. The next three samples studied drying of wood board, porous materials, and pulled through coolant continuous sheets. The last example presents freeze drying of two specimens of food. Forty-seven relevant other works are introduced shortly after the reviewed examples.

Part I is closed by a short summary of results. The basic dependences of heat transfer characteristics are presented in the form of a table where the influence parameters are arranged in order of a degree of conjugation. Such comparative information is useful in making a decision whether the conjugate solution is needed in a particular problem, or the common simple approach is enough to solve it. This question is discussed in detail and possible recommendations are formulated.

Part II incorporates two chapters consisting of applications of modern methods in fluid flow. The theory and general characteristics of those methods in both areas, peristaltic flow and direct turbulence simulation, are outlined in Chapter 5. Applications of peristaltic flow in medicine, biology and engineering and applications of direct simulation of turbulence in engineering are introduced in Chapter 6, reviewing 24 and presenting 42 as other works of original papers.

Chapter 5 starts from considering the peristaltic motion as conjugate phenomenon. Physical analysis shows that peristaltic motion adopted from creation exists due to conjugation (say interaction) between flexible walls and fluid inside tubular human organs, so that the conjugation nature is an inherent property of peristaltic flow. These considerations are confirmed by subsequent examples of human organs operating under the peristaltic flow and by explaining working principles of some devices simulating this natural motion mechanism.

The next part of Chapter 5 consists of the formulation of the conjugate model for peristaltic motion. This model is similar to that for heat transfer described in detail in Part I, and involves two subdomains with conjugate conditions on the wall/fluids interface. Conjugate relations in the case of peristaltic flow contain no-slip conditions for velocities and the balance of forces on the interface instead of equalities of temperatures and heat fluxes in the case of heat transfer. The essential difference between both conjugate models is explained, stressing that nonlinear peristaltic flow model is more complex than the linear heat transfer one. To take into account that complexity after analyzing published studies, the term "semi-conjugate model" is introduced which describes the situation when only the effect of flexible walls on the flow is investigated, ignoring more complicated impacts of flow on walls motion. The samples reviewed in Chapter 6 show that the majority of studies are of the semi-conjugate type.

The discussion of the problem solutions begins from the first simple research. The main objective of early studies was the understanding of the peristalsis mechanism in order to get insight into physical processes in the ureter, like reflux of bacteria. To simplify the problems, early authors used a linear model and assumptions of low Reynolds number and long wavelength, which are often applied up to now. Two more substantial nonlinear semi-conjugate solutions are introduced next: the analytical solution at low Reynolds number based on a perturbation series and numerical solution of a two-dimensional peristaltic flow at a moderate Reynolds number.

Two examples are analyzed to introduce fully conjugate solutions, taking into account both effects of interaction of the flexible walls and fluid inside channel. In the early paper, the equality conditions of forces on the interface are defined, employing the relatively simple approximate expressions. Conjugate conditions on the interface in another study published later are much more complicated but more realistic. These conditions are constructed using relations from the theory of thin oscillating elastic plate and two-dimensional Navier-Stokes equations and result in a system of three differential equations. Both solutions are compared with corresponding semi-conjugate data showing that the flow significantly affects the wall's behavior.

The second part of Chapter 5 presents the modern methods of direct numerical simulation of turbulence. Here, the discussion starts with a short introduction explaining the difficulties associated with extremely wide scales of turbulence eddies, which range from scales of integral length to Taylor and Kolmogorov smallest scales. In the following three sections, the new methods: direct numerical simulation (DNS), large eddy simulation (LES) and detached eddy simulation (DES) are introduced and compared.

A direct numerical simulation is a method to solve the unsteady Navier-Stokes equations in order to obtain the complete space- and time-dependent field of turbulent flow. By estimation of a number of grid points and time steps required for performing DNS, it is shown that only relatively simple engineering problems at real Reynolds numbers can be investigated by direct simulation.

A large eddy simulation is a method of reduction of the requirements for DNS in order to solve directly Navier-Stokes equations at higher Reynolds numbers. The main idea of LES proposed by Smagorinsky is to separate the treatment of large and small eddies, computing the large eddies by DNS and small eddies by Reynolds-average models. To demonstrate how the filtering procedure works providing the separation of areas with DNS and Reynolds-average models, a simple filter based on the integration is described.

The filtered form of Navier-Stokes equations are analyzed showing that this procedure gives the field of filtered large scales modified by the subgrid scales stresses (SGS). This pattern represents the interaction between large and small eddies testifying the essential role of modeling SGS in LES.

Large eddy simulation significantly widened the application of the direct solutions of the Navier-Stokes equation. However, the important engineering applications such as airfoil, ground or marine vehicle require much higher Reynolds numbers and, accordingly, greater numbers of grid points and time steps. These large requirements are caused by the near-wall region with the smallest eddies whose role increases about three times proportional to the value of Reynolds number. To reduce the number of grid points and to achieve further progress, Spalart, with co-authors, suggested detached eddy simulation. This method (DES) is a hybrid approach combining the RANS (Reynolds-average Navier-Stokes equation) for near-wall region and the LES for domain with large eddies. To provide the model behavior according to required treatment by LES or by RANS, the blending functions are used. The idea of a blending function is described showing the principle of comparing the closest to the surface distance d with the largest grid sell Δ so that the model uses RANS close to the walls, where $d << \Delta$, and works as a subgrid type pattern away from walls, where $\Delta << d$.

Some examples demonstrate the accomplishments of DES in modeling the flow separations at high Reynolds numbers, such as sub- and super-critical flows around sphere and flows past aircraft models. Nevertheless, to correct weaknesses of DES, two modifications were proposed: the delayed detached eddy simulation (DDES) and the zonal detached eddy simulation (ZDES). In these versions of DES, the treatment of the area where the model switches from RANS to LES is improved in order to get rid of the rapid decrease of the RANS eddy viscosity, which might result in strong instabilities. In DDES, to prevent this undesired depletion of the RANS strength, the switch into LES is delayed. In ZDES, this problem is resolved by introducing separated zones for RANS and LES where the regime in each zone is selected individually in line with requirements.

At the end of this chapter, a small paragraph represents the chaos theory, which studies phenomena sensitive to initial conditions, like weather, when the small variations in one location may result in widely different outcomes far away (butterfly effect). Though currently the chaos theory is not a tool for turbulence modeling, some characteristics of turbulence are of a chaotic kind, which gives hope of using the chaos theory in the future.

Chapter 6 represents applications of advanced peristaltic and turbulence models in biology, medicine and engineering. Examples of original studies are reviewed, as well as the heat transfer articles in Chapter 4, presenting problem formulation, mathematical models as systems of equations, ideas of solutions and basic results. The applications in biology and medicine are described in three sections analyzing blood flows in normal and pathologic vessels, flows in disordered human organs and biological transport processes. The first section presents flow in the arterial stenosis and flow through series of stenoses, blood flow affected by magnetic field during MRA and MRI tests, and blood flow under the hyperthermia cancer treatment. In the second section, the abnormal flows and/or irregular situations are simulated: the particle motion in ureter modeling of a bacterium or stone motility, chyme flow during gastrointestinal endoscopy and bile flow in a duct with stones. The third section describes fluid transport in the cerebral perivascular space, macromolecules transport in tumors, embryo transport modeling, and the bioheat transfer in human tissues. Twenty articles are indicated as other works.

The second part of Chapter 6 consists of applications in engineering contributed by peristaltic flow simulation (PFS), DNS, LES, and DES (4 and 9 reviewed examples of peristaltic flow and turbulence simulations, respectively). Each section of reviewed articles is followed by other works citations. 13, 15, 12, and 14 studies of PFS, DNS, LES, and DES including IDDES and ZDES, respectively, are indicated. We begin the discussion with peristaltic flow applications in engineering, and then consider the engineering applications of direct numerical turbulence simulation. The contribution of peristaltic flow in engineering is presented by four recent results obtained during the last five years, including the effects of chemical reaction, a micropumping systems optimization, the method of valve-less microfluidic peristaltic pumping design, and the construction of biomimetic swallowing robot published in 2015. These examples demonstrate the effectiveness of mathematical models in peristaltic motion applications, which cardinally changes the methods of investigation in this area.

The review of DNS examples starts from simulation of turbulent boundary layer at a relatively high Reynolds number as $Re_\theta = 2560$ published four years ago. The next two studies introduce the effects of Reynolds and Prandtl numbers in turbulent heat transfer, and a more involved recent study of exothermic gas-phase reaction in a packed bed. The last example and 15 latest results, including three articles published in 2016 cited as other works, show progress in DNS.

The next three examples demonstrate the advances in LFS via simulation of vortex and pressure fluctuation in aerostatic bearings, effects of equivalence ratio fluctuations in combustion chamber of gas turbine, and the heat transfer in pebble bed of nuclear reactor at a high temperature. Ten other works published during the last few years and the two most recent studies appearing in 2016 show in addition to reviewed articles the current situation in LES.

The last three examples display the great progress in studying the real objects characteristics by DES. The first result published 13 years ago presents patterns of sub- and super-critical flows over sphere confirming the well-known experimental data of early (82°) and much later (120°) separations in the first and second cases, respectively. Two other samples show recent investigations of complicated natural prototypes: Reentry-F flight experiment and free-surface flow around a submerged submarine fairwater. Both studies are performed at real values of Reynolds number of order $\approx 10^7$ and the Mach number about 20 in the first study and the Froude number about 0.4 in the second one. Fourteen modern articles employed DES including the latest versions DDES and ZDES cited as other works present manifold achievements of contemporary methods of direct numerical simulation of turbulence in studying the complex engineering systems.

As mentioned above, Part III serves as a subsidiary intended to help a reader to find information during studying of the basic text. Three chapters containing fundamental laws and methods compose this part: laminar and turbulent fluid flows and heat transfer (Chapters 7 and 8) and basic analytical and numerical methods in applications (Chapter 9). Chapter 7 starts with discussing two similar mechanisms of momentum, energy, and mass transport described by conservation laws. Physically grounded analysis shows that structures of Navier-Stokes, energy, and mass transfer equations are similar consisting of two groups of terms responsible for the molecular and convective transport processes.

The next several sections present different forms and properties of Navier-Stokes equations. The vector, vorticity, stream function, and irrotational invisced forms as well as the form in Einstein notations are considered. The other often used notations, Kronecker delta and Levi-Civita index are also explained. Some basic exact solutions of Navier-Stokes and energy equations

(Stokes problems, flow and heat transfer in a channel and a tube, stagnation point flow, and heat transfer in Couette flow) are analyzed. The two cases of simplified Navier-Stokes equations, the small (creeping flows) and large (boundary layer) Reynolds numbers, are presented. As an example of creeping flow, the Stokes flow around a sphere is shortly described. The derivation of boundary layer equations and dimensionless numbers are given using the comparison of the terms order in Navier-Stokes and energy equations. The merits of boundary layer approach are described. The Prandtl-Mises and Görtler forms of boundary layer equation are analyzed. The physical meaning of several dimensionless numbers is explained indicating that each number may be interpreted as a ratio of particular physical parameters. As examples of exact solutions of boundary layer equations, the Blasius, Pohlhausen, and Falkner and Skan problems are considered, showing how the initial partial differential equations are reduced to ordinary differential equations. The Karman-Pohlhausen integral method is described and some approximate solutions of boundary layer problems are analyzed.

The last section of this chapter presents the natural convection, comparing it with forced convection considered in previous sections. It is noted that a free convection occurs naturally whenever there are density differences in gravitational field in contrast to the forced one, which exists due to external force. Three examples are reviewed to show the basic features of natural convective problems. Analyzing the solution for the vertical plate reveals some characteristics of natural convection that cause this type of convection to differ from the forced one. Two examples show that in case of natural convection some additional effects should be taken into account. In particular, the radiation should often be considered along with natural convection because both heat transfer rates are usually of the same order. The other effect that is significant in that case is the flow stability as, for example, in Rayleigh-Benard free convection flow between parallel horizontal plates.

Chapter 8 presents features that differentiate turbulent flow from the laminar issue. Two parts describing averaging procedure and diverse turbulence models construct this chapter. It is explained that the process of averaging parameters developed by Reynolds leads to formulating the governing equations for turbulent flow in the form similar to that for laminar flow. Presented analysis shows how the averaging procedure yields additional unknown terms in the governing equations, called the Reynolds stresses, and finally results in an unclosed system of equations. The problem of closing this system, known as a problem of closure, is solved employing the semi-empirical or statistical turbulence models.

We begin the discussion of turbulent models from the simpler algebraic models. The first Prandtl model of this type is grounded on Boussinesq relation with unknown turbulent viscosity μ_{tb} defined through the mixing-length hypothesis. The physical interpretation of both Boussinesq and Prandtl hypotheses is followed by discussion of the structure of equilibrium turbulent boundary layers, which is the basis of the modern algebraic models. The typical velocity profile in such boundary layer consists of three standard parts: the viscous sublayer where the law of the wall holds, the defect layer with Clauser's velocity law, and the overlap logarithmic region where both laws are asymptotically valid. Three modern algebraic models, Mellor-Gibson, Cebeci-Smith, and Baldwin-Lomax, are considered. The results of modeling flows in a channel, tube, in some boundary layers, and heat transfer from surface with arbitrary temperature distribution using these models show reasonable agreement with experimental data.

The remaining part of this chapter deals with the one- and two-differential equations models. These types of models grounded on the turbulence kinetic energy equation, simulate the

turbulent flows much closer than algebraic models. A special section is devoted to the turbulence kinetic energy equation explaining the physical meaning of terms and the role of the
Kolmogorov kinetic energy equation. Some one-equation models and results of testing these
models at AFOSR conference are presented. It is noted that, according to AFOSR, the basic
shortage of the one-equation models is the absence of length scale. At the same time, it is
underlined that only two-equation models are complete models, which means that the solution
might be obtained by the model itself without using additional experimental data.

The two most popular $k - \omega$ and $k - \varepsilon$ two-equation models are considered, and both
equations defining the turbulent kinetic energy and dissipation energy rate (it serves as
length scale) are discussed. The applicability of one- and two-equation models reveals that
the turbulent flows with strong adverse pressure gradients, separated or reattachment flows,
compressible and other complex flows may be studied with reasonable accuracy only by
two-equation models since applications of more accurate new methods of direct numerical
simulation are restricted at present. At the same time, the simpler algebraic models are
preferable for the solution of problems with zero and benign pressure gradients.

Two parts of Chapter 9 present analytical and numerical mathematical methods frequently
used in applications. To illustrate the usage of considered methods, we apply mainly problems
of heat transfer in solids. Such a manner completed the set of topics important for studying the
basic text since two others, laminar and turbulent fluid flow and heat transfer, are reviewed in
previous chapters of Part III. The analytical methods are reviewed starting from error function.
It is shown that error function satisfied the unsteady one-dimensional conduction equation and
boundary conditions for infinite and semi- infinite solids and for lateral insulated thin rods.
Two examples are analyzed.

The method of separation variables is considered next. Three cases are indicated when the
general procedure of separation is possible for solution. Solutions of one-dimensional unsteady
problems applying standard technique of Fourier series are presented. A special case when the
usual Fourier series are not applicable is studied, giving an understanding of what are the eigenvalues and orthogonal eigenfunctions. The Sturm-Liouville problem is reviewed, specifying
the conditions of the existing orthogonal eigenfunctions and defining the proper series. Two
steady two-dimensional problems for Laplace equation with Dirichlet and mixed boundary
conditions are examined as well.

The Fourier and Laplace integral transforms present the next two sections. The idea of integral transform is described, and the difference between these two widely used integral methods
is explained. Four solutions for rods and rectangular sheets following this discussion show that
Fourier transform is applicable to infinite domains, whereas the Laplace transform is relevant
for semi-infinite positive variables domain.

The Green's method of analytical solution is described in the last section of this part of
Chapter 9. The idea of this approach is close to Duhamel's method: presenting a solution of
a problem with space-time dependent variables in terms of similar results for problem with
constant parameters (S.1.3.1). The general formula defining the Green function is derived for
the solution of a one-dimensional conduction problem.

In the second part of Chapter 9, we review shortly classical numerical methods Three
sections completed this review. The first section "What method is proper" shows that there is
no reason to oppose analytical and numerical methods as it becomes popular after computer
advent. In the second section, we discuss the approximate methods for solving the differential
equations. It is justified that these methods were developed and widely used many years

before they became a basis of modern numerical methods, but before they are used for entire computation domain as analytical means. As the computers came, it became possible to apply the same approximate methods for each cell of grid vastly increasing the computing accuracy and converting these simple analytical approaches into the contemporary numerical methods. We classify numerical methods according to the types of discretization of the computation domain and analytical methods used for solution. Three methods, finite-difference (FDM), finite-element (FEM), and boundary element (BEM) methods applying uniform and irregular grids are considered.

To describe the technique of employing different analytical methods, we use the weighted residual approach. The idea of this approach is explained considering solutions of a simple conduction problem governed by a one-dimensional equation. Analysis of relevant examples clarifies the features, the merits, and lack of different methods. In particular, it is explained what is the weak solution and why the boundary element method requires data only along boundaries of computing domain, whereas the other methods demand information of the whole variable field.

The final section of Chapter 9 deals with the complications in computing flow and heat transfer characteristics. Following Patankar, we discuss some ways for overcoming problems arising in computing pressure and velocity, convection-diffusion terms, and cases of false diffusion. It is shown that the difficulty in computing flow characteristics associated with the absence of explicit equations for pressure is in fact an apparent problem because the correct pressure estimation is controlled by continuity equation. Analysis indicates that usual control volume approach fails resulting in zero pressure, and to resolve the pressure computing, the staggered control volume was developed. This procedure is described, explaining that in this case, in contrast to the usual approach, the velocity components and pressure are calculated on the control volume faces. The software SIMPLE and three modified versions of it are shortly described.

The textbook is closed with a conclusion summarizing the purpose, applicability, and prediction of the feature of the contemporary methods considered in the book.

Abram Dorfman
February 2016

Acknowledgments

As always, I use this opportunity to express my appreciation to people who play a significant role in my life and work. Moreover, without contributions from some of these people, this book, as well as two previous books, could not exist, and, in fact, these people are my co-authors. First of all, my cardiac doctors: Professor R. Prager who granted me years after a heart attack in 2000, Professor M. Grossman who takes care of my heart since that time and who operated on me recently, and Professor A. Stein with his team who treated me before this operation. I am also grateful to Dr. S. Saxe and Dr. Jill Bixler who retrieved my vision and to Dr. B. Brophy who is our family doctor and helps us in different ways regularly.

My sincere thanks to Professor Massoud Kaviany from University of Michigan and my young friend Professor Alan McGaughey from Carnegie Mellon University for their interest in my work, advice and discussions improving this text.

Our dear friends Dr. Brian and Margot Schapiro and Dr. Stephen and Kim Saxe share with us our joys and sorrows, and I express my gratitude and love to them.

As usual, my family encouraged and assisted me in my work and writing. My special thanks to my computer teachers, my son-in law Isaac and my grandson Alex, for their essential contribution. My superior appreciation and love to my dearest wife Sofia who, in addition to her usual tending and serving for almost 67 years, remarkably assisted me in formatting this book.

About the Author

Abram S. Dorfman, Doctor of Science, Ph.D. was born in 1923 in Kiev, Ukraine in the former Soviet Union. He graduated from the Moscow Institute of Aviation in 1946, as an Engineer of Aviation Technology. From 1946 to 1947, he worked in the Central Institute of Aviation Motors (ZIAM) in Moscow. From 1947 to 1990, Dr. Dorfman studied fluid mechanics and heat transfer at the Institute of Thermophysics of Ukrainian Academy of Science in Kiev, first as a junior scientist from 1947 to 1959, then as a senior scientist from 1959 to 1978, and finally as a leading scientist from 1978 to 1990. He earned a Ph.D. with a thesis entitled "Theoretical and Experimental Investigation of Supersonic Flows in Nozzles" in 1952. In 1978, he received a Doctor of Science degree, which was the highest scientific degree in the Soviet Union, with a thesis and a book, *Heat Transfer in Flows around the Nonisothermal bodies*. From 1978 to 1990, he was associate editor of *Promyshlennaya Teploteknika,* which was published in English as *Applied Thermal Science* (Wiley). Dr. Dorfman was an adviser to graduate students for many years.

In 1990, he emigrated to the United States and continues his research as a visiting professor at the University of Michigan in Ann Arbor (since 1996). During this period, he has published several papers in leading American journals and two books, *Conjugate Problems in Convective Heat Transfer* (Taylor & Francis, 2010) and *Classical and Modern Engineering Methods in Fluid Flow and Heat Transfer,* (Momentum Press, 2013). He is listed in *Who's Who in America 2007*.

Dr. Dorfman has published more than 140 paper and four books in fluid mechanics and heat transfer in Russian (mostly) and in English. More than 50 of his papers published in Russian have been translated and are also available in English.

Nomenclature

$\text{Bi} = \dfrac{h\Delta}{\lambda_w}$ Biot number

$\text{Br} = \dfrac{\lambda\Delta}{\lambda_w L}\,\text{Pr}^m\text{Re}^n$ Brun number

$C = \dfrac{u'\Delta t}{\Delta x}$ Courant number

C_1, C_2 Exponents of integral universal functions

$C_f = \dfrac{\tau_w}{\rho U_\infty^2}$ Friction coefficient

$2C_f/\text{St}$ Reynolds analogy coefficient

c, c_p Specific heat and specific heat at constant pressure, J/kg K

$\hat{c} = \rho c\Delta$ Thermal capacity, J/m^2K

D, D_h Diameter and hydraulic diameter m

D_m Diffusion coefficient, m^2/s

$\text{Da} = \dfrac{k}{L^2}$ Darcy number

$\text{Ec} = \dfrac{U^2}{c_p\theta_w}$ Eckert number

$f(\xi/x)$ Influence function of unheated zone at temperature jump

$f_q(\xi/x)$ Influence function of unheated zone at heat flux jump

$\text{Fo} = \dfrac{\alpha t}{L^2}$ Fourier number

$\text{Fr} = \dfrac{U}{\sqrt{gL}}$ Froude number

$\text{Gr} = \dfrac{\beta\theta_w g L^2}{\nu^2}$ Grashof number

g_k, h_k Coefficients of differential universal functions

g Gravitational acceleration, m/s^2

h, h_m Heat and mass transfer coefficients, W/m^2K, W/m^2s

k	Specific heat ratio or turbulence energy, m^2/s^2
K_τ, K_q	Constants of rheology laws for non-Newtonian fluids
$\text{Kn} = \dfrac{l}{D_h}$	Knudsen number
l	Body length or mixing length, or free path, m
L	Characteristic length, m
$\text{Le} = \dfrac{D_m}{\alpha}$	Levis number
$\text{Ls} = \dfrac{\lambda_w h}{\rho_w^2 c_w^2 U_w^2 \Delta}$	Leidenfrost number
$\text{Lu}\ \dfrac{\rho c}{\rho_w c_w}$	Luikov number
$\text{M} = \dfrac{U}{U_{sd}}$	Mach number
M	Moisture content, kg/kg
n, s	Exponents of rheology law for non-Newtonian fluids
$\text{Nu} = \dfrac{hL}{\lambda},\ \text{Nu} = \dfrac{hL^{s+1}}{K_q U^s}$	Nusselt numbers for Newtonian and non-Newtonian fluids
p	Pressure, Pa
$\text{Pe} = \dfrac{UL}{\alpha}$	Peclet number
$\text{Pr} = \dfrac{\nu}{\alpha},\ \text{Pr} = \dfrac{\rho c_p U^{1-s} L^{1+s}}{K_q}$	Prandtl numbers for Newtonian and non-Newtonian fluids
$q,\ q_v$	Heat flux, W/m^2 or volumetric heat source, W/m^3
r/s	Exponent of isothermal heat transfer coefficient
$\text{Ra} = \dfrac{\beta \theta_w g L^3}{\nu \alpha}$	Rayleigh number
$\text{Re} = \dfrac{UL}{\nu},\ \text{Re}\dfrac{\rho U^{2-n} L^n}{K_\tau}$	Reynolds number for Newtonian and non-Newtonian fluids
$\text{Sc} = \dfrac{\nu}{D_m}$	Schmidt number
$\text{Sh} = \dfrac{h_m}{\rho c_p D_m}$	Sherwood number
$Sk = \dfrac{4\sigma T_\infty^4 L}{\lambda_\infty}$	Starks number
$\text{St} = \dfrac{h}{\rho c_p U}$ and $\text{St} = \dfrac{\omega L}{U}$	Stanton number and Strouhal number
$\text{Ste} = \dfrac{c_p \Delta T}{\Lambda}$	Stephan number

$$\text{Stk} = \frac{t_p U}{D_p}$$ Stokes number

t Time, s

T Temperature, K

u, v, w Velocity components or u, v parts of integration procedure

U Velocity on outer edge of boundary layer

U_e Velocity on outer edge of turbulent boundary layer

$u_\tau = \sqrt{\tau_w/\rho}$ Friction velocity

$u^+ = u/u_\tau, \ y^+ = yu_\tau/\nu$ Variables of wall law

x, y, z Coordinates

Greek symbols

α Thermal diffusivity, m/s^2

β Dimensionless pressure gradient or volumetric thermal expansion coefficient, 1/K

$\chi_t = \dfrac{h}{h_*}, \ \chi_p = \dfrac{h_m}{h_{m*}}$ Nonisothermicity and nonisobaricity coefficients

$\chi_f = \dfrac{C_f}{C_{f*}}$ Nonisotachicity coefficient

$\delta, \delta_1, \delta_2$ Boundary layer thicknesses, m

δ, δ_{ij} Delta function and Kronecker delta

Δ Body or wall thickness, m

ε Dissipation energy rate, m^2/s^3 or fraction of phase

κ Constant determining mixing length

λ Thermal conductivity, W/mK

Λ Latent heat, J/kg or λ_s/λ

μ Viscosity, kg/s m

ν Kinematic viscosity, m^2/s

ξ Unheated zone length, m

$\theta = T - T_\infty,$ Temperature excess, K

$\theta_w = T_w - T_\infty$ Temperature head, K

ρ Density, kg/m^3

σ Stefan-Boltzmann constant, W/m^{2}K^4

τ Shear stress, N/m^2

Φ, φ Prandtl-Mises-Görtler variables

ψ Stream function, m^2/s

ω Frequency or specific dissipation energy rate

Some of these symbols are also used in different ways as it is indicated in each case.

Subscripts		Superscipts	
av	Average	$+, -$	From both sides of interface
ad	Adiabatic	$+, ++$	Wall law
as	Asymptotic		
bl	Bulk	**Overscores**	
e	End or effective	$\bar{o}, \tilde{o}$	Dimensionless, or transformed
i	Initial; inside		
L	At $x = L$		
m	Mass average, or mean value, or moisture		
o	Outside		
p	Pressure or particle		
q	Constant heat flux		
sd	Sound		
t	Thermal		
T	Constant temperature		
tb	Turbulent		
w	Fluid-solid interface		
ξ	After jump		
∞	Far from solid		
$*$	Isothermal or special		

Part I

Applications in Conjugate Heat Transfer

Introduction

When and why Conjugate Procedure is Essential

Behavior and efficiency of natural and engineering systems and processes depend on inter-action of basic components forming those creations. It is clear that characteristics of such processes, as for example, heat transfer, blood flow, or combustion are determined by interaction of a body and fluid in the first, blood and the vessel wall in second, and fresh and burned gases in third cases, respectively. Therefore, conditions developed on the interface of basic components due to their interaction largely define the system.

However, these conditions are usually unknown in advance even if the properties of the components are specified. The conjugate procedure is an approach of determining the distribution of parameters arising on the interface as a result of elements interaction. This is achieved by solving the governing equations for each interaction element, and by following conjugation of these solutions at the interface. In fact, the conjugate procedure is essential for any problem containing at least two interaction subjects because that is only one way to find the parameters on interface and then the solution. For instance, a conjugation of the body and fluid temperature fields gives the temperature of interface and then solution of heat transfer problem. Similarly,

Applications of Mathematical Heat Transfer and Fluid Flow Models in Engineering and Medicine,
First Edition. Abram S. Dorfman.
© 2017 John Wiley & Sons Ltd. Published 2017 by John Wiley & Sons Ltd.

one gets the solutions of two other problems mentioned above conjugating at the interface forces in the blood and the wall of the vessel or the velocity and pressure distributions of the fresh and burned gases at the flame, which is an interface in combustion process.

There are countless other phenomena required by the conjugate procedure for strict investigation. In particular, such important processes in meteorology problems as energy exchange between the atmosphere and a sea or ocean or falling through air solid or liquid drops, are obviously conjugation problems requiring the study of two fluids or of drops and air interaction. The conjugate procedure is also needed in the study of drying a brick or a wood board and even in the relatively simple problem of conduction between two different solids or between subjects from the same substance but with different physical properties.

The list of such examples may be continued as long as desired because, as it will be clear from what follows, the conjugate methods are widely used in applications in different ways from complicated aerospace systems to relatively simple food processing. Moreover, there are areas where any problem is inherently conjugate. This is truth, first of all, about the heat transfer specifying, in particular, why the conjugate approach was born in this area, soon after it was recognized that such idea is realizable owning the computer advent. This gives also understanding why conjugate methods, starting with simple examples in the 1960s of the last century, now are intensively employed establishing a new avenue where the number of publications grows as avalanche.

Another phenomenon inherently conjugate is the peristaltic flow. This follows from the nature of such motion, which actually exists due to interaction between flexible wall and fluid inside channel. The Greek word peristaltikos means clasping and compressing. The corresponding English word peristalsis stands for fluid motion in a channel with flexible walls when a progressive wave propagating along the walls originates a fluid motion in a wave direction. Although there are engineering devices working on this principle, the idea of such motion was adopted from structure of human organs transporting physiological fluids, like blood vessels, urinary channels, or gastrointestinal tract. Muscular walls of such organs in the form of a tube provide consecutive narrowing and relaxing of the wall portions, which travel lengthwise the channel and results in flow movement downstream a tube.

Simulation of peristaltic motion is used in artificial organs, such as heart-lung machine or the device for hemodialysis. Other similar appliances of this type are utilized for mixing chemical reactions, transporting blood, and other biomedical clean or sterile fluids to prevent the carrying stuff from contact with parts of common pump. Similar devices are used for isolating environment from conveying corrosive and sewage fluids. In such engineering applications instead of muscular, the mechanical gears are employed. Indeed, the conjugate procedure was known long before the 1960s,when the term conjugate problem was coined. It has been a long time known that the heat transfer problems should satisfy the boundary conditions of fourth kind, which are in fact the conjugate conditions. However, before computer time, the calculation recourses permit solutions with such exact conditions only of very simple problems. As a result, the Newton's approximate expression of proportionality of the heat flux to temperature head $q_w = h(T_w - T_\infty)$, known as boundary condition of the third kind, was employed instead of exact boundary conditions. Such assumption ignores the real conditions on the solid-fluid interface and rests the accuracy of solution on a heat transfer coefficient h precision. Therefore, this approach yields acceptable solutions only when the effect of interface temperature distribution is small, and the results are close to the case of isothermal interface. Otherwise, the experimental data for heat transfer coefficient is required for successful calculation.

Nevertheless, until the last fifty years, this simple relation was only one means for calculation and study convective heat transfer. Today, the conjugate methods along with appropriate software constitute a powerful tool for heat transfer investigation and practical calculation substituting approximate methods based on heat transfer coefficient. These methods give the temperature and heat flux distributions on the interface, and then, there is no need in heat transfer coefficient, even though it can be calculated using data obtained in conjugate solution. Example of such solutions presented in suggested books (see references) show the applicability and versatility of modern conjugate methods.

A Core of Conjugation

To see the basic feature of the conjugate formulation, consider a heat transfer from a thin plate heated from one edge. Analyze the heat transfer coefficient variation along the plate in two cases of flow direction: from heated to unheated edge (first case) and in the opposite line from unheated to heated edge (second case). This simple problem clearly shows the role of interface conditions because: (i) as it follows from exact conjugate solution in two considered cases, the distributions of the heat transfer coefficient along the plate significantly differ due to contrasting interface characteristics, and (ii) according the common approach, the heat transfer coefficient distribution is the same in both cases.

Indeed, it is easy to understand that in the first case, the temperature head on the interface is maximal at the beginning, and because of that the temperature head decreases in flow direction, whereas in the second case, the starting value of temperature head on the interface is the smallest, and due to that it increases in flow direction. In the next chapters, we will see that the type of temperature head variation on the interface is the basic characteristic determining the behavior of nonisothermal or conjugate heat transfer. The value of heat transfer coefficients strongly depends on the temperature head variation, so that in the case of temperature head increasing in flow direction or in time, the heat transfer coefficients are moderately greater than those on an isothermal interface, whereas in the opposite case, when the temperature head decreases, the heat transfer coefficients dramatically lessen and might even reach zero. Below, these general features of nonisothermal heat transfer are theoretically established and are confirmed by numerous of applications including some appropriate experiments.

In particular, the exact conjugate results (Exam. 3.5) show that in two cases of flow direction along the plate heated from one edge, the heat transfer coefficients differ so much significantly that the ratio of total heat flux from the plate in both cases may reach 1.2–1.25 in turbulent and 1.5–1.6 in laminar flows. At the same time, the common approach with boundary condition of the third kind and constant or isothermal heat transfer coefficients do not show any difference for both considered cases.

Comment This example is a model of the following practical problem:

A heated from one edge object is moving through surroundings (air or water) and as a result is cooled. If the heat is supplied at the trailing edge, the temperature head increases in flow direction as in the second case considered above, whereas if the heat is delivered through the leading edge, the temperature head decreases in flow direction as in the first case of the model discussed above. As mentioned above, the exact solution of the model problem shows that in such two cases cooling effects significantly differ, so that the ratio of removing total heat may reach substantial values $\approx$20–50%.

1

Universal Functions for Nonisothermal and Conjugate Heat Transfer

1.1 Formulation of Conjugate Heat Transfer Problem

As it follows from the above discussion, the domain of any conjugate problem consists at least of two subdomains according to the interaction components. Therefore, to formulate conjugate problem, it is necessary to specify two sets of equations: initial and boundary conditions governing the problem in each of subdomains in order to further conjugation of the corresponding solutions. In the case of heat transfer, such subdomains and sets of governing equations and boundary conditions are as follows:

- **Body domain:**
  Unsteady conduction equation

$$\frac{\partial T}{\partial t} = \alpha_w \left(\frac{\partial^2 T}{\partial x^2} + \frac{\partial^2 T}{\partial y^2} + \frac{\partial^2 T}{\partial z^2} \right) + \frac{q_v}{\rho_w c_w}, \qquad \frac{\partial T}{\partial t} = \alpha_w \nabla^2 T + \frac{q_v}{\rho_w c_w} \qquad (1.1)$$

or steady conduction equations:
  Laplace's and Poisson's equations (without and with heat source q_v) (Com. 1.1)

$$\nabla^2 T = \frac{\partial^2 T}{\partial x^2} + \frac{\partial^2 T}{\partial y^2} + \frac{\partial^2 T}{\partial z^2} = 0, \qquad \nabla^2 T = \frac{\partial^2 T}{\partial x^2} + \frac{\partial^2 T}{\partial y^2} + \frac{\partial^2 T}{\partial z^2} = -\frac{q_v}{\lambda_w} \qquad (1.2)$$

or simplified conduction equations for "thin body" and "thermally thin body" (Com. 1.1)

$$\alpha_w \frac{d^2 T}{dy^2} + \frac{q_v}{\lambda_w} = 0, \qquad \frac{1}{\alpha_w} \frac{\partial T_{av}}{\partial t} - \frac{\partial^2 T_{av}}{\partial x^2} + \frac{q_{w1} + q_{w2}}{\lambda_w \Delta} - \frac{q_{v.av}}{\lambda_w} = 0 \qquad (1.3)$$

Applications of Mathematical Heat Transfer and Fluid Flow Models in Engineering and Medicine,
First Edition. Abram S. Dorfman.
© 2017 John Wiley & Sons Ltd. Published 2017 by John Wiley & Sons Ltd.

- **Fluid flow domain:**

For laminar flow: Navier-Stokes and energy equations (S. 7.1)

$$\frac{\partial u}{\partial x} + \frac{\partial v}{\partial y} + \frac{\partial w}{\partial z} = 0 \tag{1.4}$$

$$\rho\left(\frac{\partial u}{\partial t} + u\frac{\partial u}{\partial x} + v\frac{\partial u}{\partial y} + w\frac{\partial u}{\partial z}\right) = -\frac{\partial p}{\partial x} + \mu\left(\frac{\partial^2 u}{\partial x^2} + \frac{\partial^2 u}{\partial y^2} + \frac{\partial^2 u}{\partial z^2}\right) \tag{1.5}$$

$$\rho\left(\frac{\partial v}{\partial t} + u\frac{\partial v}{\partial x} + v\frac{\partial v}{\partial y} + w\frac{\partial v}{\partial z}\right) = -\frac{\partial p}{\partial y} + \mu\left(\frac{\partial^2 v}{\partial x^2} + \frac{\partial^2 v}{\partial y^2} + \frac{\partial^2 v}{\partial z^2}\right) \tag{1.6}$$

$$\rho\left(\frac{\partial w}{\partial t} + u\frac{\partial w}{\partial x} + v\frac{\partial w}{\partial y} + w\frac{\partial w}{\partial z}\right) = -\frac{\partial p}{\partial z} + \mu\left(\frac{\partial^2 w}{\partial x^2} + \frac{\partial^2 w}{\partial y^2} + \frac{\partial^2 w}{\partial z^2}\right) \tag{1.7}$$

$$\rho c_p\left(\frac{\partial T}{\partial t} + u\frac{\partial T}{\partial x} + v\frac{\partial T}{\partial y} + w\frac{\partial T}{\partial z}\right) = \lambda\left(\frac{\partial^2 T}{\partial x^2} + \frac{\partial^2 T}{\partial y^2} + \frac{\partial^2 T}{\partial z^2}\right) + \mu S$$

$$S = 2\left[\left(\frac{\partial u}{\partial x}\right)^2 + \left(\frac{\partial v}{\partial y}\right)^2 + \left(\frac{\partial w}{\partial z}\right)^2\right] + \left(\frac{\partial u}{\partial y} + \frac{\partial v}{\partial x}\right)^2 + \left(\frac{\partial u}{\partial z} + \frac{\partial w}{\partial x}\right)^2 + \left(\frac{\partial v}{\partial z} + \frac{\partial w}{\partial y}\right)^2 \tag{1.8}$$

or simplified equations for high and low Reynolds and Peclet numbers:

Boundary layer equations (S. 7.4.4.1)

$$\frac{\partial u}{\partial x} + \frac{\partial v}{\partial y} = 0 \tag{1.9}$$

$$\frac{\partial u}{\partial t} + u\frac{\partial u}{\partial x} + v\frac{\partial u}{\partial y} + \frac{1}{\rho}\frac{dp}{dx} - v\frac{\partial^2 u}{\partial y^2} = 0 \tag{1.10}$$

$$\frac{\partial T}{\partial t} + u\frac{\partial T}{\partial x} + v\frac{\partial T}{\partial y} - \alpha\frac{\partial^2 T}{\partial y^2} - \frac{v}{c_p}\left(\frac{\partial u}{\partial y}\right)^2 = 0 \tag{1.11}$$

$$-\frac{1}{\rho}\frac{\partial p}{\partial x} = \frac{\partial U}{\partial t} + U\frac{\partial U}{\partial x} \qquad -\frac{1}{\rho}\frac{dp}{dx} = U\frac{dU}{dx} \tag{1.12}$$

Creeping flow equations (S. 7.4.1)

$$\nabla \cdot \mathbf{V} = 0 \qquad \nabla p = \mu\nabla^2\mathbf{V} \tag{1.13}$$

For turbulent flow: Reynolds averaged Navier-Stokes and energy equations in Einstein notations (S. 7.1.2.2)

$$\frac{\partial u_i}{\partial x_i} = 0, \qquad \rho\frac{\partial u_i}{\partial t} + \rho u_j\frac{\partial u_i}{\partial x_j} = -\frac{\partial p}{\partial x_i} + \frac{\partial}{\partial x_j}\left[(\mu + \mu_{tb})\frac{\partial u_i}{\partial x_j}\right] \tag{1.14}$$

$$\rho\frac{\partial T}{\partial t} + \rho u_j\frac{\partial T}{\partial x_j} = \frac{\partial}{\partial x_j}\left[\left(\frac{\mu}{\mathrm{Pr}} + \frac{\mu_{tb}}{\mathrm{Pr}_{tb}}\right)\frac{\partial T}{\partial x_j}\right] + \frac{\mu + \mu_{tb}}{c_p}\left(\frac{\partial u}{\partial y}\right)^2 \tag{1.15}$$

or simplified boundary layer equations for high Reynolds and Peclet numbers (S. 8.3):

$$\frac{\partial u}{\partial t} + u\frac{\partial u}{\partial x} + v\frac{\partial u}{\partial y} + \frac{1}{\rho}\frac{dp}{dx} - \frac{\partial}{\partial x}\left[(v + v_{tb})\frac{\partial u}{\partial x}\right] = 0 \qquad (1.16)$$

$$\frac{\partial T}{\partial t} + u\frac{\partial T}{\partial x} + v\frac{\partial T}{\partial y} - \frac{\partial}{\partial y}\left[(\alpha + \alpha_{tb})\frac{\partial T}{\partial y}\right] - \frac{v + v_{tb}}{c_p}\left(\frac{\partial u}{\partial y}\right)^2 = 0 \qquad (1.17)$$

and continuity (1.9) and Bernoulli (1.12) equations.

- **Conjugate conditions** (boundary conditions of the forth kind) (Com.1.1)

$$T^+ = T^-, \quad \lambda_w\frac{\partial T}{\partial y}\bigg|^+ = -\lambda\frac{\partial T}{\partial y}\bigg|^- \qquad (1.18)$$

Comment 1.1 Body domain: (i) in equations (1.2) ∇^2 is the Laplace operator (S. 7.1.2.1) and $\nabla^2 T$ is called Laplacian of T, (ii) the first equation (1.3) is obtained from two-dimensional second equation (1.2) for thin body ($\Delta/L \ll 1$), which thermal resistance is of the same order or greater than that of coolant ($Bi = h\Delta/\lambda_w \geq 1$), and due to that longitudinal derivative may be neglected, (iii) the second equation (1.3) is derived for thermally thin body ($Bi \leq 1$) by integration of two-dimensional equation (1.1) in y-direction taken into account that transverse resistance of such body is small in comparison with that of the coolant.

Fluid flow domain: (i) the Navier-Stokes and energy equations are used in both three- and two-dimensional forms, whereas the boundary layer equations are basically employed in two-dimensional form that is appropriate for majority of applications, (ii) the creeping flow equations written in vector form (1.13) may be obtained from Navier-Stokes and energy equations after neglecting the inertia terms (left parts of equations (1.5)–(1.7) (S. 7.4.1), in this case, the continuity equation is the same equation (1.4), (iii) Navier-Stokes and energy equations are presented in instantaneous parameters, whereas the equations (1.14)–(1.17) for turbulent flow are written in averaged parameters using the same notations (S. 8.2), (iv) eddy-viscosity coefficient μ_{tb} in these equations is determined by one of turbulent models (S. 8.3) and the value of turbulent Prandtl number Pr_{tb} is usually taken to be equal or close to unit (S. 2.1.2.4).

Conjugate conditions are expressions providing continuity of the temperature fields at the interface in the form of equalities of temperatures and heat fluxes computed from both interface sides and marked by (+) and (−) for body and fluid, respectively (Exer. 1.1–1.7).

- Initial and boundary conditions for subdomains (S.7.2)

The equations just considered are used to solve the subdomain problems. The relevant initial and boundary conditions depend on the type and order of governing equation. For example, the conduction equation is of the first order in time and of the second order in space. Because of that, the solution of one-dimensional conduction problem depends on one initial and two boundary conditions given at two points. Solutions of more complicated two- or three-dimensional conduction problems also require satisfaction of one initial condition and of boundary conditions, but specified around the outline of the whole problem domain. The initial condition defines the temperature of the system as a function of coordinates at some instant $t = 0$, which is taken to be a beginning of the process, whereas the boundary conditions prescribe the values of some parameters on the boundaries of the system as the functions of time and position.

Besides the boundary conditions of the fourth kind (1.18), there are three usually employed boundary conditions that differ from each other by kind of variables assigned at some points or around boundaries of domain. The boundary condition of the first kind designates the temperature values, the second one specifies heat fluxes, and the third kind of boundary condition presents Newton's expression with known heat transfer coefficient and temperature head. For example, at some point of domain with coordinate x, three types of conditions have the form

$$T|_x = T_w, \quad or \quad q_w = -\lambda_w \left.\frac{\partial T}{\partial x}\right|_x, \quad or \quad q_w = h(T|_{w,x} - T_\infty) \tag{1.19}$$

The equations specifying the conjugate problem formulation considered above are partial differential equation of the second order. The appropriate boundary conditions for those equations depend on the type of a particular equation. It is said that such equation in its canonical form is of elliptic or hyperbolic type, depending whether it consist of a sum or a difference of two second order derivatives, whereas such canonical equation with only one second partial derivative is called a parabolic partial differential equation (Exer. 1.8). Thus, the two- and three-dimensional conduction equations, Laplace and Poisson equations as well as Navier-Stokes and energy equations for laminar flow and similar equations for turbulent flow are of elliptic type, whereas both sets of boundary layer equations for laminar and turbulent flows are the parabolic equations. The parabolic equation requires relatively simple boundary conditions specifying variable values only on a part of computational domain. For boundary layer equations, these conditions are: (i) the no-slip condition $u = v = 0$ for dynamic equation and conjugate (1.18) or one of the regular (1.19) conditions for thermal equation on the body surface ($y = 0$) and (ii) asymptotic conditions $u \to U$, $T \to T_\infty$ far from the body on the outer edge of the boundary layer ($y \to \infty$) (S. 7.4.4).

In contrast to that, the elliptic equations require boundary conditions specifying the values of parameters around the entire computational domain, and formulation of such conditions is more complicated procedure. In this case, two problem formulations with different types of boundary conditions, known as Dirichlet and Neumann problems, are usually considered. The Dirichlet problem is stated using boundary conditions of the first kind by specifying the temperature on the boundaries of domain. The Neumann problem is composed similarly by employing the boundary conditions of the second kind by specifying the derivatives of the temperatures on the boundaries domain.

The Dirichlet problem is a well-posted problem, whereas the Neumann problem is an ill-posted problem. Physically it means that the solution behavior of an elliptic equation with Dirichlet boundary conditions is regular, whereas the solution of an elliptic equation under the ill-posted Neumann boundary conditions requires satisfaction on some additional conditions. For example, the solution of the Neumann heat transfer problem demands the thermal equilibrium that a system reaches when the total heat flux inside it is zero.

Comment 1.2 Formulation of boundary conditions for the Navier-Stokes equation is associated with additional difficulties arising due to fluid nature. To understand these complications, consider a flow through a channel or past a body immersed in a parallel fluid stream. Considering the Dirichlet problem, one should specify velocities along the boundaries of domain, which for a channel or tube include the walls or body surface plus entrance and exit sections of a channel or body. However, in such a case, only zero velocity boundary conditions on the surfaces of the immersed channel or body are known and a uniform stream velocity U_∞ at $x \to \pm\infty$ far away from the immersed object. The velocity profiles at the entrance and the exit sections required for the Dirichlet problem formulation are unknown because: (i) the profile at the entrance is established due to interaction of initial uniform stream with surrounding during the way from $x \to -\infty$ to channel or body entrance, and (ii) the velocity profile at the exit is formed as a result of processes inside the flow in the channel or around the body. Because of that in practice, the experimental data or relevant assumptions are used. The same situation holds for full energy equation since this equation is of elliptic type as well, and the initial uniform temperature profiles are deformed along with the velocity profiles (S. 7.2, Exam. 7.4 and 7.5, Exer. 1.9 and 1.10).

1.2 Methods of Conjugation

Physical analysis shows that any heat transfer conjugate problem is a question of thermal interaction of a body and a fluid with unknown temperature and heat flux distribution on the body/fluid interface. This becomes clear if one looks at what is known at the beginning of the conjugate problem solution. Indeed at the beginning, we know only: (i) set of equations governing heat transfer in a body and in a fluid separately, and (ii) conjugate conditions (1.18). However, the separate boundary conditions for each subdomain are unknown since data on interface may be obtained only as a result of conjugate problem solution. Thus, the situation is deficient: to solve a particular conjugate problem we need boundary conditions for each subdomain, which may be obtained only after solution of the same conjugate problem. There are several methods for resolving this challenging problem. Here, we consider two mostly used numerical methods and analytical approach based on employing so called universal functions [120]. Examples of other procedures for solving this problem are discussed in applications.

1.2.1 Numerical Methods

One relatively simple way to realize conjugation is to apply the iterations. The idea of such a approach is that each solution for the body or for the fluid produces a boundary condition for other component of the system. The process of interactions starts by assuming that one of the boundary conditions (1.19) exists on the interface. Then, one solves the problem for a body or for fluid applying this boundary condition and uses the result of solution as a boundary condition for solving the set of governing equations for other component and so on. If this iterative process converges, it might be continued until the desired accuracy is achieved. However, the rate of convergence of the iterations highly depends on the first guessing boundary condition, and there is no way to find an appropriate condition, except using the trial-and-error approach (Exer. 1.12).

Another known numerical conjugate procedure is grounded on the simultaneous solution of a large set of governing equations for both subdomains and conjugate conditions. Patankar [306] proposed a method and software for such a solution using one generalized expression for continuous computing of the velocities and temperatures fields through the whole problem domain that includes satisfying the conjugate boundary conditions. To make sure that one generalized equation provides the correct results in different part of the whole domain, the corresponding value of physical properties for each subdomain is employed. Thus, to ensure that the velocity is zero in the solid when the velocity field is calculated, one puts a very large value of viscosity coefficient for the grids points in the body domain, whereas for the grid points in the fluid domain, the real fluid viscosity coefficient is applied. When the temperature field is computed, the real values of heat transfer characteristics for the fluid and for the body are used, which gives the actual temperature field as a result of matching the temperature distribution in both computing subdomains (Exer. 1.13).

1.2.2   Using Universal Functions

Since the boundary conditions for subdomins are unknown, the required analytical solutions for body and fluid might be obtained only applying arbitrary nonisothermal boundary conditions. In other words, it is necessary to find a solution in the form that satisfies the governing equation at any (or arbitrary) boundary conditions (Exer. 1.14). Such solutions are given, in particular, in the form of universal functions, which are called universal because they satisfied a particular equation independent of boundary conditions [120]. In the next several paragraphs, we present two forms of universal functions used in this text for investigation of nonisothermal and conjugate heat transfer, including the performance of a conjugation procedure described in Section 2.2.2.

1.3   Integral Universal Function (Duhamel's Integral)

1.3.1   Duhamel's Integral Derivation

The Duhamel's integral presents a solution of some problems with varying variables in terms of known solutions of similar problems with the same variables considered as a constant parameters. This idea is based on two principles: (i) at a small interval of a variable, the function of interest may be approximately considered as a constant, and (ii) a solution of linear differential equation is presentable as a sum of other solutions of the same equation (superposition principle) (Exer. 1.15).

   Letting the solution of the problem in question depend on some variable t according to function $F(t)$ and function $f(x, t)$ is a solution of similar problem for a different but constant t. For example, we consider the heat transfer from a body with time-dependent surface temperature $F(t)$, and function $f(x, t)$ is a known solution of the same problem with constant surface temperature. During a small interval Δt of variable t (Fig. 1.1) the given function F may be considered as an approximate constant. Then, on this small interval $\Delta F = F'(t)\Delta t$ an approximate solution of the problem in question is defined as a product $f(x, t)\Delta F$, where $f(x, t)$ is the approximate solution for constant t and ΔF is the interval. Consequently, for the first interval we have $f(x, t)F(0)$, where $F(0)$ stands for $\Delta F(0)$ at the beginning at $t = 0$ (Fig. 1.1).

   For the next interval, the approximation starts at the time $t - \Delta t$ instead of t for the first one. Thus, the approximate solution is $f(x, t - \tau_1)\Delta F(\tau_1) = f(x, t - \tau_1)F'(\tau_1)\Delta \tau_1$, where $\tau_1 = \Delta t$ is the time lag in the second interval, and the small variation of function F is determined for time τ_1 when the second interval begins (Fig. 1.1). For the third interval, one gets similar solutions

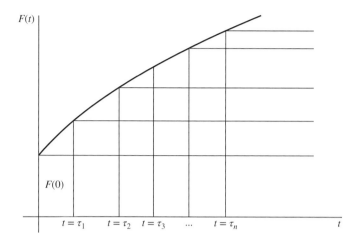

Figure 1.1 Duhamel's integral derivation: approximation of arbitrary dependence by step function

as $f(x, t - \tau_2)\Delta F(\tau_2) = f(x, t - \tau_2)F'(\tau_2)\Delta\tau_2$ with time lag τ_2 in the third interval, and so forth. The sum of those results gives the approximate solution of the considered problem, which in the limit at $\Delta t \to 0$ transforms in Duhamel's integral

$$T(x, t) = f(x, t)F(0) + \sum_{k=1}^{n} f(x, t - \tau_k)F'(\tau_k)\Delta\tau_k \tag{1.20}$$

$$T(x, t) = f(x, t)F(0) + \int_0^t f(x, t - \tau)F'(\tau)d\tau \tag{1.21}$$

Expression (1.21) presents the temperature of a body with given time-variable surface temperature $F(t)$ but in fact this is a general relation applicable to many other problems governed by linear equations. To use this integral in any particular case, it is enough to know desired function $F(t)$ and the relevant simple solution (Exer. 1.16 and 1.17).

In creating an universal function describing dependence between heat flux q_w and temperature head $\theta_w(x) = T_w - T_\infty$ in heat transfer, the role of function $F(t)$ plays temperature head $\theta_w(x)$, which is considered as a given, like a function $F(t)$. Since we are looking for relation defining the heat flux, it is clear that a heat transfer coefficient h from some solved problem should be taken instead of function $f(x, t)$. In such a case, the product $h\theta_w(x)$ determines the heat flux q_w as well as product $f(x, t)F(t)$ defines the body temperature in the example considered above. Then, the limit of a sum of products $h\theta_w(x)$, similar to the limit of the sum in equation (1.20), gives the desired universal function in the form of Duhamel's integral

$$q_w = h_1\theta_w(0) + \int_0^x h_\tau \frac{d\theta_w}{d\xi}d\xi \tag{1.22}$$

In this equation h_1 and h_τ are the heat transfer coefficients for the first and all other steps from a known solution of simple problem playing the same role as the functions $f(x, t)$ and $f(x, t - \tau)$ in equation (1.21) (Exer. 1.18).

The general expression (1.22) is applicable to any flow regime: laminar, turbulent, at zero, or non zero pressure gradient, if the heat transfer coefficients h_1 and h_τ from some relevant problem are known. It is common [201, 123] to consider as a known solution of a standard problem of heat transfer after temperature jump on a plate with an isothermal initial zone. In such a problem, the wall temperature remains constant (with isothermal heat transfer coefficient h_*) up to some point $x = \xi$ and then suddenly changes to another value resulting after temperature jump in heat transfer with coefficient h_ξ. The ratio of coefficients h_* and h_ξ defines the influence function $f(x, \xi) = h_\xi/h_*$ so called because it describes the effect of initial isothermal zone on heat transfer intensity after jump. The influence function is usually employed for putting $h_1 = h_*$ and $h_\tau = h_\xi$ which transforms equation (1.22) to the following standard form (Exer. 1.19)

$$q_w = h_* \left[\theta_w(0) + \int_0^x f(x, \xi) \frac{d\theta_w}{d\xi} d\xi \right] \tag{1.23}$$

This expression is an universal function because it determines the heat flux for arbitrary (for any) temperature head distribution $\theta_w(x)$ through integral of it derivative $d\theta_w/dx$.

1.3.2   Influence Function

Relation (1.23) is general as well as equation (1.22) and is also applicable to any flow regime, if the influence function for a specific case is known (Exer. 1.20). However, determining this function is another difficult task. For some simple cases, when the influence function depends on ratio ξ/x instead of each of those variables, an influence function was obtained using approximate methods. For the simplest case of laminar flow, Prandtl number close to one, and zero pressure gradient, the influence function was found by integral method (S. 7.6) in the form [123, 201] (Exam. 7.13)

$$f(\xi/x) = [1 - (\xi/x)^{3/4}]^{-1/3} \tag{1.24}$$

The more general result for the laminar gradient flow, the influence function was obtained for self-similar flows with power-law velocities $U = Cx^m$ (S. 7.5.2), but only for fluids with large or small Prandl numbers. It was shown that the same formula (1.24) is valid with exponents depending on velocity power m for large and small Prandtl numbers according to the first and second formula, respectively

$$f(\xi/x) = [1 - (\xi/x)^{3(m+1)/4}]^{-1/3} \qquad f(\xi/x) = [1 - (\xi/x)^{m+1}]^{-1/2} \tag{1.25}$$

The first formula was obtained assuming a linear velocity distribution in the thermal boundary layer [201, 230]. Therefore, it is applicable to fluids with large Prundtl numbers for which such assumption is close to reality because in this case the thermal boundary layer is thin relatively to velocity layer (S. 7.7). At the limit $\mathrm{Pr} \to \infty$, this approximate formula becomes exact. For another limiting case $\mathrm{Pr} \to 0$, the situation is opposite: the dynamic boundary layer thickness is so thin that velocity across the thermal boundary layer is practically equal to the external velocity $U(x)$ (S. 7.7). Since in this case, the velocity in the thermal layer is independent on y, the thermal boundary layer equation simplifies, and for self-similar flows, second formula (1.25) is derived [338].

For turbulent flows, the structure of known relations for influence function remains also the same. In particular, for this case at the same conditions as for laminar flow ($\Pr \approx 1$, zero pressure gradient), it was shown that formula (1.24) is valid with exponents 9/10 at (ξ/x) and ($-1/9$) at brackets [123, 201]. There are also several relations for turbulent flows derived from experimental data. Those are presented also in the form (1.24), but with slightly different exponents: unity at (ξ/x) and at the brackets: (-0.114) [208], (-0.12) [201, 275] and (-0.2) or with giving practically the same results exponents 39/40 at (ξ/x) and ($-7/39$) at the brackets (instead of unity and -0.2) [267].

The review shows that the known relations for influence function for laminar and turbulent flows pertain only for the simplest cases: limiting values of Prandtl numbers and basically at zero pressure gradients. Later we will see that employing the universal functions permits general expressions for influence function from which follow the particular results.

1.4   Differential Universal Function (Series of Derivatives)

This form of universal function may be obtained from the integral relation (1.23) using successive integration by parts. We start from simple case of zero pressure gradient. As we have seen above, for this simple case, the influence function depends on the ratio ξ/x, rather than of each variable separately, and integral formula (1.23) takes the form (Exer. 1.21)

$$q_w = h_* \left[\theta_w(0) + \int_0^x f(\xi/x) \frac{d\theta_w}{d\xi} d\xi \right] \tag{1.26}$$

Denoting $\zeta = \xi/x$ and applying for integration the following parts

$$u_1 = \frac{d\theta_w}{d\xi}, \quad dv_1 = xf(\zeta)d\zeta, \quad v_1 = x\left(\int_0^\zeta f(\gamma)d\gamma - 1\right), \tag{1.27}$$

we get according to formula for integration by parts an expression (Exer. 1.22)

$$uv \Big|_{\xi=0}^{\xi=x} - \int_0^x v\,du = x \frac{d\theta_w}{d\xi} \left(\int_0^1 f(\zeta)d\zeta - 1 \right) + x \frac{d\theta_w}{d\xi}\Big|_{x=0}$$

$$- x \int_0^x \left(\int_0^\zeta f(\zeta)d\zeta - 1 \right) \frac{d^2\theta_w}{d\xi^2} d\xi \tag{1.28}$$

Substitution of this result in equation (1.26) leads to modified relation for heat flux

$$q_w = h_* \left[\theta_w(0) + \frac{x}{1!} \frac{d\theta_w}{dx}\Big|_{x=0} + g_1 x \frac{d\theta_w}{dx} - x \int_0^x \left(\int_0^\zeta f(\gamma)d\gamma - 1 \right) \frac{d^2\theta_w}{d\xi^2} d\xi \right] \tag{1.29}$$

$$g_1 = \int_0^1 f(\zeta)d\zeta - 1$$

Here g_1 is a constant defined by the first integral in the right hand part of equation (1.28).

Comment 1.3 In this derivation we used several variables: x, ξ, γ, $\zeta = \xi/x$. Two of those, x and ζ, are working variables, whereas two others, ξ and γ, are so-called dummy variables, which play a subsidiary role. In this case, dummy variables are used for carry out integrals (to distinguish from upper limit of integral). Thus, γ is employed in integral (1.27) instead of ζ, and ξ is applied instead of x in the last integral in (1.28).

Relation (1.28) is further modified using the following parts for k integration

$$u_k = \frac{d^k\theta_w}{d\xi^k}, \quad v_k = \int_0^\zeta v_{k-1}d\gamma + \frac{(-1)^k x}{k!} \tag{1.30}$$

Putting here $k = 2$, we obtain the parts (1.31) for transforming the last integral in relation (1.29) via integration by parts that result in further modification of expression (1.29) for heat flux in a similar form (1.32) (Exer. 1.23)

$$u_2 = \frac{d^2\theta_w}{d\xi^2}, \quad v_2 = x\int_0^\zeta d\gamma \left(\int_0^\zeta f_2(\gamma)d\gamma - 1\right) + \frac{x}{2!}, \quad f_2(\zeta) = \int_0^\zeta d\zeta \int_0^\zeta f(\zeta)d\zeta - \zeta + \frac{1}{2} \tag{1.31}$$

$$q_w = h_* \left[\theta_w(0) + \frac{x}{1!}\frac{d\theta_w}{dx}\bigg|_{x=0} + \frac{x^2}{2!}\frac{d^2\theta_w}{dx^2}\bigg|_{x=0} + g_1 x\frac{d\theta_w}{dx} + g_2 x^2\frac{d^2\theta_w}{dx^2} + x^2\int_0^x f_2(\zeta)\frac{d^3\theta_w}{d\xi^3}d\xi \right] \tag{1.32}$$

In the first equation $f_2(\zeta) = v_2(\zeta)/x$, and in the second equation coefficient $g_2 = -f_2(1)$ is the value of this function at $\zeta = 1$. As it is seen, function $v_2(\zeta)$ arises in the last integral in equation (1.32) and is defined by relations (1.31), where for simplicity the dummy variable γ is substituted by variable $\zeta = \xi/x$ (Exer. 1.24).

Repeating the integration by applying the parts indicated by equation (1.30), we finally arrive in the following series with coefficients g_k determined by relations (1.34)

$$q_w = h_* \left[\theta_w(0) + \frac{x}{1!}\frac{d\theta_w}{dx}\bigg|_{x=0} + \frac{x^2}{2!}\frac{d^2\theta_w}{dx^2}\bigg|_{x=0} + \ldots + \frac{x^k}{k!}\frac{d^k\theta_w}{dx^k}\bigg|_{x=0} + \ldots + g_1 x\frac{d\theta_w}{dx} \right.$$

$$\left. + g_2 x^2\frac{d^2\theta_w}{dx^2} + \ldots + g_k x^k\frac{d^k\theta_w}{dx^k} \ldots + (-1)^k x^k\int_0^x f_k(\zeta)\frac{d^{k+1}\theta_w}{d\xi^{k+1}}d\xi \right] \tag{1.33}$$

$$g_k = (-1)^{k+1}f_k(1), \quad f_k(\zeta) = \int_0^\zeta d\zeta \int_0^\zeta d\zeta \ldots \int_0^\zeta f(\zeta)d\zeta + \sum_{n=1}^{n=k}\frac{(-1)^n \zeta^{k-n}}{n!(k-n)!} \tag{1.34}$$

The k times repeated integral and the sum in the last equation may be presented as follows (Exer. 1.25 and 1.26)

$$\int_0^\zeta d\zeta \int_0^\zeta d\zeta \ldots \int_0^\zeta f(\zeta)d\zeta = \frac{1}{(k-1)!}\int_0^\zeta (1-\zeta)^{k-1}f(\zeta)d\zeta \tag{1.35}$$

$$\sum_{n=1}^{n=k} \frac{(-1)^n \zeta^{k-n}}{n!(k-n)!} = \frac{1}{k!}[(\zeta - 1)^k - \zeta^k] \tag{1.36}$$

Then, the coefficients g_k according to relation (1.34) are defined as (Exer. 1.27)

$$g_k = \frac{(-1)^{k+1}}{k!}\left(k\int_0^1 (1 - \zeta)^{k-1}f(\zeta)d\zeta - 1\right) \tag{1.37}$$

Analyzing equation (1.33) for heat flux, one sees that the first sum, starting with $\theta_w(0)$, represents an expansion of function $\theta_w(x)$ as a Taylor series at $x = 0$. Therefore, if at $k \to \infty$, the last integral in equation (1.33) (which is a remainder) goes to zero, this expression turns into infinite series determining the heat flux in terms of derivatives of temperature head distribution $\theta_w(x)$ in the following form (Exer. 1.28)

$$q_w = h_*\left(\theta_w + \sum_{k=1}^{\infty} g_k x^k \frac{d^k \theta_w}{dx^k}\right) \tag{1.38}$$

As shown in [101] and repeated in [119, p. 55], this relation is an exact particular result for zero pressure gradient obtained from more general exact solution of thermal boundary layer equation. At the same time, we just derive this relation from integral formula for heat flux (1.26) using the exact procedure of integration by parts. These two facts show that both relations—the integral (1.26) and differential formula (1.38)—are two equivalent exact expressions for heat flux in a flow with a zero pressure gradient. These expressions are universal because they both describe the dependence between heat flux and arbitrary (say any) surface temperature head in two forms: as integral consisting of the first derivative of temperature head or as a series of derivatives of it (Exer. 1.29).

1.5 General Forms of Universal Function

The more general expression for heat flux than relation (1.38), derived in [101] and mentioned above, is an exact solution of thermal boundary layer equation for self-similar flows with power-low external velocity distribution $U = Cx^m$ (S. 7.5.2). This solution is the same series (1.38) but written in Görtler variable Φ (S. 7.4.4.2) (see [119])

$$q_w = h_*\left(\theta_w + \sum_{n=1}^{\infty} g_k \Phi^k \frac{d^k \theta_w}{d\Phi^k}\right) \qquad \Phi = \frac{1}{\nu}\int_0^x U(\xi)d\xi \tag{1.39}$$

In the case of zero pressure gradient ($U = const$), series (1.39) transforms in (1.38) since for constant U variable $\Phi = \mathrm{Re}_x$, and due to that Φ is proportional to x (Exer. 1.30). Because series (1.38) and integral relation (1.26) are equivalent, it is clear that substituting the Görtler variable Φ for x in integral (1.26) yields expression also valid for self-similar flows with power-low external velocity distribution $U = Cx^m$ (Exer. 1.31 and 1.32)

$$q_w = h_*\left[\theta_w(0) + \int_0^\Phi f(\xi/\Phi)\frac{d\theta_w}{d\xi}d\xi\right] \tag{1.40}$$

The two equivalent universal functions (1.39) and (1.40) are solutions of boundary layer equations because, as it noted at the beginning of this section, expression (1.39) is obtained from thermal boundary layer equation. It is known, that solutions of that type are applicable to majority of practically important applications. Two facts ensure this conclusion: (i) the properties (kinematic viscosity and thermal diffusivity) of the essential technical fluids, such as air, water, oil, and liquid metals, are small, and due to that the corresponding Reynolds and Peclet numbers are large, and (ii) as it was shown by Prandtl, the Navier-Stokes equations simplifies to boundary layer equations for high Re and Pe because in this case, the viscosity effects are significant only in a thin layer adjacent to the body surface (S. 7.4.4.1).

To employ universal functions (1.39) and (1.40) for calculations, the series coefficients g_k and influence function $f(\xi/\Phi)$ for the integral are required. Examples considered above in Section 1.3.2 show that all known formulae for influence function have the same structure with different values of exponents. Proceeding from that fact of similarity, we use analogous expression for general form of influence function with unknown exponents C_1 and C_2

$$f(\xi/\Phi) = [1 - (\xi/\Phi)^{C_1}]^{-C_2} \tag{1.41}$$

In the next sections, the coefficients g_k and exponents C_1 and C_2 are calculated for laminar, turbulent flows, and several other regimes and conditions.

Exercises

It is assumed (see Preface) that to perform some exercises a reader gets additional knowledge from Part III using references indicated in text.

1.1 Explain how the conjugate problem differs from other heat transfer problems.

1.2 Why do conjugate heat transfer problems contain at least two subdomains? Name these subdomains.

1.3 What is the difference between Laplace and Poisson equations? In what cases and why are these equations simplified? Explain physically the difference between "thin" and "thermally thin" bodies.

1.4 Compare Navier-Stokes and full energy equations with simplified boundary layer and creeping equations. Explain what and why part of equation terms may be neglected? Are the simplified equations exact or approximate? Think: why the set of equations for creeping flow does not include the equation for turbulent flow?

1.5 Study Sections 8.2.3 to understand the Reynolds averaging and arising Reynolds stresses in averaged Navier-Stokes equations. Think: why are these stresses so much greater than molecular ones? Explain the differing between physical nature of coefficients μ and μ_{tb}, Compare Navier-Stokes (1.4)–(1.7) and Reynolds (1.14) equations to better understand the Einstein notations.

1.6 What is the essential difference between boundary conditions of the third and forth kinds?

1.7 Is the problem of atmosphere and ocean interaction a conjugate one? Explain your answer. What are the subdomains in such a problem? What equations are relevant?

1.8 Study some features of partial differential equation of second order using, for example, *Advanced Engineering Mathematics Course*, to understand what is a canonical equation form? How do canonical equations differ from each other? Learn or recall why the Navier-Stokes equation is difficult to solve. What is nonlinearity? How do nonlinear and linear equations differ?

1.9 Discuss with other students or colleagues the difficulties arising in formulating the boundary conditions for Navier-Stokes and full energy equations. Explain why it is easier to formulate boundary conditions for boundary layer equations.

1.10 Compare well- and ill-posed problems. What causes the additional difficulties in formulating Dirichlet problem for Navier-Stokes and energy equations?

1.11 Explain the term "deficient situation". Why do such situations occur in conjugate problem statements? How can this difficulty be resolved?

1.12 What is the basic idea of iteration method of conjugation? What is the method of trial-and-error? (see article "Trial and Error" on Google, on Wikipedia)

1.13 Explain how the problem of different physical properties of body and fluid is resolved in the conjugation by one equation for entire domain.

1.14 What is the universal function? Why is such a function required for an analytical solution of the conjugate problem?

1.15 The idea of Duhamel's integral is based on a superposition principle. Read about the superposition method to understand why it is applicable for linear, and is not appropriate for nonlinear equations (see Exer. 1.8).

1.16 In the development of Duhamel's integral, we used the first ordinate of each interval as a value of function $F(t)$. Think: will the result (1.20) be the same if we use the middle or final ordinate of each interval instead of the first one? Explain your answer.

1.17 What is the time lag? Remind or read about this term, for example, on Wikipedia. Think: what is the difference between time legs in the two equations (1.20) and (1.21)? Can the first one be transformed in the second?

1.18 Repeat the derivation of equation (1.22) from Duhamel's integral (1.21) to answer why: (i) function $\theta_w(x)$ and heat transfer coefficient represent in this case functions $F(t)$ and $f(x,t)$ in initial integral (1.21), and (ii) heat transfer coefficients h_1 and h_τ relate to functions $f(x,t)$ and $f(x, t - \tau)$.

1.19* Describe the problem of heat transfer after a temperature jump on the plate with an isothermal initial zone, which is usually used as a standard problem with a known solution. Draw a graph showing the temperature variation along the plate. Obtain equation (1.23).

1.20 What is an influence function? How does it relate to Duhamel's integral?

1.21 Discuss with your friend or colleague the benefits of universal function for conjugate heat transfer problems. Think about other examples where universal functions might be useful. What is the basic characteristic that distinguishes this type of relations from others?

1.22* Repeat the first integration by parts and derive equation (1.29) from universal function (1.26). Hint: note that the integration is performed by variable ξ, and because of that, x is considered as a constant parameter in process of this integration.

1.23* Obtain parts for third integration from general relation (1.30). Hint: begin from deriving the parts for first and second integrations using relations (1.30) to understand the procedure.

1.24 Extend the series (1.32) by adding two next terms without farther calculations. Compare your results with formulae (1.33) and (1.34). Hint: first, analyze the rules according to which the existing terms are constructed and then proceed using these rules.

1.25* Expression (1.35) for k-times repeated integral is obtained by series of integration by parts. Show that this is true for $k = 2$. Hint: take the integral with function $f(\zeta)$ as one part (u) and $d\zeta$ as another part (dv).

1.26 Check equation (1.36) for k $= 2$ and $k = 3$ to see that this equation is correct.

1.27 Derive equation (1.37) from relation (1.34) using expressions (1.35) and (1.36). Hint: take into account that $k!$ is a product of (k-1)! and k.

1.28 Recall or study the Taylor series to understand the procedure of transforming equation (1.33) into series (1.38).

1.29 At the end of this section (S. 1.4) it is stated that the integral with the first derivative of temperature head and a series with successive derivatives of it are universal functions, because they describe the dependence between heat flux and arbitrary (say any) surface temperature head. Explain why this statement is true or in other words, why a function describing a dependence of some arbitrary variable is universal? In what sense is such function universal?

1.30 Show that in the case of constant external flow $U = const$, the Görtler variable Φ becomes Reynolds number, and expression (1.39) transforms into relation (1.38).

1.31 Prove that in the case of constant external flow the expression (1.40) transforms into equation (1.26). Explain why it is possible to write $f(\xi/\Phi) = f(\xi/x)$ in this case. Hint: think about the connection between dummy and working variables explained in analyzing examples in Comment 1.2. Are the dummy variables ξ the same in both influence functions in equations (1.26) and (1.40)?

1.32 Recall why the arbitrary external velocity corresponds to arbitrary pressure gradient. What equation from a set of relations given in the beginning of this chapter tells us about this fact? What is the name of this equation? In what type of flow does such connection between velocity and pressure hold? Hint: see Section 7.1.2.5.

1.6   Coefficients g_k and Exponents C_1 and C_2 for Laminar Flow

In this section we discuss the basic features of coefficients g_k that determine the differential universal function for laminar flows and show how the exponents C for the general form (1.41) of influence function may be estimated using known coefficients g_k. Then, in the next sections, similar coefficients and exponents are evaluated for turbulent, compressible flows and for some other cases.

1.6.1   Features of Coefficients g_k of the Differential Universal Function

We discussed in the previous section how the exact solution of the thermal boundary layer equation is obtained for self-similar laminar flows with power-law external velocity distribution $U = cx^m$. Such distribution occurs on the wage with open angle $\pi\beta$ (Fig. 1.2) streamlined by potential flow. The exponent m in velocity distribution and open wage angle β are connected by relation $m = \beta/(2 - \beta)$. The first four coefficients g_k of series (1.39) calculated for this case are plotted in Figures 1.3 and 1.4 as functions of Prandl number for different external flow velocities (different β).

These data are obtained numerically (details in [119]). For limiting cases $\text{Pr} \to 0$ and $\text{Pr} \to \infty$, the corresponding simplified equations are solved analytically leading to following

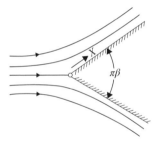

Figure 1.2   Flow past a wage. At the leading edge, the potential velocity is $U = cx^m$

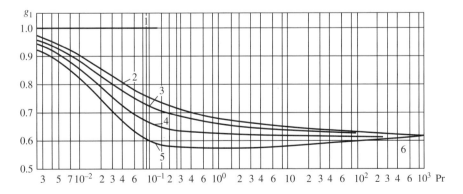

Figure 1.3   Coefficient $g_1(\text{Pr}, \beta)$ of universal function (1.39) for laminar boundary layer. Asymptotes: $1 - \text{Pr} = 0$, $6 - \text{Pr} \to \infty$; β: 2-1 (stagnation point), $3 - 0.5$ (favorable pressure gradient), 4-0 (zero pressure gradient), 5-(−0.16) (preseparation pressure gradient) (S. 7.5.2)

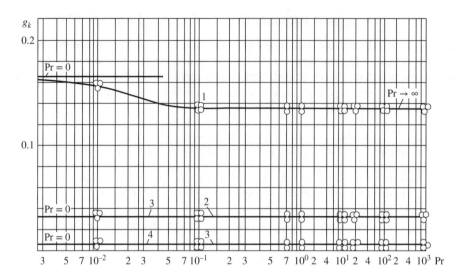

Figure 1.4   Coefficients $g_k(\mathrm{Pr})$ of universal function (1.39) for laminar boundary layer. $1 - (-g_2)$, 2-g_3, $3 - (-g_4)$, ∘ − numerical integration

formulae and numerical results [119] ($\Gamma(j)$ -is gamma function, Exer. 1.33)

$$g_k = \frac{(-1)^{k+1}}{k!(2k-1)} \quad \mathrm{Pr} \to 0, \quad g_k = \Gamma\left(\frac{2}{3}\right) \sum_{i=0}^{i=k} \frac{(-1)^{k+i}\Gamma\,[(4i/3)+1]}{(k-i)!\,i!\,\Gamma\,[(4i/3)+2/3]} \quad \mathrm{Pr} \to \infty \tag{1.42}$$

$$g_1 = 1,\ g_2 = -1/6,\ g_3 = 1/30,\ g_4 = -1/168,\ g_5 = 1/1080,\ g_6 = -1/7920 \quad \mathrm{Pr} = 0 \tag{1.43}$$

$$g_1 = 0.6123,\ g_2 = -0.1345,\ g_3 = 0.0298,\ g_4 = -0.0057 \qquad \mathrm{Pr} \to \infty \tag{1.44}$$

Data from Figures 1.3 and 1.4 and limiting values (1.43) and (1.44) yield the following basic features of coefficients g_k of universal function (1.39): (i) coefficient g_1 depends on the external velocity (via β) and on the Prandtl number; this dependence is more significant for small Prandtl numbers ($\mathrm{Pr} < 0.5$); for $\mathrm{Pr} \to 0$ and $\mathrm{Pr} \to \infty$, the values of g_1 for all β tend to the greatest $g_1 = 1$ and to the lowest $g_1 = 0.6123$ values, respectively, (ii) coefficient g_2 is practically independent of the external velocity and depends slightly only on the Prandtl number in the region of small Prandtl numbers; for $\mathrm{Pr} \to 0$ and $\mathrm{Pr} \to \infty$ the values of g_2 also tend to the greatest absolute value $|g_2| = 1/6$ and to the lowest absolute value $|g_2| = 0.1345$, respectively; (iii) coefficients g_3 and g_4 are independent of both the external velocity and the Prandtl number, so that numerically obtained values for whole diapason of Prandtl numbers practically coincide with the limiting values $g_3 = 1/30$ and 0.0298 and $g_4 = -1/168$ and -0.0057, (iv) therefore, for $k \geq 3$, coefficients g_k may be estimated using simple formula (1.42) for $\mathrm{Pr} \to 0$, (v) coefficients g_k for universal function (1.39) rapidly decrease with the number of terms, so that using first two or three coefficient usually gives acceptable results (Exer. 1.34).

Comment 1.4 The isothermal heat transfer coefficient h_* required for employing both universal functions (1.39) and (1.40) may be estimated by any of known methods reviewed, for example, in [369]. These methods are well tested during many years when the isothermal heat transfer coefficient was used for practical calculations, in particular, as a part of the boundary conditions of the third kind.

Analysis of just considered features of coefficients g_k shows that the exact for power-law external flows universal function (1.39) provides practically accurate approximate results for arbitrary external flow velocity distribution. The following facts specified this statement: (i) coefficient g_1 slightly depends on external velocity (say on β), whereas other coefficients are practically independent of the external velocity, (ii) if coefficients g_1 were also independent of β, the relation (1.39) would be an exact relation for arbitrary external velocity $U(x)$, (iii) in reality, according to Figure 1.3, the effect of β (i.e., external velocity) on the first coefficient reaches the maximum of $\pm12\%$ from average value $g_1 \approx 0.675$ in vicinity of $\mathrm{Pr} = 0.1$ and then decreases to zero in both limiting cases of Prandtl number at $\mathrm{Pr} \to 0$ and $\mathrm{Pr} \to \infty$.

Thus, universal function (1.39) in general case of arbitrary external flow (or pressure gradient) with average values of coefficients g_k provides the calculation results with inaccuracy less than $\pm12\%$ which is comparable with accuracy of other existing approximate methods (S. 7.6) (see also [338]). Accuracy may be increased by estimating the value of β. In some cases, β is known, for example, it is clear that for the flow past plate $\beta = 0$ as well as for transverse flow past circular cylinder or other body with blunt nose $\beta = 1$ close to the stagnation point, whereas for the rest part of the surface of such body the value of β may be approximately considered as zero. In other cases, the parameter β may be estimated using some simple relation, for example, formula

$$\beta = 2(1 - \Phi/\mathrm{Re}_x) \tag{1.45}$$

which results from assumption that average velocities of considering distribution $U(x)$ and power-law flow distribution $U = cx^{\beta/(2-\beta)}$ are equal (Exer. 1.35).

Comment 1.5 To understand why Görtler variable extends applicability of universal function (1.38) obtained for zero pressure gradients, consider Görtler variable in the form $\Phi = U_{av}x/\nu = \mathrm{Re}_{(av)x}$ (Exer. 1.36). Here, $U_{av}(x)$ is the average external flow velocity for the interval from the leading edge ($x = 0$) to point with coordinate x. It follows from this presentation of Görtler variable that function Φ (1.39) takes into account the flow history. Physically, it means that the flow characteristics at point x are determined not only by local parameters at this point but also by those at other points along the whole considering interval $(0, x)$. Thus, Görtler variable takes into account, in particular, the variation of pressure gradients along the considering interval.

Comment 1.6 In general, the characteristics of some point of interest are governed by: (i) in the simplest case, local data only (at this point); for example, the coefficients of friction and heat transfer at some point in a flow past wage with given surface and free stream temperatures are specified only by local values of Reynolds and Nusselt numbers (S. 7.5.2), (ii) local and historical data (at this and behind points) as, for example, in the same problem for wage but with nonisothermal surface when to get characteristics of some point, the surface temperature at this and behind points is required in addition (S. 2.1.1), and (iii) local, historical and future data (at this, behind and advanced points) as, for example, in any Dirichlet problem for elliptical equation when for characteristics of any point, the information is necessary from whole boundary of domain (Exer. 1.37).

1.6.2 Estimation of Exponents C_1 and C_2 for Integral Universal Function

Relation (1.37) establishes the connection between coefficients g_k and influence function $f(\zeta)$ where for general case $\zeta = \xi/\Phi$. Substituting expression (1.41) into relation (1.37), expanding $(1 - \zeta)^{k-1}$ via a binominal formula, and introducing a new variable $r = \zeta^{C_1}$ leads to an equation

$$g_k = \frac{(-1)^{k+1}}{k!} \left[\frac{k}{C_1} \sum_{m=0}^{k-1} (-1)^m \frac{(k-1)!}{m!(k-m-1)!} B\left(\frac{m+1}{C_1}, 1 - C_2\right) - 1 \right] \tag{1.46}$$

determining the dependence between coefficients g_k and exponents C_1 and C_2 required for integral universal function (1.40). Here, $B(i,j)$ is beta function, which is specified through combination of gamma functions (Exer. 1.38)

$$B(i.j) = \int_0^1 r^{j-1}(1-r)^{j-1} dr = \Gamma(i)\Gamma(j)/\Gamma(i+j) \tag{1.47}$$

It is easy to calculate coefficients g_k employing relation (1.46) and knowing exponents of some influence function. For example, since relation (1.46) is a result of exact solution for self-similar flows, this equation may be used to check the accuracy of influence function (1.24) and others of this type approximate equations. Substitution $C_1 = 3/4$ and $C_2 = 1/3$ in equation (1.46) gives: $g_1 = 0.612$, $g_2 = -0.131$, $g_3 = 0.03$, $g_4 = -0.0056$. Those are practically the same as (1.44) obtained from exact solution for the limiting case $Pr \to \infty$. This tells us that function (1.24) is correct (Exer. 1.38). Similarly, equation (1.46) with exponents $C_1 = 1$ and $C_2 = 1/2$ gives coefficients g_k that are in agreement with the values (1.43) obtained from exact solution for $Pr \to 0$ (details in [119]).

More complicated is the inverse problem of determining exponents in influence function (1.41) knowing coefficients g_k. Two facts cause the difficulties in solving this problem: (i) relation (1.46) is transcendental, and due to that could not be solved for exponent C_1 or C_2 (Exerc. 1.39), and (ii) there are only two unknowns C_1 and C_2, but countless known coefficients g_k. While the first difficulty is a technical question that may be resolved applying a graphic approach, or software based on trial and error or on other numerical method, the second problem is a fundamental complexity, known as an overdetermined system, when a number of equations exceeds the number of unknowns. In this case, such situation produces numerous results because each pair of coefficients g_k after substitution in equation (1.46) gives two equations defining unknown C_1 and C_2. However, as we have seen, this particular set of coefficients g_k consists of only two weighty coefficients g_1 and g_2, whereas the others are comparatively negligible. Due to that, it may be expected that exponents C_1 and C_2 obtained from system of two equations with coefficients g_1 and g_2 would be appropriate (Exer. 1.40 and 1.41).

This is confirmed by calculation results plotted in Figure 1.5 showing that all particular cases of laminar flow considered in Section 1.3.2 follow from data of this pattern. It is seen from Figure 1.5 that for whole interval of Prandtl number, exponents C_1 and C_2 vary slightly from 1 to 3/4 and from 1/2 to 1/3, respectively, as it should be according to relations (1.25).

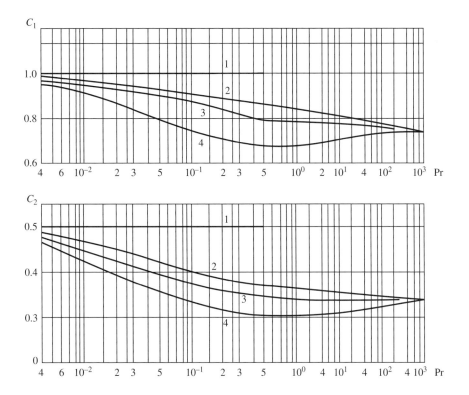

Figure 1.5   Exponents C_1 and C_2 for laminar boundary layer. $1 - \mathrm{Pr} = 0, 2 - \beta = 1, 3 - 0, 4\text{-}(-0.16)$

These data tells us as well as data of a slight dependence of g_k on pressure gradient (on β from Figs 1.3 and 1.4) identifies that the Görtler variable Φ takes into account the flow history (Com. 1.5). This property of variable Φ results in independence of the exponents in influence function (1.41) written in variables (ξ/Φ) on pressure gradients. That means that the values of C_1 and C_2 remains the same $3/4$ and $1/3$ at $\mathrm{Pr} \to \infty$ as well as 1 and $1/2$, at $\mathrm{Pr} \to 0$ for arbitrary external velocity distribution, as those in the known simple influence functions (1.25) for the plate at zero pressure gradient ($m = 0$) presented in variables (ξ/x).

This is also in line with data from Figure 1.5 showing that C_1 and C_2 are independent on β in both limiting cases being the same for arbitrary external velocity as just indicated values $3/4$ and $1/3$ at $\mathrm{Pr} \to \infty$ and 1 and $1/2$, at $\mathrm{Pr} \to 0$. Moreover, it is easy to check that the functions (1.25) for self-similar flows follow from relation (1.41) because in the case of $U = cx^m$, one gets: $\xi/\Phi = (\xi/x)^{m+1}$ (Exer. 1.42).

Now we have a full set of constants for using universal functions for general case of laminar flow. The same universal function (1.39) and (1.40) are applicable to other flow regimes and situations, however, with proper coefficients g_k and exponents C. Next sections present those constants for several other cases.

Comment 1.7   As we will see in applications, employing two forms of universal function provides accurate results of calculations. This is achieved by using the differential form with several first terms, when the series converges fast, and employing the integral form if the achieved by series accuracy is not satisfactory.

1.7 Universal Functions for Turbulent Flow

To get coefficients g_k and exponents C for the two forms of universal functions in the case of turbulent flow, the solution of thermal boundary layer equation for this case, analogous to that given in [101] for laminar flow, was obtained in [106] using the same Görtler variable (1.39).

Comment 1.8 This solution as well as some others considering in this and in the next chapters are presented more detailed in author's monograph [119]. Therefore, below we indicate in the brackets only the relevant page or section of this book without repeating citation [119]. For the case of turbulent flow solution such citation is [Chapter 4] instead of [119, Chapter 4].

The thermal boundary layer equation (1.17) for turbulent flow differs from similar equation (1.11) for laminar flow by last two terms containing additional dynamic v_{tb} and thermal α_{tb} turbulent transfer coefficients. In contrast to analogous v and α laminar coefficients, which are physical properties, the turbulent coefficients are complex functions depending on flow characteristics, and because of that a turbulence model is required for their estimation. Usually, to estimate turbulent thermal diffusivity, the turbulent Prandtl number $Pr_{tb} = v_{tb}/\alpha_{tb}$ similar to physical Prandtl number is introduced. Then, coefficient v_{tb} is defined, and turbulent diffusivity is found as ratio $\alpha_{tb} = v_{tb}/Pr_{tb}$ using Pr_{tb} which usually is taken to be close or equal unity (S. 2.1.2.4, Exer. 1.43).

Solution of turbulent thermal boundary layer equation requires also the velocity profiles in boundary layer. In solution derived in [106], the Mellor-Gibson turbulence model [260] is used to calculate both transfer coefficients and velocity profiles. This model is one of the modern algebraic models based on dependences valid for equilibrium turbulent boundary layers, which are as well as the self-similar laminar boundary layers flows with constant dimensionless pressure gradient (S. 8.3.2 and S. 8.3.3)

$$\beta = \frac{\delta_1}{\tau_w} \frac{dp}{dx} \tag{1.48}$$

where δ_1 and τ_w are displacement thickness and skin friction stress (S. 7.5.1.1). As we have seen in Section 1.6.1, the parameter β determines the wage angle $\pi\beta$ and in the case of laminar self-similar flows is connected with the exponent in velocity power law $U = cx^m$. For this case, it is easy to check that the complex (1.48) is a constant (Exer. 1.44). In the case of the equilibrium turbulent boundary layer that can be demonstrated as well by more complicated analysis [422].

It is shown [106 or p. 99, Com. 1.8] that solution of thermal turbulent boundary layer equation for heat flux on arbitrary nonisothermal surface may be presented by the same two universal functions in slightly different Görtler variables (S. 8.3.6.3) with specific coefficients g_k of series (1.39) and exponents C of influence function in integral (1.40). In this case, these specific parameters depend on pressure gradient via β and on Prandtl number as in the case of laminar flow and in addition on Reynolds number. Calculations were performed for $\beta = -0.3$ (flow at stagnation point), $\beta = 0$ (zero pressure gradient flow), $\beta = 1$ and $\beta = 10$ (flows with weak and strong adverse pressure gradients) for the following Prandtl and Reynolds numbers:

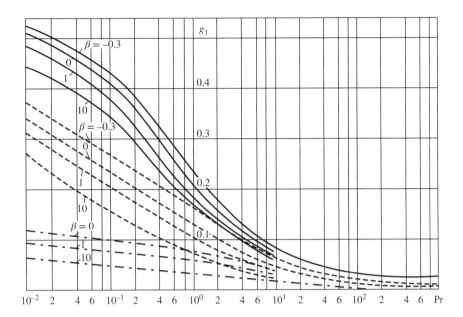

Figure 1.6 Coefficient g_1 for turbulent boundary layer: ____ $\mathrm{Re}_{\delta_1} = 10^3$, _ _ _ 10^5, _._._ 10^9

Pr $= 0.01, 0.1, 1, 10, 100, 1000$ $\mathrm{Re}_{\delta_1} = 10^3, 10^5, 10^9$. Other details of turbulent boundary layer computing may be found in [106, p. 102]. The results are plotted in Figures 1.6 and 1.7 for coefficients g_k and in Figures 1.8 and 1.9 for exponents C as functions of Prandtl number for different values of β and Re_{δ_1}.

The following conclusions are formulated analyzing these data: (i) the values of the coefficients g_k rapidly decrease with increasing k as well as in the case of laminar flow; hence, one may use only a few of the first terms in series (1.39) to get satisfactory accuracy, (ii) as in the case of laminar flow, the coefficients g_k decrease with increasing Prandtl number (Exer. 1.45); however, in contrast to the case of laminar flow, where at large Prandtl numbers the value of coefficients become independent of Pr, but remain finite, in the case of turbulent flow they tends to zero with increasing Pr, so that starting with some value of Prandtl number (say $\approx 10^2$), the effect of nonisothermicity becomes negligible, (iii) for this case, coefficients g_k are smaller than the corresponding coefficients for laminar flow, and they decrease with increasing Reynolds number indicating that the nonisothermicity affects the heat transfer in turbulent flows relatively lesser (Exer. 1.46), (iv) coefficients g_1 and g_2 depend slightly on β, whereas the others are practically independent of β, this allows one to use the universal function (1.39) for turbulent flows with an arbitrary pressure gradient as for laminar flows estimating value of β by the same approach, (v) exponents C_1 and C_2 increase with decreasing Prandtl and Reynolds numbers, (vi) exponent C_2 increases with decreasing pressure gradient, whereas the exponent C_1 is practically independent of pressure gradient.

It follows from Figures 1.8 and 1.9 that $C_1 = 1$, $C_2 = 0.18$ for zero pressure gradient flow when Pr $= 1$ and $\mathrm{Re}_{\delta_1} = 10^3$, but under the same conditions and $\mathrm{Re}_{\delta_1} = 10^5$, the same figures give $C_1 = 0.84, C_2 =$ ⁀hese computed results are close to exponents in well-known influence function cons⁀ ⁀ection 1.3.2: the first exponents are almost the same as $C_1 = 1$

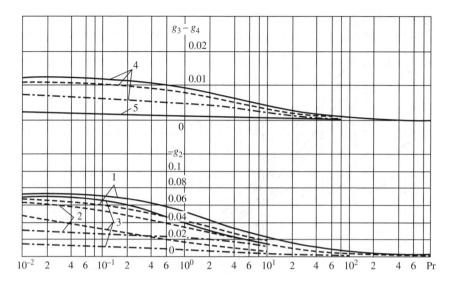

Figure 1.7 Coefficients g_k for turbulent boundary layer: g_2 : $1 - \beta = -0.3$, $2 - \beta = 0$, $3 - \beta = 1$, $4 - g_3$, $5 - g_4$, ___ $\mathrm{Re}_{\delta_1} = 10^3$, _ _ _ 10^5, _._._ 10^9

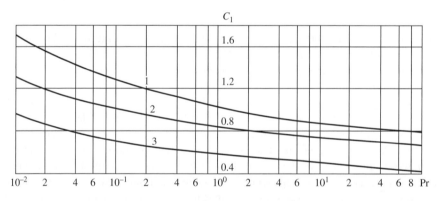

Figure 1.8 Exponent C_1 for turbulent boundary layer: $\mathrm{Re}_{\delta_1} = 10^3$, $2 - 10^5$, $3 - 10^9$

and $C_2 = 0.2$ derived theoretically and confirmed by measuring data at relatively low Reynolds number $\mathrm{Re}_x = 5 \cdot 10^5$ [201], and the second values are close to $C_1 = 9/10$ and $C_2 = 1/9$ obtained experimentally at greater Reynolds number $\mathrm{Re}_x = 10^8$ in [267] (Exer. 1.47).

Comment 1.9 In the case of turbulent flow, the isothermal heat transfer coefficient h_* is required also for applying both universal functions (1.39) and (1.40). For this purpose, a special investigation was provided for wide range of Reynolds and Prandtl numbers [107] using the same Mellor turbulent model. We present those results and comparison with available experimental data in applications in Section 2.1.2.3.

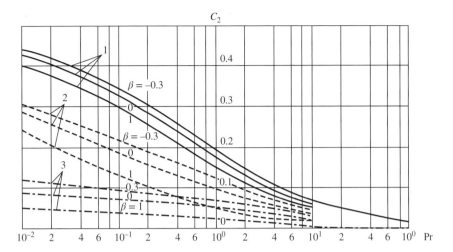

Figure 1.9 Exponent C_2 for turbulent boundary layer: $\mathrm{Re}_{\delta_1} = 10^3, 2 - 10^5, 3 - 10^9$

Exercises

1.33 Read about gamma function in some *Advanced Engineering Mathematics Course* and calculate limiting values (1.43) and (1.44) using formulae (1.42). Hint: for Gamma function estimation use some handbook or Mathcad.

1.34 Estimate a rate of convergence of series (1.39) by comparing the first four values of terms using limiting coefficients for Pr $\rightarrow$ 0 or Pr $\rightarrow$ ∞ and assuming that all products $\Phi^k(d^k\theta_w/d\Phi^k)$ are of the same order.

1.35 Derive relation (1.45) for estimating β. Hint: equal results of integration of two flows velocity distributions: given $U(x)$ and self-similar $U = cx^{\beta/(2-\beta)}$, which is used for estimating β.

1.36 Show that the Görtler variable Φ (1.39) is the average value of Reynolds number on interval $(0, x)$.

1.37 Think about other examples of problems and different types of information required to understand the role of local, historical, and future data in determining point characteristics of interest.

1.38 Read about beta function (see Exer. 1.33) and calculate coefficients g_k using relation (1.46) and exponents from influence function (1.24).

1.39 Recall or read about a term "transcendental expression" using, for example, Wikipedia to understand the difficulties in solving the inverse problem and methods to overcome those.

1.40* Think how the inverse problem may be solved graphically. Hint: (i) calculate two functions $g_1(C_1, C_2)$ and $g_2(C_1, C_2)$ using equation (1.46) and varying C_1 from 1 to $3/4$ and

C_2 from $1/2$ to $1/3$, (ii) plot these functions and choose coefficients g_1 and g_2 from Figures 1.3 and 1.4 for the same value of Prandtl number, (iii) draw lines $g_1 = const.$ and $g_2 = const.$ on each of graphs to get two relations between exponents C_1 and C_2, (iv) find the intersection point of two curves plotted according to data for C_1 and C_2 from (iii).

1.41* Create a software for solving the inverse problem applying trial and error method. Hint: (i) choose coefficients g_1 and g_2 from Figures 1.3 and 1.4 for the same value of Prandtl number, (ii) take two equations (1.46) for g_1 and g_2 and guess a value of C_1 or C_2, (iii) by varying another than guessed exponent satisfy one of equation (1.46) to get the corresponding coefficient g, (iv) by varying another exponent satisfy another equation (1.46) to get the second coefficient g, (iv) continue the procedure until desired accuracy of both refined coefficients g_1 and g_2 will be achieved, (vi) if the process would be poor converged, change the value of the first guessing coefficient.

1.42 Show using the Görtler variable that for self-similar flows with velocity $U = cx^m$, the formula (1.25) is valid. Hint: calculate variable Φ for $U = cx^m$, and substitute ξ/Φ for ξ/x in equation (1.24).

1.43 What means physically $Pr_{tb} = 1$? Recall, what is the Reynolds analogy? Explain physically why such an analogy exits. Hint: think about similarity of transport processes (S. 7.1.1).

1.44 Prove that for laminar boundary layer the parameter $\beta = (\delta_1/\tau_w)(dp/dx)$ is constant. Hint: use the external velocity distribution $U = Cx^{\beta/(2-\beta)}$, second equation (1.12) for dp/dx, and equations for δ_1 and τ_w from Section 7.5.1.

1.45 Explain physically why coefficients g_k decrease with a Reynolds or Prandtl number increasing. Hint: think about relation between dynamic and thermal boundary layer thicknesses (Exam. 7.8).

1.46 Explain why effect of nonisothermicty (and coefficients g_k) in turbulent flow is less than in laminar flow. Hint: think about different nature of transport process in both cases (S.8.2.1).

1.47 Calculate exponents C_1 and C_2 for several examples using laminar and turbulent data for g_1 and g_2 and graphical approach or software described in Exercises 1.41 and 1.42, respectively.

1.8   Universal Functions for Compressible Low

The results obtained above for incompressible flows are applicable to the compressible flows with variable density in the case of zero pressure gradient. This follows from the fact that boundary layer equations for compressible flow past plate in Dorodnizin or Illingworth-Stewartson variables (according Russian or English literature) has the same form as the boundary layer equations for incompressible flow in physical variables (Exer. 1.48). Therefore, the solution of thermal boundary layer equation for heat flux on a plate with arbitrary temperature distribution in compressible flow in such variables may be presented as universal function (1.38), but written in differences of enthalpy instead of temperature head

[111, p. 78]

$$q_w = \frac{q_{w*}C_x}{\sqrt{C}}\left(i_{0w} + \sum_{k=1}^{\infty} g_k x^k \frac{d^k i_{0w}}{dx^k}\right), \quad i_{0w} = \frac{J_w - J_{ad}}{J_\infty} = i_w - \frac{r}{2}(k-1)\,\mathrm{M}^2 \quad (1.49)$$

$$C = \sqrt{\frac{T_{w.av}}{T_\infty} \frac{T_\infty + T_S}{T_{w.av} + T_S}}$$

Here, q_{w*} is the heat flux on an isothermal surface with average temperature head of considering nonisothermal plate (like h_* in relation (1.38)), J_w and J_{ad} are wall and adiabatic wall enthalpy, i_{0w} is stagnation enthalpy difference (Exer. 1.49), C and C_x are coefficients of proportionality in approximation law for gas viscosity $\mu/\mu_\infty = C\,(T/T_\infty)$ calculated using average $T_{w.av}$ or local T_w surface temperature, r is a recovery factor (S. 1.14), M is Mach number, and T_S is Sutherland gas constant. Equation (1.49) is an exact solution of a thermal boundary layer equation for the plate with an arbitrary temperature distribution. The well-known Chapman-Rubesin solution for polynomial surface temperature follows from this relation as a particular case.

The equivalent integral universal function for compressible flow past plate is obtained from equation (1.26) after similar substitution of stagnation enthalpy difference i_{0w} and product $q_{w*}C_x/\sqrt{C}$ before brackets for the temperature head and isothermal heat transfer coefficient h_*, respectively

$$q_w = \frac{q_{w*}C_x}{\sqrt{C}}\left[i_{0w}(0) + \int_0^x f(\xi/x)\frac{di_{0w}}{d\xi}d\xi\right] \quad (1.50)$$

1.9   Universal Functions for Power-Law Non-Newtonian Fluids

The term "non-Newtonian fluid" belongs to fluids which rheology (a science of flow of a matter) behavior is different from that of Newtonian fluids. In particular, the Newtonian fluids viscosity depends only on the temperature, whereas there are countless other fluids and materials (like polymers) whose viscosity is governed by more complex laws depending on spatial or/ and on time deformation (Com. 5.8). The universal function was obtained for non-Newtonian and non-Fourier power laws fluids [351, p. 83] which characteristics obeys the following expressions of power types

$$\tau = K_\tau \left(\frac{1}{2}I_2\right)^{\frac{n-1}{2}} \mathbf{e}, \quad \mathbf{q} = K_q\left(\frac{1}{2}I_2\right)^{\frac{s}{2}} \mathrm{grad}\ T, \quad I_2 = 4\left(\frac{\partial u}{\partial x}\right)^2 + 4\left(\frac{\partial v}{\partial y}\right)^2 + 2\left(\frac{\partial u}{\partial y} + \frac{\partial v}{\partial x}\right)^2$$
$$(1.51)$$

Here, τ and $\mathbf{e}$ are the stress and rate of deformation tensors, $\mathbf{q}$ is the heat flux vector, K_τ and K_q are constants and I_2 is a second invariant of rate of deformation tensor. Non-Newtonian power laws satisfactory describe the behavior of a group of substances, like suspensions, polymer solutions and melts, starch pastes, clay mortars.

Comment 1.10 Tensor is a general term defining the other quantities by order of its degree. A tensor order depends on array of numerical values (or indices) that determines the tensor. In this terminology, the scalar is a tensor of zero order, whereas the vector is the tensor of order one since three values (three coordinate) is necessary to define a vector. Similarly, the tensor of second order is an array of nine values.

Relations (1.51) simplifies for boundary layer flows for which they become

$$\tau = K_\tau \left(\frac{\partial u}{\partial y} \right)^n, \qquad q = K_q \left(\frac{\partial u}{\partial y} \right)^s \frac{\partial T}{\partial y} \tag{1.52}$$

The Newton and Fourier laws follow from (1.52) as well as from (1.51) at $n = 1$ and $s = n - 1 = 0$ (Exer. 1.50 and 1.51).

It is shown that universal functions (1.39) and (1.40) are valid also for power law non-Newtonian fluids, but only for such, for which the condition $s = n - 1$ is satisfied. This equality means that viscosity K_τ and heat conductivity K_q (called apparent parameters for non-Newtonian fluids, Exer. 1.52) are proportional to each other as for Newtonian fluids, and due to that, the analogous exact solution of thermal boundary layer equation may be obtained. [111, p. 84] In that, more general case, the Görtler variable and parameter β, defined the self-similar flows, are determined as follows (Exer. 1.53)

$$\Phi = \frac{\rho}{K_\tau} \int\limits_0^x U^{2n-1}(\xi) d\xi \qquad \beta = \frac{(n+1)m}{(2n-1)m+1} \qquad m = \frac{Ux}{x \int\limits_0^x U(\xi) d\xi} - 1 \tag{1.53}$$

These equations indicate that for power-law fluids, the pressure gradient parameter β is depended not only on exponent m of the external velocity $U = cx^m$ as for Newtonian fluid, but also on exponent n in rheology law (1.51) or (1.52). Because of that, the pressure gradient is characterized not by β, which depends also on n, but by using directly the exponent m applying the third formula (1.53) instead of relation (1.45) for β in the case of Newtonian fluid (Exer. 1.54).

The coefficients g_k were calculated in the same way only for large Prandtl numbers $Pr = 10, 100, 1000$ typical for non-Newtonian fluids, exponents n from 0.2 to 1.8, and for $m = 0$ (zero pressure gradient), $m = 1/3$ (negative pressure gradient) and $m = 1$ (stagnation point flow). The results given in Figure 1.10 indicate that: (i) the dependences of coefficients g_k on Prandtl number and pressure gradient are similar to those for Newtonian fluids (Fig.1.3), (ii) for $Pr > 10$ and relatively small pressure gradients $m = 0$ and $m = 1/3$, coefficients g_k are practically independent on Pr as well as for Newtonian fluid, (iii) as the pressure gradient increases ($m = 1$), this dependence becomes more significant, but still remains slight, (iv) functions $g_1(n)$ and $g_2(n)$ (for $m = 1/3, m = 1$ and $Pr = 100$ they merge in one curve 2) present the effect of non-Newtonian behave of fluid (exponent n) on heat transfer intensity showing that nonisotermicity effect increases markedly with growing exponent n, (v) the greatest coefficient g_1 becomes larger from two to more than three times as the exponent n increases from 0.2 to 1.8.

In Figure 1.11 are plotted the results obtained for coefficient g_0 which defines the isothermal heat transfer coefficient h_* for non-Newtonian fluids according to formula for Nusselt number [111, p. 84]

$$\frac{Nu_*}{Re^{\frac{n}{n+1}}} = g_0 \left(\frac{C_f}{2} Re^{\frac{1}{n+1}} \right)^{\frac{2n-1}{2n}} \left(\frac{\Phi}{Re} \right)^{\frac{1}{2(n+1)}} \tag{1.54}$$

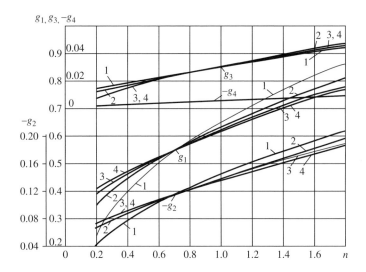

Figure 1.10 Coefficients g_k for non-Newtonian fluids: $s = n - 1$, $1 - m = 1$, $\text{Pr} = 10$, $2 - m = 1$, $\text{Pr} = 100$ and $m = 1/3$, $\text{Pr} > 10$, $3 - m = 1$, $\text{Pr} > 1000$, $4 - m = 0$, $\text{Pr} > 10$

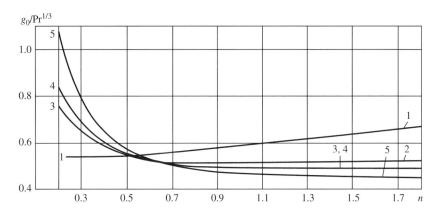

Figure 1.11 Quotient $g_0/\text{Pr}^{1/3}$ of the coefficient g_0 and Prandtl number $\text{Pr}^{1/3}$ for non-Newtonian fluids: $s = n - 1$, $1 - m = 0$, $\text{Pr} > 10$, $2 - m = 1/3$, $\text{Pr} > 10$, $3 - m = 1$. $\text{Pr} > 1000$, $4 - m = 1$, $\text{Pr} = 100$, $5 - m = 1$, $\text{Pr} = 10$

where Nu and Re are special numbers for non-Newtonian fluids given in nomenclature. The friction coefficient C_f containing in (1.54) may be estimated using special literature, for example [351]. Figure 1.11 presents the value $g_0/\text{Pr}^{1/3}$ (instead of g_0) as a function of exponent n for different values of pressure gradient (different m) and Prandtl number. It is seen that for large Pr, this quotient slightly depends on Prandl number. This indicates that the heat transfer coefficient for an isothermal surface for power-law fluids is practically proportional to $\text{Pr}^{1/3}$ as well as for Newtonian fluids (S. 7.5.1).

Although the coefficients C_1 and C_2 here are not given, they may be calculated by the same way using data for coefficients g_k (Exer. 1.55). For the general case of arbitrary exponents n and s (not connected by condition $s = n - 1$) for laws (1.51) only approximate similar solutions for universal function have been obtained [104].

1.10 Universal Functions for Moving Continuous Sheet

The systems in which a continuous material goes out of a slot and moves through surrounding coolant are used in a number of industrial processes, such as forming of synthetic films and fibers, the rolling of metals, glass production, and so on. Due to viscosity of the surrounding, on the surface of such moving sheet a boundary layer similar to that on a streamlined or a flying body forms as it schematically is shown on Figure 1.12. Despite both boundary layers are similar, at the middle of the last century it was shown [333] that the boundary layer on continuous sheet differs from the well-known boundary layer existing on streamlined bodies. As it is clear from Figure 1.12 in this case, the boundary layer grows in the direction of the motion, whereas on the moving or streamlined body, the boundary layer develops in an opposite to the moving direction.

It can be shown that in coordinate system attached to the moving surface, the boundary layer equations are unsteady (Exer. 1.56), and hence, they differ from the usual steady equations for a plate, but the boundary conditions in the moving frame are identical with those for streamlined plate. At the same time, in a frame attached to the slot, the problem of the moving sheet is steady, and both boundary layer equations coincide, however, in this case, the boundary conditions differ because the flow velocity on a sheet relatively to a slot is not zero.

The first calculations revealed that for the moving continuous sheet, the friction coefficient and the isothermal heat transfer coefficient at $Pr = 0.7$ are greater than those for streamlined plate by 34% and 20%, respectively [333, 395]. Exact solution for arbitrary nonisothermal surface in the same form of universal functions (1.38) and (1.26) for zero pressure gradient was obtained in [109] for stationary and blowing surrounding coolant with different ratio $\phi = U_\infty / U_w$, where U_w and U_∞ are velocities of surface and coolant. The coefficients g_k and the exponents C_1 and C_2 are given in Figure 1.13 and 1.14. The coefficient g_0 plotted in Figure 1.15

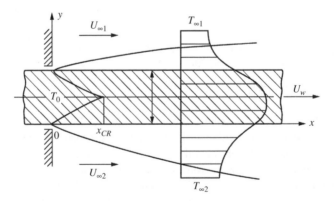

Figure 1.12 Schematic pattern of boundary layer on moving continuous sheet for symmetric ($U_1 = U_2, T_{\infty 1} = T_{\infty 2}$) and asymmetric flows

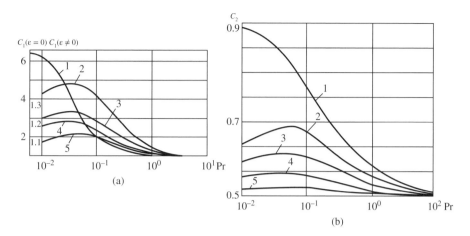

Figure 1.13   Coefficients $g_1(a)$ and $g_2(b)$ as a function of Prandtl number and a ratio $\phi = U_\infty/U_w$ for moving continuous sheet: $1 - \phi = 0, 2 - 0.1, 3 - 0.3, 4 - 0.5, 5 - 0.8$, 6-streamlined plate, $7(b)$-g_3

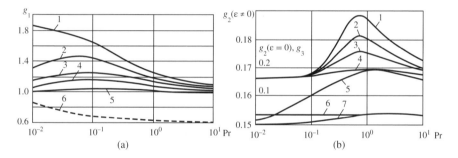

Figure 1.14   Exponent C_1 and C_2 as a functions of Prandtl number and ratio $\phi = U_\infty/U_w$ for a plate moving through surrounding. $1 - \phi = 0, 2 - 0.1, 3 - 0.3, 4 - 0.5, 5 - 0.8$

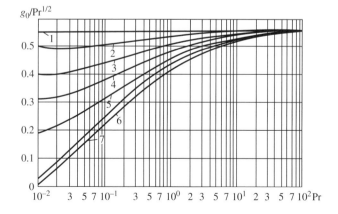

Figure 1.15   Quotient $g_0/Pr^{1/2}$ as functions of Prandtl number and ratio $\phi = U_\infty/U_w$ for moving isothermal continuous sheet. $1 - \phi = 1, 2 - 0.8, 3 - 0.5, 4 - 0.3, 5 - 0.1, 6 - 0.7, 7 - (-0.05)$

for different ratio $\phi = U_\infty/U_w$ defines the isothermal heat transfer coefficient according to formula $h_* = g_0\sqrt{U_w/vx}$ (Exer. 1.57).

The coefficients g_k in this case are markedly greater than those for streamlined plate (Figs. 1.3 and 1.4), but decrease rapidly as well with coefficient number increasing. For instance, in the case of stationary surrounding, the first coefficient g_1 for the moving sheet is twice greater which indicates that the effect of nonsothermcity in this case is significantly larger than that for streamlined stationary or moving plate. As well as for usual case, the coefficients g_k for $k \geq 3$ are practically independent of the Prandtl number and blowing parameter ϕ and may be calculated by the same formula (1.42) for Pr $\rightarrow$ 0 as for universal functions (1.38).

1.11   Universal Functions for a Plate with Arbitrary Unsteady Temperature Distribution

An exact solution of unsteady thermal boundary layer equation for a plate with arbitrary unsteady temperature distribution $T_w(t, x)$ shows that in this case, a differential universal function has the form similar to series (1.38) [117, p. 75] (Exer. 1.58)

$$q_w = h_*\left(\theta_w + g_{10}x\frac{\partial\theta_w}{\partial x} + g_{01}\frac{x}{U}\frac{\partial\theta_w}{\partial t} + g_{20}x^2\frac{\partial^2\theta_w}{\partial x^2} + g_{02}\frac{x^2}{U^2}\frac{\partial^2\theta_w}{\partial t^2} + g_{11}\frac{x^2}{U}\frac{\partial^2\theta_w}{\partial x\partial t} + \dots\right)$$
$$(1.55)$$

Series (1.55) contains three types of terms with derivatives depending: on coordinate only, on time only, and on both coordinate and time with coefficients g_{k0}, g_{0i} and g_{ki}, respectively. Coefficients g_{k0} are the same as g_k in universal function (1.38) for steady temperature distribution. The two others depend on Prandtl number as well as g_k and on time (Exer. 1.59). For the case of zero pressure gradient and Pr = 1, the first four coefficients $g_{ki}(i \neq 0)$ are computed numerically. They are plotted on Figure 1.16, which shows that the coefficients g_{ki} are similar to coefficients g_k rapidly decreasing with growing number ki. Due to that, it is possible to obtain satisfactory accurate calculations using only first several terms of the series (1.55) as well as in the case of employing steady state universal function (1.38).

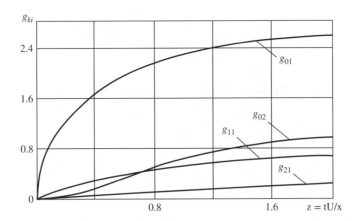

Figure 1.16   Coefficients $g_{ki}(i \neq 0)$ as functions of z for zero pressure gradient and Pr = 1

It follows from Figure 1.16 that the coefficients g_{ki} gradually grow with time and finally attain the values of $(g_{ki})_{t \to \infty}$ that coincide with those obtained by Sparrow in a similar problem without an initial conditions [370]. The ratio $g_{ki}(z)/(g_{ki})_{t \to \infty}$ is about 0.99 when dimensionless time $z = Ut/x$ (see Exer. 1.59) becomes $z > 2.4$. Hence, for $z > 2.4$, the coefficients g_{ki} are practically independent of time. Comparing these coefficients with coefficients g_k of series (1.38) for zero pressure gradient and $Pr = 1$ reveals that coefficients g_{ki} for large time are much greater. In particular, the first coefficient $g_{01} \approx 2.4$ is four times larger than corresponding value of steady state $g_1 \approx 0.6$ (Fig.1.3). This specifies that the nonisothermicity effect caused by time temperature gradient $\partial \theta_w / \partial t$ is four times greater than the effect of nonisothermicity produced at the same conditions by spatial temperature gradient $\partial \theta_w / \partial x$.

The following integral universal function corresponds to unsteady differential universal function (1.55)

$$
q_w = h_* \left[\theta_w(t, 0) + \int_0^x f[(\xi/x,) 0, z] \frac{\partial \theta_w}{\partial \xi} d\xi + \int_0^t f[0, (\eta/t), z] \frac{\partial \theta_w}{\partial \eta} d\eta \right.
$$

$$
\left. + \int_0^t d\eta \int_0^x f\left[(\xi/x), (\eta/t), z\right] \frac{\partial^2 \theta_w}{\partial \xi \partial \eta} d\xi \right] \tag{1.56}
$$

Applying the same technique of repeated integration by part as in the case of steady-state heat transfer described in Section 1.4, one can show that this expression is identical with the differential form in series (1.55).

1.12   Universal Functions for an Axisymmetric Body

According to Stepanov and Mangler (in conformity with Russian and English literature), the problem for an axisymmetric body is transformed to two-dimensional problem by using variables which in case of non-Newtonian power law fluids are [351]

$$
\tilde{x} = \int_0^x R^{n+1}(\xi) d\xi, \quad \tilde{y} = Ry, \tag{1.57}
$$

where R is a cross section radius (Exer. 1.60). It follows from this result that universal functions (1.39) and (1.40) are valid for flow past axisymmetric body in Görtler variable transformed according to relation (1.57). Such variable for the case of the power law non-Newtonian fluids is obtained after substituting $R^{n+1} dx$ for dx in the corresponding Görtler variable for non-Newtonian fluids (1.53). The heat flux q_w obtained for plane two-dimensional problems should be multiplied in this case by R^n to get

$$
\Phi = \frac{\rho}{K_\tau} \int_0^x U^{2n-1}(\xi) R^{n+1}(\xi) d\xi, \quad \tilde{q}_w = R^n q_w \tag{1.58}
$$

In the case of Newtonian fluid ($n = 1$), the first expression (1.58) in conformity with equations (1.57) transforms in the Görtler variable Φ (1.39) modified by multiplying the integrand by R^2. The second equation (1.58) in this case reduces to $\tilde{q}_w = Rq_w$ where q_w is the heat flux obtained by universal function (1.39) for plane problem, which also is in line with the second equation (1.57) (Exer. 1.61).

Comment 1.11 Special variables play a significant role in simplifying the initial form of equations. So far, we consider those three variables: Görtler variable, which transforms the solution for zero pressure gradient to form applicable for flows with arbitrary pressure gradient, Dorodnizin-Illingworth-Stefartson variable transforming the equations for compressible fluid in an incompressible form, and the Stepanov-Mangler variable which reduces the axisymmetric equation to the equivalent two-dimensional equation. There are other specific variables transforming equations, in particular, as later we will see the Falkner-Skan, Blasius, and other similarity variables, which change the partial boundary layer equations in ordinary differential equations (S. 7.5.1 and 7.5.2), Prandtl-Mises variables converting boundary layer equation in the form, which is close to one-dimensional conduction equation (S. 7.4.4.2). These examples are taken from boundary layer theory. However, the equations transforming by applying new variables is a general method widely used in mathematics.

1.13 Inverse Universal Function

The inverse universal function is obtained as a result of solution of the inverse problem when the surface heat flux distribution is specified and the corresponding temperature head distribution should be found. We consider two such problems: first, the general inverse universal functions for surface with given arbitrary heat flux distribution is obtained and then, the specific inverse problem determining the universal function for recovery factor is solved.

1.13.1 Differential Inverse Universal Function [112]

Because we are seeking a temperature distribution, we solve equation (1.39) for temperature head to obtain

$$\theta_w + \sum_{k=1}^{\infty} g_k \Phi^k \frac{d^k \theta_w}{d\Phi^k} = \frac{q_w}{h_*} = \theta_{w*}(\Phi) \tag{1.59}$$

The right hand part of this equation is a known function of x and hence, of Görtler variable Φ since in the problem in question the heat flux distribution is given. Physically the ratio q_w/h_* determines the temperature head which would be established by given heat flux if the considering surface were isothermal. Therefore, we use for this function the notation $\theta_{w*}(\Phi)$ (Exer. 1.62). In fact, equation (1.59) is a differential equation defining unknown function $\theta_w(\Phi)$, which solution may be presented in terms of known function $\theta_{w*}(\Phi) = q_w/h_*$ as a series similar to universal function (1.39)

$$\theta_w = \theta_{w*} + \sum_{n=1}^{\infty} h_n \Phi^n \frac{d^n \theta_{w*}}{d\Phi^n}, \tag{1.60}$$

where h_n denotes coefficients similar to g_k. To prove this, we show that relation (1.60) satisfies equation (1.59) which we transform by replacing the right hand part to the left

$$\theta_w - \theta_{w*}(\Phi) + \sum_{k=1}^{\infty} g_k \Phi^k \frac{d^k \theta_w}{d\Phi^k} = 0 \tag{1.61}$$

Substitution of equation (1.60) into this equation leads to following equation (Exer. 1.63)

$$\sum_{n=1}^{\infty} h_n \Phi^n \frac{d^n \theta_{w*}}{d\Phi^n} + \sum_{n=1}^{\infty} g_k \Phi^k \frac{d^k \theta_{w*}}{d\Phi^k} + \sum_{n=1}^{\infty} g_k \Phi^k \frac{d^k}{d\Phi^k} \sum_{n=1}^{\infty} h_n \Phi^n \frac{d^n \theta_{w*}}{d\Phi^n} = 0 \qquad (1.62)$$

After changing indices n to k and performing differentiation in the last term, we modify this equation by assembling terms containing the same groups $\Phi^k (\partial^k \theta_{w*}/\partial \Phi^k)$ for $k = 1, 2, 3 \ldots$. The expression obtained in such a way is a summation of the partial sums of groups $\Phi^k (\partial^k \theta_{w*}/\partial \Phi^k)$ for different numbers k, which according to (1.62) should be equal zero. Since those partial sums of different groups are independent of each other, the required condition of zero may be satisfied only if each sum of groups would be equal zero. The equalities obtained in such procedure by setting each of partial sums to zero contain known k coefficients g_k and $k - 1$ coefficients h_k. Therefore, they define coefficients h_k in sequence so that the first equality for $k = 1$ specifies h_1, the second one gives h_2 via known h_1, and so on, resulting in the following equations (Exer. 1.64)

$$h_1 + g_1(h_1 + 1) = 0, \quad h_2 + g_1(2h_2 + h_1) + g_2(2h_2 + 2h_1 + 1) = 0,$$

$$h_3 + g_1(3h_3 + h_2) + g_2(6h_3 + 4h_2 + h_1) + g_3(6h_3 + 6h_2 + 3h_1 + 1) = 0,$$

$$h_4 + g_1(4h_4 + h_3) + g_2(12h_4 + 6h_3 + h_2) + g_3(24h_4 + 18h_3 + 6h_2 + h_1)$$

$$+ g_4(24h_4 + 24h_3 + 12h_2 + 4h_1 + 1) = 0 + \ldots \qquad (1.63)$$

Figures 1.17 and 1.18 present the values of the first four coefficients h_k for laminar and turbulent flows. For the limiting Prandtl numbers the coefficients h_k are

$$h_1 = -1/2, \quad h_2 = 3/16, \quad h_3 = -5/96, \quad h_4 = 35/1968 \qquad \text{Pr} \to 0 \qquad (1.64)$$

$$h_1 = -0.38, \quad h_2 = 0.135, \quad h_3 = -0.037, \quad h_4 = 0.00795 \qquad \text{Pr} \to \infty \qquad (1.65)$$

Like coefficients g_k, the first few coefficients h_k are a weak functions of β and the others are practically independent of β and Pr.

1.13.2 Integral Inverse Universal Function [112]

The integral inverse universal function is obtained in the same way by considering the universal function (1.40) as an equation for temperature head determination to get

$$\int_0^\Phi f(\xi/\Phi) \frac{d\theta_w}{d\xi} d\xi = \frac{q_w(\Phi)}{h_*(\Phi)} \qquad \text{or} \qquad \int_0^\Phi [1 - (\xi/\Phi)^{C_1}]^{-C_2} \frac{d\theta_w}{d\xi} d\xi = \frac{q_w(\Phi)}{h_*(\Phi)} \qquad (1.66)$$

It is seen that in this case, the unknown function θ_w is located under sign of an integral with variable limit. To find such unknown function, one should solve the second equation (1.66), which is Volterra integral equations (Exer. 1.65). Because there is no standard approach for solving integral equations, the solution of such a problem is a hard task. In this specific case, applying new variables: $\Phi^{C_1} = z$, $\xi^{C_1} = \zeta$, and $q_w(\Phi)/h_*(\Phi) z^{C_2} = F(z)$ converts second equation (1.66) into Abel integral equation (1.67), of which the known solution for this

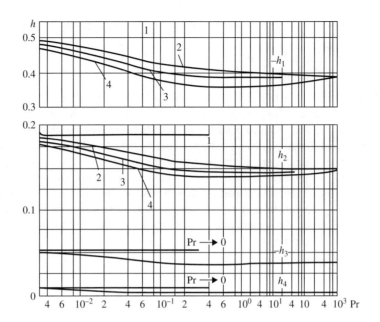

Figure 1.17   Coefficients h_k for laminar boundary layer. Asymptotes $1 - \text{Pr} = 0$, β : $2 - 1$, $3 - 0$, $4 - (-0.16)$

particular case is shown by second expression (1.67) (Exer. 1.66)

$$\int_0^z \frac{d\theta_w}{d\zeta} \frac{d\zeta}{(z - \zeta)^{C_2}} = F(z) \frac{d\theta_w}{dz} = \frac{1}{(C_2 - 1)!(-C_2)!} \frac{d}{dz} \int_0^z \frac{F(\zeta)d\zeta}{(z - \zeta)^{1-C_2}} \qquad (1.67)$$

Returning to variables Φ and ξ yields a relation for the temperature head (Exer. 1.67)

$$\theta_w = \frac{C_1}{\Gamma(1 - C_2)\Gamma(C_2)} \int_0^\Phi \left[1 - \left(\frac{\xi}{\Phi} \right)^{C_1} \right]^{C_2 - 1} \left(\frac{\xi}{\Phi} \right)^{C_1(1 - C_2)} \frac{q_w(\xi)}{h_*(\xi)\xi} d\xi \qquad (1.68)$$

For the case of zero pressure gradient, this expression coincides with known relation

$$\theta_w = \frac{0.623}{\lambda} \text{Re}_x^{-1/2} \text{Pr}^{-1/3} \int_0^x \left[1 - \left(\frac{\xi}{x} \right)^{3/4} \right]^{-2/3} q_w(\xi)d\xi \qquad (1.69)$$

obtained in [201] using another approach (Exer. 1.68).

1.14   Universal Function for Recovery Factor

The recovery factor determines a part of fluid mechanical energy that is recovered as thermal energy. This process is important for estimation of the wall adiabatic temperature (S. 7.3.5).

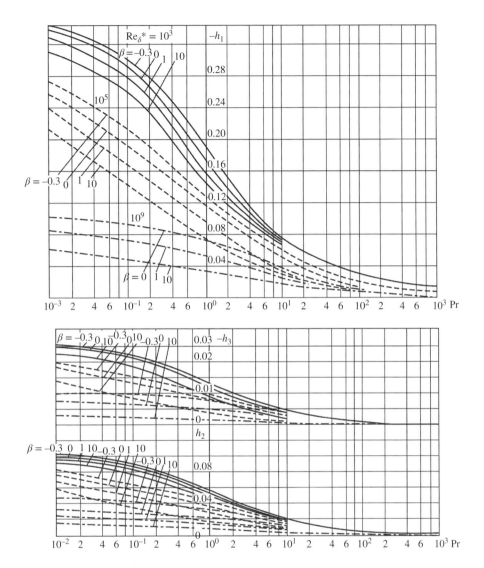

Figure 1.18 Coefficients h_k for turbulent boundary layer ___ $Re_{\delta_1} = 10^3$,_ _ _ 10^5,_._ _. 10^9

Actually, the recovery factor calculation is an inverse problem of the temperature head determining under condition of zero heat flux (Exer. 1.49) and significant mechanical energy dissipation. Solution of such a problem is similar to that considered in the previous section, but in this case in addition, the effect of mechanical energy dissipation should be taken into account.

It is shown [119, p. 57] that the exact solution of the laminar thermal boundary layer equation for the case with significant mechanical energy dissipation differs from universal function (1.39) by the term $g_d(U^2/c_p)$ where g_d is a special coefficient similar to coefficients g_k. Taking this into account, we solve equation (1.39) with such additional term (the first equation (1.70)) for temperature head θ_w in the same way as in Section 1.13.1 to obtain the second equation (1.70) instead of similar equation (1.59)

$$q_w = h_* \left(\theta_w + \sum_{n=1}^{n=k} g_k \Phi^k \frac{d^k \theta_w}{d\Phi^k} - g_d \frac{U^2}{c_p} \right), \qquad \theta_w + \sum_{k=1}^{\infty} g_k \Phi^k \frac{d^k \theta_w}{d\Phi^k} = \frac{q_w}{h_*} + g_d \frac{U^2}{c_p} \qquad (1.70)$$

Considering the last equation (1.70) as the differential equation defining temperature head as a sum $(q_w/h_* + g_d U^2/c_p)$ instead of θ_{w*}, we get the same solution (1.60) for the problem with significant mechanical dissipation in which $\theta_{w*} = q_w/h_*$ is substituted by the sum $(q_w/h_* + g_d U^2/c_p)$ regarding the effect of dissipation

$$\theta_w = q_w/h_* + g_d U^2/c_p + \sum_{n=1}^{\infty} h_n \Phi^n \frac{d^n(q_w/h_* + g_d U^2/c_p)}{d\Phi^n} \qquad (1.71)$$

After putting here $q_w = 0$, this expression gives the universal function for recovery factor

$$r = \frac{T_{ad} - T_\infty}{U^2/2c_p} = \frac{T_w - T_{ad}}{U^2/2c_p} = 2g_d \left(1 + \sum_{k=1}^{\infty} h_k \frac{\Phi^k}{U^2} \frac{d^k U^2}{d\Phi^k} \right) \qquad (1.72)$$

where T_{ad} is adiabatic wall (i.e., isolated) temperature, and T_w is the temperature of the usual non-isolated wall defined by equation (170) which wall has if it would be nonisolated. It follows from the last two relations that the adiabatic may be determined as a temperature of isolated wall or as a wall temperature cooled due to dissipation mechanical energy (Exer. 1.69 and 1.70).

Accordingly, integral universal function for recovery factor is obtained from relation (1.68) by the same substitution of the sum $(q_w/h_* + g_d U^2/c_p)$ for $\theta_{w*} = q_w/h_*$ in its integrand and following putting $q_w = 0$ (Exer. 1.71)

$$r = \frac{T_w - T_{ad}}{U^2/2c_p} = \frac{2g_d C_1}{\Gamma(1 - C_2)\Gamma(C_2)U^2} \int_0^\Phi \left[1 - \left(\frac{\xi}{\Phi} \right)^{C_1} \right]^{C_2-1} \left(\frac{\xi}{\Phi} \right)^{C_1(1-C_2)} \frac{U^2(\xi)}{\xi} d\xi \qquad (1.73)$$

Relations (1.72) and (1.73) present recovery factor for flows with arbitrary external flow $U(x)$. For zero pressure gradient with $U = const.$ both expressions (1.72) and (1.73) give the well-known result $r = 2g_d$ (Exer. 1.72).

The effect of dissipation, which is proportional to the square of velocity $U^2(x).$, is minor for incompressible fluids due to typical relatively small incompressible flows velocities. In such a case, the effect of dissipation becomes significant only for large Prandtl numbers when the last term in thermal boundary layer equation (1.11), which is proportional to Prandtl number, is comparable with other terms (Exer.1.73). The large Prandtl numbers are common, in particular, for non-Newtonian fluids. Because of that, the coefficient g_d was calculated for non-Newtonian $(n = 0.6 - 1.2)$ including Newtonian $(n = 1)$ fluids only for large Prandtl numbers, $\text{Pr} > 1000$, for zero pressure gradient $(m = 0)$ and for stagnation point flow $(m = 1)$. The same approach as for coefficients g_k calculation (S. 1.6.1) was used (Exer. 1.74). The results are plotted in Figure 1.19. It is seen from Figure 1.19 that the recovery factor and hence, the effect of dissipation increases with pressure gradient and decreases as the exponent n increases. This implies that the effect of dissipation for non-Newtonian fluids with $n > 1$ (dilatant fluids) is greater and limited for with $n < 1$ (pseudoplasic fluids) is smaller than that for Newtonian fluid with $n = 1$.

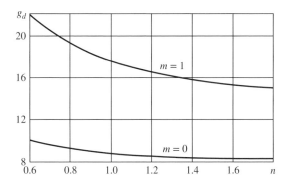

Figure 1.19 Coefficients g_d for Newtonian ($n = 1$) and non-Newtonian fluids ($s = n - 1$, S.1.6), Pr >1000

Comment 1.12 One merit of using the adiabatic wall temperature is that the substitution of the adiabatic temperature head defined as $\theta_{ad} = T_w - T_{ad}$ for usual temperature head $\theta_w = T_w - T_\infty$ in known relations yields the expressions valid for the case with heat dissipation. For example, equation (1.49) for compressible fluid written using adiabatic stagnation enthalpy difference i_{0w} has the same form as the relation (1.39) written in terms of usual temperature head θ_w for incompressible fluid. This is achieved due to applying adiabatic stagnation enthalpy difference according to the second equation (1.49) (Exer. 1.75). Applying adiabatic temperature is also important to understand some heat exchange processes. For instance, as it shown in Section 2.1.4.3, the fluid cools the wall until it reaches the adiabatic temperature, and then, the heat flux changes its direction so that the fluid heats the wall.

Exercises

1.48* The basic idea of transforming boundary layer equation for compressible fluid to the form of incompressible fluid is to take into account the variability of gas density using integral of density $\eta = \eta_0 \int_0^y \frac{\rho(\xi)}{\rho_0} d\xi$. That gives the transverse variable η, which converts the compressible boundary layer equation to the form of incompressible boundary layer equation. Note, that expression for η is of the same type as Görtler variable Φ defined by equation (1.39). Compare these relations, thinking of their similarity and dissimilarity. Are the considerations about different kinds of information and flow history outlined in comments 1.5 and 1.6 applicable to relation for η defined as integral of density? Think about dummy variables in this relation and in Görtler variable (1.39). Compare both these dummy variables denoted by the same letter ξ. Are they actually the same? If no, explain why do you think so, and how differ these dummy variables from each other?

1.49 Recall what is the adiabatic wall temperature (S.7.3.5) and enthalpy to understand what is the stagnation enthalpy difference i_{0w} and how is it calculated? Compare relations (1.49) and (1.50) for compressible fluids with analogous relations (1.38) and (1.26) for incompressible flows.

1.50 Show that for the boundary layer flows, the relations (1.52) follow from general expressions (1.51). Hint: compare magnitude order of terms as in derivation boundary layer equations (S. 7.4.4.1) and use Newton's law for shear stress.

1.51 Obtain Newton's and Fourier's laws from relations (1.52) for Newtonian fluids ($n = 1$) and prove that in this case the proportionality between viscosity and conductivity exists because of $s = n - 1$. Hint: note that in this case the deformation tensor $\mathbf{e}$ (see (1.51)) equals $\partial u / \partial y$.

1.52* Show the same from relations (1.52) for non-Newtonian fluid (at any n) for the case when $s = n - 1$. Hint: modify the relations (1.52) to the forms $\tau = K_\tau \left(\dfrac{\partial u}{\partial y} \right)^{n-1} \dfrac{\partial u}{\partial y}$ and $q = K_q \left(\dfrac{\partial u}{\partial y} \right)^{n-1} \dfrac{\partial T}{\partial y}$ which are similar to those obtained for Newtonian fluid in example 1.51. Compare the Newton's and Fourier's laws for Newtonian fluid with similar relations for non-Newtonian fluids to see the difference between viscosity μ and conductivity λ in the first case and factors $K \left(\dfrac{\partial u}{\partial y} \right)^{n-1}$ at velocity and temperature derivatives in the second case. Think: why those coefficients for non-Newtonian fluids are called "apparent" viscosity and conductivity (see Com. 5.8)?

1.53* Show that in the case of constant external flow $U = const$, the Görtler variable (1.53) becomes Reynolds number in a special form for non-Newtonian fluids (see nomenclature) as in similar case for Newtonian fluid (Exer. 1.30).

1.54 Derive the third formula (1.53) for exponent m using approach described in Exer. 1.35.

1.55* Calculate exponents C_1 and C_2 for several examples using coefficients g_k from Figure 1.10 by graphical approach or software described in Exer. 1.41 and 1.42, respectively; see also Exer. 1.47.

1.56 Think and explain why the problem of a moving continuous sheet is unsteady in the frame attached to the sheet, but is steady in unmoved frame located out of sheet. Hint: consider what the observer sees looking at the boundary layer when he or she: (i) moves along with the sheet and (ii) sit out of the moving sheet.

1.57* Compare coefficients g_0, g_1 and g_2 for different values of parameter $\phi = U_\infty / U_w$ on Figures 1.13 and 1.15. Think: why coolant blowing increases the first coefficient, but reduces two others showing that the cooling effect of isothermal surface grows, whereas the effect of nonisithermicity decreases as coolant velocity increases? What physically causes those opposite effects? Hint: analyze the mechanism of such two phenomena.

1.58 Continue series (1.55) for $i = 0, k > 0$, $k = 0, i > 0$, $k > 0, i > 0$ knowing that the general term is $g_{ki}(x^{k+i}/U^i)(\partial^{k+i}\theta_w/\partial x^k \partial t^i)$. Hint: first check the given terms.

1.59* Show that time and mixed derivatives are in fact derivatives with respect to dimensionless time $z = Ut/x$ so that, for example, the third, fifth and sixth terms in equation (1.55) actually are $\partial\theta_w/\partial z$, $\partial^2\theta_w/\partial z^2$, $\partial^2\theta_w/\partial x\partial z$.

1.60* Compare variable (1.57) with others of this type, Görtler variable (1.39) and Dorodnizin-Illingworth-Stewartson variable, considered in Exercise 1.48. Answer the same questions about variable (1.57): are the considerations about different kinds of information outlined in comment 1.3 applicable to this variable defined as integral of cross section radius? What variable represents the same dummy variable ξ in this integral?

1.61 Show that in the case of Newtonian fluid, equation (1.58) transforms in both Görtler variable for plane two-dimensional problem and for axisymmetric body that are relation (1.39) and modified relation (1.39), respectively.

1.62 Explain physically why the ratio q_w/h_* presents the temperature head on an isothermal surface and why the notation $\theta_{w*}(\Phi)$ is proper for this ratio.

1.63 Obtain expression (1.62) by substituting equation (1.60) into equation (1.61). Hint: the first sum in (1.62) represents the difference $\theta_w - \theta_{w\bullet}$, which is found from equation (1.60), the second and the third terms are obtained by substituting θ_w defined by the same equation (1.60) into derivatives in the last term of equation (1.61).

1.64* Derive relations (1.63) following directions from text and calculate coefficients h_k for limiting cases. Explain why the summation of the sums of groups with the same complexes $\Phi^k(\partial^k\theta_{w*}/\partial\Phi^k)$ equals zero it follows that each these sums equals zero.

1.65 Read the article about integral equations on Wikipedia (at least the beginning part of definitions) to understand the difference between Volterra and Fredholm integral equations. Read also on Wikipedia about young mathematician Abel who became famous despite he died at the age of 26.

1.66* Show that substitution of new variables indicated in the text transforms the second integral equation (1.66) into integral Abel equation (1.67). Hint: note that ζ is the integration variable, whereas z is considered as a parameter (see Exercise 1.22).

1.67 Obtain the inverse universal function (1.68) from solution of Abel equation (1.67) by returning to physical variables. Hint: take care of difference between functions of variables Φ and ξ. Note also, that factorial of noninteger numbers is determined by gamma function as $\gamma! = \Gamma(\gamma + 1)$.

1.68* Prove that expression (1.68) becomes (1.69) in the case of zero pressure gradient. Hint: as in previous example, be careful in using variable Φ, which in this case is proportional to x, and variable ξ. To calculate gamma function apply known formulae, for example, $\Gamma(x)\Gamma(1 - x) = \pi/\sin \pi x$.

1.69* Show that the first and second expressions (1.72) for recovery factor follow from equations (1.71) and (1.70), respectively. Verify that that both definition of adiabatic temperature given in text follow from these equations.

1.70 Derive differential universal function (1.72) for recovery factor using relation (1.60) as described in the text. Comparing sums in functions (1.60) and (1.72), one sees that the second sum may be obtained directly from the first one by substituting U^2 for θ_{w*}. Explain physically why such substitution yields true result.

1.71 Obtain integral universal function (1.73) for recovery factor using relation (1.68). Holds the same physical explanation from the previous exercise in this case?

1.72 Show that in the case of zero pressure gradient, relations (1.72) and (1.73) yield well-known result for recovery factor $r = 2g_d$. Hint: in integrand (1.73) take a new variable $z = (\xi/x)^{3/4}$ and transform integral (1.73) to beta function defined by equation (1.47).

1.73   Prove that that only the last term in thermal boundary layer equation (1.11) is proportional to Prandl number. Explain physically why high Prandtl number leads to significant effect of mechanical energy dissipation.

1.74*   What parameters should be changed to make relations (1.72) and (1.73) applicable for turbulent flows?

1.75   Derive the second equation (1.49) for adiabatic stagnation enthalpy difference i_{0w} at zero pressure gradient from relation (1.72) knowing that Mach number is $M = U/U_{sd}$ where speed of sound is $U_{sd} = \sqrt{kRT_\infty}$.

2

Application of Universal Functions

2.1 The Rate of Conjugate Heat Transfer Intensity

In this section, we investigate the effect of different factors on conjugate heat transfer intensity considering the conjugate problem as a case of heat transfer from a surface with variable (nonisothermal) temperature or heat flux. Such an approach is founded on the conception (see Introduction) that a variable temperature (or temperature head) of a body/fluid interface is one of the basic characteristics of any conjugate problem. The results are obtained analyzing universal functions and are supplemented with relevant examples.

2.1.1 Effect of Temperature Head Distribution

The universal functions structure shows that there are three factors determining the effect of temperature head variation on the heat transfer intensity: (i) signs and values of temperature head derivatives, (ii) signs and values of coefficients g_k depending on Prandtl number for laminar and on Prandtl and Reynolds numbers for turbulent flows, and (iii) pressure gradient. The rate and a type (favorable or adverse) of the temperature head effect are defined basically by two first factors, whereas the pressure gradient specifies the independent variable (x or Φ) of derivatives (Exer. 2.1).

2.1.1.1 Effect of Temperature Head Gradient and Higher Derivatives

The results of calculation presented in Table 2.1 show that the first coefficient g_1 is significantly larger than others. Even in comparison with second one, which is the greatest among others, the first coefficient is from 3 (unsteady laminar flow) to 10 (turbulent flow) times larger. If all derivatives of the temperature head are of the same order, this means that the first derivative that defines the temperature head gradient basically specifies the effect of nonisotermicity (Exer. 2.2).

It is seen that the first coefficient g_1 is positive for all universal functions (except for inverse function, as indicate positions 3 and 4 in Table 2.1). Due to that, the positive temperature head gradients (the first derivative) lead to an increasing of heat flux, whereas the negative

Applications of Mathematical Heat Transfer and Fluid Flow Models in Engineering and Medicine,
First Edition. Abram S. Dorfman.
© 2017 John Wiley & Sons Ltd. Published 2017 by John Wiley & Sons Ltd.

Table 2.1 Relation between coefficients g_1 and g_2 of the universal functions. Laminar layer: arbitrary $\theta_w - 1 - \mathrm{Pr} \to 0$, $2 - \mathrm{Pr} \to \infty$, arbitrary $q_w - 3 - \mathrm{Pr} \to \infty$, $4 - \mathrm{Pr} \to 0$, unsteady laminar layer: $5 - \mathrm{Pr} = 1$, turbulent layer: $6 - \mathrm{Pr} \to 0$, $\mathrm{Re}_{\delta_1} = 10^3$, $7 - \mathrm{Re}_{\delta_1} = 10^9$, $8 - \mathrm{Pr} = 1$, $\mathrm{Re}_{\delta_1} = 10^3$, non-Newtonian fluid: $9 - n = 1.8$, $10 - n = 0.2$, 11-moving sheet: $\mathrm{Pr} \approx 1$, $\varepsilon = 0$.

	g_1	g_2	$\dfrac{|g_2|}{|g_1|}$
1	1	$-1/6$	1/6
2	0.6123	-0.1345	0.22
3	-0.380	0.135	0.36
4	$-1/2$	3/16	3/8
5	2.4	-0.8	1/3
6	≈ 0.5	≈ -0.05	≈ 0.1
7	≈ 0.1	≈ -0.01	≈ 0.1
8	≈ 0.2	≈ -0.04	≈ 0.2
9	≈ 0.8	≈ -0.2	≈ 0.25
10	≈ 0.4	≈ -0.06	≈ 0.15
11	1.25	-0.15	0.12

gradients cause a decreasing of the heat flux. This results in a greater than an isothermal heat transfer coefficients in the case of increasing temperature heads (positive first derivative) and in smaller than an isothermal heat transfer coefficients in the opposite case of decreasing temperature heads (negative first derivative). However, the calculations show that the same value of increasing and decreasing of the temperature head yields significantly different changes in the heat transfer coefficients. Physically, the reason of this is that the same absolute change in the falling and growing temperature heads yields much greater relative variation in the first case, when the temperature head itself is small, than in the second one when the temperature head increases. These considerations give the understanding why growing temperature heads results in only relatively modern increasing of heat transfer intensity, whereas the falling temperature heads lead to extremely small heat transfer coefficients reaches sometimes even zero.

The universal functions show that the third, fifth and higher odd derivatives cause qualitatively a similar effect, increasing the heat flux for positive and decreasing it for negative derivatives. This follows from the fact that all odd coefficients g_k as well as first one g_1 are positive. An opposite effect is produced by even derivatives because the corresponding even coefficients g_k are negative, resulting in decreasing and increasing heat flux for positive and negative derivatives, respectively.

The other feature that one may see from universal functions structure is that the total change in heat flux determines not by derivatives only, but rather by the products $x^k(d^k\theta_w/dx^k)$ or $\Phi^k(d^k\theta_w/d\Phi^k)$ of coordinate x or Görtler variable Φ with corresponding derivatives for zero pressure gradient or for general case, respectively. This means that, in particular, the second term of series $x(d\theta_w/dx)$ or $\Phi(d\theta_w/d\Phi)$ may be significantly large even in the case of small temperature head gradient if the body is enough long. For example, at the large distance from the leading edge on a plate, the heat transfer coefficient may be small (and even become zero) not only due to significant negative temperature head gradient $d\theta_w/dx$, but also at relatively minor gradient through large coordinate x.

Comment 2.1 It is easy to see that if the temperature head is proportional to coordinate $\theta_w = cx$, the temperature head gradient is independent on coordinate and hence, the last conclusion may be confusing. Exercise 2.4 helps to understand that this is a special case, whereas in the other, ordinary cases, the just-analyzed relation between temperature head gradient and corresponding product of gradient and coordinate presents the real situation.

The formulated above basic general properties of the heat transfer behavior in the case of variable temperature head are discussed in detail and justified, considering the examples of flows past nonisothermal surfaces, conjugate heat transfer, and different applications. We start with laminar flows, and then show some examples with turbulent flows as well as more complicated models.

■ Example 2.1: Laminar Flow Past Plate with Linear Temperature Head

The calculation data for this case for the fluids with Pr > 0.5 are plotted in Figure 2.1. The reason of choosing this case is that the coefficient $g_1 = 0.6123$ is the lowest for such fluids and is practically constant for $\beta = 0$ (zero pressure gradient) (Fig. 1.3). At the same time, these results of the smallest nonisothermicity effects are qualitatively valid for any laminar flow, including the limiting case Pr → 0 with the greatest value of coefficient $g_1 = 1$ when the effect of nonisothermicity at the same other conditions is maximum.

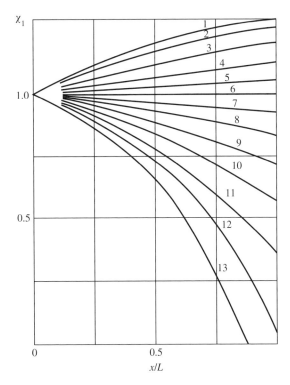

Figure 2.1   Effect of nonisothermicity for linear temperature head distribution on a plate. $1 - \theta_{we}/\theta_{wi} = 2$, $2 - 1.75$, $3 - 1.5$, $4 - 1.25$ $5 - 1.1$, $6 - 1.0$, $7 - 0.9$, $8 - 0.8$, $9 - 0.7$, $10 - 0.6$, $11 - 0.5$, $12 - 0.4$, $13 - 0.3$

The results in Figure 2.1 are presented using nonisothermicity coefficient obtained after dividing both sides of universal functions (1.39) and (1.40) by heat flux $q_{w*} = h_* \theta_w$

$$\chi_t = \frac{h}{h_*} = 1 + \sum_{k=1}^{\infty} g_k \frac{\Phi^k}{\theta_w} \frac{d^k \theta_w}{d\Phi^k}, \quad \chi_t = \frac{h}{h_*} = \frac{1}{\theta_w} \left[\int_0^{\Phi} f\left(\frac{\xi}{\Phi}\right) \frac{d\theta_w}{d\xi} d\xi + \theta_w(0) \right] \quad (2.1)$$

Substituting x for Φ gives corresponding equations for zero pressure gradient

$$\chi_t = \frac{h}{h_*} = 1 + \sum_{k=1}^{\infty} g_k \frac{x^k}{\theta_w} \frac{d^k \theta_w}{dx^k}, \quad \chi_t = \frac{h}{h_*} = \frac{1}{\theta_w} \left[\int_0^{x} f\left(\frac{\xi}{x}\right) \frac{d\theta_w}{d\xi} d\xi + \theta_w(0) \right] \quad (2.2)$$

It follows from those equations that the nonisothermicity coefficient shows how much the heat transfer coefficient in some variable temperature head is more or less than that for an isothermal surface.

In Figure 2.1, each curve relates to a fixed value of ratio θ_{we}/θ_{wi} determining the coefficient K in a linear dependence

$$\theta_w/\theta_{wi} = 1 - (1 - \theta_{we}/\theta_{wi})(x/L) = 1 - K(x/L), \quad (2.3)$$

where θ_{wi} and θ_{we} denote the initial and ending temperature heads. It is seen how strongly the effects for rising ($\theta_{we}/\theta_{wi} > 1$) and falling ($\theta_{we}/\theta_{wi} < 1$) temperature heads differ. For example, an increasing of the temperature head in $1.5 - 2$ times leads to growing of heat transfer coefficient of about 20 to 30%, whereas the same decreasing of temperature head results in lessening of heat transfer coefficient in $1.5 - 2.5$ times as compared to that for isothermal surface. If the decreasing of temperature head becomes more than three times, the heat flux at the end of the plate reaches zero. For small Prandtl numbers, the effect of nonisothermicity is greater because coefficient g_1 grows as Prandtl number decreases (Fig. 1.3). In particular, the double decreasing temperature head for $Pr = 0.01$ yields six times less heat transfer coefficient at the plate end and for $Pr \to 0$ reduces heat flux to zero (Exer. 2.5).

■ **Example 2.2: Laminar Flow Past Cylinder with Linear Temperature Head**

This case requires taking into account the nonzero pressure gradient applying the formulae (2.1). To perform this, it is necessary to have the function $\theta_w(\Phi)$, which corresponds to linear distribution $\theta_w(x)$ (2.3). Such function we construct as follows:

(i) adapt from [369] experimentally established function

$$U/U_\infty = 3.631(x/D) - 3.275(x/D)^3 - 0.168(x/D)^5 \quad (2.4)$$

which gives the external velocity distribution around a cylinder, (ii) compute the function $\Phi(x/D)$ using equation (2.4) and formula (1.39) for Görtler variable, (iii) derive the inverse to $\Phi(x/D)$ dependence for x/D as a function of Φ (Exer. 2.6)

$$x/D = 0.74(\Phi/Re)^{1/2} + 0.1(\Phi/Re), \quad \theta_w = 1 - K \left[0.74(\Phi/Re)^{1/2} + 0.1(\Phi/Re)\right] \quad (2.5)$$

and (iv) substitute first function (2.5) into linear distribution (2.3) $\theta_w/\theta_{wi} = 1 - K(x/D)$ to get the desired function $\theta_w(\Phi)$ in the form of the second equation (2.5).

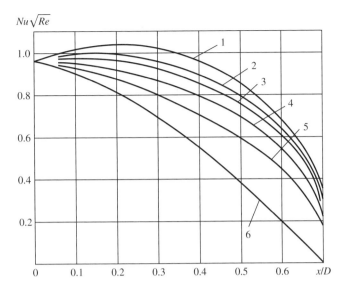

Figure 2.2 Heat transfer from a cylinder with linear temperature head in transverse flow of air $(\text{Pr} = 0.7)$. $1 - \theta_{we}/\theta_{wi} = 2, 2 - 1.5, 3 - 1.25\ 4 - 1.0, 5 - 0.75, 6 - 0.5$

The dimensionless heat transfer characteristic for cylinder with linear temperature head distribution is plotted in Figure 2.2. This result is found as a product $(\text{Nu}_*/\sqrt{\text{Re}})\chi_t$

$$\frac{\text{Nu}}{\sqrt{\text{Re}}} = \frac{\text{Nu}_*}{\sqrt{\text{Re}}}\left\{1 - \frac{K(\Phi/\text{Re})^{1/2}[0.37(g_1 + g_2/2) + 0.1g_1(\Phi/\text{Re})^{1/2}]}{1 - K(\Phi/\text{Re})^{1/2}[0.74 + 0.1(\Phi/\text{Re})^{1/2}]}\right\} \qquad (2.6)$$

where $\text{Nu}_*/\sqrt{\text{Re}}$ is similar heat transfer characteristic for isothermal cylinder, which may be calculated by one of known methods [338, 369] (Exer. 2.7). The nonisothermicity coefficient is estimated using the first formula 2.1 with only two first derivatives. The derivatives with respect to Φ are computed applying second relation (2.5).

In Figure 2.2, the same significant effect of nonisothermicity as in example 2.1 is observed. Comparing the curves for isothermal (curve 4) and nonisothermal cylinders shows that for $\theta_{we}/\theta_{wi} < 1, (K > 0)$ when the temperature head decreases in flow direction, the Nusselt number falls much intensely than that in the case of isothermal surface. Thus, for $\theta_{we}/\theta_{wi} = 0.5$, the heat flux becomes almost zero at the point close to separation flow. Vice versa, an increasing temperature head slows the falling in the Nusselt number. Therefore, in this case, at close to unity ratio $\theta_{we}/\theta_{wi} > 1, (K < 0)$ (for example $\theta_{we}/\theta_{wi} = 1.25$), the Nusselt number decreases slower than on an isothermal cylinder, whereas for greater nonisothermities, the heat transfer intensity even increases first (as for example for $\theta_{we}/\theta_{wi} = 2$) and only then goes down.

2.1.1.2 Effect of Pressure Gradient

As follows from the above discussion, the effect of the pressure gradient is specified by the Görtler variable. This conclusion becomes clear from the second term of series with the first temperature head derivative presented in the form

$$\Phi \frac{d\theta_w}{d\Phi} = \frac{U_{av}}{U} x \frac{d\theta_w}{dx} \tag{2.7}$$

This expression indicates that the ratio U_{av}/U is a measure of the effect of the pressure gradient on heat transfer intensity from nonisothermal surface. Because according to Görtler variable, U_{av} is an average velocity on the interval $(0, x)$, one concludes that relations $U_{av}/U < 1$ and $U_{av}/U > 1$ correspond to accelerating and decelerating flows, respectively (Exer. 2.8). Therefore, the second term (2.7) of universal function becomes lesser in the first and greater in the second cases, which leads to decreasing or increasing the effect of the temperature head as the flow accelerates (pressure reduces) or decelerates (pressure grows). This means that in a case of reducing pressure ($U_{av}/U < 1$ at accelerating velocity) the heat transfer coefficient decreases if the first derivative $d\theta/dx$ is positive and increases in the opposite case at negative temperature head derivative $d\theta/dx$. Consequently, in another case of growing pressure ($U_{av}/U > 1$ at decelerating velocity) heat transfer coefficient increases if the first derivative $d\theta/dx$ is positive and decreases in the opposite case at negative derivative $d\theta/dx$. (Exer. 2.9).

As a simple example, consider the self-similar flows when $U = cx^m$ (S. 7.5.2), and hence, the ratio $U_{av}/U = 1/(m+1)$ decreases the nonisothermicity effect in the case of $m > 0$ and diminishes it in the opposite case when $m < 0$. In particular, for the flow near stagnation point ($m = 1$), the second term (2.7) of universal function is a half as much as in the case of the zero pressure gradient when $U_{av}/U = 1$ and additional effect of pressure gradient is zero.

Comment 2.2 We considered a qualitative analysis of pressure gradient effect. To estimate the pressure gradient effect for a given external velocity $U(x)$, one gets a nonisothermicity coefficient χ_t applying the first formula (2.1) with two or three terms or using the second formula (2.1) if the accuracy of the first formula is unacceptable. The dependence of $\chi_t = h/h_*$ on $\Phi = \int_0^x U(\xi)d\xi$ shows the effect of pressure gradient.

2.1.2   Effect of Turbulence

2.1.2.1   Comparative Effects of Turbulent and Laminar Flows

Coefficients g_k for turbulent flow are less than those the laminar flow. The higher are the Reynolds and Prandtl numbers the less are coefficients g_k (Figs. 1.6 and 1.7) and, correspondingly, is less the effect of nonisothermicity. Nevertheless, the qualitative effect of the temperature head gradient on the heat transfer intensity is the same as in the case of laminar flow discussed in previous section. The quantitative results might be seen on Figure 2.3 where a comparison of nonisothermicity coefficients for laminar and turbulent flows is given.

It is seen that in spite of smaller coefficients g_k, nonisothermicity strongly affects the heat transfer intensity in turbulent flows if the temperature head decreases. In that case, the effect of nonisotermicity is not as high as in laminar flows, but if a surface is long enough, the heat flux finally also reaches zero. This is true for all cases except the turbulent flows of fluids with large Prandl numbers (say Pr > 100) for which the effect of nonisothermicity is negligible due to very small coefficients g_k, (Figs. 1.6 and 1.7).

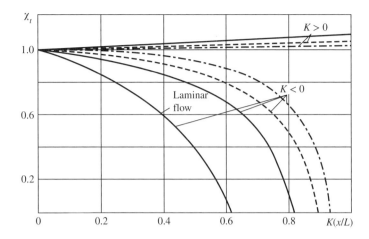

Figure 2.3 Nonisothermicity coefficients for turbulent and laminar flows. -------- $Re_{\delta_1} = 10^3$, - - - - $Re_{\delta_1} = 10^5$, -. -. -. $Re_{\delta_1} = 10^9$

2.1.2.2 Comparison of Calculations with Experimental Data

■ **Example 2.3: Linear and Exponential Temperature Head and Heat Flux [106]**

The effect of nonisothermicity was studied experimentally in zero pressure gradient turbulent flow of air by Leontev *et al.* [224]. They obtained data for the increasing and decreasing linear temperature heads and for exponential variations of temperature head and heat flux. Figure 2.4 presents the comparison of calculation results with these data.

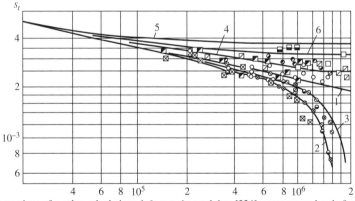

Comparison of results: calculations-1-6, experimental data [224]: temperature heads θ_w:
7 —○— 158.6; 8 —⊘— [222—150(x/L)]; 9 —⊗— [137—81(x/L)]; 10 —◐— [204—140(x/L)];
11 —●— [44+170(x/L)]; 12 —□— 0.19 exp 7(x/L); 13 —⊠— [34+159(x/L)]; 14 —⊠— [159–100(x/L)]; 15 —◪— [3.8+110(x/L)]; 16 —■— $q_{CT} = 3.4$ exp 5.8(x/L);

Figure 2.4 Comparison between calculation [106] and experimental [224] results for different temperature heads and heat flux variations for turbulent flow

In Figure 2.4, the line 1 corresponds to isothermal zero pressure gradient flow and is computed by well-known first equation (2.8) [201]

$$St_* = 0.0295 \, Re_x^{-0.2} \, Pr^{-0.4}, \quad St = St_* \left[1 - \frac{g_1}{(Re/KRe_x) - 1} \right] \tag{2.8}$$

The rest curves are obtained using universal functions. For zero pressure gradient flow and linear temperature head (2.3), the series (2.2) with two first terms was used in the form of second equation (2.8), where $Re/Re_x = L/x$.

Experiments were performed under low Reynolds numbers. According to Figure 1.6 for $Pr \approx 1$ and low Reynolds number ($Re_{\delta_1} = 10^3$), the first coefficient is estimated as $g_1 = 0.2$. Experimental data for linear temperature heads were obtained for the following inverse values of coefficient K in (2.3): $1/K = 1.46, 1.49, 1.59, 1.68$ for decreasing and $1/K = 0.215, 0.258, 0.346$ for increasing temperature heads. Curves 2, 3, 4, and 5 are calculated by second formulae (2.8) for two limiting values of coefficient K for decreasing (curves 2 and 3) and for increasing (curve 4, two curves coincide) temperature heads. Line 5 represents exponential increasing temperature head $\theta_w = \theta_{wi} \exp[K(x/L)]$ with $K = 7$. For this case with the large value of K instead of series, which converges slowly, the integral universal function (2.2) with exponents $C_1 = 1$ and $C_2 = 0.2$ ($Pr \approx 1$, $Re_{\delta_1} = 10^3$) is used in the form

$$St = St_* \left\{ \exp\left[-K(x/L) \right] + K(x/L)^{0.2} \int_0^{x/L} \xi^{-0.2} \exp(-K\xi) d\xi \right\} \tag{2.9}$$

To calculate curve 6 for exponential temperature head distribution $q_w = q_{wi} \exp[K(x/L)]$ with $K = 5.8$, equation (1.68) with the same value of $C_1 = 1$ and $C_2 = 0.2$ and the first equation (2.8) are used to get the expression (Exer. 2.10).

$$\frac{1}{St} = 6.35 Re^{0.2} \, Pr^{0.6} \int_0^{x/L} \xi^{-0.8} \exp(K\xi) d\xi \tag{2.10}$$

Figure 2.4 shows that there is a reasonable agreement between both results. The points representing the experimental data are close to corresponding theoretical results: curves 2 and 3 to points 8, 9, 14 (linear decreasing temperature head), curve 4 to points 11, 13, 15 (linear increasing temperature head), curve 6 to point 12 (exponential growing temperature head), and curve 5 to point 16 (exponential growing heat flux).

■ **Example 2.4: Stepwise Temperature Head [111]**

Moretti and Kays [275] experimentally studied heat transfer in the case of stepwise temperature head variations. In Figure 2.5 the comparison between calculations and Moretti and Kays experimental results is shown.

The experimental conditions differ from each other or by temperature head variation or by free stream velocity distribution. The free stream velocity along the surface increases: or gradually (Figs. 2.5a, e), or stepwise (Fig. 2.5b, c), or first increases and then decreases (Fig. 2.5d).

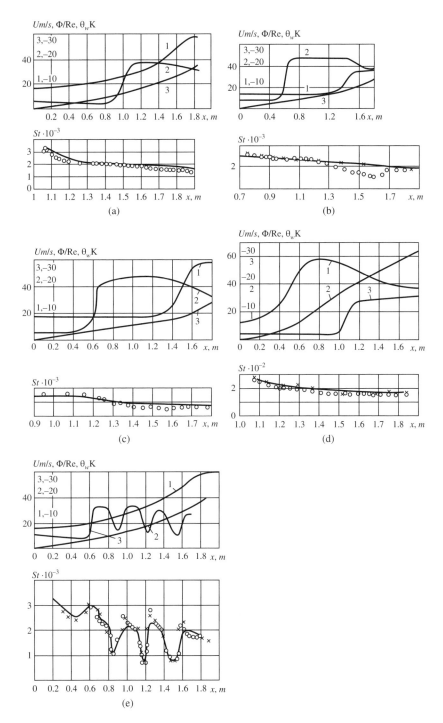

Figure 2.5 Comparison of results for different stepwise temperature heads and various pressure gradients obtained by calculation with experimental data [275] for turbulent flow; ×-numerical integration (b) [304], (e) [362]; $1 - U$, $2 - \theta_w$, $3 - \Phi/Re$

The temperature head variations contain one jump, as in four cases (Fig. 2.5a, b, c, and d), or several jumps (as in Fig. 2.5e). Since the temperature head distributions contain jumps and pressure gradients are not zero, the calculations are performed using general form of integral universal function (2.1) with Görtler variable. For example, in the case presented in Fig. 2.5a, this equation transformed to Stanton number has the form (Exer. 2.11)

$$
St = \frac{St_*}{\theta_w} \left\{ 1 + \left[1 - \left(\frac{\Phi_1}{\Phi} \right)^{C_1} \right]^{-C_2} \Delta\theta_w + \int\limits_{\Phi_2}^{\Phi} \left[1 - \left(\frac{\xi}{\Phi} \right)^{C_1} \right]^{-C_2} \frac{d\theta_w}{d\xi} d\xi \right\} \tag{2.11}
$$

Here, Φ_1 and Φ_2 relate to the start and to end points of the temperature head jump $\Delta\theta_w$.

It is seen that results giving by integral universal function (2.1) are in agreement with the experimental data not only for increasing temperature heads but, unlike the other analytical methods, also in the cases with decreasing temperature heads (Fig. 2.5e). Some more differences between both data in the cases of sharp increasing free stream velocity (cases 2.5b, c) are associated with changes in boundary layer structure, which correction requires additional information [275].

2.1.2.3   Reynolds Analogy and Heat Transfer of an Isothermal Surface [107]

The investigation of Reynolds analogy and heat transfer from an isothermal surface in turbulent flow was performed applying the same approach as used for coefficients g_k estimation (S. 1.7). Solutions of the thermal boundary layer equation (1.17) with the turbulent transport characteristics defined by Mellor-Gibson turbulence model were taken to compute the Reynolds analogy coefficient $2St_*/C_f = g_0$, where g_0 is a coefficient defining isothermal heat transfer intensity in the form of Reynolds analogy coefficient, which is similar to coefficients g_k. Calculations were performed for the same range of parameters as for coefficients g_k estimation: $\beta = -0.3$ (stagnation point flow), $\beta = 0$ (zero pressure gradient), $\beta = 1$ and $\beta = 10$ (flows with weak and strong adverse pressure gradients) and for following Prandtl and Reynolds numbers:

$$
Pr = 0.01, 0.1, 1, 10, 100, 1000 \quad Re_{\delta_1} = 10^3, 10^5, 10^9
$$

The results obtained in [107] (presented also in Figs. 4.1–4.4 [119]) show that the coefficient of Reynolds analogy: (i) equals unity for zero pressure gradient and $Pr = 1$ for all Reynolds numbers as it should be according to well-known Reynolds analogy, (ii) increases above unity and decreases below unity for $Pr < 1$ and $Pr > 1$, respectively, (iii) decreases at $Pr < 1$ and increases at $Pr > 1$ as the Reynolds number grows, (iv) changes its value much intensively with Re growing at $Pr < 1$ than that at $Pr > 1$, and (v) changes its increasing/decreasing point depending of pressure gradient: at $Pr = 1$ for zero pressure gradient ($\beta = 0$) according to Reynolds analogy, at $Pr \approx 0.5 - 0.7$ for favorable pressure gradient ($\beta = -0.3$), at $Pr \approx 3$ for weak ($\beta = 1$) and at $Pr \approx 250$ for strong ($\beta = 10$) adverse pressure gradients.

It follows from these data that the Reynolds analogy holds at zero pressure gradient and $Pr = 1$ but violates in other cases so that, depending on pressure gradient, Prandtl, and Reynolds numbers, the Stanton number St is greater or lower than half-friction coefficient $C_f/2$. Stanton number is considerably greater than the half-friction coefficient at small Prandtl and Reynolds numbers. However, as Prandtl number grows, the value of ratio $2St_*/C_f$ reduces significantly at all Reynolds numbers and becomes asymptotically zero at $Pr \to \infty$.

Reynolds analogy data are used in [107] for creating relations determining the isothermal heat transfer rate in turbulent flows. The four equations are derived by the computing results approximation

$$St_* = Pr^{-1.35}(C_f/2)^{1-0.3\log Pr}(1 < Pr < 50), \quad St_* = 0.113\,Pr^{-3/4}(C_f/2)^{1/2} \ (Pr > 50) \tag{2.12}$$

$$(Nu_x)_*^{-0.023} = 1.04 - 0.0335\lg Pe_x \quad St_* = cPe_\Phi^n \tag{2.13}$$

$$c = 0.282, 0.036, 0.00575 \ \text{and} \ n = -0.38, -0.2, -0.1$$

$$Pe_\Phi = \Phi\,Pr = 10^3 - 10^5, \ 10^5 - 5\cdot10^8, \ 5\cdot10^8 - 2.5\cdot10^{12}$$

The first two equations are applicable for zero pressure gradient and $Pr > 1$. The two others describe the data for $Pr < 1$. The first relation (2.13) is applicable for whole range $0 < Pr < 1$, but only for zero pressure gradient, whereas the second relation (2.13) is suitable for both zero and nonzero pressure gradients, but with different constants for three intervals of Peclet number $Pe_\Phi = \Phi\,Pr$ defined through Görtler variable.

Figures 2.6 through 2.9 compare calculations with experimental data confirming the applicability of equations (2.12) and (2.13). Figure 2.6 compares the second formula (2.12) with experimental data from [209]. The curves $St_*\sqrt{2/C_f} = f(Pr)$ computed for different Reynolds numbers, which merge into one for large Prandtl numbers, are continued into region $Pr > 10^3$

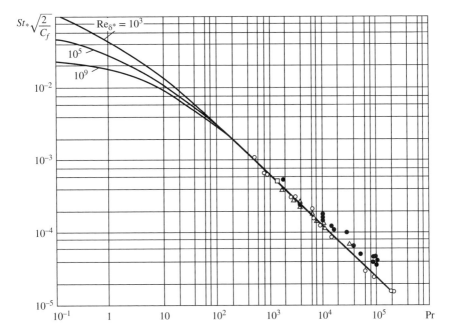

Figure 2.6   Comparison between calculation (second equation (2.12)) and experimental data [209] for zero pressure gradient flow and large Prandtl numbers $Pr > 50$

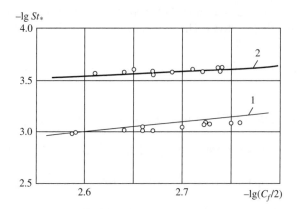

Figure 2.7 Comparison between calculation (first equation (2.12)) and experimental data [444] for zero pressure gradient flow and large Prandtl numbers. 1-water (Pr = 5.5), 2-oil (Pr = 55)

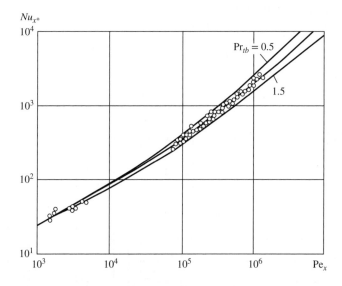

Figure 2.8 Comparison between first equation (2.13) for Pr < 1 and experimental data for air [308] and liquid metal [136]

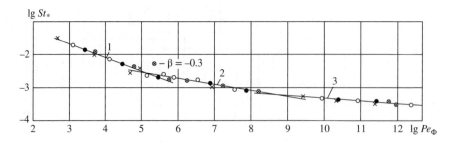

Figure 2.9 Dependence between St_* and Pe_Φ for Pr < 1 and zero and nonzero pressure gradients flows: $\otimes - \beta = -0.3$, $\bullet - \beta = 0$, $\circ - \beta = 1$, $\times - \beta = 10$, 1, 2, 3-second equation (2.13)

by calculating the slope of the tangent at the point $Pr = 10^3$. Good agreement is seen between the computed and experimental results: the coefficient 0.113 in the second equation (2.12) determined by calculation is close to value 0.115 determined from experimental data in [209]. Comparison between other equation (2.12) for region $1 < Pr < 50$ and experimental data from [444] presented in Figure 2.7 also shows good agreement.

The conformity of the first relation (2.13) for Nusselt number at $Pr < 1$ with experimental results from [308] for air and from [136] for liquid metals is shown in Figure 2.8. This relation also well agrees with other experimental results obtained in [444] and [321]. The reliance of the second relation (2.13) follows from Figure 2.9 showing that the data related to flows with zero and nonzero pressure gradients at $Pr < 1$ form a single curve approximated by three power-law function for Stanton number defined through Görtler variable (Exer. 2.12).

Calculations show that for zero pressure gradient and favorable gradients at large Reynolds numbers the error of the second formula (2.13) is less than 5%. For low Reynolds numbers, adverse gradients and Prandtl numbers close to 1, the error increases and reaches 35% for $Pr = 1$, $Re_{\delta_1} = 10^3$ and $\beta = 10$. In these cases, the approximate estimation of isothermal heat transfer coefficient in turbulent flow may be obtained using well-known formula 2.8 in the form based on Görtler variable $St_* = 0.0295\Phi^{-0.2} Pr^{-0.6}$.

Comment 2.3 Formulae (2.12) and (2.13) along with the last equation for Stanton number covered the whole diapason of parameters determining the heat transfer coefficients of isothermal surfaces in turbulent incompressible flow.

2.1.2.4   Effect of Turbulent Prandtl Number

The same approach is employed in [114] for studying the effect of turbulent Prandtl number on calculation accuracy of heat transfer characteristics. Computations have been performed for zero pressure gradient, two values of turbulent Prandtl numbers $Pr_{tb} = 0.5$ and $Pr_{tb} = 1.5$, four values of Prandtl numbers $Pr = 10^{-1}$, 1, 10^2, 10^3 and three Reynolds numbers $Re_{\delta_1} = 10^3 (Re_x = 2.95 \cdot 10^5)$, $10^5 (7.93 \cdot 10^7)$, $10^9 (2.56 \cdot 10^{12})$. The results in the form of the ratio $St_*/(St_*)_{Pr_{tb}=1}$ as a function of the Prandtl number are given in Figure 2.10 where St_* and $(St_*)_{Pr_{tb}=1}$ are values of Stanton number for turbulent Prandtl values $Pr_{tb} \neq 1$ and for $Pr_{tb} = 1$. It follows from Figure 2.10 that an increase in turbulent Prandtl yields a reduction, whereas the decrease in Pr_{tb} leads to an increase in heat transfer compared to that at $Pr_{tb} = 1$.

An increase in Pr_{tb} to 1.5 yields less reduction in the Stanton number than the corresponding increase in the Stanton number caused by the decrease of Pr_{tb} to 0.5. The most significant effect of turbulent Prandtl number is observed when the physical Prandtl number is close to unity. For $Pr_{tb} = 0.5$, the maximum increase in the Stanton number in comparison with that at $Pr_{tb} = 1$ is 67% at $Pr = 1$ and $Re_{\delta_1} = 10^9$. The maximum decrease of this effect for $Pr_{tb} = 1.5$ is also at $Pr = 1$ and $Re_{\delta_1} = 10^9$, but it is by about 25% less. These differences decrease with increasing Pr and for $Pr > 10^2$ practically vanish becoming independent of Pr as well as of Re (Fig. 2.10).

Comment 2.4 Comparatively small influence of Pr_{tb} at large Prandtl numbers is a result of a thin thermal boundary layer lying basically in the viscous sublayer. The effect of Pr_{tb} lessens also as Prandtl number decreases, which is a result of significant molecular heat conduction at small Prandtl numbers.

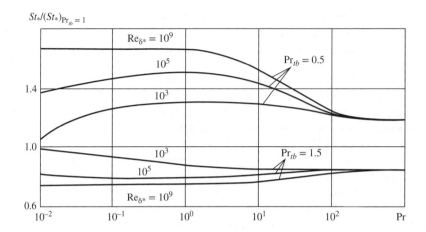

Figure 2.10   Dependence between $St_*/(St_*)_{\mathrm{Pr}_{tb}=1}$ and Pr for different Re_{δ_1} and Pr_{tb}

The calculation results for $\mathrm{Pr}_{tb} = 0.5$ and $\mathrm{Pr}_{tb} = 1.5$ are compared with measuring data. In particular, the value 0.115 of the coefficient in the second equation (2.12) found in [209] from experimental data (Fig. 2.6) was compared with coefficients 0.096, 0.113, and 0.136 computed for $\mathrm{Pr}_{tb} = 0.5$, $\mathrm{Pr}_{tb} = 1$, $\mathrm{Pr}_{tb} = 1.5$. It is seen that the value 0.113 obtained for $\mathrm{Pr}_{tb} = 1$ practically coincides with the experimental value 0.115, whereas two others for $\mathrm{Pr}_{tb} = 0.5$ and $\mathrm{Pr}_{tb} = 1.5$ differ from this result markedly. The other result of comparison is presented in Figure 2.8 where the calculations of Nusselt number for the same three values of $\mathrm{Pr}_{tb} = 0.5$, 1, 1.5 are plotted together with data from [308] and [136]. The same conclusion follows from this pattern: the curve obtained for $\mathrm{Pr}_{tb} = 1$ agrees much better with measuring data than the two others lying up and under of experimental points (Exer. 2.13).

Comment 2.5   For a long time, the contradictory experimental data (e.g., [209, 225, 243, 274]) leads to a lengthy discussion about dependence of the turbulent Prandtl number on transverse coordinate, in particular, at the wall vicinity. The more relable results are obtained by direct numerical calculations that we consider below in Chapter 6 (Exam. 6.18). These modern data confirm the derived here conclusion of $\mathrm{Pr}_{tb} = 1$, which is in line with practice becoming common to take the turbulent Prandtl number as a constant either close to or equal to one [304, 338, 422, 444].

2.1.3   Effect of Time-Variable Temperature Head

It follows from Table 2.1 that the first coefficient in the differential universal function is the highest at the time-dependent temperature head. According to Figure 1.16, the coefficient in equation (1.55) at the first derivative with respect to time is $g_{01} \approx 2.4$ for the dimensionless time $z > 2.4$ when the coefficients of series (1.55) are independent on time. The coefficient in this series at the first derivative with respect to space as well as the coefficient g_1 in series

(1.38) for the same case of zero pressure gradient and $Pr \approx 1$ is $g_{10} \approx 0.6$ (Fig. 1.3). Thus, if the both derivatives are of the same order, effect induced by time variation temperature head is $g_{01}/g_{10} \approx 4$ times greater than that caused by space temperature head variation.

■ Example 2.5: Linear Time-Dependent Temperature Head

Figure 2.11 shows the time variation of heat flax and nonisothermicity coefficient obtained by equation (1.55) and first equation (2.2) for a linear decreasing temperature head $\theta_w = a_0 - a_1 t$. In this case, these equations take the form (Exer. 2.14)

$$\frac{q_w}{h_* a_0} = 1 - C_t \left[Z + g_{01}(z) \frac{x}{L} \right], \quad \chi_t = 1 - \frac{C_t g_{01}(z)(x/L)}{1 - C_t z}, \quad C_t = \frac{a_1 L}{a_0 U}, \quad Z = \frac{tU}{L} \quad (2.14)$$

where $z = tU/x$ and coefficient $g_{01}(z)$ is given in Figure 1.16.

The following specific features of unsteady nonisothermal heat transfer may be derived from equations (2.14) and Figure 2.11: (i) the heat flux and the heat transfer coefficient (as well as nonisothermicity coefficient) depend not only on time, but also on the coordinate x, despite the surface temperature depends only on time $\theta_w = a_0 - a_1 t$ (Exer. 2.15), (ii) it follows from

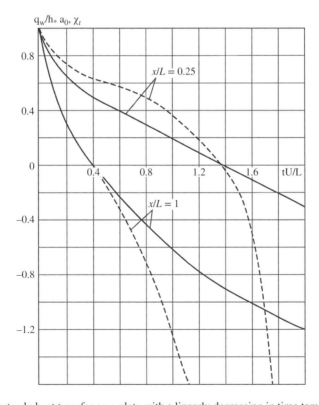

Figure 2.11 Unsteady heat transfer on a plate with a linearly decreasing in time temperature head ___ $q_w/h_* a_0$, $- - - \chi_t$

equation (1.55) that this property is always true for unsteady heat transfer because the terms with time and mixed derivatives depend also on coordinate x and on velocity U, (iii)) due to a large coefficient g_{01}, the heat transfer characteristics (q_w and χ_t) decrease much faster than that in steady-state regime, (iv) these characteristics quickly become zero, reaching it the sooner the greater the distance from leading edge is; in particular, it is seen from Figure 2.11 that in the case in question, this occurs at $z = 0.4$ and at $z = 1.4$ for $x/L = 1$ and $x/L = 0.25$, respectively.

2.1.4 Effects of Conditions and Parameters in the Inverse Problems

Here, we consider the effects arising in the problems and the solution, which requires the use of inverse universal functions (1.60) or (1.68) determining the temperature head at a given heat flux distribution. In a general case with given arbitrary functions $q_w(\Phi)$ and $h_*(\Phi)$, these problems are solved numerically applying integral function (1.68). At the same time, there are practical important inverse problems associated with relatively simple functions $q_w(\Phi)$ and $h_*(\Phi)$, which enable analytical solutions. Below, several of such solutions are analyzed showing effects of different parameters on the heat transfer rate.

2.1.4.1 Effect of Boundary Conditions

First, the connection between the most often employed simple boundary conditions $q_w = const.$ and $T_w = const.$ is studied. Functions (1.60) and (1.68) after dividing both sides by q_w give the relation between heat transfer coefficients h_q and h_* on surfaces with $q_w = const.$ and $T_w = const.$ in two (serial and integral) forms

$$\frac{1}{h_q} = \frac{1}{h_*} + \sum_{k=1}^{\infty} h_k \Phi^k \frac{d^k(1/h_*)}{d\Phi^k}$$

$$\frac{1}{h_q} = \frac{C_1}{\Gamma(C_2)\Gamma(1 - C_2)} \int_0^\Phi \left[1 - \left(\frac{\xi}{\Phi}\right)^{C_1} \right]^{C_2-1} \left(\frac{\xi}{\Phi}\right)^{C_1(1-C_2)} \frac{d\xi}{h_*(\xi)\xi} \qquad (2.15)$$

Both relations are valid for laminar and turbulent flows with arbitrary pressure gradient and Prandtl numbers, but in the case of arbitrary function $h_*(\Phi)$, the second equation requires also numerical evaluation of the integral. Two examples when function $h_*(\Phi)$ is given or may be approximated by polynomial or power-law function are considered next. In the first case, the known polynomial with coefficients a_i for $1/h_*$ produces polynomial for $1/h_q$ with coefficients b_i determined according to the first equation (2.15) via the beta functions (1.47) (Exer. 2.16)

$$\frac{1}{h_*} = \sum_0^{i=k} a_i \Phi^i, \qquad \frac{1}{h_q} = \sum_0^{i=k} b_i \Phi^k, \qquad b_i = a_i B \left[\frac{C_1(1 - C_2) + i}{C_1}, C_2 \right] / B(C_2, 1 - C_2) \quad (2.16)$$

In the second case, when for isothermal heat transfer coefficient is used in the customary form $h_* = c\Phi^{-n}$, both equations (2.15) show that ratio h_*/h_q is independent on variable Φ being determined as (Exer. 2.17)

$$\frac{h_*}{h_q} = 1 + \sum_{k=1}^{k=\infty} h_k n(n-1)(n-2) \ldots (n-k=1),$$

$$\frac{h_*}{h_q} = \frac{\Gamma\{[C_1(1-C_2)+n]/C_1\}}{\Gamma\{C_2+[C_1(1-C_2)+n]/C_1\}\Gamma(1-C_2)} \tag{2.17}$$

■ **Example 2.6: Heat Transfer From Cylinder in Transverse Flow at $q_w = const$ and $T_w = const$**

The last relation (2.16) and data of heat flux for cylinder with linear temperature head variation presented in Figure 2.2 are used to calculate the heat transfer rate from transverse streamlined cylinder with $q_w = const$. The function $Nu_*(x/D)/\sqrt{Re}$ for isothermal cylinder giving by curve with $\theta_{we}/\theta_{wi} = 1$ on Figure 2.2 is approximated by polynomial in inverse form $\sqrt{Re}/Nu_*(\Phi)$ required for using equations (2.16)

$$\sqrt{Re}/Nu_* = 1.04 + 0.75(\Phi/Re) - 0.83(\Phi/Re)^2 + 3.4(\Phi/Re)^3 \tag{2.18}$$

The coefficients b_i of corresponding polynomial determining $\sqrt{Re}/Nu_q$ for cylinder with $q_w = const$. are obtained by equation (2.16). Since the values of exponents C slightly depend on β, they are estimated from Figure 1.5 approximately: $C_1 = 0.92$ and $C_2 = 0.4$ for the region near the stagnation point ($\beta \approx 1$) and $C_1 = 0.9$ and $C_2 = 0.38$ for the rest part of cylinder ($\beta \approx 0$). Then, the corresponding polynomial is computed (Exer. 2.18)

$$\sqrt{Re}/Nu_q = 1.04 + (0.44 \div 0,46)(\Phi/Re) - (0.39 \div 0.42)(\Phi/Re)^2 + (1.4 \div 1.5)(\Phi/Re)^3 \tag{2.19}$$

Here, a sign $\div$ is used to indicate two coefficients obtained for two values of β (different C_1 and C_2). Calculation shows that both coefficients yield practically the same data plotted in Figure 2.12 confirming that Görtler variable takes into account the flow history (Com. 1.5) leading to slight effects of parameter β on the final calculation results.

Comparison with data for isothermal cylinder giving by equation (2.18) and plotted on the same figure presents the effect of boundary conditions of the rate of heat transfer.

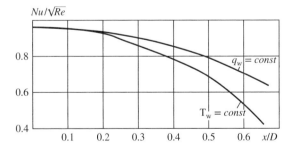

Figure 2.12 Effect of boundary conditions on heat transfer from a cylinder in transverse flow

■ **Example 2.7: Effects of Prandtl Number and Pressure Gradient**

The effects of Prandtl number and pressure gradient on ratio h_*/h_q are estimated by equations (2.17). As mentioned above, these equations do not depend on Φ, but they contain the exponent n from customary formula for heat transfer coefficient $h_* = c\Phi^{-n}$. To estimate this exponent, consider the self-similar flows with external flow $U = cx^m$. Because for isothermal surface $Nu_{x*} \sim \sqrt{Re_x}$ (see (2.33)), we have from this relation $h_*x/\lambda \sim \sqrt{Ux/v}$, which for self-similar flows gives $h_* \sim x^{(m-1)/2} \sim \Phi^{(m-1)/2(m+1)}$ (since $\Phi \sim x^{m+1}$). Thus, for zero pressure gradient ($m = 0$) and for stagnation point ($m = 1$), we have $n = 1/2$ and $n = 0$, respectively. Then, the second equation (2.17) with powers C_1 and C_2 from Figure 1.5 yields for zero pressure gradient with $n = 1/2$ the well-known data: $h_*/h_q = 0.74$ ($C_1 = 3/4, C_2 = 1/3$) for $1 < Pr < \infty$ (here, exponents are practically constant), and $h_*/h_q = \Gamma(1)/\Gamma(3/2)\Gamma(1/2) = 2/\pi \approx 0.64$ for $Pr \to 0$. Close (about 10% over) results $h_*/h_q \approx 0.78$ (instead of 0.74) and $h_*/h_q \approx 0.7$ (instead of 0.64) provides the first equation (2.17) with first several coefficients h_k (1.65) and (1.64), respectively (in this case, series converges poorly).

For the second case of stagnation point with $n = 0$, both equations (2.17) show that for any values of C_1 and C_2 (which means for any Prandtl number and any pressure gradient) both heat transfer coefficients are equal to each other, so that $h_*/h_q = 1$. This is obvious, because in this case, the sum in the first equation (2.17) is zero resulting in unity, and the second equation gives in this case $\Gamma(1 - C_2)/\Gamma(1)\Gamma(1 - C_2) = 1$ (Exer. 2.19).

The following conclusions of boundary conditions effects may be formulated: (i) for zero and positive (adverse) pressure gradients ($0 \geq \beta > -0.199$), the ratio h_*/h_q increases from 0.64 to 0.74 as the Prandtl number growth from $Pr \to 0$ to $Pr \to \infty$, so that for any Prandtl number and zero or possible positive gradient, the heat transfer rate at $q_w = const.$ is greater than that at $T_w = const.$ of 36% for large and 57% for small Prandtl numbers, (ii) for negative pressure gradients ($0 < \beta \leq 1$) for whole diapason of Prandtl numbers from $Pr \to 0$ to $Pr \to \infty$, the ratio $h_*/h_q = 1$, so that for any Pr and negative gradients, both heat transfer rate at $q_w = const.$ and $T_w = const.$ are equal to each other, (iii) it follows from these data that the effect of boundary conditions on heat transfer rate significantly varies from remarkable influence, like in zero and positive pressure gradient flows at any Prandtl number or in flow past cylinder (Fig.2.12), when the heat transfer at $q_w = const.$ is fairly greater up to 57% than that on the isothermal surface, to no effects at all in flows with negative pressure gradients at arbitrary Prandtl numbers, as in stagnation point flow, when the heat transfer rate for both boundary conditions $q_w = const.$ and $T_w = const.$ is the same (Exer. 2.20 and 2.21).

2.1.4.2 Heat Transfer in the Case of Heat Flux Jump

To derive the influence function for the case of the heat flux jump, similar to the function $f(\xi/\Phi)$ in equation (1.40) for an unheated isothermal zone at the temperature jump, we integrate by parts the integral function (1.68) by setting

$$u(\xi) = q_w, \quad v(\xi, \Phi) = \frac{C_1}{\Gamma(C_2)\Gamma(1 - C_2)} \int_\xi^\Phi \left[1 - (\zeta)^{C_1}\right]^{C_2-1} (\zeta)^{C_1(1-C_2)} \frac{d\zeta}{h_*(\zeta)\zeta} \qquad (2.20)$$

Here, $\zeta = \xi/\Phi$ is a new dummy variable. The last function (2.20) at $\xi = \Phi$ equals zero $v(\Phi, \Phi) = 0$ and according to the second equation (2.15) at $\xi = 0$ becomes $v(0, \Phi) = 1/h_q$.

Due to that the result of integration takes the form (Exer. 2.22)

$$\theta_w = \frac{q_w(0)}{h_q} + \int_0^\Phi v(\xi, \Phi) \frac{dq_w}{d\xi} d\xi \qquad (2.21)$$

A comparison shows that this equation for temperature head has the same structure as equation (1.40) for heat flux with influence function $f(\xi/\Phi)$. To see this, note that: (i) the ratio $q_w(0)/h_q$ in equation (2.21) determines the temperature head θ_{wq} on the surface with $q_w = $ const. as the product $\theta_w(0)h_*$ determines the heat flux q_{w*} on an isothermal surface in equation (1.40), (ii) both integrals (2.21) and (1.40) specify the increments: the first one, of the temperature head, and the second, of the heat flux, whereas (iii) the expressions $(dq_w/d\xi)d\xi$ and $(d\theta_w/d\xi)d\xi$ ascertain the corresponding increments of the heat flux and of the temperature head in equations (2.21) and (1.40), respectively. Consequently, the equation (2.21) is similar to equation (1.40) and the structure of this equation indicates that function $v(\xi, \Phi)$ in front of the derivative in (2.21) plays the same role as the influence function $f(\xi/\Phi)$ in front of the derivative in (1.40). This implies that: (i) the function $v(\xi, \Phi)$ gives the reciprocal ratio $h_*/h_{q\xi}$ of the heat transfer coefficients after the heat flux jump $h_{q\xi}$ and on the initial isothermal zone h_* as well as the influence function $f(\xi/\Phi) = h_\xi/h_*$ specifies the ratio of the heat transfer coefficients after the temperature jump h_ξ and on the initial isothermal zone h_* and (ii) the function $v(\xi, \Phi)$ is a reciprocal influence function of the heat flux jump as well as the function $f(\xi/\Phi)$ is the influence function of the temperature head jump.

Consequently, the equation (2.21) is similar to equation (1.40) and the structure of this equation indicates that function $v(\xi, \Phi)$ in front of the derivative in (2.21) gives the reciprocal ratio $h_*/h_{q\xi}$ of the heat transfer coefficient $h_{q\xi}$ after the heat flux jump and the heat transfer coefficient h_* on the initial isothermal zone as well as the influence function $f(\xi/\Phi) = h_\xi/h_*$ in front of the derivative in (1.40) specifies the ratio of the heat transfer coefficient h_ξ after the temperature jump and the heat transfer coefficient h_* on an initial isothermal zone. This implies that the function $v(\xi, \Phi)$ is a reciprocal influence function of the heat flux jump as well as the function $f(\xi/\Phi)$ is the influence function of the temperature head jump.

Using for the influence function of the heat flux jump the notation $f_q(\xi, \Phi)$ similar to $f(\xi/\Phi)$, and applying the second equation (2.20) determining the function $v(\xi, \Phi)$, we arrive to the following expression for reciprocal influence function $1/f_q(\xi, \Phi)$, where $\zeta = \xi/\Phi$ is dummy variable (Exer. 2.23)

$$\frac{1}{f_q(\xi, \Phi)} = \frac{h_*}{h_{q\xi}} = \frac{C_1 h_*}{\Gamma(C_2)\Gamma(1 - C_2)} \int_{\xi/\Phi}^1 (1 - \zeta^{C_1})^{C_2-1} \zeta^{C_1(1-C_2)-1} \frac{d\zeta}{h_*(\Phi\zeta)} \qquad (2.22)$$

Because h_* is known, this expression determines the heat transfer coefficient after heat flux jump $h_{q\xi} = q_w/\theta_w$ where q_w and θ_w are heat flux jump and corresponding temperature head caused by this jump. Therefore, since the heat flux q_w is given, relation (2.22) determines actually the distribution of the temperature head after heat flux jump.

Equation (2.22) is written in Görtler variables and due to that is valid for arbitrary gradient flows. To obtain the expression for zero pressure gradient flow, notice that in this case, $\Phi = Re_x, h_* = cRe_x^{-1/2}$, and hence, $h_*(\Phi\zeta) = h_*(x)\zeta^{-1/2}$. Then, after using a new variable

$\sigma = 1 - \zeta^{C_1}$ and applying the incomplete beta function $B_\sigma(i,j)$, equation (2.22) takes the well-known form [321]

$$f_q\left(\frac{\xi}{x}\right) = \frac{h_{q\xi}}{h_*} = \frac{B(C_2, 1 - C_2)}{B_\sigma\{C_2, [C_1(1 - C_2) + 1/2]/C_1\}}, \quad B_\sigma(i,j) = \int_0^\sigma r^{i-1}(1 - r)^{j-1}dr \quad (2.23)$$

The incomplete beta function becomes beta function (1.47) at $\sigma = 1$. For $C_1 = 3/4$ and $C_2 = 1/3$ equation (2.23) transforms in formula for temperature head after the heat flux jump in the simplest case of zero pressure gradient and $\mathrm{Pr} > 1$ (Exer. 2.24 and 2.25).

$$\theta_w = 0.276 \ (q_w/h_*)B_\sigma(1/3, 4/3) \tag{2.24}$$

Comment 2.6 Although both influence functions for temperature head and for heat flux jumps are similar, the latter $f_q(\xi, \Phi)$, unlike the previous $f(\xi/\Phi)$, depends not on ratio ξ/Φ only but on each of variables ξ and Φ separately. This difference is caused by more complex integrand in expression (2.22), which depends not only on variable $\zeta = \xi/\Phi$ but also on the isothermal heat transfer coefficient $h_*(\Phi\zeta)$.

Comment 2.7 General formulae for laminar flow, like considered in this section, are applicable to turbulent flows and to other cases with relevant values of coefficients g_k, h_k and exponents C_1 and C_2 as well as formulae of type (2.17) (see Exer. 2.21).

■ **Example 2.8: The Heat Flux Jump on a Cylinder in Transverse Flow and on a Plate**

The solution for this problem is given by function (2.22) containing isothermal heat transfer coefficient $h_*(\Phi\zeta)$. Therefore, in general case of arbitrary pressure gradient flows, the solution requires numerical evaluation of the integral as in the second formula (2.15). For the polynomial presentation (2.18) for reciprocal isothermal coefficient $1/h_*(\Phi\zeta)$ with coefficients a_i, which we used in Example 2.6, equation (2.22) for $1/f_q$ yields polynomial solution with coefficients b_i defined by formula (2.16) with the incomplete beta functions (2.23) instead of complete beta functions (1.47), and the solution of the problem in question takes the form (Exer. 2.26)

$$\frac{1}{h_{q\xi}} = \frac{1}{B(C_2, \ 1 - C_2)} \sum_{i=0}^{i=n} a_i \Phi^i B_\sigma[(1 - C_2 + i \ /C_1), C_2] \tag{2.25}$$

where $\sigma = [1 - (\xi/\Phi)^{C_1}]$.

It is assumed that the jump occurs at central angle $30°(x/D = \pi/12)$ before which the unheated isothermal zone exists. The results for $\mathrm{Pr} > 1$ are plotted in Figure 2.13. The data for a plate calculated by the first equation (2.23) for zero pressure flow is plotted on the same figure. It is seen that in this case, the pressure gradient significantly affects the variation of the heat transfer coefficient along the surface showing remarkable increasing of the heat transfer intensity caused by pressure gradients on a cylinder.

Comment 2.8 The examples just considered show two types of heat transfer behavior at the leading edge. A body with blunt entrance, like cylinder in a transverse flow, reveals a finite heat

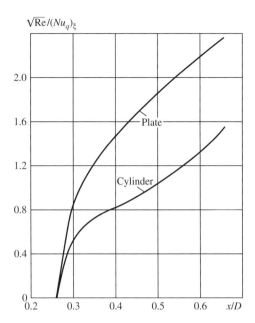

Figure 2.13   Heat transfer from a cylinder in transverse flow and from a plate after heat flux jump

transfer coefficient at the stagnation point (Figs. 2.2 and 2.12), whereas at the leading edge of a plate or at the starting point in the cases specifying by influence function of heat flux jump $f_q(\xi, \Phi)$, the heat transfer coefficient is infinite (Fig 2.13) because it is proportional to $x^{-1/2}$ in laminar or to $x^{-1/5}$ (equations (2. 33)) in turbulent flow. The cause of this infinite result is the parabolic type of boundary layer equations, which could not satisfy boundary conditions at the leading edge of this type, which in order to satisfy requires Dirichlet problem formulation (S. 7.2).

2.1.4.3   Heat Transfer on an Adiabatic Surface in an Impingent Flow

This problem is of the same type as previous one and is solved similarly. If an isothermal section precedes the thermally insulated adiabatic surface, at the entrance of it, the heat flux in impingent flow drops abruptly to zero, whereas the temperature head decreases gradually, becoming practically equal to zero only at a some distance from the entrance section. To determine the variation of this temperature head along the adiabatic surface, the universal function (1.68) is again integrated by parts putting

$$u(\xi) = \frac{q_w}{h_*} = \theta_{w*}, \quad v\left(\frac{\xi}{\Phi}\right) = \frac{C_1}{\Gamma(C_2)\Gamma(1 - C_2)} \int\limits_{\xi}^{\Phi} [1 - (\zeta)^{C_1}]^{C_2-1}(\zeta)^{C_1(1-C_2)}\frac{d\zeta}{\zeta} \quad (2.26)$$

The function $v(\xi/\Phi)$ at $\xi = \Phi$ equals zero $v(1) = 0$ and at $\xi = 0$ becomes unity $v(0) = 1$ because the integral in (2.26) at $\xi = 0$ equals $[\Gamma(C_2)\Gamma(1 - C_2)]/C_1$ (compare to a factor in front of integral). Hence, relation for temperature head (1.68) after integration becomes

$$\theta_w = \theta_{w*}(0) + \int_0^\Phi v\left(\frac{\xi}{\Phi}\right) \frac{d\theta_{w*}}{d\xi} d\xi, \tag{2.27}$$

where $\theta_{w*}(0)$ is the temperature head on the isothermal section at the entrance of the adiabatic surface (Exer. 2.27).

Comparing as in previous section relation (2.27) with equation (1.40) for the case of temperature jump, one concludes that: (i) the integral in (2.27) determines the temperature head change $\theta_{w*}(0) - \theta_w$ from temperature head $\theta_{w*}(0)$ on an isothermal section to some value θ_w on an adiabatic surface similar to integral (1.40), which in case of temperature jump determines the heat flux change $(q_w - q_{w*})$, and (ii) the product $(d\theta_{w*}/d\xi)d\xi$ in (2.27) specifies the corresponding change of temperature head on the isothermal section as well as similar product $(d\theta/d\xi)d\xi$ in (1.40) specifies the change of the temperature head corresponding to heat flux change. Therefore, the function in front of derivative in (2.27) may be defined as $v(\xi/\Phi) = (\theta_{w*} - \theta_w)/\theta_{w*}$ analogously to influence function in front of derivative in (1.40) $f(\xi/\Phi) = [(q_w - q_{w*})/\theta]/h_* = h_\xi/h_*$.

Considering the expression (2.26) for function $v(\xi/\Phi) = (\theta_{w*} - \theta_w)/\theta_{w*}$ as an equation, and solving it for temperature head θ_w on the adiabatic surface leads to desired solution (Exer. 2.28)

$$\theta_w = \theta_{w*}(0)\left[1 - \frac{B_\sigma\left(C_2, 1 - C_2\right)}{B(C_2, 1 - C_2)}\right], \qquad \sigma = 1 - (\xi/\Phi)^{C_1} \tag{2.28}$$

Here, $B_\sigma(i,j)$ is incomplete beta function defined by equation (2.23). Equation (2.28) gives the temperature head variation in an impingent flow for general case of arbitrary pressure gradient and Prandtl number. For zero pressure gradient and $\text{Pr} > 1$, the well-known relation follows from equation (2.28)

$$\theta_w = \theta_{w*}(0)\left[1 - \frac{B_\sigma(1/3, 2/3)}{B(1/3, 2/3)}\right], \qquad \sigma = 1 - (\xi/x)^{3/4} \tag{2.29}$$

2.1.5   Effect of Non-Newtonian Power-Law Rheology Fluid Behavior

It was indicated in Section 1.9 that universal functions (1.39) and (1.40) obtained for Newtonian fluids are valid as well for power-law non-Newtonian fluids for which the condition $s = n - 1$ is satisfied. The analogous exact solution of thermal boundary layer equation exists for self-similar flows in special Görtler variable and pressure gradient parameter β defined by expressions (1.53). Consequently, the effect on the heat transfer intensity may be estimated using formula (1.54) for an isothermal surface and function (1.39) or (1.40) for nonisothermal effect estimation. Coefficients g_0 and g_k are given in Figures 1.10 and 1.11, whereas the exponent C_1 and C_2, which are not present here, may be calculated by the same way using coefficients g_k as it described for Newtonian fluid in Section 1.6.2 and relevant exercises.

■ **Example 2.9: Cylinder in Transverse Flow of Non-Newtonian Power-Law Fluid**

In this case, the heat transfer from nonisothermal cylinder was studied for a theoretical external velocity distribution given by a sinusoidal function [338]

$$U/U_\infty = 2 \sin 2(x/D) \tag{2.30}$$

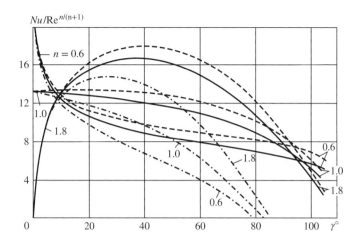

Figure 2.14 Heat transfer from nonisothermal cylinder in transverse flow of power law non-Newtonian fluid. ——— $\theta_{we}/\theta_{wi} = 1$, – – – $\theta_{we}/\theta_{wi} = 1 + x/D$, –.–.– $\theta_{we}/\theta_{wi} = 1 - x/D$ (notations in Example 2.1) Pr = 1000

and three temperature head variations: increased, decreased, and constant (Fig. 2.14). Equations (1.54) and (1.40) are used for estimation of heat transfer rate on isothermal cylinder and for computing the nonisothermicity effect. The results plotted in Figure 2.14 show that the value of heat flux and it variation along cylinder strongly depend on the type of the fluid determined by an exponent n in the rheology law (1.51) or (1.52). In particular, the heat transfer coefficient is a finite value at the stagnation point for Newtonian fluids, whereas for non-Newtonian fluids it becomes zero for dilatant ($n > 1$) or tends to the infinite for pseudoplastic ($n < 1$) fluids.

The temperature head variation also considerably affects the heat transfer intensity. In this case as well as in the others discussed above, the distribution of Nusselt number along the cylinder differs highly for decreasing and increasing temperature heads. Thus, in the first case, the heat flux becomes zero on cylinder at angle $\gamma = 80°$, whereas in the second case at the same location, heat flux value is close to that in the case of constant temperature head. As a result, the variation of Nusselt number along the cylinder changes from parabolic curve for dilatant fluids in the case of constant or growing temperature head to a s-shaped curve for pseudoplastic fluids and falling temperature head (Fig. 2.14) (Exer. 2.29).

2.1.6 Effect of Mechanical Energy Dissipation

As mentioned in Section 1.14, the effect of dissipation is proportional to square of the velocity (see equation (1.70)). Because of that, for incompressible fluid flows that are usual relatively low-velocity flows, the effect of dissipation is significant only in the case of large Prandtl numbers. Therefore, the effect of dissipative heat was studied for non-Newtonian fluids for which high Prandtl numbers are typical.

■ **Example 2.10: Dissipation Effect on a Cylinder in Transverse Non-Newtonian Flow**

The results were obtained using equation (1.70) and coefficients g_d from Figure 1.19 for theoretical external sinusoidal flow (2.30) of two types of non-Newtonian fluids as in previous example and Pr = 1000.

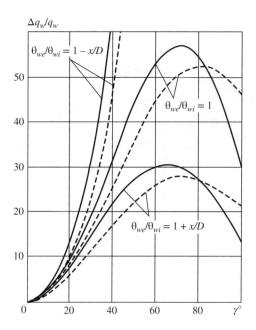

Figure 2.15 Additional heat flux caused by energy dissipation on a cylinder transversally streamlined by non-Newtonian fluid, $Ec = U_\infty^2/c_p\theta_w = 1$, $s = n - 1$, $Pr = 1000$, ------- $n = 0.6$, - - - - $n = 1.8$

The results in Figure 2.15 are given in the form of the ratio $\Delta q_w/q_w$ where Δq_w is additional energy dissipation, when the Eckert number equals unity, to heat flux q_w, calculated without regard to dissipation. Because the additional dissipation energy is proportional to Eckert number, and the velocity of incompressible flows usually is small, the heat flux associated with dissipation is small as well. Nevertheless, in some cases, the effect of dissipation is appreciable. In particular, this is true in the case of decreasing temperature head when the heat flux without regard to dissipation is small (Fig. 2.15). For constant and increasing temperature head, the dissipation energy is usually small. For example, if the Eckert number is 1/500, the addition heat flux in the case given in Figure 2.15 is about 10% for a constant temperature head and about 5% for increasing one at the point with angle $\gamma \approx 70°$ where the effect of dissipation reaches maximum.

2.1.7   Effect of Biot Number as a Measure of Problem Conjugation

Analysis of universal function (1.39) and others of this type as well as examples considered above show that the character of heat transfer behavior is basically determined by variation of temperature head. We have seen that the heat transfer coefficient increases relatively not much if the temperature head increases in flow direction or in time and vigorously decreases, becoming even zero sometimes, in the opposite case when the temperature head decreases in flow direction or in time.

The second basic parameter defining the intensity of conjugate heat transfer is the Biot number, which is the ratio of thermal resistances of the body and fluid flowing around or inside it. In the case of the given temperature head variation, the Biot number largely specifies the absolute value of heat transfer rate caused by the nonisothermicity of the surface or of the body-fluid interface. This can be shown using conjugate conditions and the universal function defining a heat flux. Taken into account that nonisothermicity effect is basically determined by the second term of a series (1.39) (S. 2.1.1 1) and using the conjugate conditions (1.18), one gets the following relation (Exer. 2.30)

$$\lambda_w \frac{\partial T}{\partial y}\bigg|_{y=0} = h_* g_1 \Phi \frac{\partial \theta_w}{\partial \Phi} \quad or \quad \frac{1}{Bi_*} \frac{\partial T}{\partial (y/\Delta_{av})}\bigg|_{y=0} = g_1 \Phi \frac{\partial \theta_w}{\partial \Phi} \quad Bi_* = \frac{h_* \Delta_{av}}{\lambda_w} \qquad (2.31)$$

where Δ_{av} is the average body thickness.

This relation tells us that the value of temperature head gradient $\partial \theta_w / \partial \Phi$ is inversely proportional to Biot number for an isothermal surface. Because relations (2.31) is obtained from conjugate conditions, this connection between Bi and $\partial \theta_w / \partial \Phi$ means that Biot number specifies the degree of problem conjugation indicating that the larger the Biot number is the smaller the gradient $\partial \theta_w / \partial \Phi$ establishes in a conjugate problem. Analysis of equation (2.31) yields the following conclusions: (i) in both limiting cases $Bi_* \rightarrow \infty$ and $Bi_* \rightarrow 0$, the conjugate problem degenerates, because in these situations, only one resistance is finite, whereas another is either infinite as in the first case or becomes zero as in the second one, (ii) in the first case, a conjugate problem transforms in a problem with isothermal surface, since according to equation (2.31) the temperature gradient there is zero; this case corresponds to situation when the body of infinite thickness (or negligible conductivity) is streamlined by the fluid with finite heat transfer coefficient, or when a body of finite thickness and conductivity is streamlined by a fluid with infinite heat transfer coefficient, (iii) in the other case, a conjugate problem transforms into a problem in which the temperature head changes in a stepwise manner, because according to equation (2.31), the temperature head gradient is infinite; this case corresponds to situation when the body of finite thickness and conductivity thermally interacts with a non-conducting fluid ($h_* \rightarrow 0$), or the body with infinite conductivity (or negligible thickness) is streamlined by fluid with finite heat transfer coefficient (Exer. 2.31), (iv) because in both limiting cases the conjugate problem degenerates, one deduces that the greatest effect of nonisothermicity should occurs when the both resistances are of the same order, and the Biot number (2.31) is close to unity; such problems, characterized by close to unity Biot number, should be considered as conjugate.

In what follows we will see that the results of this analysis are supported by numbers of conjugate problems considering in the next chapters and in its applications.

Comment 2.9 In deriving equation (2.31) we apply for heat flux only the second term of series with the first derivative instead of full universal function. Such simplification is possible because, as mentioned above, the effect of nonisothermicity is defined basically by the second term of series. This conclusion follows from two facts: (i) the first series term $h_* \theta_w$ defines the only isothermal heat flux, and (ii) coefficient g_1 is the largest, whereas all the others are comparatively small (Table 2.1) (Exer. 2.32).

Comment 2.10 Some authors use other criteria for characterizing relation of body-fluid thermal resistances. For example, Luikov, in an early work, suggested Brun number $Br = (\Delta/x)(\lambda/\lambda_w) Pr^{1/3} Re_x^{1/2}$ or later, Cole proposed criterion $(\lambda/\lambda_w)(Pe)^{1/3}$. In fact, both those criteria are Biot numbers (Exer. 2.33).

Exercises

2.1   Analyze any of universal functions for heat flux to list the factors determining the effect of temperature head variation. Compare your results with conclusion indicated in the text.

2.2   Estimate the relation g_2/g_1 using graphs and numerical data from Chapter 1 for some cases different from those given in Table 2.1. Compare your results with values from Table 2.1.

2.3   A simple example helps to understand better the difference of effects at growing and falling temperature heads. Add and subtract 5 from 20 and estimate the percentage of relative changes in comparison with new amounts in the first $5/25 = 20\%$ and in second $5/15 \approx 33\%$ cases. Increasing the absolute changes from 5 to 8 gives $\approx 28\%$ and $\approx 67\%$ respectively, resulting in increasing between relative changes in both cases from $33 - 20 = 13\%$ to $67 - 28 = 39\%$. Finally, if we add and subtract 10, we will have $\approx 33\%$ in the first and 100% (similar to zero heat transfer coefficient) in the second cases.

2.4   Consider some temperature head dependency other than proportionality, for example, $\theta_w = Kx^{1/3}$. Calculate the derivative $d\theta_w/dx$ and the product $x(d\theta_w/dx)$ for different values of coordinate x. Plot both dependencies of the derivative and of the product on coordinate x to see that the product increases with x growing despite the significant lessening of the derivative.

2.5   Compute and draw a pattern 2.1 for linear temperature head variation 2.1 for the limiting case $Pr \to 0$ and zero pressure gradient ($\beta = 0$) when the value of coefficient $g_1 = 1$ is maximum. Compare your results with data from Figure 2.1 to estimate the difference between greater and minor effects of surface nonisothermicity. Think: is there only a quantitative difference or there is a qualitative diversity, similar to distinction between cases of increasing and decreasing temperature heads?

2.6*   Obtain function $\Phi(x/D)$ using formula (1.39) for Görtler variable and the equation (2.4) for velocity distribution. Find the inverse function (2.5). Hint: because the first term of given function $\Phi(x/D)$ is $\Phi(\bar{x}) = a\bar{x}^2$, where $\bar{x} = x/D$, the first term of the inverse function is $\bar{x}(\Phi) = [(1/a)\Phi]^{1/2}$. Then, it is reasonable to assume that the second term is $b(\Phi^{1/2})^2 = b\Phi$. To estimate coefficient b, plot the difference $\Delta(\Phi) = \bar{x}(\Phi) - [(1/a)\Phi]^{1/2}$, and approximate the curve $\Delta(\Phi)$ by the line $\Delta\Phi = b\Phi$. Note: the function $\bar{x}(\Phi)$ in the equation for $\Delta\Phi$ is obtained using known function $\Phi(x/D)$, but is plotted as a curve $\bar{x} = x/D$ vs. Φ.

2.7   Compute the dependence $Nu/\sqrt{Re}(x/D)$ using values of $Nu_*/\sqrt{Re}$ for isothermal cylinder from Figure 2.2 (the curve 4 for $\theta_{we}/\theta_{wi} = 1$)) and first equation (2.1) with two derivatives for calculation of the nonisothermicity coefficient χ_t. Compare your results with data from Figure 2.2.

2.8   Obtain equation (2.7) and show that inequalities $U_{av}/U < 1$ and $U_{av}/U > 1$ are true for accelerating and decelerating flows leading to changes of the second term (2.7) of universal function described in the text. Hint: note that average velocity for interval $(0, x)$ lies above the local value of velocity at point x on a falling curve and below the local value of velocity at point x on a growing curve.

2.9   Repeat the analysis of pressure gradient effect on heat transfer coefficient in accelerating and decelerating flows presented in the text. Calculate the nonisotermicity coefficient χ_t using first formula (2.1) for self-similar flows with $U = K_1 \bar{x}^m$ and linear temperature head $\theta_w = 1 + K_2 \bar{x}$, where $\bar{x} = x/L$. Consider positive and negative K_2, compare results, and formulate conclusions.

2.10   Derive the second formula (2.8), or formulae (2.9), or (2.10) from example 2.3 as indicated in the text. Compute the corresponding curve from Figure 2.4 to check the approximate agreement between calculation and experimental data.

2.11* Derive equation (2.11) and calculate some curves for Stanton number from Figure 2.5 to practice employing the Görtler variable and to see how it works. Hint: the heat flux in the case of temperature head jump is estimated by the expression, similar to second equation (2.1), with a sum of products $f(\xi_k/\Phi)\Delta\theta_{wk}$ where $f(\xi_k/\Phi)$ is the influence function and Φ is the Görtler variable at the beginning of the temperature head jump $\Delta\theta_{wk}$ at the point k.

2.12   Obtain Figure 2.9 using the first formula (2.13) for Nusselt number. Approximate the results by the second relation (2.13) applying coordinates $\lg \mathrm{St}_*$ and $\lg \mathrm{Pe}_\Phi$ and estimate the coefficients c and exponents n. Hint: plot the calculation results as a function $\lg \mathrm{St}_* = f(\lg \mathrm{Pe}_\Phi)$ and draw three lines to determine coefficients c and exponents n.

2.13* It is shown that calculations for $\mathrm{Pr}_{tb} = 1.5$ and $\mathrm{Pr}_{tb} = 0.5$ give respectively lower and higher results compared to measured data. The same result is observed in Figure 2.8. At the same time, the results for corresponding coefficients for second formula (2.12) are opposite: it is higher for the first case and lower in the second then experimental value. Think: what is the reason of such seeming contradiction?

2.14   Obtain equations (2.14) from series (1.55) and compute curves plotted in Figure 2.11 for different values of x/L. Indicate the terms in series (1.55), which show that the heat transfer characteristics depend on coordinate x also, despite the fact that the given temperature head depends only on time.

2.15* Think: what physically means that the terms containing time-derivatives are multiplying by ratio x^{n_1}/U^{m_2} with exponents corresponding to the number of relevant derivative? Hint: see Exercise 1.59.

2.16* Derive formula (2.16) for polynomial coefficients b_i by substituting polynomial $1/h_* = a_i \Phi^i$ into integral (2.15) and applying relation (1.47) for beta function. Hint: use a new variable $\zeta = (\xi/\Phi)^{C_1}$ and polynomial for $1/h_*$ expressed in this variable.

2.17   Obtain expressions (2.17) from equations (2.15) for the case of $h_* = c\Phi^{-n}$. Hint: use a new variable, similar to variable in previous exercise to transform (2.15) to beta function.

2.18 Calculate coefficients of the polynomial (2.18) proceeding from the polynomial with coefficients a_i and formula (2.16) for coefficients b_i. Hint: for estimating Γ-function use tablets or Mathcad.

2.19 Compute ratio h_*/h_q for different cases considered in Example 2.7 using formulae (2.17) and coefficients h_k or exponents C_1 and C_2. Compare your results with values given in the text.

2.20 Analyze effects of boundary condition and other factors on the intensity of heat transfer using your data from previous exercise.

2.21 To understand the effect of boundary condition in turbulent flows, compute ratio h_*/h_q for turbulent flow using the second formula (2.17) and exponents C_1 and C_2 (Fig. 1.8 and 1.9) for $Pr = 1$ and 10^{-2}, $Re_{\delta_1} = 10^3$ and 10^5 and $\beta = -0.3$ (negative pressure gradient) and $\beta = 1$ (positive pressure gradient). Compare the results with data for laminar flow from Exercise 2.19. Formulate conclusions. Think: why there are no large Prandtl and Reynolds numbers among just suggested parameters? What are the physical reasons of such a situation?

2.22 Perform the integration by parts indicated in the text to derive the expression (2.21).

2.23* Discuss with a friend or college the proof that the function $v(\xi, \Phi)$ is the reciprocal influence function of the heat flux jump to understand the details of this analysis.

2.24 Derive formula (2.23) from expression (2.22) following directions given in the text.

2.25 Obtain formula (2.24) from equation (2.23). See additional information about incomplete beta function on Wikipedia.

2.26 Obtain equation (2.25) from relation (2.22) as it is described in the text.

2.27* Show that the integral in equation (2.26) for function $v(\xi/\Phi)$ at $\xi = 0$ is equal $[\Gamma(C_2)\Gamma(1 - C_2)]/C_1$ and derive the relation (2.27) from universal function (1.68) by performing integration by parts specified by relations (2.26). Explain why the function $v(\xi/\Phi)$ in this case depends on ratio ξ/Φ, while a similar function considered in the previous section depends on each variable ξ and Φ separately.

2.28* Repeat the reasoning leading to expression for function $v(\xi/\Phi)$ and finally to the derivation of equations (2.26) and (2.27). Think to understand why $[(q_w - q_{w*})/\theta]/h_*$ equals h_ξ/h_*.

2.29 Explain the behavior of Newtonian and non-Newtonian fluids at the stagnation point observed on Figure 2.14 and described in the text. Hint: analyze the behavior of heat flux at stagnation point for different n using equations (1.52).

2.30 Obtain relation (2.31) following explanation given in the text. Hint: define the heat flux in conjugate boundary conditions (1.18) by term with first derivative only, which basically determines the heat transfer rate, and then divide both side of first equation (2.31) by h_* and use variable y/Δ_{av}.

2.31 Repeat the analysis of equation (2.31) to understand details. Think: what type of processes analyzing in text are quenching and lumped models?

2.32 Think: is it always true that only the second term instead of the full universal function practically defines the heat flux due to large coefficient g_1?

2.33* Show that the Brun number is actually a Biot number. Hint: use formula for the Nusselt number. Show that criterion proposed by Cole becomes a Brun number after including the dimensionless body thickness. Think: in what cases a parameter without body thickness may be used as conjugate parameter?

2.2   The General Convective Boundary Conditions

The universal function (1.39) and other function of this type in the form of series might be considered as a sum of the surface temperature distribution perturbations or a sum of perturbation of boundary conditions. The case when all derivatives are zero, and in series retains only the first term, corresponds to the isothermal surface. Such series is considered as undisturbed boundary conditions since it does not consist of any derivative. The series containing the first term and the second one with first derivative presents the linear boundary condition. The series with two derivatives describes the quadratic boundary condition, and so on. In general, the series consisting infinite number of derivatives describes an arbitrary boundary conditions (Exer. 2.34).

It follows from such considerations that universal function (1.39) and others of this type are general boundary conditions describing different types of boundary conditions. A series with only the first term is the well-known boundary condition of the third kind, which gives the solution of the first approximation. A better solution of the same problem yields using the series with two first terms as a boundary condition. The more terms retained in the series, the more accurate the data may be obtained resulting, in principle, in an exact solution in the case of employing the series with infinite number of terms as boundary condition (Exer. 2.35).

Those step-by-step refinements make it possible to estimate the result accuracy at any step by comparing two consecutive approximations. In particular, comparing data obtained by boundary conditions with one and two first terms gives an accuracy of solution with boundary condition of the third kind. More coarse estimation of that accuracy may be obtained in simpler way if one estimates the value of the second term of series using the solution given by boundary condition of the third kind. Then, comparing this value of the second term with this solution presents an estimation of accuracy of applying the boundary condition of the third kind (Exer. 2.36).

2.2.1   Accuracy of Boundary Condition of the Third Kind

As indicated in the introduction, the common simple approach based on boundary condition of the third kind ignores the real thermal conditions on the body/fluid interface. In such a case, the accuracy of results depends basically on experimentally estimated heat transfer coefficient. Two reasons are responsible for using this primitive approach during past centuries and even now. The first one is the absence of alternative practical realizable methods until the computers came to wide use. The second reason is that there are problems with almost every isothermal body/fluid interface for which common approach gives practical acceptable accuracy. Thus, in such a case, there is no sense to employ more complicated conjugate methods.

Below are presented several examples showing that the just-described estimation of second term of the series (1.39) helps to understand what approach a simple common or a conjugate one is reasonable to use in a particular case.

∎ Example 2.11: The Overall Heat Transfer Coefficient in Flow Past Both Sides of a Plate

We start from one of the basic problems considering heat transfer between two fluids separated by a thin plate. The conjugate solution of this problem is presented in the next chapter. Here, we analyze a common solution based on the third kind of boundary condition and isothermal heat transfer coefficients. In this case, the dimensionless temperature head on one side of the plate is determined by estimating the overall heat transfer coefficient as a sum of thermal resistances

$$\theta_{w1} = \frac{T_{\infty 1} - T_{w1}}{T_{\infty 1} - T_{\infty 2}} = \frac{q_w}{h_{*1}(T_{\infty 1} - T_{\infty 2})} = \frac{1}{1 + h_{*1}/h_{*2} + h_{*1}\Delta/\lambda_w} \tag{2.32}$$

where $T_{\infty 1}$ and $T_{\infty 2}$ are fluid temperatures far from the plate and the isothermal heat transfer coefficients are given for laminar and turbulent flows by well-known formulae

$$\mathrm{Nu}_{x*} = 0.332\,\mathrm{Pr}^{1/3}\,\mathrm{Re}_x^{1/2} \quad \mathrm{Nu}_{x*} = 0.0295\,\mathrm{Pr}^{0.4}\,\mathrm{Re}_x^{0.8} \tag{2.33}$$

The accuracy of formula (2.32) is estimated as described above by computing the second term $g_1 x(d\theta_{w1}/dx)$ of universal function (1.38) using the result (2.32). If the flow regimes on both sides of a plate are the same, the second denominator term h_{*1}/h_{*2} in (2.32) does not depend on coordinate x. The next term in this sum, which is the Biot number $\mathrm{Bi}_* = h_*\Delta/\lambda_w$, according to equations (2.33) changes along the plate as a power function x^{-n}. Thus, the denominator in (2.32) is the function $D_1 + D_2 x^{-n}$, where D_1 and D_2 are constants. Then, expression (2.32) and relative error of this result defined as a ratio of second series term $g_1 x(d\theta_{w1}/dx)$ and value of θ_{w1} defined by (2.32) are

$$\theta_{w1} = \frac{1}{D_1 + D_2 x^{-n}}, \quad \sigma = g_1 \frac{x}{\theta_w}\frac{d\theta_w}{dx} = \frac{g_1 n D_2}{D_1 x^n + D_2} \tag{2.34}$$

Analysis shows that the maximal error $\sigma_{max} = g_1 n$ occurs at $x = 0$ when the denominator takes minimum. Hence, the greatest error is $g_1/2$ and $g_1/5$ for laminar and turbulent flows, respectively. For laminar flow at zero pressure gradient and $\mathrm{Pr} > 0.5$, the coefficient $g_1 = 0.62$ is practically independent of Pr (Fig. 1.3), so in this case, the maximum error is $\approx 30\%$. For turbulent flow with zero pressure gradient and $\mathrm{Pr} = 0.5$, the coefficient $g_1 = 0.22$ (Fig. 1.6). Because the value of g_1 decreases as the Prandtl number increases, in this range of Pr, the maximum error is $\sigma_{max} \approx 4\%$. However, for $\mathrm{Pr} = 0.01$ $g_1 = 0.52$, and, hence, $\sigma_{max} \approx 10\%$. (Exer. 2.37). Thus, for laminar flow the error may be moderate, whereas for turbulent flow, when $\mathrm{Pr} > 0.5$, the use of the common overall heat transfer coefficient does not lead to significant errors.

These estimates are in agreement with corresponding conjugate problem solutions. For laminar flow such solution gives the maximum error of $20 - 25\%$ (Exam. 4.1), for turbulent flow the conjugate solution gives the maximum error of about 7% (Exer. 2.37 and 2.38).

Comment 2.11 The errors in this problem are relatively small because the temperature head increases in flow direction on both sides of the plate. Usually in problems with increasing temperature head the nonisothermicity effect is moderate, and the heat transfer coefficients are not much different from isothermal coefficients. However, there are problem where this rule does not work because sometimes the combination of the isothermal heat transfer coefficient behavior and temperature head variations results in unusual notable nonisothermicity effect. One such example is the heat transfer from the plate heated from one end (second case) that we considered in introduction. In Section 3.1.3, we consider this question more precisely analyzing other problems with decreasing temperature heads when conjugate solutions are necessary independent of flow regime.

■ **Example 2.12: A Thermally Thin Plate at Given Temperature at One End**

A steel plate of length $0.25\ m$ and of thickness $0.01m$ is streamlined by air flow of temperature $300\ K$ with velocity $3\ m/s$. The left-hand end is insulated, and the temperature of the right-hand end is maintained at T_{wL}.

The steady-state differential equation (1.3) for a thermally thin plate in variables $\bar{x} = x/L$ and $\theta_w = (T - T_\infty)/(T_{wL} - T_\infty)$ and its solution satisfying given boundary conditions of the third kind with average h_* may be written in the form

$$\frac{d^2\theta_w}{d\bar{x}^2} + 2\mathrm{Bi}_{*L}\theta_w = 0, \quad \theta_w = \frac{T_w - T_\infty}{T_{wL} - T_\infty} = \frac{1}{ch(\sqrt{2\mathrm{Bi}_{*L}})}ch\left(\sqrt{2\mathrm{Bi}_{*L}}\frac{x}{L}\right) \tag{2.35}$$

Here, $\mathrm{Bi}_{*L} = h_*L^2/\lambda_w\Delta = \mathrm{Bi}_*(L/\Delta)^2$ is a modified Bio number, which takes into account effect of relative body length. This complex of parameters appears in thermally thin equation (1.3) in dimensionless form (2.35) (Exer. 2.39).

The accuracy of result (2.35), obtained by common approach, is estimated as in the previous example by differentiating expression (2.35) and apprising a value of the ratio of the second series term and temperature head itself (2.35)

$$\sigma = g_1\frac{x}{\theta_w}\frac{d\theta_w}{dx} = g_1\sqrt{2\mathrm{Bi}_{*L}}\frac{x}{L}th\left(\sqrt{2\mathrm{Bi}_{*L}}\frac{x}{L}\right) \quad \sigma_{max} = g_1\sqrt{2\mathrm{Bi}_{*L}} \tag{2.36}$$

For calculation of the Biot number, the relation (2.33) for laminar zero pressure gradient of air flow (Pr = 0.7) and $\mathrm{Re}_L = 5 \cdot 10^4$ are used giving $\mathrm{Nu}_{*L} = 66$, and then, the Bio number and maximal error for $g_1 = 0.62$ (Pr > 0.5) are counted as $\mathrm{Bi}_{*L} = \mathrm{Nu}_{*L}L\lambda/\lambda_w\Delta = 0.66$ (at $\lambda = 0.267 \cdot 10^{-2}$ and $\lambda_w = 65\mathrm{W/mK}$) and $\sigma_{max} \approx 0.7$. This result is obtained using Nusselt number $\mathrm{Nu}_{*L} = 66$ estimated at the end of a plate. Usually when a common approach is used, the average heat transfer coefficient or average Nusselt number would be $\mathrm{Nu}_{*av} = 2\mathrm{Nu}_{*L}$. Then, the error will be $\sigma_{max} \approx 0.7 \cdot \sqrt{2.} \approx 1$. Hence, the solution of this problem with the boundary conditions of the third kind is unacceptable despite that the temperature head increases in flow direction. As follows from Exercise 2.40, the same result yields the estimation error for this problem in the case of turbulent flow. The conjugate solution of this problem is given in Example 3.1.

Comment 2.12 The majority examples above and below consider the heated bodies cooled by the flow with lower temperature. In such a case, the temperature head increases in flow direction if the body temperature increases along the interface, and vice versa, it decreases in flow direction if the body temperature decreases along the interface. However, the situation is opposite if the flow temperature is higher than the body temperature, like in the problem considered in this example, when the flow heated the body, and the temperature head decreases in flow direction as the body temperature increases along the interface, and vice verse, it increases in flow direction if the body temperature decreases along the interface.

■ **Example 2.13: Thermally Treated Polymer Sheet Passed Through Surrounding**

The sheet of polymer at temperature T_0 is extruded from a die and passed at the velocity U_w through water (Pr = 6.1) at temperature T_∞. This problem is solved using boundary condition of the third kind in Example 4.14. The result is presented in the form $\theta_w(\overline{x})$, where $\overline{x} = x\alpha_w/\Delta^2 U_w$, $\theta_w = .(T_w - T_\infty)/(T_0 - T_\infty)$, and calculation is made for ratio of plate-coolant thermal resistances $\gamma = (c_p\rho\lambda)_w/c_p\rho\lambda = 8.51$, and quiescent ($U_\infty = 0$) surrounding. The error is estimated for the same values of parameters. Numerical differentiation of the curve $\theta_w(\overline{x})$ (Fig. 4.9) gives the derivative $d\theta_w/d\overline{x}$. Then, using this derivative and the first coefficient for moving sheet $g_1 = 1.3$ (Fig. 1.13), one obtains the accuracy of this solution as proposed by computing a ratio of the second term of universal function (1.38) and temperature head $\sigma = \overline{x}(d\theta_w d\overline{x})/\theta_w$. The results reveal that the error .grows as the distance from the die increases and finally reaches $\sigma_{max} \approx 2.6$. It is evident that this problem must be solved as conjugate. The reason of that is again the decreasing along the sheet temperature head (Exer. 2.41).

Comment 2.13 The evaluation of the second term of universal function provides satisfactory estimations of common approach accuracy in agreement with conjugate problem solutions. Examples show that in the case of temperature head decreasing, the problem should be considered as a conjugate one. In another case, when the temperature head increases, the proposed error estimation helps to understand whether the simple solution of a particular problem is acceptable or the conjugate analysis is required.

2.2.2   Conjugate Problem as an Equivalent Conduction Problem

It is clear from the above discussion that the solution of a conjugate problem is a result of coupling two other solutions obtained for a body and for a fluid under an arbitrary temperature head distribution on their interface. At the same time, the universal function (1.39) is a solution of thermal boundary layer equations for a fluid at arbitrary surface temperature head distribution, which in fact is one of the just-mentioned two solutions required for coupling in a conjugate problem. Because this solution determines the heat flux distribution on the interface, in the case of using universal function, there is no need to solve the boundary layer equation for fluid, and only the conduction equation for a body subjected to the boundary condition on the interface in the form of universal function remains to be solved (like in one of numerical methods of conjugation, S. 1.2). In other words, application of universal function transforms the conjugate problem to an equivalent conduction problem for a body with boundary condition in the form of universal function (1.39) or others of this type defining the heat flux on the interface.

We have seen at the beginning of this section that the universal function may be considered as a general boundary condition. Therefore, if the heat conduction equation is solved using this condition with first term only, an approximate solution of the conjugate problem is obtained as that in the case of applying the boundary condition of the third kind. By retaining the first two terms in equation (1.39) or (1.38) and solving the heat conduction equation, one finds a more accurate solution of the conjugate problem. This process of refining, in principle, might be continued by retaining more terms in universal function. However, this entails difficulties posed by the calculation of higher-order derivatives, and therefore, the integral form of general boundary condition (1.40) or (1.26) should be used for high-order approximations.

In practical calculations, it is convenient to retain the first few terms of the series and to calculate the error arising by neglecting the rest terms of it. If the first three terms of the series are retained, and the error is determined using the integral universal function, the conjugate problem is transformed to a heat conduction equation for a solid with the boundary condition in the form

$$q_w = h_* \left[\theta_w + g_1 \Phi \frac{d\theta_w}{d\Phi} + g_2 \Phi^2 \frac{d^2\theta_w}{d\Phi^2} + \varepsilon(\Phi) \right], \quad \varepsilon(\Phi) = \frac{1}{h_*}(q_w^{int} - q_w^{diff})$$

$$\frac{d\theta_w}{d\Phi} = \frac{v}{U}\frac{d\theta_w}{dx}, \quad \frac{d^2\theta_w}{d\Phi^2} = \frac{v^2}{U^2}\frac{d^2\theta_w}{dx^2} - \frac{v^2}{U^2}\frac{dU}{dx}\frac{d\theta_w}{dx}$$

$$\text{(2.37)}$$

This form of general boundary condition for arbitrary pressure gradient simplifies in the case of zero pressure gradient to (Exer. 2.42)

$$q_w = h_* \left[\theta_w + g_1 x \frac{d\theta_w}{dx} + g_2 x^2 \frac{d^2\theta_w}{dx^2} + \varepsilon(\Phi) \right], \quad \varepsilon(\Phi) = \frac{1}{h_*}(q_w^{int} - q_w^{diff}) \qquad \text{(2.38)}$$

Quantities q_w^{int} and q_w^{diff} are defined by integral relation (1.40) and by the first equation (2.37) in general case and by similar relations (1.26) and first equation (2.38) in the case of zero pressure gradient. The first approximation is found by solving the conduction equation for a body employing the first equation (2.37) or (2.38) as a boundary condition and assuming $\varepsilon(\Phi) = 0$. Computing the error $\varepsilon(\Phi)$ using the first approximation data, one incorporates it in equations (2.37) or (2.38) and obtains the next approximation in the same way, applying one of these equations as boundary condition. By continuing this process, the solution with the desired accuracy can be achieved.

Retaining in equations (2.37) and (2.38) terms with first two derivatives only leads to differential equations of the second order in any approximation. Methods of solutions of this type of equations are well investigated. Therefore, in this case, the reducing of a conjugate problem to conduction equation with universal function as boundary condition results in differential equations of which solutions are known. If a solid is thin or thermally thin, and one-dimensional conduction equation (1.3) is applicable, the proposed approach leads to ordinary differential equation of the second order. Examples of such problems are presented in the following chapters of applications. In the more complicated two- or three- dimensional conduction equation for a body, the solution of a conjugate problem is reduced to corresponding two- or three-dimensional resulting equations. Numerical solution of those equations are similar to well-developed means for solving conduction equations with boundary condition of the third kind. In some cases of simple form of solid, like a plate, the two-dimensional problem may be solved analytically. Such problem is considered in Example 4.7. Several examples of numerical solutions of two-dimensional conjugate problems illustrating usage of this approach are considered in applications as well.

The method of reducing a conjugate problem to an equivalent conduction problem is applicable to any linear convective heat transfer problem, but not to nonlinear problems for which the principle of superposition is unacceptable (Exer. 2.43).

2.3 The Gradient Analogy

Some investigators [14, 224, 445] pointed out that the temperature head gradient qualitatively affects the heat transfer coefficient as the free stream velocity gradient affects the friction coefficient. In author's papers [101, 120] it is shown that such analogy holds not only for the both gradients, which corresponds to the first derivatives, but for all other subsequent derivatives in both cases. To reveal that the variable, the velocity, and the temperature head affect similar corresponding coefficients, we should consider universal functions for the friction coefficient analogous to functions (1.39) and (1.40) for heat transfer characteristic in order to compare those dynamic and thermal functions.

Since the dynamic boundary layer equation is nonlinear, in this case, the exact solution could not be presented in the form of a sum of subsequent derivatives like series (1.39) for heat flux. However, using a linearization of the dynamic boundary layer equation as it shown in [120, 119], an approximate solution in analogous form can be obtained

$$c_f = c_{f*} \left(1 + b_1 \frac{\Phi}{P_w} \frac{dP_w}{d\Phi} + b_2 \frac{\Phi^2}{P_w} \frac{d^2 P_w}{d\Phi^2} + b_3 \frac{\Phi^3}{P_w} \frac{d^3 P_w}{d\Phi^3} . + .. \right) \tag{2.39}$$

Here, $c_{f*} = 0.664/\sqrt{\mathrm{Re}_{av}} = 0.664\Phi^{-1/2}$ (* denotes constant U, like it denotes constant temperature head for universal functions) is the average within interval $(0, x)$ skin friction coefficient defined for average within this interval Reynolds number $\mathrm{Re}_{av} = U_{av}x/v$ with average external flow velocity U_{av} and $P_w = \rho U^2/2$ is the dynamic pressure of the external flow, which remains unchanged across boundary layer (S. 7.4.4.1).

Comparing shows that: (i) equation (2.39) presents the friction coefficient in a series of derivatives of $P_w = \rho U^2/2$ as well as universal function (1.39) determines in series of derivatives of temperature head the heat transfer coefficient $h = q_w/\theta_w$ (after dividing both parts of (1.39) by θ_w), (ii) the dynamic pressure P_w as well as temperature head θ_w is independent on transverse coordinate y, (iii) structure of both series (2.39) and (1.39) is the same, therefore (iv) the corresponding to (1.39) integral relation has the same form as integral universal function (1.40) (Exer. 2.44)

$$c_f = \frac{c_{f*}}{P_w} \left[P_w(0) + \int_0^\Phi f(\xi/\Phi) \frac{dP_w}{d\xi} d\xi \right], \quad f(\xi/\Phi) = \left[1 - \left(\frac{\xi}{\Phi} \right)^{C_1} \right]^{-C_2} \tag{2.40}$$

This implies that relations (2.39) and (2.40) are universal function defining friction coefficient for arbitrary pressure gradient via dynamic pressure $P_w(\Phi)$ or through external velocity $U(x)$ because both variables P_w and Φ are functions of $U(x)$.

Coefficients b_k and exponents C are computed in the same way as coefficients g_k and exponents C for heat transfer problems (S. 1.6). As one may observe from Table 2.2, they also only slightly depend on β due to using the variable Φ defined by integral (1.39), which takes into account the flow history. Physically, this means that the value of Φ at the point x is determined not only by data at point x but rather by data of whole interval from $x = 0$ to x of the considered point (Com. 1.5 and 1.6). Because of a slight dependency on β, the accuracy of it estimation barely affects the final calculation results Therefore, in this case, β also may be evaluated intuitively or using some approximate formula, for example, relation (1.45) (Exam. 2.6).

Let us introduce a nonisotachicity coefficient $\chi_f = c_f/c_{f*}$ (an isotach is a line of constant velocities, like isotherm is a line of constant temperatures) similar to the nonisothermicity

Table 2.2 Coefficients b_k and exponents C for universal functions (2.39) and (2.40)

β	b_1	b_2	b_3	b_4	C_1	C_2
0	2.28	−0.30	0.058	−0.0096	0.52	0.57
0.5	1.97	−0.26	0.050	−0.0089	0.50	0.54
1	1.85	−0.24	0.047	−0.0085	0.48	0.52
average	2.0	−0.25	0.05	−0.009	0.50	0.54

coefficient, which shows how much the friction coefficient in a flow with a variable external velocity is more or less than that in a flow with constant velocity. From expressions (2.39) and (2.40), we get relations for nonisotachicity coefficient analogous to equations (2.1) for nonisothermicity coefficient

$$\chi_f = 1 + \sum_{k=1}^{\infty} b_k \frac{\Phi^k}{P_w} \frac{d^k P_w}{d\Phi^k} \qquad \chi_f = \frac{1}{P_w} \left[\int_0^{\Phi} f\left(\frac{\xi}{\Phi}\right) \frac{dP_w}{d\xi} d\xi + P_w(0) \right] \tag{2.41}$$

■ Example 2.14: Friction in Self-Similar Flows

In this case: $U^2 \sim x^{2\beta/(2-\beta)}$, $\Phi \sim x^{2/(2-\beta)}$, $P_w \sim U^2 \sim \Phi^\beta$ (S. 1.6.1). Thus, these equations lead to the following expressions for nonisotachicity coefficient

$$\chi_f = 1 + b_1 \beta + b_2 \beta(\beta - 1) + ... \qquad \chi_f = \beta \int_0^1 [1 - \zeta^{C_1}]^{-C_2} \zeta^{\beta-1} d\zeta \tag{2.42}$$

The second formula is obtained from the last equation (2.41) after defining the derivative $dP_w/d\Phi = \beta \Phi^{\beta-1}$ and employing the variable $\zeta = \xi/\Phi$. For integer β, the first formula gives: $\chi_f = 1$ and $\chi_f = 2.85$ for $\beta = 0$ and $\beta = 1$. Hence, $c_f = 0.664/\sqrt{\mathrm{Re}_x}$ and $c_f = 2.649\sqrt{\mathrm{Re}_x}$ for the first and the second cases, respectively. The second result is 7% larger than the exact value. For not integer β, the first expression (2.42) diverges (Exer. 2.45). The second one yields: $\chi_f = 1.56, 2.17, 2.89$, and 3.42 for $\beta = 0.2, 0.5, 1.0$, and 1.6, respectively (Exer. 2.46). These values are in reasonable agreement with data for self-similar flows[338, Fig. 9.1] (Exer. 2.47).

■ Example 2.15: Friction in Howarth Flow

The boundary layer flow with linear external velocity distribution $U = U_0 - ax$ is known as the Howarth flow because Howarth was the one to first consider this case. Using dimensionless variables $\bar{x} = ax/U_0$ and $\bar{\Phi} = a\Phi/U_0^2$, one gets $\bar{\Phi} = \bar{x} - \bar{x}^2/2$, $P_w = (\rho U_0^2/2)(1 - 2\bar{\Phi})$, $dP_w/d\bar{\Phi} = -2(\rho U_0^2/2)$. Then, equations (2.41) become (Exer. 2.48)

$$\chi_f = 1 - \frac{2b_1 \bar{\Phi}}{1 - 2\bar{\Phi}}, \qquad \chi_f = \frac{1}{1 - 2\bar{\Phi}} \left[1 - 2 \int_0^{\bar{\Phi}} \left[1 - \left(\frac{\xi}{\bar{\Phi}}\right)^{C_1} \right]^{-C_2} d\xi \right] \tag{2.43}$$

Both relations with average values of $b_1 = 2$ and exponents $C_1 = 0.5$ and $C_2 = 0.54$ lead to the practically same numerical results: $\chi_f = 1.33, 1.18, 0.78, 0.53$ for $\bar{x} = -0.1, -0.05, 0.05$, and 0.1, which are also in reasonable agreement with the Howarth data.[338, Fig. 9.8] The coordinate of separation is found using $b_1 = 2.31$ and $C_1 = 0.53, C_2 = 0.58$. These quantities are estimated by extrapolation of data from Table 2.2 to the close of separation value $\beta = -0.199$. Both equations (2.43) with those coefficients yield the separation coordinate $\bar{x} \approx 0.16$, whereas the exact value is $\bar{x} = 0.12$ (Exer. 2.49).

Comment 2.14 Estimation of the separation coordinate from the first equation (2.43) leads to quadratic equation $\bar{x} - \bar{x}^2/2 = 1/6.62$. Solution of this equation yields small differences of two mach larger values, which usually results in poor accuracy of the final data. In this case, the change in the value of square root from 5.5299 to 5.5300 leads to variation in final data from 0.18 to 0.16. Better results may be obtained using iterations. Presenting the quadratic expression in the form $\bar{x} = 1/6.62 + \bar{x}^2/2$, one gets the first approximation $\bar{x} = 1/6.62$ by neglecting the second term. Then, the neglecting term is estimated using this result $\bar{x}^2/2 = (1/6.62)^2/2$, and the second approximation is calculated: $1/6.62 + \bar{x}^2/2 = 1/6.62 + (1/6.62)^2/2 = 0.17$.

■ **Example 2.16: Friction in Transverse Flow Past Cylinder**

The potential velocity is given by equation (2.30): $U = 2U_\infty \sin(x/R) = 2U_\infty \sin \bar{x}$. Then, we have: $\bar{\Phi} = 1 - \cos \bar{x}$, where $\bar{\Phi} = \Phi/Re$, $\bar{x} = x/R$, and $Re = 2RU_\infty/\nu$. In these notations, the dynamic pressure is defined as $P_w = 2\rho U_\infty^2 (2\bar{\Phi} - \bar{\Phi}^2)$. To compare the results with known data, we modified equation (2.39) applying relations $c_f = \tau_w/\rho U^2$ and $c_{f*} = 0.664/\sqrt{\bar{\Phi} Re}$. After that procedure and determining the derivatives of P_w in terms of $\cos \bar{x}$, the dimensionless expression (2.39) with first two derivatives only is presented in the form applicable for comparison with known data (Exer. 2.50).

$$\bar{\tau}_w = \frac{\tau_w}{\rho U_\infty^2} \sqrt{\frac{U_\infty R}{\nu}} = 0.940[2(1 + b_1)(1 - \cos \bar{x})^{1/2} - (1 + 2b_1 + 2b_2)(1 - \cos \bar{x})^{3/2}]$$

(2.44)

The condition $\tau = 0$ gives the following equation for the coordinate of a separation point $\cos \bar{x} = -(1 - 2b_2)/(1 + 2b_1 + 2b_2)$ ($\bar{x}$ is an arc length) and the according value of central angle $\gamma = 108.9°$ estimated applying constant $b_1 = 2.31$ and $b_2 = -0.31$ adapted from Example 2.15. This result is in conformity with other approximate predictions 109.5° and 108.8° [338] given by integral method (S. 7.6 [338, p. 215]) and by Blasius series approach[338, p. 168], respectively, whereas the more accurate data found by numerical solution of boundary layer equations is 104.5°. The results for shearing stress distribution $\bar{\tau}_w(\gamma°)$ around a cylinder, obtained by equation (2.44) and average values of coefficients $b_1 = 2$ and $b_2 = -0.25$ agreed with other approximate data and with numerical solution for entire cylinder surface, except a small region near separation point [338, Fig. 10.7] (Exer. 2.51).

Comment 2.15 The average error of universal functions (2.39) and (2.40) of about 10% is a common value for approximate methods. At the same time, the comparison with others inexact methods (for example, from [338]) reveals that estimating the friction coefficient by universal functions is much simpler because this procedure does not required of any differential equation solution.

Comment 2.16 Coefficients b_k and exponents C in Table 2.2 are applicable for laminar flow. For turbulent flow they can be determined analogously using the same procedure as that for evaluation of coefficients g_k (S. 1.6) and exponents C (S. 1.6.2).

Now, we return to analysis of gradient analogy stated at the beginning of this section. Comparing equations (2.39) and (1.39), we observe that such similarity actually exists because: (i) these equations are similar in form, and for each dynamic term with derivative of pressure P_w (which is proportional to velocity square U^2) in the first equation, there is a corresponding analogous thermal term with derivative of temperature head θ_w in the second one, and (ii) it is clear from equation (2.39) that friction coefficients are greater in favor ($dU/dx > 0$) and lesser in adverse ($dU/dx < 0$) pressure gradient flows that that in zero pressure gradient flows ($dU/dx = 0$); in the same way, as it follows from equation (1.39), the increasing ($d\theta_w/dx > 0$) or decreasing ($d\theta_w/dx < 0$) temperature head in flow direction affects the heat transfer coefficient leading to growing or lessening its value in comparison with that for an isothermal surface.

It is known that the different influence of the favor and adverse velocity gradients on the friction coefficient is induced by deformation of velocity profiles in a dynamic boundary layer. Similarly, the different effects of increasing and decreasing temperature head on the heat transfer coefficient is caused by deformation of the temperature profiles in a thermal boundary layer.

To understand physically this process, consider first the case when the surface temperature is higher than the temperature of flowing fluid. If the wall temperature increases in the flow direction, the adjoining to the wall descending layers of fluid come in contact with increasingly hotter wall. Because of the fluid inertness, these layers warm up gradually in vertical direction. As a result, the cross-sectional temperature gradients near a wall turn out to be greater than in the case of constant wall temperature, which leads to the higher than for isothermal surface heat transfer coefficients (according to a slope of the tangent to the temperature profile at the wall). Similarly, in the case of decreasing surface temperature in the flow direction, the temperature cross-sectional gradients near a wall and the coincident heat transfer coefficients are less than those for an isothermal surface. The same situation exists in the case when a fluid heats a cooler surface. The difference is only that in this case, the absolute values of the falling temperature head and lesser heat transfer coefficients correspond to increasing in the flow direction wall temperature, and inversely, the growing absolute values of the temperature head and higher heat transfer coefficients correspond to the decreasing in the flow direction wall temperature (Com. 2.12 and Exer. 2.52).

Considering the second terms of the universal functions (2.39) and (1.39), one concludes that since coefficients g_2 and b_2 are negative, the effect of the second terms is opposite to just considered effect of gradients, defined by the first derivatives: the positive second derivatives lead to a reduction of the friction and heat transfer coefficients, and a negative derivatives yield an increasing of those coefficients. The effects of the third, fifth, and other odd derivatives is of the same kind as that of the first derivative, whereas the effects of even derivatives is of the same kind as that of the second derivative. That is because all odd coefficients of both universal functions are negative, whereas all even coefficients are positive.

Despite equations (2.39) and (1.39) are similar, they are indeed significantly different due to nonlinearity of dynamic boundary layer equation in contrast to linear thermal boundary layer equation. As a result, the coefficients g_k in universal function (1.39) are constant or depend weakly on Pr and β, but they do not depend on the temperature head θ_w that is the basic variable in equation (1.39). Although the coefficients b_k in universal function (2.39) also only slightly depend on β, in contrast to previous case, where β is independent on temperature head, here, the parameter β depends on the pressure P_w (or on external velocity U). Consequently, in universal function (2.39), β, and therefore, the coefficients b_k as well, strongly speaking, depend on the basic variable P_w in this equation. The reason for this is the nonlinearity of the dynamic boundary layer equation and using linearization of this equation (Exer. 2.43).

Comment 2.17 Coefficient b_1 are greater than coefficient g_1, and as a result, the pressure gradient affects the friction characteristics more significantly than the same temperature head gradient changes the heat transfer rate. This may be confusing, since according to Newton and Fourier laws both friction and heat transfer coefficients are defined through similar derivatives $(\partial u/\partial y)_{y=0}$ and $(\partial \theta/\partial y)_{y=0}$. Nevertheless, this diversity occurs because function (1.39) determines the temperature head θ_w, whereas function (2.39) deduces the pressure P_w that is proportional to U^2, rather than to U (Exer. 2.53).

2.4 Heat Flux Inversion

Analysis of universal functions (2.39) and (1.39) shows that for certain relations between terms of series, the friction or heat transfer coefficient as well as corresponding local friction stress or local heat flux vanishes. Usually, when the first series term plays a primary role, this may occur at negative gradients, in which case the external velocity or temperature head decreases along a surface. It is well known that vanishing of the friction is accompanied by separation of the boundary layer and is associated with a deformation of the velocity profiles in the boundary layer of flows with adverse pressure gradient [338]. Analogously, the vanishing of the heat flux in a flow with decreasing temperature head is associated with the deformation of the

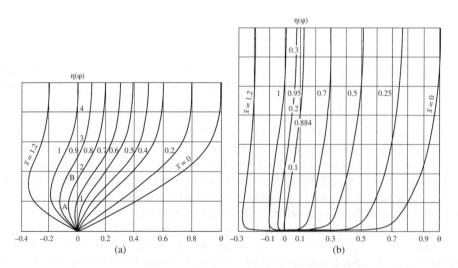

Figure 2.16 Deformation of the excess temperature profile for the linear decreasing temperature head $\theta_w/\theta_{wi} = 1 - \bar{x}$, with $\bar{x} = x/L$ and $\mathrm{Pr} = 0.7$, for laminar (a) and turbulent $\mathrm{Re}_{\delta_1} = 10^3$ (b) boundary layers

temperature profiles in the thermal boundary layer. Examples of such flows are considered in [65, 123] and [153].

Figure 2.16 presents the schematic pattern of the excess temperature $\theta = T_w - T$ deformation in laminar and turbulent boundary layers for the case of linear temperature head decreasing on the plate at zero pressure gradient [111, 119].

It is seen how a usual initial temperature profile deforms first into a profile with an inflection point and then converts into a profile with a vertical tangent at the wall. Although the temperature head is finite at this point, the local heat flux vanishes and changes its direction. The coordinate of this point known as a point of heat flux inversion is found from equation (1.38) or (2.2), determining heat flux or nonisotrmicity coefficient. In the case of linear temperature head $\theta_w/\theta_{wi} = 1 - x/L = 1 - \bar{x}$, equation (2.2) becomes $\chi_t = 1 - g_1\bar{x}/(1 - \bar{x})$. Then, one gets from equation $\chi_t = 0$ the coordinate of the inversion point $(x/L)_{inv} = 1/(1 + g_1)$, which gives $(x/L)_{inv} = 0.62$ and $(x/L)_{inv} = 0.834$ for laminar ($g_1 = 0.62$, Pr > 1) and turbulent ($g_1 = 0.2$, Pr = 1, $Re_{\delta_1} = 10^3$) boundary layers, respectively (Exer. 2.54 and 2.55).

Comment 2.18 An inflection point is a point of a curve at which a curvature changes from convex to concave or vice versa. The vertical tangent at the wall indicates that the heat flux at the wall is zero. This follows from Fourier law because the derivative in normal direction is zero.

The pattern of the temperature profile deformation in thermal boundary layer shown in Figure 2.16 is analogous to a well-known picture of velocity profile deformity in dynamic boundary layer, which leads to separation of the boundary layer. Nevertheless, these phenomena are fundamentally different. Separation leads to restructuring of the flow, to the appearance of the reverse streams, and results in actual destruction of the boundary layer flow, so that after separation the completely different flow structure establishes, and boundary layer equations are no longer valid beyond the separation. In contrast with this situation, the thermal boundary layer equations remain valid beyond the point of zero heat flux, since only the direction of the heat flux changes at this point, but the flow structure remains the same. Beyond the inverse point of the heat flux in the region before the point $\bar{x} = 1$, at which the temperature head vanishes, the heat flux is directed from the liquid to the wall (because the heat flux changed his direction), despite (as before) the fact that the wall temperature is higher than that of the fluid outside of boundary layer.

We present physical analysis [153] of that deformation process considering the same situation as in the previous section when the surface temperature is higher than that of fluid. Because in this case, the temperature head gradient decreases in flow direction, the hot wall temperature decreases as well, and the adjoining to the wall descended layers of fluid come into contact with cooler parts of the wall. As a result, the temperature difference between the wall and the layers of fluid near the wall decreases and in time becomes zero at the inverse point with the coordinate $\bar{x}_{inv} = 1/(1 + g_1)$. After this point with vertical tangent at the wall, the temperature of the fluid near the wall turns out to be above the wall temperature, because the temperature of the wall continues to decrease (the temperature head gradient remains the same). Thus, before the inverse point, the heat flux is directed from the wall to the fluid, whereas after this point, close to the surface the heat flux direction changes, so that the heat flux near the wall up to the point A in Figure 2.16 (a) is directed from the fluid to the wall. Because of this, the thermal boundary layer near the wall is divided vertically by point A, at which the heat flux vanishes, into two regions. In the region adjacent to the wall, the heat flux is directed toward the wall, and in the other region, above A until point B, the heat flux is directed away from the wall.

At the end of this region, at the point $\bar{x} = 1$, the flow temperature outside the boundary layer and the surface temperature become equal, and the temperature head vanishes. Nevertheless, the heat flux at this point does not vanish, so the functions $h(x)$ and $\chi_t(x)$ go to infinity, get discontinuity, and the concept of the heat transfer coefficient (and of nonisothermicity coefficient) becomes meaningless, as it first indicated by Chapman and Rubesin [65]. After the point $\bar{x} = 1$, where the temperature head reaches zero, it changes direction for whole boundary layer (like before until point B), so that both the temperature head and heat flux, which changed the direction before at inversion point $\bar{x}_{inv}$ continue to increase to infinity as $x \to \infty$ (Exer. 2.56).

Comment 2.19 The pattern of the temperature head deformation presented in Figure 2.16 could not be obtained using common relation of proportionality of heat flux and temperature head. In the case of a complicated temperature field and basic heat transfer characteristics, the hypothesis of proportionality is not applicable. In particular, this is obvious at the specific points such as A, B, or point with coordinate $\bar{x} = 1$ where the heat flux or temperature head vanish, whereas the other component remains finite. Physically, this means that one of the components changes its direction, which results or in two zones of flow as at the point A and B where the heat flux changes direction or in meaningless infinite heat transfer coefficient as at third point ($\bar{x} = 1$) where the temperature head becomes zero (Exers. 2.56).

2.5   Zero Heat Transfer Surfaces

The feature of friction coefficient lessening in flows with decreasing external velocity and becoming zero at the point of separation is used in so-called preseparated diffusers exhibiting small energy losses. The shape of such diffuser is designed so that in each channel section the flow is close to separation. This provides small friction coefficients resulting in low total losses [146].

Since the nature of heat transfer coefficient in the case of decreasing temperature head is similar, it makes possible to create a surface with theoretically zero heat losses [123, 201]. The temperature head along such surface should vary so that the condition of heat flux inversion is satisfied in each section of flow. In such a case, the temperature profile in the fluid layer adjoining the wall has a vertical tangent at the surface. Because the vertical tangent at the surface ensures zero heat transfer coefficient (Com. 2.18), the adjoining the wall fluid layer with vertical tangent serves as an insulating coating. Providing such a condition of vertical tangent of the temperature profile in each section of flow, one obtains a surface with theoretically zero heat losses.

In self-similar flows with $U = cx^m$ and power law temperature head $\theta_w = c_1 x^{m_1}$, the corresponding distribution of heat flux along the surface turns out to be also of power law $q_w = c_2 x^{m_2}$ with exponent $m_2 = m_1 + (1/2)(m + 1)$. This result follows from formula (2.33) for isothermal heat transfer coefficient $\mathrm{Nu}_x = c_3 \mathrm{Re}_x^{1/2}$. Then, it is easy to find that the heat flux along the surface becomes constant if the exponent in the temperature head distribution is $m_1 = -(1/2)(1 + m)$ yielding $m_2 = 0$ in formula for heat flux (Exer. 2.57).

Proceeding from this result, we assume that in general, the zero heat flux may be obtained also in the case of power law temperature head distribution specified in Görtler variable as

$\theta_w = K(\Phi/\text{Re})^s$ with unknown exponent s. Substituting this relation in the universal function (1.40) and using a new dummy variable $\sigma = (\xi/\Phi)^{C_1}$, we find

$$q_w = h_* \left[\frac{Ks(\Phi/\text{Re})^s}{C_1} \int\limits_0^1 (1 - \sigma)^{-C_2} \sigma^{(s/C_1)-1} d\sigma + \lim_{\Phi\to 0} K\left(\frac{\Phi}{\text{Re}}\right)^s \right] \tag{2.45}$$

Here, the last term presents the temperature head $\theta_w(0)$ from equation (1.40) in the form of limit at $\Phi \to 0$. Such presentation is necessary because in the case of negative unknown exponent s, this term becomes infinite. The same occurs with the integral in (2.45) where the integrand at negative exponent s also becomes infinite at $\sigma \to 0$. To understand whether this integral (called improper) is finite (converges) or infinite (diverges), the limiting cases should be considered. Expression (2.45) after performing integration and substituting integral limits 1 and 0 becomes (Exer. 2.58)

$$\lim \frac{q_w}{h_*} = K(\Phi/\text{Re})^s \left(\lim_{\sigma\to 1} \frac{s(1 - \sigma)^{-C_2+1}}{C_1(1 - C_2)} - \lim_{\sigma\to 0} \sigma^{s/C_1} \right) + \lim_{\Phi\to 0} K\left(\frac{\Phi}{\text{Re}}\right)^s \tag{2.46}$$

Comment 2.20 Improper is an integral for which the interval of integration is not finite.

It follows from Figures 1.5, 1.8, and 1.9 that for laminar and turbulent flows both exponents C_1 and C_2 are positive and $C_2 < 1$. Therefore, for positive exponent ($s > 0$), all three limiting terms are zero, and hence, we have $\lim(q_w/h_*) = 0$, which shows that relation (2.45) converges. For negative exponent ($s < 0$), the first term is still zero, but two others go to infinite as $\Phi \to 0$, Nevertheless, in this case, the limit (2.46) for q_w/h_* again is zero because both infinite terms are equal having opposite signs. This becomes clear if the second term is modified applying $\sigma = (\xi/\Phi)^{C_1} = \zeta^{C_1}$ in order to present this term as $\lim_{\sigma\to 0} \sigma^{s/C_1} = \lim_{\zeta\to 0} \zeta^s$ (Exer. 2.59).

Thus, analysis reveals that relation (2.45) converges at any exponent s and due to that may be used for farther consideration. From previous discussion we know that integrals of type (2.45) are expressed via beta function (or through combination of gamma functions) (1.47). Applying this approach, we arrive at the relation from which follows that the temperature head distribution providing zero heat flux along a surface is a power law function (as it was expected) with showing below exponent

$$q_w = \frac{q_{w*}s}{C_1} \frac{\Gamma(1 - C_2)\Gamma(s/C_1)}{\Gamma(1 - C_2 + s/C_1)}, \quad \theta_w = K\left(\frac{\Phi}{\text{Re}}\right)^{-C_1(1-C_2)} \tag{2.47}$$

The value of this exponent is obtained from first equation (2.47) by satisfying condition $q_w = 0$. This can be achieved if $s = 0$ or if the denominator in (2.47) goes to infinity $\Gamma(1 - C_2 + s/C_1) \to \infty$. It is clear that the first case corresponds to isothermal surface. The second case means that $1 - C_2 + s/C_1 = 0$ and hence, $s = -C_1(1 - C_2)$.

Simple calculation shows that for laminar flow (Fig. 1.5) the exponent in the final result (2.47) is practically the same for whole range of Prandtl number and pressure gradients (different β) and equals $s = -1/2$. For turbulent flow (Figs. 1.8 and 1.9), the exponent in (2.47) varies from $s = -1$ for small to $s = -0.4$ for large Prandtl and Reynolds numbers, respectively, and is almost independent of pressure gradient (Exer. 2.60). It is known [242] that for self-similar flows, the heat flux along the surface is zero if the temperature head varies according to power law with exponent $s = -(m + 1)/2$. This particular result follows from general formula (2.47) since in this case external velocity distribution $U = cx^m$ leads to the Görtler variable $\Phi \sim x^{m+1}$ and, therefore, for laminar flow the zero heat flux is obtained at temperature head $\theta_w \sim x^{-(m+1)/2}$ (Exer. 2.61).

The temperature distribution (2.47) is difficult to implement in reality because according to this relation, the surface temperature becomes infinite at the starting point at $\Phi \to 0$ [123]. Therefore, only laws close to (2.47) might be realized practically. For instance, one such temperature distribution, which gradually approaches the relation (2.47) as the distance from the starting point increases, is as follows

$$\theta_w = K\left(K_1 + \frac{\Phi}{\mathrm{Re}}\right)^s, \quad q_w = q_{w*}\left\{(1-z)^s + s\int_0^z \left[1 - \left(\frac{\zeta}{z}\right)^{C_1}\right]^{-C_2} \frac{d\zeta}{(1-z+\zeta)^{1-s}}\right\}$$

(2.48)

Here, $\zeta = \xi/(K_1 + \Phi/\mathrm{Re})$ and z is the value of ζ when $\zeta = \Phi$. As $\Phi \to \infty$, the last expressions coincides with equations (2.47) and heat flux becomes zero, whereas in preceding sections of flow, the heat flux is close to zero.

Comment 2.21 In principle, these results may be used to minimize the heat losses in the car or other vehicles if the heat sources inside a wall are distributed so that heat releases according to equation (2.48). In such a case, almost total heat remains inside the car because the outer side of the wall is almost isolated.

2.6   Optimization in Heat Transfer Problems

The universal functions indicate that the heat transfer intensity is determined by an isothermal heat transfer coefficient and the temperature head distribution. This follows from the structure of universal functions defining the heat flux as a product of heat transfer coefficient h_* and series (1.39) or integral (1.40), which both are specified by temperature head distribution θ_w. Therefore, by appropriate selection of the temperature head distribution, one can ensure that a given particular system satisfied the desired conditions, for example, a maximum or minimum rate of heat transfer.

In the general case of gradient flow over an arbitrary nonisothermal surface, the two integral universal functions (1.40) and (1.68) defining the heat flux or temperature head may be applied for solution of such problems. Mathematically, the optimization problem is a question of extreme (maximum or minimum) of one of those integrals. The solution of such problem in general is difficult, especially in the cases when the conjugate effects are important. It is much simpler to investigate different possible versions of interest using integral (1.40) or (1.68) and select the best comparing the results. Although such approach does not capture all capabilities of the optimization, the results are instructive and useful for applications.

To illustrate the proposed approach, we consider solutions of three simple optimization problems [116]. As usually in statement of optimization problems, the considered examples of solution start with detailed problem formulation indicating the specific conditions of optimization and applying assumptions. To see the effect of flow regime, the same problems are solved for laminar and turbulent flows at zero pressure gradients and for laminar flow near stagnation point. In addition, the second and third problems are considered also for jet wall flow at zero pressure gradient. The optimal results show significant effects reaching up to 30% for laminar and 20% for turbulent flows.

2.6.1   Problem Formulation

It is required to install on a plate several heat sources or sinks, for instance, electronic components with linear varying strengths. How should these be arranged to ensure the minimum of the maximal plate temperature?

2.6.1.1   Zero Pressure Gradient Flow

For linear varying heat flux $q_w = K + K_1(x/L)$ and zero pressure gradient flow, equation (1.68) may be presented in following form

$$\theta_w = \frac{1}{h_*}[KI_1 + K_1I_2(x/L)] \tag{2.49}$$

$$I_i = \frac{C_1}{\Gamma(C_2)\Gamma(1-C_2)}\int_0^1 (1-\zeta^{C_1})^{C_2-1}\zeta^{C_1(1-C_2)+n+i-2}d\zeta$$

where n is the exponent in the expression $\mathrm{Nu}_{x*} = c\mathrm{Re}_x^{1-n}$, $\zeta = \xi/x$, and $i = 1, 2$. This integral can be expressed using gamma functions like other similar integrals above:

$$I_i = \frac{\Gamma[1 - C_2 + (n+i-1)/C_1]}{\Gamma(1 - C_2)\Gamma[1 + (n+i-1)/C_1]} \tag{2.50}$$

Expressing coefficients K and K_1 in terms of difference $\Delta q = q_{max} - q_{min}$ and total heat flux Q_w, one presents first relation (2.49) for temperature head in the form (Exer. 2.62)

$$\theta_w = \frac{1}{h_*}\left\{I_1q_{av} \pm \Delta q\left[\frac{1}{2}I_1 - I_2(x/L)\right]\right\} \tag{2.51}$$

where $q_{av} = Q_w/BL$ is an average heat flux, B is the width of the plate and signs plus and minus correspond to decreasing and increasing heat flux along the plate, respectively.

For zero pressure gradient flow $h_* \to \infty$ as $x \to 0$. Hence, the temperature head at the beginning becomes zero. Then, it follows from equation (2.51) that the maximum temperature head for increasing heat flux is located at the end of the plate, at $x \to L$. This is not obvious in the case of decreasing heat transfer. To find the location of temperature head maximum in this case, one differentiates equation (2.51), taking into account that $h_*(x) = h_*(L)(x/L)^{-n}$. Equating the result to zero, gives this coordinate

$$x_m/L = \frac{I_1 n[1 + 0.5(\Delta q/q_{av})]}{I_2(1 + n)(\Delta q/q_{av})} \tag{2.52}$$

Setting in this expression $x_m = L$, one finds the limiting value of ratio $\Delta q/q_{av}$ which corresponds to case when the maximum is located at the end of the plate

$$\left(\frac{\Delta q}{q_{av}}\right)_{\lim} = \frac{I_1 n}{I_2(1+n) - 0.5 I_1 n} \tag{2.53}$$

For all ratios $\Delta q/q_{av}$ lower than this limiting value, the maximum of temperature head is located at the end of the plate, and for all those which are higher than the limiting value, it is placed at $x < L$ (Exer. 2.63).

In the case of laminar flow at $\mathrm{Pr} \geq 1$: $C_1 = 3/4$, $C_2 = 1/3$ (Fig.1.5) and $n = 1/2$; in the case of turbulent flow at $\mathrm{Pr} \approx 1$: $C_1 = 1$, $C_2 = 0.18$ at $\mathrm{Re}_{\delta_1} = 10^3$ (Re $= 5 \cdot 10^5$), $C_1 = 0.84$, $C_2 = 0.1$ at $\mathrm{Re}_{\delta_1} = 10^5$ (Re $= 10^8$) (Figs. 1.8 and 1.9) and $n = 1/5$; for wall jet of laminar flow at $\mathrm{Pr} \approx 1$: $C_1 = 0.42$, $C_2 = 0.35$ and $n = 3/4$ [148]. At these values of the exponents equation (2.49) yields: for laminar flow $I_1 = 0.73$ and $I_2 = 5/9$, for turbulent flow $I_1 = 0.93$ and $I_2 = 0.8$ at $\mathrm{Re}_{\delta_1} = 10^3$; $I_1 = 0.96$ and $I_2 = 0.88$ at $\mathrm{Re}_{\delta_1} = 10^5$; for laminar wall jet $I_1 = 0.55$ and $I_2 = 0.43$ (Exer. 2.64).

Calculation results obtained by equation (2.49) using these values of integrals are plotted in Figure 2.17 where the difference $\Delta\theta_{w.\max} = |\theta_{w.\max.in} - \theta_{w.\max.de}|$ of the maximum temperature heads in the cases of increasing $\theta_{w.\max.in}$ and decreasing $\theta_{w.\max.de}$ heat fluxes, referred to the higher of those two is plotted as a function of $\Delta q/q_{av}$. Estimating from this figure the difference $\Delta\theta_{w.\max}$ for the value of ratio $\Delta q/q_{av}$ of given heat sources or sinks yields the data of how much the maximum of the temperature head in the case of increasing heat elements is greater than in the opposite situation with decreasing units. It follows from these data that the maximum surface temperature at the same heat flux (same $\Delta q/q_{av}$) could be significantly reduced by arranging the cooled sources in the row with decreasing and the heated sinks in the chain of increases strength in flow direction, respectively. It is seen from Figure 2.17 that the value of reduced surface temperature is approximately proportional to ratio $\Delta q/q_{av}$ at small values of this parameter and approaches about 30% at $\Delta q/q_{av} \approx 0.5$ increasing farther remarkable for laminar and mach slower for turbulent flows as $\Delta q/q_{av}$ grows.

2.6.1.2 The Flow Near Stagnation Point

The external velocity in this case is proportional to coordinate $U = cx$, and the heat transfer coefficient from an isothermal surface is independent of x ($n = 0$). Thus, the Görtler variable is $\Phi = cx^2/2$, and hence, the linear dependence for heat flux in Görtler variable turns to $q_w = KI_1 + K_1 I_2 \sqrt{(2/c)}\Phi$. Using this result and knowing that an isothermal heat transfer coefficient is independent on coordinate, one gets two formulae similar to relations (2.49) for temperature head θ_w and for integrals I_i. This leads to expression (2.50) in gamma functions for integrals I_i, where $(n + i - 1)$ is replaced by $(i - 1)/2$, and results in a formula for temperature head similar to (2.51)

$$I_i = \frac{\Gamma[1 - C_2 + (i-1)/2C_1]}{\Gamma(1 - C_2)\Gamma[1 + (i-1)/2C_1]}, \quad \theta_w = \frac{1}{h_*}\left\{q_{av} \pm \Delta q\left[\frac{1}{2} - 0.73\,(x/L)\right]\right\} \tag{2.54}$$

Here, the last expression is obtained from equation (2.51) after substituting the values of integrals $I_1 = 1$ and $I_2 = 0.73$ computed applying the same as for zero pressure gradient exponents $C_1 = 3/4$ and $C_2 = 1/3$.

It is ease to understand that in this case the temperature head is maximal at $x = 0$ under decreasing heat fluxes, whereas the temperature head is the highest at $x = L$ in the case of

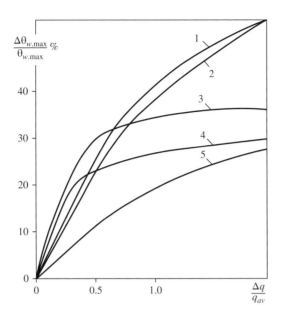

Figure 2.17 Absolute value of difference between maximum temperature heads under linear decreasing and increasing heat fluxes: 1- wall jet, laminar flow, 2-plate, laminar flow, 3 and 4-turbulent flow $Re = 10^8$ and $Re = 5 \cdot 10^5$, 5- stagnation point

increasing heat fluxes. This is obvious because at an invariable heat transfer coefficient, the temperature head changes according to heat flux variation yielding in increasing or in decreasing of θ_w as heat flux increases or decreases. Computing based on this information shows that, unlike the situation in previous case of zero pressure gradient, the value of maximum temperature head at increasing heat flux is smaller then that in opposite case at decreasing heat flux. Therefore, in order to reduce the maximal surface temperature, the strength of the sources should be increased in the flow direction of cold coolant, whereas the strength of sinks should be decreased in flow direction of a hot stream. It is seen from Figure 2.17 that in this case the effect of reducing the maximal surface temperature in smaller than that for the case of a plate (Exer. 2.65).

2.6.2 Problem Formulation

It is given allowable maximal surface temperature. Find the mode of temperature head variation providing a maximal removed (or supplied) total heat flux.

2.6.2.1 Zero Pressure Gradient Flow

If the sought temperature head is approximated by quadratic polynomial, the heat flux is reasonable to determine by differential universal function since in such a case only two terms are retained in the series

$$\theta_w = a_0 + a_1(x/L) + a_2(x/L)^2$$

$$q_w = h_*[a_0 + a_1(1 + g_1)(x/L) + a_2(1 + 2g_1 + 2g_2)(x/L)^2]$$

$$(2.55)$$

After integrating the last equation and using $h_* = h_*(L)(x/L)^{-n}$, one gets a total heat flux

$$Q_w = h_*(L)LB \left[\frac{a_0}{1-n} + \frac{a_1(1+g_1)}{2-n} + \frac{a_2(1+2g_1+2g_2)}{3-n} \right] \tag{2.56}$$

Two cases should be considered:

(i) The maximal temperature head is located at a leading or at traveling end of the plate. In this case first equation (2.55) has one of two forms:

$$\theta_w = \theta_{w.max} + a_1(x/L) + a_2(x/L)^2, \quad \theta_w = \theta_{w.max} + a_1[1 - (x/L)] + a_2[1 - (x/L)]^2 \tag{2.57}$$

Because the first terms present the maximal value of the temperature head, the sum of the two other terms should be negative at any x in the whole range $0 \leq x \leq L$.:

$$a_1(x/L) + a_2(x/L)^2 \leq 0, \quad a_1[1 - (x/L)] + a_2[1 - (x/L)]^2 \leq 0 \tag{2.58}$$

If $a_2 \leq 0$, then in order to satisfy the first of these inequalities at $x \to 0$ or the second at $x \to 1$, it is necessary that $a_1 \leq 0$. Then, it follows from equation (2.57) that the maximum heat flux is attained at $a_1 = a_2 = 0$ and $a_0 = \theta_{w.max}$, that is, when a plate is heated uniformly to the specified maximum temperature. If $a_2 > 0$, then should be $a_1 < 0$. Satisfaction of the first inequality (2.58) at $x = L$ or the second at $x = 0$ requires $|a_1| > a_2$. It easy to check that under these conditions, the sum of the two last terms in equation (2.57) for laminar (e.g., $Pr \approx 1$, $g_1 = 0.62$, $g_2 = -0.135$, and $n = 1/2$) or for turbulent (e.g., $Pr \approx 1$, $Re_{\delta_1} = 10^3$, $g_1 = 0.2$, $g_2 = -0.05$ and $n = 1/5$) flows at zero pressure gradient is negative, and hence, the total heat flux is again maximum at $a_1 = a_2 = 0$ and $a_0 = \theta_{w.max}$.

(ii) The maximum temperature head occurs at $0 < x < L$. In this case at $a_2 < 0$, the first parabola (2.55) is open down toward the negative ordinate. Therefore, after changing the sign to a minus at the last term in equation (2.55), only the case with positive a_2 should be studied. In such a case, the coordinates of parabola vortex in the frame (θ_w vs x) are determined as $x_m/L = a_1/2a_2$ and $\theta_{wmax} = a_0 + a_1^2/4a_2$. Considering these relations as a system of two equations, one finds a_0 in terms of two other coefficients, and after substitution the result in equation (2.56) gets the expression for total heat flux

$$Q_w = h_*LB \left[\frac{\theta_{w.max}}{1-n} + a_2 F \left(\frac{a_1}{a_2} \right) \right],$$

$$F \left(\frac{a_1}{a_2} \right) = -\frac{1}{4(1-n)} \left(\frac{a_1}{a_2} \right)^2 + \frac{1+g_1}{2-n} \frac{a_1}{a_2} - \frac{1+2g_1+2g_2}{3-n} \tag{2.59}$$

The discriminant of the last quadratic trinomial defined as

$$\Delta = \frac{1+2g_1+2g_2}{(1-n)(3-n)} - \left(\frac{1+g_1}{2-n} \right)^2 \tag{2.60}$$

is positive for laminar and turbulent flows as well as for laminar wall jet over the plate, which means that trinomial F (2.59) has no roots. It follows from this result that the trinomial $F(a_1/a_2)$ is negative at all ratios a_1/a_2 because it is negative at $a_1/a_2 = 0$. Thus, we arrive again at the same conclusion that the maximum heat flux is obtained when the plate is heated uniformly to the specified maximum temperature (Exer. 2.66).

At first sight, the solution of this problem is worthless. In fact, this is not true because analysis shows that in this case, the result depends on the relation between values of coefficients g_1 and g_2, which govern the effect of nonisothermicity, and exponent n in relation for the heat transfer coefficient on an isothermal wall. Two examples in which a nonuniform distribution of temperature head is optimal are given below (Exer. 2.67).

2.6.2.2   The Flow Near Stagnation Point

Although the temperature head is approximated by the same first polynomial (2.55), in this case, the integral universal function for computing heat flux is convenient to apply. The reason of this is that for flow with pressure gradient the Görtler variable should be used, which is easier to perform by the integral equation (1.40), required only the first derivative, than series in derivatives. Substituting a first derivative of temperature head (2.55) with respect to Görtler variable $\Phi = cx^2/2$ in equation (1.40), gives for heat flux formula similar to expression (2.55) $q_w = h_*[a_0 + a_1(I_1/2)(x/L) + a_2I_2(x/L)^2]$ with $I_1 = 2.9$ and $I_2 = 1.62$. Integration of this expression and taken into account that $n = 0$ for stagnation point, one gets for removed total heat similar to (2.59) relation (Exer 2. 68)

$$Q_w = h_* LB\left[\theta_{w.\max} + a_2 F\left(\frac{a_1}{a_2}\right)\right], \quad F\left(\frac{a_1}{a_2}\right) = -\frac{1}{4}\left(\frac{a_1}{a_2}\right)^2 + 0.73\frac{a_1}{a_2} - 0.533 \quad (2.61)$$

The discriminant of this trinomial is $\Delta \approx 0$, and the root is 1.46. Therefore, the first equation (2.61) shows that $Q_{w.\max} = h_*\theta_{w.\max}$ is achieved at any a_2 as long as ratio $a_1/a_2 = 1.46$, and hence, $F(a_1/a_2) = 0$. In such a case, the temperature head distribution (2.55) and corresponding heat flux are (Exer. 2.69)

$$\theta_w = \theta_{w.\max} - a_2[0.53 - 1.46(x/L) + (x/L)^2]$$
$$q_w = h_*\{\theta_{w.\max} - a_2[0.53 - 2.12(x/L) + 1.62(x/L)^2]\} \quad (2.62)$$

The first relation (2.62) presents a nonuniform temperature head distribution providing the same total removed or supplied heat (2.61) as the uniform heating to a maximally permissible plate temperature, but at lower than that temperature for whole plate, except one point. Distribution of temperature head (solid curves) and local heat fluxes (dashed curves) (2.62) are plotted in Figure 2.18 for different values of $a_2/\theta_{w.\max}$. It is seen that the ratio $\theta_w/\theta_{w.\max} < 1$ on whole plate, except point $x/L = 0.8$ where the temperature is equal to the permissible maximal temperature. The temperature head decreases as the value of $a_2/\theta_{w.\max}$ grows, and becomes zero at the beginning of the plate at maximum value $a_2/\theta_{w.\max} = 1.88$ (Fig. 2.18). The physical explanation for this effect is that despite the temperature head is lower than the maximal, it increases along the basic part of the plate resulting in greater then isothermal heat transfer coefficient on the uniformly heated plate, and this compensates the decrease in the temperature head (Exer. 2.70).

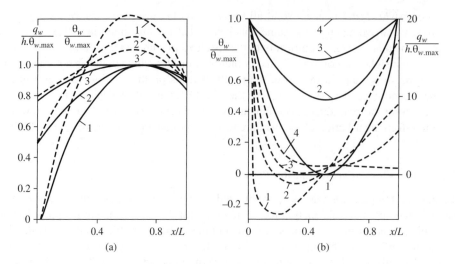

Figure 2.18 Different distributions of the temperature head (solid curves) and the corresponding heat fluxes distributions (dashed curves) providing the same total heat removed from surface a) stagnation point $a_2/\theta_{w.max}$: $1 - 0.4, 2 - 1.0, 3 - 1.88$ b) jet wall: $1 - 0, 2 - 1.0, 3 - 3, 4 - 4$

2.6.2.3   The Jet Wall Flow at Low Prandtl Numbers and Zero Pressure Gradient

In this case, $g_1 = 4$, $g_2 = -0.03$ and $n = 3/4$ [148]. It is ease to check that under these values of parameters, the sum of two last terms in equation (2.56) at any value of coefficients $a = a_2 = -a_1$ is close to zero ≈ 0.027. If for simplicity this small value is neglected, the removed total heat is determined only by coefficient $a_0 = \theta_{w.max}$, and distributions of temperature head and local heat fluxes take the forms similar to (2.55)

$$\theta_w = \theta_{w.max} - a(x/L)[1 - (x/L)], \qquad q_w = h_*\{\theta_{w.max} - a(x/L)[5 - 9(x/L)]\} \qquad (2.63)$$

Figure 2.18 (b) shows defined by the first equation nonunifom temperature head distributions for several values of $a/\theta_{w.max}$ providing at lower surface temperature the same removed total heat as in the case of uniformly heated plate at maximal surface temperature. The same effect of reducing temperature is seen. However, the effect is less than that in the case of the stagnation point flow (compare, for example, both curves for $a/\theta_{w.max} = 1$) because here, the temperature head distributions are symmetrical so that the parts of surface with growing and lessening temperature heads are equal.

Comment 2.22 The calculation shows that despite the distribution of the temperature head being symmetrical in this case, the average heat transfer coefficient is larger than an isothermal coefficient, and it increases as $a/\theta_{w.max}$ grows also (Exer. 2.71 and 2.72).

2.6.3   Problem Formulation

It is known the total heat flux which is necessary to remove (or supplied). Find the pattern of this heat flux variation providing the smallest maximal plate temperature.

The desired heat flux distribution we approximate by a quadratic polynomial (2.64), similar to the temperature head distribution (2.55)

$$q_w = a_0 + a_1(x/L) + a_2(x/L)^2, \quad q_{w.\text{un}} = \frac{Q_w}{BL} = a_0 + \frac{a_1}{2} + \frac{a_2}{3}, \tag{2.64}$$

Integrating this polynomial gives the total removed heat Q_w, and then, the uniform local heat flux $q_{w.\text{un}}$ (2.64) is obtained. The corresponding temperature head distribution is determined as before in example (2.17) by equation (1.68). Applying the first relation (2.64) and knowing that for zero pressure gradient $n = 1/2$, one gets

$$\theta_w = \frac{1}{h_*}[a_0 I_1 + a_1 I_2(x/L) + a_2 I_3(x/L)^2] \tag{2.65}$$

Here, integrals $I_i (i = 1, 2, 3)$ are analogous to integrals (2.49) and are computed by equation (2.50) through gamma functions for laminar, turbulent flows on a plate, and for jet wall flow for the same values of parameters as in Example 2.17 (Exer. 2.73).

The coordinate of the temperature head maximum is found by standard procedure of differentiating expression (2.65) for θ_w and equating it to zero, which gives an equation

$$\left(\frac{x_m}{L}\right)^2 + \frac{I_2 a_1(n+1)}{I_3 a_2(n+2)}\left(\frac{x_m}{L}\right) + \frac{I_1 n}{I_3(n+2)}\left(\frac{q_{w.\text{un}}}{a_2} - \frac{a_1}{2a_2} - \frac{1}{3}\right) = 0 \tag{2.66}$$

To analyze this quadratic equation, we consider again two cases:

(i) Equation (2.66) has no roots. In this case, the discriminant of equation (2.66) is negative so that the following inequality is valid

$$\left[\frac{I_2 a_1(n+1)}{2 I_3 a_2(n+2)}\right]^2 - \frac{I_1 n}{I_3(n+2)}\left(\frac{q_{w.\text{un}}}{a_2} - \frac{a_1}{2a_2} - \frac{1}{3}\right) < 0 \tag{2.67}$$

Since there are no roots, the sign of expression (2.66) does not change, and hence, the maximum of temperature head is attained at the beginning or at the end of a plate. So, if expression (2.66) is positive at $x = 0$, it is positive for whole interval $0 < x/L < 1$. It follows from equation (2.66) that this occurs if the last term is positive because the two other terms vanish at $x = 0$. Calculation results presented in Table 2.3 show that in the cases considered here, this is true because $q_{w.\text{un}} > 0$ and $a_1/a_2 < 0$. Then, it is clear that under positive last term and negative ratio a_1/a_2, the trinomial (2.66) decreases along the plate, and consequently, the maximal temperature head is located at the plate end at $x = L$, because the positive function could not decrease along the plate starting from zero (Exers. 2.74 and 2.75). In this case, the maximal temperature head is defined as

$$\frac{\theta_{w.\text{max}}}{\theta_{w.\text{max.un}}} = 1 + \frac{a_2}{q_{w.\text{un}}}\left[\frac{a_1}{a_2}\left(\frac{I_2}{I_1} - \frac{1}{2}\right) + \frac{I_3}{I_1} - \frac{1}{3}\right] \tag{2.68}$$

Here, $\theta_{w.\text{max}}$ is obtained from equation (2.65) at $x/L = 1$ after using the second equation (2.64) for eliminating coefficient a_0. The uniform temperature head $\theta_{w.\text{max.un}}$ is found applying universal function (1.68) for the case $q_w = cont.$ resulting in equations similar to (2.49) and (2.50) (Exer. 2.76).

Table 2.3 Maximal temperature head reducing at optimal pattern of removed heat flux

Kind of flow	I_1	I_2	I_3	n	a_1/a_2	$q_{w.un}/a_2$	$\theta_{w.max}/\theta_{w.max.un}$
Laminar flow	0.73	0.56	0.48	½	−2.14	1.18	0.79
Turbulent flow $\mathrm{Re}_{\delta_1} = 10^3$	0.93	0.80	0.74	1/5	−2.31	3.23	0.88
Turbulent flow $\mathrm{Re}_{\delta_1} = 10^5$	0.96	0.88	0.84	1/5	−2.38	3.33	0.86
Laminar wall jet	0.55	0.43	0.37	¾	−1.97	0.69	0.67

The optimal values of a_1/a_2 correspond to minimum of $\theta_{w.max}$. To find this minimum, it is necessary to get an equation by setting to zero the derivative of equation (2.68) with respect to a_1/a_2. Thus, differentiation of equation (2.68), yields (Exer. 2.77)

$$\frac{q_{w.un}}{a_2}\left(\frac{I_2}{I_1}-\frac{1}{2}\right)-\frac{d}{d(a_1/a_2)}\left(\frac{q_{w.un}}{a_2}\right)\left[\frac{a_1}{a_2}\left(\frac{I_2}{I_1}-\frac{1}{2}\right)+\frac{I_3}{I_1}-\frac{1}{3}\right]=0 \qquad (2.69)$$

This equation consists of two unknown: $q_{w.un}/a_2$ and a_1/a_2. Therefore, it is necessary to express the ratio $q_{w.un}/a_2$ in terms of a_1/a_2. This is done using an equation that is obtained from condition (2.67) by changing the sign of inequality to equality sign. After substituting the result of solving this equation for $q_{w.un}/a_2$ along with corresponding derivative of this term in equation (2.69), we arrive at following expression

$$\frac{n+2}{n}\left(\frac{I_2}{I_1}-\frac{1}{2}\right)\frac{I_3}{I_1}\left[\frac{I_2(n+1)}{2I_3(n+2)}\right]^2\left(\frac{a_1}{a_2}\right)^2+2\left(\frac{I_3}{I_1}-\frac{1}{3}\right)\frac{(n+2)I_3}{n}\frac{I_3}{I_1}\left[\frac{I_2(n+1)}{2I_3(n+2)}\right]^2\frac{a_1}{a_2}$$

$$+\frac{1}{2}\frac{I_3}{I_1}-\frac{1}{3}\frac{I_2}{I_1}=0 \qquad (2.70)$$

which determines the optimum values of a_1/a_2. The results of solution of this quadratic equation are listed in Table 2.3 (Exer. 2.78).

Comment 2.23 To understand the way it is possible to use an equation similar to inequality (2.67), consider such equation as a limiting case of condition (2.67). The inequality sign in (2.67) tells us that its left side is less than zero, but this sign does not specify how much it is less than zero. Therefore, this inequality is valid at any difference between the right inequality side and zero including small amounts up to limiting case of infinite small difference from zero. But such inequality (2.67) of which the left side differs from zero by an infinite small amount in fact is an equation that we used (Exer. 2.79).

Table 2.3 shows that an appropriate selection of the heat flux pattern allows us to reduce the maximal temperature on 12-14% for turbulent, 20% for laminar flows over plate, and up to 30% for jet wall flow. The corresponding optimal heat flux distribution providing the required amount of the removed (supplied) heat is defined by the first equations (2.64). We present this equation in two forms, which may be used depending of calculation convenience (Exer. 2.80)

$$q_w = a_2 \left(\frac{q_{w.un}}{a_2} - \frac{a_1}{2a_2} - \frac{1}{3} + \frac{a_1}{a_2}x + x^2 \right)$$

$$= q_{w.un} \left\{ 1 + \frac{a_2}{q_{w.un}} \left[\frac{a_1}{a_2} \left(\frac{x}{L} - \frac{1}{2} \right) + \left(\frac{x}{L} \right)^2 - \frac{1}{3} \right] \right\} \qquad (2.71)$$

(ii) The equation (2.66) has roots. We have showed that for the case considered, there are no roots of equation (2.66). To understand the way we consider the opposite case, take into account that we study a problem of removing heat. Therefore, if there are roots, they should be special, such that distribution (2.71) could not change its sign along the surface. Otherwise, after the sign changes from minus to plus, the heat will be supplied to the plate instead of removing it, and vice versa in the case of supplying heat when plus will change to minus resulting in heating instead of supplying. To avoid such a situation, the following inequality should be satisfied ensuring that only imaginary rots are possible

$$\frac{1}{4} \left(\frac{a_1}{a_2} \right)^2 - \left(\frac{q_{w.un}}{a_2} - \frac{1}{2} \frac{a_1}{a_2} - \frac{1}{3} \right) < 0 \qquad (2.72)$$

This relation is a discriminant of quadratic trinomial (2.71), and a satisfaction of inequality (2.72) means that the heat flux distribution (2.71) does not have points of maximum or minimum at which the change of sign occurs (Exer. 2.81).

Thus, if desired distribution (2.71) has special roots, the two conditions should be true: equation (2.66) and inequality (2.73). Substituting the last sum in parentheses from equation (2.66) in condition (2.72), we obtain inequality

$$\frac{1}{4} \left(\frac{a_1}{a_2} \right)^2 + \frac{I_3}{I_1} \frac{n+2}{n} (x_m/L)^2 + \frac{I_2}{I_1} \frac{a_1}{a_2} \frac{n+1}{n} (x_m/L) < 0 \qquad (2.73)$$

which only one should be satisfied if both conditions (2.66) and (2.72) are true.

It is clear that at $a_1/a_2 > 0$ inequality (2.73) is not satisfied for $x_m/L > 0$. This is not apparent in the case when $a_1/a_2 < 0$. However, one can easily check by direct calculation that at $0 < x_m/L < 1$ inequality (2.73) is not satisfied for the studying cases also at $a_1/a_2 < 0$. So, analysis indicates that there are no any roots of equation (2.66) for the problem in question showing that the statement at the beginning of the case (ii) is not a true (Exer. 2.82).

Exercises

2.34 What means the term "perturbation"? Think about examples of perturbation other then perturbation boundary conditions. Hint: read some examples from the article about "perturbation" on Wikipedia.

2.35 What is the general boundary condition? What benefits does such a condition offer?

2.36   Explain how the estimation of the value of the second series term may be used to evaluate the accuracy of the boundary condition of the third kind.

2.37   Derive relation (2.34) for estimating error caused by common approach and obtain corresponding evaluations for laminar and turbulent flows. Show that in this problem the temperature head on both sides of a plate increases in flow direction. Hint: consider how the boundary layer thickness and the corresponding difference $T_w - T_\infty$ change along the plate as the distance from the starting point increases.

2.38*   Solve the problem from Example 2.11 for the case when the flow on a side 1 is the same, that is, laminar, and on a side 2, the flow is turbulent. Compare your results with data from previous exercise. Hint: consider the expression for error taking $h_1 = c_1 x^{-n_1}$ and $h_2 = c_2 x^{-n_2}$ to see that: (i) if index 1 denotes side with laminar flow, the result qualitatively is the same as in Example 2.11 when flow on both sides is laminar or turbulent, and (ii) if index 1 denotes side with turbulent flow, error becomes negative, which does not fit consider situation. Think why this awkward result happens.

2.39*   Obtain equation and solution (2.35). Prove that this solution satisfies the equation (2.35) and given boundary conditions.

2.40   Solve the problem from Example 2.12 for turbulent flow at average $Re_x = 5 \cdot 10^5$. Compare your results with data from previous exercise.

2.41   Estimate numerically the derivative $d\theta_w/dx$ of curve $\theta_w(x)$ in Figure 4.9 (curve 7) and calculate the error caused by the boundary condition of the third kind in the case of thermally treatment polymer sheet in Example 2.13.

2.42   Obtain relations (2.36) and show that equation (2.37) follows from (2.36) in the case of zero pressure gradient. What causes the error determined by the second equations (2.36) and (2.37)? Why is this error estimated by such difference?

2.43   Recall or study in *Advanced Engineering Mathematics* the distinction between linear and nonlinear differential equations. What is the superposition principle? Show that superposition approach is unacceptable to nonlinear equation. Hint: consider simple linear and nonlinear equations and substitute a sum of two simple solutions (e.g., two constants assuming that those are solutions) and see the results after substitution. Are the results in both cases (of linear and nonlinear equations) as sums of two separate similar solutions?

2.44   Show that relations (2.39) and (2.40) have exactly the same form as universal functions (1.39) and (1.40). Hint: take into account a note in parentheses or consider the total skin friction $\tau = c_f P_w = c_{f_w}(\rho U^2/2)$ to compare it with heat flux $q_w = h\theta_w$.

2.45   Derive the first expression (2.42) for coefficient χ_f and compute skin friction coefficients for $\beta = 0$ and $\beta = 1$ using formula $c_f = 0.664\Phi^{-1/2}\chi_f$ as it done in text Show that this series (2.42) diverges for not integer values of β. Hint: use some not integer value of β, for example, $\beta = 1/2$ or $\beta = 1/3$.

2.46* Obtain the second equation (2.42) from integral formula (2.41) following the explanation given in the text. Repeat the calculation presented in the text. Hints: (i) substitute the derivative in integrand (2.42) using a dummy variable to get $dP_w/d\zeta = \rho\beta\zeta^{\beta-1}$ instead of $dP_w/d\Phi = \rho\beta\,\Phi^{\beta-1}$, (ii) to calculate integrals use software, for example, Mathcad.

2.47 Compare the results obtained in previous exercise with data for self-similar flows given in [338, Fig. 9.1]. Hints: (i) measure the slope of the tangent to the corresponding curve at the surface on Fig. 9.1 in [338], and compute the quotient by dividing this result by 0.664. The values of χ_f gained by calculations compare with coincident quotients. For example, the slope of the tangent to the curve $\beta = 1.6$ is about ≈ 2, then the quotient is $2/0.664 = 3.01$, whereas the computing value is 2.85, which is $\approx 6\%$ less.

2.48 Derive equations (2.43) and repeat calculation for the same values of $\bar{x}$. See hints in the last exercise.

2.49 Calculate the coordinate of separation point using both relations (2.43). Hints: (i) solve the quadratic equation defining the point of separation by the first relation (2.43) and perform calculation to see the problem of small difference arising in this case, (ii) solve this equation by iteration as explained in Comment 2.14, (iii) in using the second formula (2.43) compute χ_f for several values of $\overline{\Phi}$, plot the curve $\chi_f(\overline{\Phi})$, and extrapolate this curve to abscissa axis to find the point of intersection, which gives the point coordinate of separation where $\chi_f = 0$.

2.50 Obtain equations (2.44) and expression for point separation as it described in the text. Calculate the coordinate of point separation. Hint: use relations for c_f, c_{f*} and $P_w = 2\rho U_\infty^2(2\overline{\Phi} - \overline{\Phi}^2)$ knowing that $\overline{\Phi} = 1 - \cos\bar{x}$. Compare your result with other approximate data indicated in text and estimate the errors of your and other results with value of the separation angle 104.5° gained numerically.

2.51 Calculate and plot dependence $\bar{\tau}_w(\gamma°)$ using coefficients b given in the text. Compare your results with the data from Figure 10.7 in [338]. Hint: use equation (2.44), relation for $\cos\bar{x}$, and dependence between arc length $\bar{x}$ and central angle $\gamma°$.

2.52 Repeat the physical analysis of temperature profile deformation to understand this process in details that gives insight into such type of phenomena.

2.53 Consider Fourier and Newton laws for heat flux and friction stress and equations (2.39) and (1.39) defining pressure P_w and temperature head θ_w to see the similarity of both laws and difference of both equations explained in comment 2.17.

2.54 Derive the expression for the coordinate of heat flux inversion points using relation (1.38) for linear temperature head $\theta_w/\theta_{wi} = 1 - (x/L)$ as explained in text. Find the coordinate of inversion point for laminar flow for Pr = 0.01 taken value of coefficient g_1 from Figure 1.3. Draw curve $\chi_t(\bar{x})$ for interval $0.5 < \bar{x} < 0.7$ to see how heat flux changes its direction resulting in negative χ_t and hence, in negative heat transfer coefficient.

2.55 Find the coordinate of heat flux inversion point for turbulent flow for $Pr = 0.01$, $Re_{\delta_1} = 10^3$ and for $Pr = 0.01$, $Re_{\delta_1} = 10^5$. Compare your results with data for the cases considered in the text and in the previous exercise. Describe how Prandtl and Reynolds numbers affect the location of heat inversion point. Hint: take values of coefficient g_1 from Figure 1.6.

2.56 Repeat the analysis of temperature profile deformation to understand the details, in particular, the role of points A and B on Figure 2.16 and heat flow directions in different zones.

2.57 Show that in the self-similar flows with $U = cx^m$ and power law temperature head $\theta_w = c_1 x^{m_1}$, the heat flux is also distributed along surface according to power law with exponent $m_2 = m_1 + (1/2)(m + 1)$. Hint: use formulae $Nu_x = c_3 Re_x^{1/2}$ and $Nu_x = q_w x/\theta_w \lambda$. Find the value of exponent m_1 at which the heat flux is constant and compare the result with obtained in the text.

2.58 Study in *Advanced Engineering Mathematics* (simpler explanation is in the Kreyszig's book) the improper integral to understand the derivation of equation (2.46) and obtain this equation. Hint: construct expression (2.46) from (2.45) as two limits after integration as described in the text and replace ratio s/C_1 from factor before integral into integrand.

2.59 Show that expression (2.45) is finite at any value of exponent s (positive, negative, and zero) to understand the analysis from text. Explain why this result is important. What happens in an opposite case? Hint: use equation (2.46).

2.60 Prove that the second equation (2.47) follows from the first one and gives the solution of the problem in question. Estimate the value of exponent s in this equation for laminar flow using several values of exponents C from Fig. 1.5 to see that s slightly depends on Pr and β being about (-1/2). Compute this exponent for turbulent flows taken the values of C_1 and C_2 from Figures 1.8 and 1.9. Compare your data with that from the text.

2.61 Show that the known result of temperature head distribution providing the zero heat flux in self-similar flows follows as a particular case from formula (2.47).

2.62 Derive the second relations (2.49) from integral universal function (1.68) for the case of linear law $q_w = K + K_1(x/L)$ and present the first relation (2.49) for this case in terms of total heat flux $Q_w = q_{av}BL$ and difference $\Delta q = q_{max} - q_{min}$ as expression (2.51). Hint: first obtain the equation (2.50) using beta function (1.47).

2.63 Obtain equation (2.52) and (2.53) as described in the text and explain the results of analysis. Hint: first consider simple examples when $\Delta q/q_{av}$ is lower or higher than limiting value (2.53).

2.64* Calculate dependence $\Delta\theta_{w.max}/\theta_{w.max} = f(\Delta q/q_{av})$ for laminar flow at low Prandtl number and for turbulent flow at low Prandtl and Reynolds numbers using equations (2.50)–(2.53). Draw curves similar to plotted in Figure 2.17 in the same form and perform analysis analogous to given in the text. Compare your results with data from Figure 2.17. Hint: first calculate integrals I_i for $i = 1$ and 2 by equation (2.50) and then obtain formula for $\Delta\theta_{w.\max}/\theta_{w.\max}$ using relation (2.51). To find the points with maximal surface temperature apply equations (2.52) and (2.53).

2.65   Explain why in the case of stagnation point, the sources and the sinks should be arranged in the opposite manners than that in the case of a plate?

2.66   Recall the role of discriminant in studying algebraic equations from calculus to understand the analysis of case (ii) in example 2.18.

2.67*   The analysis of zero pressure gradient case in example 2.18 shows that the largest heat flux is removed if the plate temperature is uniform and equal to the specified maximal temperature. Nevertheless, it is clamed that this is not always the case because in general the result depends on nonisotermicity effect and on the value of isothermal heat transfer coefficient. Explain physically why these facts indicate that there is another, nonuniform optimal distribution of temperature head. Hint: take into account that the rate of heat transfer depends on two components as it is explained at the beginning of this section.

2.68*   Derive relation for local heat flux q_w and expression (2.61) for total removed heat Q_w. Hint: to obtain derivative of temperature head (2.55) with respect to Görtler variable use chain rule $\frac{d\theta_w}{d\Phi} = \frac{d\theta_w}{dx}\frac{dx}{d\Phi}$ (see calculus), where the last derivative is found applying the Görtler variable specified for the stagnation point. Follow explanations from the text.

2.69   Perform the following text analysis of trinomial (2.61) leading to relations (2.62) to understand why the same total removed heat is obtained at any value of coefficient a_2.

2.70   Find coordinates in which all curves of temperature head (2.62) coincide and estimate the coordinate of the maximum of the unified curve.

2.71   Estimate the sum of two last terms in equation (2.56) to show that this sum is close to zero and explain the derivation of equations (2.63). Calculate the ratio h/h_* for $a/\theta_{w.\max} = 1$ using these relations to see that the heat transfer coefficients on the part of surface with increased temperature head are significantly higher than those on another surface part with decreasing temperature head. Show that the average heat transfer coefficient is greater than isothermal one and explain why this is the reason of reducing the surface temperature. Hint: use relation $h = q_w/\theta_w$ and calculate the average heat transfer coefficient by integrating this relation.

2.72   Perform the same calculation for some value of parameter $a/\theta_{w.\max}$ from a range $1.5 < a/\theta_{w.\max} < 4$. You will see that on surface with decreasing temperature head the heat transfer coefficient is negative at one or more points. Draw the graph h/h_* vs x to see that the curves close to these points (exactly at x where $h/h_* = 0$ and this ratio changes the sign) become discontinuous. Explain these effects physically. Also calculate the average heat transfer coefficient as in the previous example to show that despite there being negative coefficients, the average heat transfer coefficient is greater than an isothermal one. Moreover, the average value of ratio h/h_* significantly grows as the parameter $a/\theta_{w.\max}$ increases. Hint: for physical considerations, recall what is heat flux inversion (S. 2.4).

2.73 Explain how the equations (2.64) and (2.65) are obtained, and what values of parameters in formula (2.50) should be used for estimating three integrals in this case.

2.74 Obtain equations (2.66) and (2.68) and inequality (2.67). Repeat an analysis of the case (i) when there is no root of equation (2.66) to understand that the temperature head maximum in this case is located at $x = L$.

2.75 Explain why two facts that (i) the sign of expression (2.66) does not change at $0 < x/L < 1$ and (ii) the sign of this expression is positive at $x = 0$ are enough to conclude that the temperature head maximum is located at the end of a plate. Hint: analyze the change of some monotonic function $f(x)$ with growing the coordinate x.

2.76 Derive equation (2.68) following the directions from the text. Hint: on the derivation of expression for $\theta_{w.max.un}$, consider that in this case: $i = 1$, heat flux q_w is constant, and temperature head may be presented as $\theta_{w.max.un} = q_w/h_*$.

2.77 Obtain equation (2.69) by differentiating equation (2.68). Hint: present equation (2.68) in the form $1 + f(a_1/a_2)/(q_{w.un}/a_2)$ where $f(a_1/a_2)$ is the expression in the brackets in equation (2.68).

2.78 Find the expression for $q_{w.un}/a_2$ as it described in the text and deduce the derivative of this term with respect to a_1/a_2. Use these results to get equation (2.70) determining the optimum values of a_1/a_2. Obtain some optimal values of a_1/a_2 by solving quadratic equation (2.70). Take required numerical data of integrals from Table 2.3. Compare your results for optimal a_1/a_2 with corresponding data from Table 2.3.

2.79* Think of Comment 2.22 to understand the validation of the equation obtained from inequality (2.67). Compute using relations (2.65) and (2.64) the temperature head θ_w as a function of coordinate x/L and draw corresponding graph to see that the maximum temperature head indeed is located at the end of the plate. Hint: (i) find a_0 from equation (2.64) and substitute the result in (2.65), (ii) present heat transfer coefficient in the form $h_* = h(1)x^{-n}$ and draw the graph in coordinates $\theta_w h(1)/a_2 I_1$ vs x/L. Compare the calculated value at the plate end with according data from Table 2.3.

2.80 Obtain equation (2.71) from two equations (2.64) using the second equation for eliminating a_0 from the first equation. Calculate distributions described by obtained equation (2.71) applying the data given in Table 2.3. Draw corresponding curves in coordinates $q_w/q_{w.un}$ vs x/L to compare the heat flux distributions providing the same removed total heat in different cases. Explain why curves intersect.

2.81 Explain what fact is proved in the case (ii) and why this proof is needed. Deduce inequality (2.72). Hint: to obtain inequality (2.72) present trinomial in the form of the first equation (2.64).

2.82 Obtain inequality (2.73) as described in the text. Explain why it is clear that this inequality is not satisfied for positive values of a_1/a_2 and $x/L > 0$. Calculate the left part of inequality (2.73) for some negative values of a_1/a_2 using data for integrals from Table 2.3 to see that this inequality is positive in interval $0 < x/L < 1$. Explain why it follows from these results that there are no roots of equation (2.66) for problem in question, and because of that the starting initial assumption in the case (ii) is false. Hint: study carefully the discussion about the last question.

3

Application of Conjugate Heat Transfer Models in External and Internal Flows

In the next two chapters, the applications of conjugate heat transfer problems are considered. First in Chapter 3, generally, without concrete usage, the flows past plates, around the bodies (external flows), and inside the channels and tubes (internal flows) are analyzed. Then in Chapter 4, the specific applications of conjugate heat transfer problems in industrial and technology areas are discussed. Analysis of examples include problem formulation and models as a basis of equations, short description of solution, and the most important results. An interested reader may get more detailed information using cited original papers. Besides analyzing examples, other related publications are shortly reviewed to give the reader extended knowledge of specific literature.

Comment 3.1 Although the choosing examples present different methods of solution and subjects of applications, it is clear that: (i) the most of relevant published solutions are outside of introduced here survey, and (ii) the choice of examples is random and depends at least on preferences and background of a reviewer. Nevertheless, author hopes that the considered results give a reader a primary understanding of the situation in modern applications in conjugate heat transfer.

3.1   External Flows

3.1.1   Conjugate Heat Transfer in Flows Past Thin Plates

In this section, the conjugate heat transfer in flows past thin plates at different situations is investigated. The majority problems are considered for thermally thin plates when the second conduction equation (1.3) is valid. Such simplification is justified because as we will see results

Applications of Mathematical Heat Transfer and Fluid Flow Models in Engineering and Medicine,
First Edition. Abram S. Dorfman.
© 2017 John Wiley & Sons Ltd. Published 2017 by John Wiley & Sons Ltd.

obtained in this way are qualitatively representative for exact solutions and in many cases are acceptable even quantitatively for the most part of the plate, except relatively small regions at the leading edge (S. 4.1.2). On the basis of these facts, we developed the charts for a simple solution for the conjugate problems of flows past thermally thin plates. Examples show the wide applicability of this approach and physically comprehensible results.

3.1.1.1 Temperature Singularities on the Solid-Fluid Interface

In the early works [246, 361], it was shown that in the case of laminar flow past thermally thin plate, the wall temperature distribution is not an analytic function near the leading edge rather it is presented at $x = 0$ as a series of variable $x^{1/2}$. Indeed, as it shown in [105] the temperature distribution on a thermally thin plate at $x = 0$ is in general not an analytic function of coordinate x. Analysis reveals that the temperature of a thermally thin plate at $x = 0$ is a singular function presented in series of variable $x^{1/s}$ where s is the denominator of exponent in relation for an isothermal heat transfer coefficient in the form (index L denotes the plate end) (Exer. 3.1)

$$h_* = h_{*L}(x/L)^{-r/s} \tag{3.1}$$

Comment 3.2 An analytic function at some point x_0 may be presented in a neighborhood of this point by Taylor series in powers of variable x. Otherwise, it is said that the function is singular at this point (S. 7.1.2.5). Most functions encountered in applications are analytic at all x or are singular at some specific points. One example of such function is analyzed in considering the boundary layer equation in Prandtl-Mises form (S.7.4.4.2).

Basic Equations
Consider a thermally thin plate of finite length with given boundary conditions at the edges. For generality, it is assumed that the plate is streamlined past both sides by laminar or turbulent, gradient or zero pressure gradient flows of Newtonian or power law non-Newtonian fluids. The flow may be symmetrical or asymmetrical as two streams with different characteristics. Examine first the case when exponents r/s in equation (3.1) for both sides are the same (e.g., both streams are laminar or turbulent), but the temperatures of flow far away from a body and the isothermal heat transfer coefficients at the plate end h_{*L} (see equation (3.1)) are different. For such a case, substitution of universal function (1.38) defining heat fluxes for both streams into second equation (1.3) yields the following dimensionless equation for the plate temperature (Exer. 3.2)

$$\sum_{k=0}^{\infty} D_k \zeta^k \frac{d^k\theta}{d\zeta^k} - \zeta^{r/s}\frac{d^2\theta}{d\zeta^2} - \text{Bi}_{*L2} - \zeta^{r/s}\overline{q}_v = 0$$

$$\theta = \frac{T_w - T_{\infty 1}}{T_{\infty 2} - T_{\infty 1}}, \quad \text{Bi}_{*L} = \frac{h_{*L}L^2}{\lambda_w \Delta} = \text{Bi}_*\left(\frac{L}{\Delta}\right)^2, \quad \overline{q}_v = \frac{q_{v.av}L^2}{\lambda_w(T_{\infty 2} - T_{\infty 1})} \tag{3.2}$$

Here, θ is the average across plate thickness temperature head (Com. 1.1), coefficients of sum $D_k = g_{k1}\text{Bi}_{*L1} + g_{k2}\text{Bi}_{*L2}$ and $D_0 = \text{Bi}_{*L1} + \text{Bi}_{*L2}$ depend on Biot numbers and on coefficients g_k, where k denotes the number a term of universal function (1.38), indices 1 and 2 denote plate sides, and $\zeta = x/L$ is dimensionless coordinate. Equation (3.2) is valid for the case of equal fluid temperatures as well if the scale in definition of θ is changed to T_∞ for both sides, and

Bi_{*L2} in equation (3.2) is omitted. For a symmetric streamlined plate, one should set in addition $Bi_{*L1} = Bi_{*L2}$ and $g_{k1} = g_{k2}$.

If the heat sources $\bar{q}_v$ near the origin is an analytic function of x, the solution of equation (3.2) at $x = 0$ is given as a series in the variable $x^{1/s}$

$$\bar{q}_v = \sum_{i=0}^{\infty} b_i \zeta^i, \qquad \theta = \sum_{i=0}^{\infty} a_i \zeta^{i/s} \tag{3.3}$$

This follows from the fact that substituting series (3.3) into equation (3.2) yields the relations determining the series coefficients a_i under knowing coefficients b_i

$$\sum_{i=0}^{\infty} [D_0 + D_1(i/s) + D_2(i/s)(i/s - 1) + \ldots + D_k(i/s)(i/s - 1) \ldots (i/s - k + 1) + \ldots] a_i \zeta^{\frac{i}{s}}$$

$$-\sum_{i=0}^{\infty} (i/s)(i/s - 1) a_i \zeta^{\frac{i+r}{s}-2} - Bi_{*L2} - \sum_{i=0}^{\infty} b_i \zeta^{i+\frac{r}{s}} = 0 \tag{3.4}$$

Here, the first, second, and the last sums are obtained from corresponding terms of equation (3.2) as follows: (i) the first sum by differentiating the second series (3.3) to find the k derivatives of temperature θ, (ii) the second sum by using the second derivative of the same temperature θ to get the second term of equation (3.2), and (iii) the last sum by differentiating the first series (3.3) to find the derivatives of source $\bar{q}_v$ (Exer. 3.3).

The sums in equation (3.4) contain variable ζ with different exponents. To collect the terms with the same exponents from these sums, we change the indices from i to $i - (2s - r)$ in the first and to $(i/s) - 2$ in the last sums, respectively. That transforms indices in the first and the third sums to the same index $(i + r)/s - 2$ as in the second sum. Then, because indices must be positive integers, it follows from the new indices in the first sum that the inequality: $i > 2s - r$ should be satisfied (otherwise, i becomes negative). Hence, the first $2s - r$ coefficients a_i are zero except coefficients a_0 and a_s, which are free because the terms in the second sum in equation (3.4) at $i = 0$ and $i = s$ vanish independent of the values of those coefficients. With regard to these reason, the collection of terms with new indices in equation (3.4) leads to the following equations for coefficients a_i of series (3.3) (Exer. 3.4).

$$D_0 a_0 - (r/s - 1)(r/s - 2) a_{2s-r} - Bi_{*L2} = 0, \quad \text{for } i = 2s - r$$

$$[D_0 + D_1(j - 1) + D_2(j - 1)(j - 2) + \ldots + D_k(j - 1)(j - 2) \ldots (j - k) + \ldots] a_{s(j-1)}$$

$$-(i/s)(i/s - 1) a_i - b_{(i/s)-2} = 0, \quad j = (i + r - s)/s, \quad \text{for } i > 2s - r \tag{3.5}$$

It is seen that coefficients a_i are calculated one after another in terms of a_0 and a_s starting from the first one a_{2s-r}, which is defined by the first equation (3.5). Then, the second coefficient is defined through the first one, the third though the second and so on, whereas the coefficients a_0 and a_s are determined from the given boundary conditions at the plate ends (see examples in next section). The coefficients b_i ($i = 0, 1, 2 \ldots$) from the first equation (3.3) are taken into account only when the exponents of variable ζ in terms of second series (3.3) are the same integers 0, 1, 2 ... as in corresponding terms with b_i in the first equation (3.3). This happens when in the second series (3.3) indices are $i = 0, s, 2s, 3s \ldots$, and the corresponding coefficients are $a_0, a_s, a_{2s} \ldots$.

In the case of asymmetric flow, the exponents s/r might be different on two sides. This happened, for instance, when the flow on one side is laminar and on another side turbulent, or when the pressure gradients or the exponents in the power law non-Newtonian fluids are different on two sides. For this case, basic equation (3.2) becomes

$$\frac{d^2\theta}{d\zeta^2} - \text{Bi}_{*L1}\zeta^{-\frac{r_1}{s_1}}\sum_{k=0}^{\infty}g_{1k}\zeta^k\frac{d^k\theta}{d\zeta^k} - \text{Bi}_{*L2}\zeta^{-\frac{r_2}{s_2}}\left(\sum_{k=0}^{\infty}g_{2k}\zeta^k\frac{d^k\theta}{d\zeta^k} - 1\right) + \overline{q}_v = 0 \qquad (3.6)$$

The temperature head near $\zeta = 0$ is presented by the same series (3.3) in integer power of variable $\zeta^{1/s_1 s_2}$ instead of $\zeta^{1/s}$. If the liquids are numbered so that $r_1 s_2 < r_2 s_1$, then the first $(2s_1 s_2 - r_2 s_1)$ coefficients a_i are zero, except a_0 and $a_{s_1 s_2}$, which are found from the boundary conditions. The rest of the coefficients are obtained analogously from equation, which is like equation (3.5) (Exer. 3.5 and 3.6).

$$(r_2/s_2 - 1)(r_2/s_2 - 2)a_{(2s_1 s_2 - r_2 s_1)} + (1 - a_0)\text{Bi}_{*L2} = 0, \quad \text{for } i = 2s_1 s_2 - r_2 s_1$$

$$\text{Bi}_{*L1}[1 + g_{11}(j_1 - 1) + g_{21}(j_1 - 1)(j_1 - 2) + \dots + g_{k1}(j_1 - 1) \dots (j_1 - k) + \dots]a_{s_1 s_2(j_1 - 1)}$$

$$+\text{B.i}_{*L2}[1 + g_{12}(j_2 - 1) + g_{22}(j_2 - 1)(j_2 - 2) + \dots + g_{k2}(j_2 - 1) \dots (j_2 - k) + \dots]a_{s_1 s_2(j_2 - 1)}$$

$$-(i/s_1 s_2)[(i/s_1 s_2) - 1]a_i - b_{(i/s_1 s_2) - 2} = 0, \quad j_1 = (i + r_1 s_2 - s_1 s_2)/s_1 s_2,$$

$$j_2 = (i + r_2 s_1 - s_1 s_2)/s_1 s_2, \quad \text{for } i > 2s_1 s_2 - r_2 s_1 \qquad (3.7)$$

Examples of Singular Series

- *Laminar flow at zero pressure gradient.* In this case $s = 2$, and the wall temperature is presented as a series (3.3) in powers of variable $x^{1/2}$. The first three $(2s - r = 3)$ coefficients are zero except a_0 and a_s. Therefore, $a_1 = 0$, a_0 and $a_s = a_2$ are determined from the boundary conditions. The coefficient a_3 is found from the first equation (3.5) in terms of a_0. The rest coefficients are obtained from the second equation (3.5) starting from $a_4 = 0$ because $j = (i + r - s)/s = 3/2$ and $a_{s(j-1)} = a_1 = 0$. Then, a_5 is determined through $a_{s(j-1)} = a_2 = a_s$ since the next $j = 2$. These three coefficients establish three groups with spacing $2s - r = 3$: $a_0, a_3, a_6 \dots$, $a_1, a_4, a_7 \dots$, $a_2, a_5, a_8 \dots$ in which following coefficients are defined one after another in terms of knowing a_0, $a_1 = 0$, and $a_2 = a_s$, respectively. The coefficients b_i are taken into account only when indices of a_i are proportional to $s = 2$, which means for all even indices: 0, 2, 4 $\dots$ Finally, the resulting series (3.3) contains only terms $a_{3i}\zeta^{(3/2)i}$ and $a_{3i+2}\zeta^{(3/2)i+1}$ ($i = 0, 1, 2 \dots$) because coefficients of the second group are zero due to $a_1 = 0$.
- *Laminar flow at the stagnation point.* It follows from equation (3.1) that the wall temperature at $x = 0$ is an analytic function of the longitudinal coordinate only when the exponent in this equation is an integer. This occurs, in particular, for laminar flow at the stagnation point for which the external velocity is proportional to x, and an isothermal heat transfer coefficient is independent of coordinate ($r/s = 0$). In this case, there is no singularity, and the wall temperature is presented as a series in powers of x in form (3.3) by terms $a_i\zeta^i$.
- *Turbulent flow at zero pressure gradient.* In this case, $r/s = 1/5$, hence, the temperature head is presented as a series in power of $\zeta^{1/5}$. The first nine coefficients are zero except a_0 and a_5. Therefore, seven group starting by these seven coefficients from a_1 to a_8 (except a_5) with spacing $2s - r = 9$ (for example, $a_1, a_{10}, a_{19} \dots$) are zero. So, two groups of terms

$a_{9i}\zeta^{(9/5)i}$ and $a_{9i+5}\zeta^{(9/5)i+1}(i = 0, 1, 2 \dots)$ construct the final series (3.3) in this case. The coefficients b_i should be counted only when indices of a_i are proportional to $s = 5$, which means for indices: 0, 5, 10, 15 ... but these according to above analysis are zero except a_0 and a_5 (Exer. 3.7).

- *Laminar gradient self-similar flow with external velocity $U = cx^m$.* The heat transfer coefficient for isothermal surface is defined as $h_* = h_{*L}(x/L)^{(m-1)/2}$. This relation follows from formula (2.33) for laminar flow $Nu_x = c_1 Re_x^{1/2}$ after substituting U. The values of s and r are determined after simplifying the fraction $(1 - m)/2$. For example, if $m = 1/5$ or $m = 1/3$, one gets $s = 5$, $r = 2$ or $s = 3$, $r = 1$. In the latter case, the first five $(2s - r)$ coefficients a_1, a_2 and a_4 (except a_0 and a_3) are zero. Analogously, these five coefficients establish five groups with spacing $2s - r = 5$ (e.g., $a_0, a_5, a_{11}, a_{16}, a_{21}$) three of which are zero. Therefore, only two groups of terms $a_{5i}\zeta^{(5/3)i}$ and $a_{5i+3}\zeta^{(5/3)i+1}$ make up the solution series. The coefficients b_i should be regarded only when indices of a_i are proportional to $s = 3$, which means for indices: 0, 3, 6, 9 ...

- *Zero pressure gradient flow of power law non-Newtonian fluid.* According to equation (1.54), isothermal heat transfer coefficient is defined as $h_* = h_{*L}(x/L)^{-n/(n+1)}$. If n is an integer, then $s = n + 1$ and $r = n$. If n is a fraction as $n = n_1/n_2$, then $s = n_1 + n_2$ and $r = n_1$. So, if for example, $n = 2$, then $s = 3$, $r = 2$, whereas in the case of $n = 3/5$, one obtains $s = 8$, $r = 3$. It is easy to estimate in the same way as in above examples that the solution in the last case is formed by terms $a_{13i}\zeta^{(13/8)i}$ and $a_{13i+8}\zeta^{(13/8)i+1}$ (Exer. 3.8).

- *Asymmetric laminar-turbulent flow.* Consider the case when the flow is laminar $(r_2/s_2 = 1/2)$ on one side and is turbulent $(r_1/s_1 = 1/5)$ on another side. Then, $1/s_1 s_2 = 1/10$, and the temperature head distribution is described by series in power of $\zeta^{1/10}$. Since $r_1 s_2 = 2 < r_2 s_1 = 5$, the first 13 coefficients are zero because $2s_1 s_2 - r_2 s_1 = 15$, and a_0 and a_{10}, are determined from the boundary conditions. Correspondingly, the final series includes two terms $a_{15i}\zeta^{(3/2)i}$ and $a_{15i+10}\zeta^{(3/2)i+1}$. The exponents consist of fraction $3/2$ instead of $15/10$ (Exer. 3.9–3.12).

Below the second series (3.3) and technique of analyzing its structure are used for conjugate problems solution.

3.1.1.2 Conjugate Problems Solution by Charts

Series (3.3) determines the temperature head close to the leading edge up to some value $\zeta > 0$. Then, the numerical integrating of equation (3.2) or (3.6) gives the solution for the rest part of a plate. Because usually these equations are used with only several first derivatives, the numerical integration could be performed by standard methods, for example, by Runge-Kutta method. In some cases, the basic equations (3.2) or (3.6) might be integrated analytically using well-investigated equations. One such case we employed to create charts for solving conjugate heat transfer problems for flows past thin plates.

Chart Creating
Equation (3.2) with two first derivatives can be transformed to a hypergeometric equation, of which the solution is presented in the following form

$$\theta = C_1 x F(\alpha, \beta, \gamma, D_2 x^{2-r/s}) + C_2 F(\alpha - \gamma + 1, \beta - \gamma + 1, 2 - \gamma, D_2 x^{2-r/s}) + \sigma_{Bi} + \vartheta_q \quad (3.8)$$

$$\alpha + \beta = \frac{D_1 + D_2}{D_2(2 - r/s)}, \quad \alpha\beta = \frac{D_1 + D_0}{D_2(2 - r/s)^2}, \quad \gamma = \frac{3 - r/s}{2 - r/s}, \quad \sigma_{Bi} = \frac{Bi_{*L2}}{Bi_{*L1} + Bi_{*L2}} \quad (3.9)$$

Here, two hypergeometric functions F are solutions of a homogeneous (S. 9.2.1) equation (3.2) for the case when γ is not integer (as in (3.9)), α and β are roots of the quadratic equation, σ_{Bi} is a ratio of thermal resistances of the two coolants and ϑ_q is a particular solution of an inhomogeneous equation (3.2) with source $\overline{q}_v$ (Exer. 3.13).

Equation (3.8) takes into account the singularity of the temperature distribution at the leading edge at $x = 0$ studied above. To see this, note that two hypergeometric functions (3.8) depend on variables $x^{2-r/s} \cdot x$ and $x^{2-r/s}$, respectively. It is easy to check that the corresponding Taylor series created by these two functions is actually the same second series (3.3). For example, in the case of $r/s = 1/2$ for the laminar flow at zero pressure gradient, one finds that the Taylor series of solution (3.8) is formed by two terms with the same variables $x^{(2-1/2)i} = x^{(3/2)i}$ and $x^{(2-1/2)i+1} = x^{(3/2)i+1}$ ($i = 0, 1, 2 \ldots$) as the second series (3.3). Analogously, one gets that the Taylor series of solution (3.8) is in line with series (3.3) in other cases (Exer. 3.14).

Functions F in relation (3.8) are independent on boundary conditions, whereas the constants C_1 and C_2 are determined by boundary conditions specifying the particular problem. Therefore, the hypergeometric functions F are universal in that respect and hence, can be tabulated. We consider the case when coefficients g_k for both flows around a plate are the same. In this case, equation (3.2) is transformed to the following

$$\sum_{k=0}^{\infty} g_k z^k \frac{d^k\theta}{dz^k} - z^{r/s}\frac{d^2\theta}{dz^2} - \sigma_{Bi} - z^{r/s}\frac{\overline{q}_v}{z_L^2} = 0 \quad (3.10)$$

$$z = (Bi_{*L1} + Bi_{*L2})^{1/(2-r/s)}(x/L) \quad z_L = (Bi_{*L1} + Bi_{*L2})^{1/(2-r/s)} \quad (3.11)$$

According to this equation, the temperature head θ in the dimensionless form (3.2) for thermally thin plate depends only of single variable z. Due to that, the charts can be created in the form of two hypergeometric function (3.8) (Exer. 3.15)

$$\vartheta_1 = F\left(\alpha - \gamma + 1, \ \beta - \gamma + 1, \ 2 - \gamma, \ g_2 z^{2-r/s}\right) \quad \vartheta_2 = zF\left(\alpha, \ \beta, \ \gamma, \ g_2 z^{2-r/s}\right) \quad (3.12)$$

For laminar and turbulent flows at zero pressure gradient these functions and their first two derivatives are calculated using simple initial conditions

$$\vartheta_1(0) = 1, \quad \vartheta_1'(0) = 0, \quad \vartheta_2(0) = 0, \quad \vartheta_2'(0) = 1 \quad (3.13)$$

The results plotted in Figures 3.1 and 3.2 construct the basic part of charts for solving homogeneous equation (3.10). The other part of these charts consisting of functions ϑ_3 and ϑ_4 gives the particular solutions of inhomogeneous equation (3.10) for linear source $\overline{q}_v = A + B(x/L)$ in the form (Exer. 3.16)

$$\vartheta_q == \frac{AL^2}{\lambda_w\left(T_{\infty 2} - T_{\infty 1}\right)z_L^2}\vartheta_3 + \frac{BL^3}{\lambda_w\left(T_{\infty 2} - T_{\infty 1}\right)z_L^3}\vartheta_4 = \overline{A}\vartheta_3 + \overline{B}\vartheta_4 \quad (3.14)$$

In a more general case, when there is no closed form of solution, the data for chart can be computed by series (3.3) for starting values of coordinate at $x = 0$ and subsequent numerical solution of equation (3.2) for the rest part of a plate. In some cases, the series (3.3) is applicable for the whole problem domain, like in Example 3.4, which is considered below.

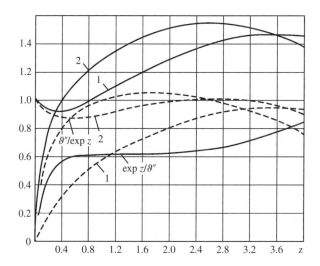

Figure 3.1 Chart functions ϑ_1 and ϑ_2 for laminar flow, $Pr > 0.5$, ------ϑ_1, - - - - ϑ_2, $1 - \vartheta / \exp z$, $2 - \vartheta' / \exp z$, $\exp z / \vartheta''$, $\vartheta'' / \exp z$

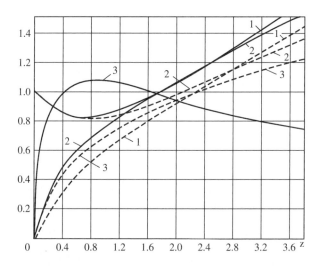

Figure 3.2 Chart functions ϑ_1 and ϑ_2 for turbulent flow, $Pr = 0.7$, $Re = 10^6...10^7$, ------ϑ_1, - - - - $-\vartheta_2$, $1 - \vartheta / \exp(3z/4)$, $2 - \vartheta' / \exp(3z/4)$, 3------$\exp(3z/4)/\vartheta_1''$, - - - - - $-\vartheta_2'' / \exp(3z/4)$

Examples of Using Charts

Here, we consider in details several conjugate problem solutions [111] by using charts to show the applicability and simplicity of this procedure. The solution starts from defining constants in equation (3.8) applying chart functions (3.12) in the form

$$\theta = C_1\vartheta_1 + C_2\vartheta_2 + \sigma_{Bi} + \vartheta_q \tag{3.15}$$

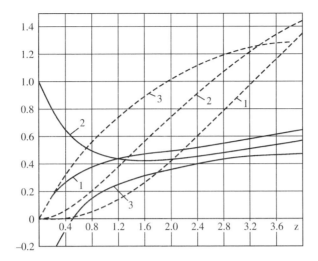

Figure 3.3 Chart functions ϑ_3 and ϑ_4 for laminar flow, $Pr > 0.5$, ------ϑ_3, - - - - ϑ_4, $1 - \vartheta/\exp(3z/4)$, $2 - \vartheta'/\exp(3z/4)$, $3 - \vartheta''/\exp(3z/4)$

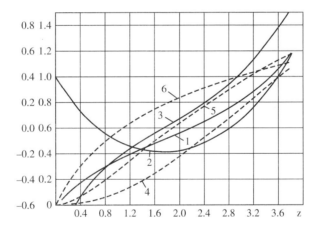

Figure 3.4 Chart functions ϑ_3 and ϑ_4 for turbulent flow. $Pr = 0.7$, $Re = 10^6...10^7$, ----- ϑ_3, - - - ϑ_4,, $1 - \vartheta_3/2$, $2 - \vartheta'_3$, $3 - \vartheta''_3$, $4 - [-\vartheta_4/\exp(3z/4)]$, $5 - [-\vartheta'_4/\exp(3z/4)]$, $6 - [-\vartheta''_4/\exp(3z/4)]$

In this relation, θ is the temperature head in the form (3.2) applicable for the case of different flow temperatures on both sides of a plate. For symmetrical flow, the scale $T_{w2} - T_{w1}$ in this definition should be changed to the temperature head $T_{w0} - T_\infty$ at the leading edge at $x = 0$. Two last terms σ_{Bi} and ϑ_q in equation (3.15) are the same as in equation (3.8) specifying the inhomogeneous part of equation (3.2). Two other basic characteristics: the dimensionless local heat flux $\bar{q}_w$ from a plate and heat flux along the plate $\bar{q}_x$ are determined through derivatives of temperature head θ' and θ'' with respect to coordinate x (or z). The heat flux $\bar{q}_w$ from a plate is determined by the steady-state second equation (1.3) and according this equation is proportional to second derivative, whereas the heat flux along the plate $\bar{q}_x$ is proportional to the first derivative. For the symmetrical flow with temperature head scale $T_{w0} - T_\infty$, both relations are (Exer. 3.17)

$$\bar{q}_w = \frac{q_w L^2}{\lambda_w \Delta \left(T_{w0} - T_\infty\right) z_L^2} = \frac{\theta''}{2\theta_0}, \quad \bar{q}_x = -\frac{q_x L}{\lambda_w \left(T_{w0} - T_\infty\right) z_L} = \frac{\theta'}{\theta_0} \quad (3.16)$$

■ **Example 3.1:** A Steel Plate of a Length $L = 0.25\ m$ and Thickness $\Delta = 0.01\ m$ Is In the Air Flow of Velocity 3 m/s. The left end of the plate is isolated, the temperature of another end is T_{wL}. The air temperature is 300 K. Obtain the heat transfer characteristics.

In this case, it is convenient to use the dimensionless temperature head in the form $\theta = (T_w - T_\infty)/(T_{wL} - T_\infty)$, based on the given temperature T_{wL} instead of leading edge temperature T_{w0}. The reason of this is that in this case, the variable domain becomes simple: from zero ($T_{w0} = T_\infty$) at the leading to one at trailing edge, respectively. Then, applying the homogeneous equation (3.15) with tabulated functions ϑ_1, ϑ_2 and given boundary condition at left $q_x(0) = 0$ (isolated end) and at right $\theta(z_L) = 1$ (known temperature), one gets two equations

$$q_x = C_1 \vartheta_1'(0) + C_2 \vartheta_2'(0) = 0, \qquad \theta(z_L) = C_1 \vartheta_1(z_L) + C_2 \vartheta_2(z_L) = 1 \quad (3.17)$$

The problem solution is obtained from these relations taking into account that: (i) from initial conditions (3.13) it is known that $\vartheta_1'(0) = 0$, (ii) then, it follows from the first relation (3.17) that $C_2 = 0$, and (iii) therefore, from second equation (3.17) that $C_1 = 1/\vartheta_1(z_L)$. Finally, relation (3.15) gives the solution $\theta = \vartheta_1(z)/\vartheta_1(z_L)$.

Estimation of Reynolds number Re $\approx 5 \cdot 10^4$ shows that the flow is laminar. Hence, the Nusselt number for isothermal plate is (see (2.33)) $\mathrm{Nu}_{*L} = 0.332 \cdot 0.7^{1/3} \sqrt{\mathrm{Re}_L} = 66$. The corresponding Biot number (3.2) and variable (3.11) at the plate end are obtained as

$$\mathrm{Bi}_{*L} = \frac{h_{*L} L^2}{\lambda_w \Delta} = \frac{\mathrm{Nu}_{*L} \lambda L}{\lambda_w \Delta} = \frac{66 \cdot 2.58 \cdot 10^{-2} \cdot 0.25}{65 \cdot 10^{-2}} = 0.656, \quad z_L = (2\mathrm{Bi}_{*L})^{\frac{1}{2-r/s}} = 1.2$$
$$(3.18)$$

According to equations (3.16) one gets: local heat flux from a plate $\bar{q}_w = \vartheta_1''(z)/2\vartheta_1(z_L)$, and along the plate $\bar{q}_x = -\vartheta_1'(z)/\vartheta_1(z_L)$. These relations are obtained proceeding from the solution $\theta = \vartheta_1(z)/\vartheta_1(z_L)$ and taking into account that the temperature at the leading edge is $\theta_0 = 1/\vartheta_1(z_L)$ because $\vartheta_1(0) = 1$ (see (3.13)). The numerical results are presented in the Table 3.1 (Exer. 3.18).

Comment 3.3 We considered this problem in Chapter 2 (Exam. 2.12) when estimated the errors caused by boundary condition of the third kind. For estimating the error, we used solution of this problem with boundary condition of the third kind. Now, we compare the result of this approximate common solution with just obtained data of conjugate solution. A maximal difference between both results occurs at the starting section of a plate where according to the common solution (2.35) the temperature head is $\theta_w = 1/ch\sqrt{2\mathrm{Bi}_{*av}} = 1/ch\sqrt{4\mathrm{Bi}_{*L}} = 1/ch\sqrt{4.0.656} = 0.381$, whereas the conjugate solution gives the value $\theta_w = 0.278$ (Table 3), which is 0.73 times less showing that the error estimated in Example 2.12 is in line with this result.

Table 3.1 Heat transfer characteristic of a plate heated from one end (Example 3.1)

z	x/L	$\vartheta_1(z)$	$\vartheta_1'(z)$	$\vartheta_1''(z)$	θ	$-\bar{q}_x$	$\bar{q}_w$
0	0	1	0	∞	0.278	0	∞
0.2	0.167	1.12	0.949	2.75	0.311	0.264	0.382
0.4	0.334	1.37	1.48	2.66	0.388	0.411	0.369
0.6	0.501	1.72	2.04	3.02	0.478	0.567	0.419
0.8	0.668	2.19	2.70	3.62	0.608	0.750	0.503
1.0	0.835	2.81	3.50	4.42	0.780	0.969	0.614
1.2	1	3.60	4.48	5.12	1	1.244	0.753

■ **Example 3.2:** A Copper Plate of Length 0.5 m and 0.02 m in Thickness Is Streamlined On One Side By Air at Temperature 313 K with Velocity 30 m/s. Another side of the plate is isolated. The temperatures of the plate edges are maintained at $T_{w0} = 593$ K and $T_{wL} = 293$ K. Find the local temperature and heat flux distributions.

If the dimensionless temperature is used in the form based of the leading edge temperature $\theta = (T_w - T_\infty)/(T_{w0} - T_\infty)$, the boundary conditions are: $\theta(0) = 1$ and $\theta(L) = \theta_L$. Then, according to conditions (3.13), the equation $C_1\vartheta_1(0) + C_2\vartheta_2(0) = 1$ for leading edge gives $C_1 = 1$, and after that, the similar condition for another edge $\vartheta_1(z_L) + C_2\vartheta_2(z_L) = \theta_L$ yields $C_2 = [\theta_L - \vartheta_1(z_L)]/\vartheta_2(z_L)$. The Reynolds number Re $= 0.88 \cdot 10^6$ shows that the flow is turbulent. Therefore, $Nu_{*L} = 0.0255 Re^{4/5} = 1453$, $Bi_{*L} = 2.53$, and $z_L = 2.53^{5/9} = 1.67$. Taken from the charts the values of $\vartheta_1(z_L)$ and $\vartheta_2(z_L)$, one finds $C_2 = -1.22$ and the solution $\theta = C_1\vartheta_1(z) + C_2\vartheta_2(z) = \vartheta_1(z) - 1.22\vartheta_2(z)$.

The calculation results are plotted in Figure 3.5. It is seen that the heat transfer coefficient h obtained in conjugate solution significantly differs from the isothermal coefficient h_* (dotted line). Whereas the former sharply decreases and reaches zero at $x \approx 0.4$ m, the latter remains almost constant at this part of the plate. This happened because in this example the temperature head decreases in the flow direction (Exer. 3.19).

■ **Example 3.3:** Consider a Similar Problem For Aluminum Plate of Length 0.3 m and 0.002 m in Thickness Past a Flow of Air of a Velocity 250 m/s On an Altitude of 20 km. The air temperature is $T_\infty = 223$ K, kinematic viscosity is $v = 1.65 \cdot 10^4 m^2/s$. The trailing isolated edge is at stagnation flow temperature $T_{\infty L} = T_{ad.L} = 223 + (250^2/2000) = 254$ K, and the leading edge temperature is mentioned at $T_{w0} = 323$ K.

Because the Mach number is M $= U/U_{sd} = 250/(20.1 \cdot \sqrt{223}) = 0.833$ (U_{sd} is a speed of sound) is close to unity, the compressibility effect should be taken into account. This is achieved by using the adiabatic enthalpy difference instead of the temperature head (S. 1.8), or the adiabatic temperature difference (ignoring the dependence $c_p(T)$) if the Mach number is not very close to the unity. In the last case, the dimensionless temperature head is used in usual form $\theta = (T_w - T_{ad.L})/(T_{w0} - T_{ad.L})$, in which the fluid temperature T_∞ is substituted by adiabatic flow temperature at the trailing edge $T_{ad.L}$ (Exer. 3.20).

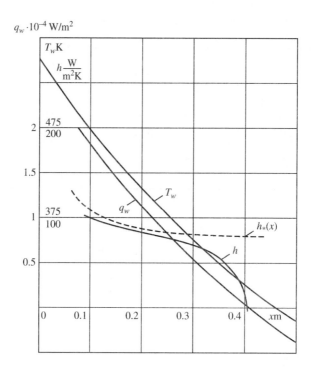

$q_w \cdot 10^{-4}$ W/m^2

Figure 3.5   Heat transfer characteristics for a plate streamlined on one side by turbulent flow

The boundary conditions at the plate ends $\theta(0) = 1$ and $\theta(L) = 0$ along with initially relations (3.13) give the constants $C_1 = 1$ and $C_2 = -\vartheta_1(z_L)/\vartheta_2(z_L)$. Then, the solution is obtained
according to (3.15) $\theta = \vartheta_1(z) - \vartheta_2(z)\vartheta_1(z_L)/\vartheta_2(z_L)$. To find a value of the parameter z_L, a Biot
number should be multiplied by $C_x/\sqrt{C}$. This becomes clear if one writes a steady-state
equation (1.3) for thermally thin plate past compressible flow using relation (1.49) for heat
flux to get an equation similar to relation (3.6) (Exer. 3.21)

$$\frac{d^2\theta}{d\zeta^2} - \frac{C_x}{\sqrt{C}} \mathrm{Bi}_{*L1} \zeta^{-\frac{r_1}{s_1}} \sum_{k=0}^{\infty} g_{1k} \zeta^k \frac{d^k\theta}{d\zeta^k} - \frac{C_x}{\sqrt{C}} \mathrm{Bi}_{*L2} \zeta^{-\frac{r_2}{s_2}} \left(\sum_{k=0}^{\infty} g_{2k} \zeta^k \frac{d^k\theta}{d\zeta^k} - 1 \right) + \bar{q}_v = 0$$

(3.19)

From this equation is clear that the ratio $C_x/\sqrt{C}$ may be considered as a factor of a Biot number. To estimate this ratio, the Chapman-Rubesin formula (1.49) for coefficient C in viscosity
law $\mu/\mu_\infty = C(T/T_\infty)$ is used. Calculation shows that for the case in question, the values of
both coefficients are close so that $C_x/\sqrt{C} \approx \sqrt{C} = 0.975$.

Reynolds value $\mathrm{Re} = 4.55 \cdot 10^5$ tells us that the flow is laminar, and corrected (multiplied
by $\sqrt{C}$) Biot number is $\mathrm{Bi}_{*L} = 2.91$, and hence, $z_L = 2.04$. Knowing z_L, we estimate the
ratio $\vartheta_1(z_L)/\vartheta_2(z_L)$ by charts and after that get the solution for temperature head employing gained above formula $\theta = \vartheta_1(z) - 1.69\vartheta_2(z)$. The relations for heat fluxes are found from
equations (3.16) after substituting the difference $T_{ad} - T_{wL}$ for the scale $T_w - T_\infty$. The results
are summarized in Table 3.2. Observe that in this case as well as in the former example, the
local heat flux becomes zero (here at $x/L \approx 0.64$) and heat flux inversion (S. 2.4) occurs.

Table 3.2 Heat transfer characterizes of the plate streamlined by compressible flow (Example 3.3)

z	x/L	$\vartheta_1(z)$	$\vartheta_1'(z)$	$\vartheta_1''(z)$	$\vartheta_2(z)$	$\vartheta_2'(z)$	$\vartheta_2''(z)$	θ	$\bar{q}_x$	$\bar{q}_w$
0	0	1	0	∞	0	1	0	1	1.6	∞
0.4	0.196	1.37	1.48	2.66.	0.446	1.29	1.18	0.656	0.584	0.772
0.8	0.392	2.19	2.70	3.62	1.08	1.94	2.10	0.462	0.404	0.260
1.2	0.588	3.60	4.48	5.42	2.06	3.02	3.37	0.304	0.352	0.028
1.6	0.784	5.89	7.16	8.14	3.58	4.71	5.17	0.152	0.376	−0.132
2.04	1	11.0	11.7	12.6	6.26	7.61	8.04	0	0.476	−0.264

It follows from Table 3.2 that the reason of this is the same- decreasing of temperature head (Exer. 3.22).

■ **Example 3.4:** Air Flows (Re $= 5 \cdot 10^4$) Over One Side of a Thin ($\Delta/L = 1/600$) Radiating Plate ($\lambda/\lambda_w = 0.135 \cdot 10^{-4}$) with Uniform Internal Heat Source ($\bar{q}_v = 5.1$). The front end is at the free stream temperature T_∞. Another side of the plate is isolated. The radiation is taken into account by parameter $N = \sigma \varepsilon T_\infty^4 / \lambda_w \Delta = 0.07$, where σ and ε are Stefan-Boltzmann constant and emissivity.

This problem was solved by Sohal and Howell [360] using numerical integration of integro-differential equation (see Exam. 3.14). We consider solution of this conjugate problem as an example of applicability of series (3.3) in variable $x^{1/s}$ for whole problem domain. In the case of radiating plate, the proper form of dimensionless plate temperature is $\theta = T_w/T_\infty$ because according to Stefan-Boltzmann law the heat of radiation is proportional to T^4 rather than to the temperature head, as in the case of convective heat transfer. The basic equation (3.19) for a plate streamlined on one side with additional radiating term and assumption $C_x/\sqrt{C} \approx 1$ takes the following form

$$\zeta^{r/s}\frac{d^2\theta}{d\zeta^2} - \mathrm{Bi}_{*L}\left(\theta - 1 + \sum_{k=1}^{\infty} g_k \zeta^k \frac{d^k\theta}{d\zeta^k}\right) - \zeta^{r/s}N(\theta^4 - 1) + \zeta^{r/s}\bar{q}_v = 0 \qquad (3.20)$$

To solve this equation in series (3.3), the additional radiating term should be expanded in similar series in terms $d_i\zeta^{i/s}$. After multiplying this series by $\zeta^{r/s}$ (in line with relation (3.20)), we change the index i to $i - 2s$ in the exponent $(i + r)/s$ of resulting variable. That transforms the radiating terms to the form $d_{i-2s}\zeta^{(i+r)/s-2}$ with the same variable exponent as in other terms, like in equation (3.4). After adding terms $Nd_{i-2s}\zeta^{(i+r)/s-2}$ to the last equations (3.5) and taking into account that the flow past one plate side is an asymmetric issue, the equations for coefficients a_i are obtained (Exer. 3.23)

$$(r/s - 1)(r/s - 2)a_{2s-r} + (1 - a_0)\mathrm{Bi}_{*L} = 0, \quad \text{for} \quad i = 2s - r$$

$$\mathrm{Bi}_{*L}[1 + g_1(j - 1) + g_2(j - 1)(j - 2) + \ldots + g_k(j - 1)(j - 2) \ldots (j - k) + \ldots]a_{s(j-1)}$$

$$-(i/s)(i/s - 1)a_i + Nd_{i-2s} - \bar{q}_v = 0, \quad j = (i + r - s)/s, \quad \text{for} \quad i > 2s - r \qquad (3.21)$$

Coefficients d_i are calculated applying the chine rule for differentiating function $(\theta^4 - 1)$ with respect to the variable $\zeta^{1/s}$. The complicated expressions of this procedure are significantly simplified due to their dependency on the coefficients a_i of series (3.3) several of which are zero. The first coefficients d_i required for farther steady are: $d_0 = d_1 = d_3 = 0, d_2 = 4a_2, d_4 = (1/2)a_2^2 - (1/6)\overline{q}_v$ (Exer. 3.24).

From foregoing discussion we know that the series coefficients a_i are determined through coefficients a_0 and a_s, whereas the two last should be estimated from boundary conditions. In the problem in question, the only one condition is given $\theta(0) = 1$ at the left edge because the plate is considered as a semi-infinite. In such a case, when there is no real trailing edge, another condition is obtained from the following reasoning. As the distance from the leading edge increases, the heat transfer intensity diminishes due to growing boundary layer, and the plate temperature head gradually decreases along the plate becoming asymptotically constant far away from starting edge, theoretically at the infinity. This asymptotically constant value θ_{as} is found from algebraic equation

$$\text{Bi}_{*L}(\theta_{as} - 1) - N(\theta_{as}^4 - 1) + \overline{q}_v = 0 \tag{3.22}$$

which is deduced from equation (3.20) by setting to zero all temperature head derivatives. Equation (3.22) may be solved graphically or by trial-and-error method. Because the flow is laminar, the estimation of Biot number gives: $\text{Nu}_{*L}\lambda L / \lambda_w \Delta = 66 \cdot 0.135 \cdot 10^{-4} \cdot 600 = 0.535$. Then, solution of equation (3.22) yields $\theta_{as} = 2.79$ (Exer. 3.25).

At some point, where the temperature head becomes practically constant (with desired accuracy, for example, with 1% changes), may be considered as an end of a plate at which $x/L = 1$ and where the asymptotic value $\theta_{as} = 2.79$ is achieved. That may be done numerically using, for example, Runge-Kutta method. Retaining in the sum, as before, only two first derivatives, one gets the differential equation (3.20) of the second order, which requires for solution the initial values of θ and θ' at $x = 0$. The first one is known: $\theta(0) = 1$, whereas the second is found by trial and error. This procedure starts by guessing some value of $\theta'(0)$ and solving equation (3.20) to get the value of θ at $\zeta = 1$. Repeating the solution of this equation for different initial values of $\theta'(0)$, one finds among gained data the correct value of $\theta'(0)$, that corresponds to $\theta_{as} = 2.79$. Because of singularity of equation (3.20) at $x = 0$, the series (3.3) with first several terms is used to find θ and θ' at $\zeta > 0$ close to $x = 0$. These values define the coefficients $a_0 = \theta(0)$ and $a_s = \theta'(0)$, which serve as initial data for numerical solution (see S. 3.1.1.1) (Exer. 3.26).

Much simpler such problem is solved using series (3.3) with several first terms for entire domain in the case of acceptable series convergence. The trial-and-error approach helps to estimate how many terms should be retained to obtain the desired accuracy. We begin retaining the first terms up to a_5. The flow is laminar (Re $= 5 \cdot 10^4$), and hence, $r/s = 1/2$, $a_0 = \theta(0) = 1$, $a_s = a_2$. The further coefficients of series are given by equations (3.21). From the first one we get $a_3 = 0$ since $a_0 = 1$. Then, we have: $j = 3/2$, $d_0 = 0$, $a_{s(j-1)} = a_1 = 0$, and $a_4 = -\overline{q}_v/2 = -2.55$. The next is: $j = 2$, $d_3 = 0$, $a_{s(j-1)} = a_2$, and $a_5 = 0.432\,\text{Bi}_{*L}a_2 = 0.231\,a_2$. So the series is: $\theta = 1 + a_2\zeta - 2.55\zeta^2 + 0.231\,a_2\,\zeta^{5/2}$. Assuming that this expression describes the problem satisfactorily for the entire domain up to $\zeta = 1$ where the asymptotic value $\theta_{as} = 2.79$ is attained, we find a_2 from the equation $2.79 = 1.231a_2 - 1.55$ to get $a_2 = 3.526$. The second approximation is obtained by adding the next term: $i = 6$, $j = 5/2$, $a_{s(j-1)} = a_3 = 0$, $d_2 = 4a_2$, and $a_6 = (2/3)Na_2$. The series becomes $\theta = 1 + a_2\zeta - 2.55\zeta^2 + 0.231a_2\zeta^{5/2} + 0.0467a_2\zeta^3$, which gives a new value of

Table 3.3   Dimensionless temperature $\theta = T_w/T_\infty$ along the radiating plate (Example 3.4)

$\zeta = x/L$	0	0.1	0.2	0.3	0.4	0.5	0.6	07	0.8	0.9	1.0
θ	1.00	1.31	1.58	1.82	2.01	2.18	2.31	2.45	2.57	2.68	2.79

unknown coefficient as $a_2 = 3.397$. The relative difference between two results is about 4%. We calculate two terms more arriving in the following final expression

$$\theta = 1 + a_2\zeta - 2.55\zeta^2 + 0.231a_2\zeta^{5/2} + 0.0467a_2\zeta^3$$
$$- 0.314\zeta^{7/2} + (0.0226a_2 + 0.0351a_2^2 - 0.0596)\zeta^4 \qquad (3.23)$$

which yields $a_2 = 3.642$ or $a_2 = 3.330$ when one used the only first or both additional terms, respectively. Comparing data for a_2 shows that there are two pair of values close to each other: the first 3.526 and the third 3.642, and the second 3.397 and fourth 3.330. The accuracy for these pairs is about 3% or 2%, accordingly. Table 3.3 presents the data obtained with the last value $a_2 = 3.330$. These well agreed with results given in [360] on Figure 5 (Exer. 3.27 and 3.28).

3.1.1.3   Investigation of Conjugate Heat Transfer in Flows Past Plates

We continue using the charts for solving conjugate problems of flows past thin plate in order to continue showing the basic features of conjugate heat transfer. Here, much less attention is given to solution details considered above rather focusing the major interest in results of investigation. The most of examples are adopted from the book [111], some are taken from other publications indicated below.

■ **Example 3.5: A Plate Heated From One End in a Symmetrical Flow**

The book begins with the qualitative analysis of this problem as a typical example of conjugate problem, which gives a reader an understanding of the core of conjugation. Recall that this example clearly demonstrates the role of the temperature head variation because the surface temperature decreases in flow direction, if the plate is passed from the heated end, and increases when the flow runs in opposite direction, starting from unheated edge, whereas in both cases everything else remains the same.

To solve this conjugate problem, assume that the temperature head of heated end is θ_h and the other edge is isolated. Determining the constants in the basic equation (3.15) applying known conditions yields the temperature head in both cases

$$\frac{\theta}{\theta_h} = \vartheta_1 - \frac{\vartheta_1'(z_L)}{\vartheta_2'(z_L)}\vartheta_2, \qquad \frac{\theta}{\theta_h} = \frac{\vartheta_1}{\vartheta_1(z_L)} \qquad (3.24)$$

The local heat fluxes from and along the plate are obtained from relations (3.16). The total heat flux from a plate is found by integration of the first equation (3.16) for local heat flux. Using

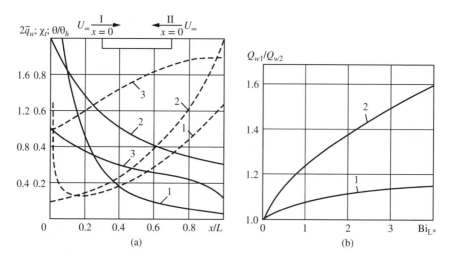

Figure 3.6   Heat transfer characteristics for the plate heated from one end. (a) Local characteristics $\mathrm{Bi}_{*L} = 1.4$, I ------ first case, II- - - - second case, $1 - 2\bar{q}_w$, $2 - \theta/\theta_h$, $3 - \chi_t$; (b) Ratio of total heat fluxes removed from plate 1- turbulent flow and 2-laminar flow

this expression and taking into account that in the first case $\theta'(z_L) = 0$ and in the second case $\theta'(0) = 0$, one obtains the ratio of total fluxes (Exer. 3.29)

$$Q_w = \frac{2L}{\lambda_w(T_h - T_\infty)z_L\Delta}\int_0^L q_w dx = \frac{\theta'(z_L) - \theta'(0)}{\theta_h}, \qquad \frac{Q_{w1}}{Q_{w2}} = \frac{\vartheta_1(z_L)}{\vartheta'_2(z_L)} \tag{3.25}$$

The results for laminar flow are plotted in Figure 3.6 (a). It is seen that heat transfer characteristics in both cases differ substantially. In the first case when the temperature head decreases, the heat transfer coefficients are significantly less than the isothermal coefficients (small non-isothermicity coefficients χ_t, curve 3), and the heat flux sharply decreases along the plate (curve 1), so that the situation is close to inversion at the plate end. In that case at the end, the heat transfer coefficient is 4.5 time less than an isothermal one. In another case, the temperature head increases (dashed curve 2), and according to this, the heat transfer coefficients are greater than an isothermal coefficients, but not more than 1.8 times (dashed curve 3). Nevertheless, the total heat flux in this case is less than that in the first case where the local heat transfer coefficients are much lower (curves 3). To physically understand this result, note that: (i) in the first case, the large temperature heads at the beginning of the plate coincide with high isothermal heat transfer coefficients at this part of the plate resulting in high local and total heat fluxes, despite the small nonisothermicity coefficients, whereas (ii) in the second case, the high starting values of isothermal coefficients at the beginning are accompanied with small temperature heads and vice versa, which leads to partly decreasing local heat fluxes (dashed curve 1) and finally to relatively smaller total heat flux from a plate (Exer. 3.30).

The ratio of total heat fluxes Q_{w1}/Q_{w2} presented in Figure 3.6 (b) depends on Biot number and in the case of laminar flow riches significant values. For instance, for a steel plate with $\Delta/L = 1/10$ past air ($\mathrm{Bi}_{*L} = 0.8$) or water ($\mathrm{Bi}_{*L} = 4.5$) this ratio is 1.2 and 1.65, respectively. In the case of turbulent flow, the difference between total flaxes is smaller, not more than

1.2. However, the distributions of the local heat fluxes along the plate in two opposite flow directions differ in essence similar to pattern for laminar flow given in Figure 3.6(a).

Comment 3.4 Problems of such type with significantly variable surface temperature cannot be analyzed via a common approach based on proportionality between heat flux and temperature head because the distribution of heat transfer coefficient is unknown a priori.

■ Example 3.6: A Plate Streamlined on One Side and Isolated on Another

Actually, this is the problem from example 3.5 with negligible radiation effects. Due to radiation absence, this problem can be solved using charts. Such solution shows that employing charts significantly simplifies the conjugate heat transfer investigation.

We consider this problem assuming that the temperature θ_0 and dimensionless heat flux $\bar{q}_x = \bar{q}_0$ are given at the starting end of a plate. These boundary conditions along with the basic equation (3.15) and second equation (3.16) yield constants $C_1 = \theta_0$ and $C_2 = \theta_0 \bar{q}_0$. The solution of this problem given below (first equation (3.26)) is valid for some other similar problems with different boundary conditions if the corresponding value of $\bar{q}_0$ at the starting end is first properly specified. In particular, in the case when the temperature θ_L or heat flux $\bar{q}_x = \bar{q}_L$ at the trailing edge (instead of temperature head θ_0) is given, the proper value of $\bar{q}_0$ is defined by second or third expression (3.26) (Exer. 3.31).

$$\frac{\theta}{\theta_0} = \vartheta_1 + \bar{q}_0 \vartheta_2, \qquad \bar{q}_0 = \frac{\theta_L/\theta_0 - \vartheta_1(z_L)}{\vartheta_2(z_L)}, \qquad \bar{q}_0 = \frac{\bar{q}_L - \vartheta_1'(z_L)}{\vartheta_2'(z_L)} \qquad (3.26)$$

Figure 3.7 shows the variation of the nonisothemicity coefficient and of the temperature head for laminar (a) and turbulent (b) flows in three cases: $\bar{q}_0 = 10, 0,$ and (-2). In the

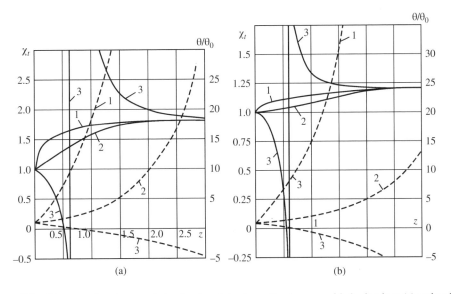

(a)   (b)

Figure 3.7   Heat transfer characteristics for the plate streamlined on one side by laminar (a) and turbulent (b) flows ------ χ_t, - - - - θ/θ_0, $1 - \bar{q}_0 = 10, 2 - \bar{q}_0 = 0, 3 - \bar{q}_0 = -2$

first two cases, the temperature head increases along the plate. In the third case, it first decreases, reaches zero, and then its absolute value increases. The same character of the heat transfer rate variation as in other examples is seen. For an increasing temperature head, the heat transfer coefficients are greater than those for an isothermal surface but not more than 75% to 80% in the case of laminar flow and not more than 20% to 25% for turbulent flow. In the third case, in which the temperature head partly decreases, these coefficients are so much smaller that in the points where the temperature head turns to zero, the heat transfer coefficient becomes meaningless, and corresponding curve $\chi_t(z)$ undergoes discontinuity.

■ **Example 3.7: A Plate Streamlined by Turbulent Flow on Both Sides at Different Temperatures and Heated Leading Edge**

In this case, the scales in relations (3.16) should be substituted by $T_{\infty 2} - T_{\infty 1}$. Assuming that both initial values of temperature head θ_0 and heat flux q_0 are given at leading edge as well as in previous example, we have the same constants $C_1 = \theta_0$ and $C_2 = \theta_0 \bar{q}_0$, and hence, the same solution (3.26) for temperature head with proper scale $\theta = \theta_0[(T_w - T_\infty)/(T_{\infty 2} - T_{\infty 1})]$. For heat fluxes from the plate we need the new equations because relation (3.16) for $\bar{q}_w$ is obtained for symmetrical flow. From equation (3.10) we find that in this case (when $q_v = 0$), the sum of derivatives equals $d^2\theta/dz^2 + z^{-r/s}\sigma_{Bi}$. This gives an expression for heat flux on one side of the plate, since a sum of derivatives determines the heat flux. Then, after transforming equation (3.6), which is valid for asymmetrical flow over a plate, to variable z, one obtains an equation similar to (3.10) and finds the heat flux on the other side of a streamlined plate (Exer. 3.31)

$$\bar{q}_{w1} = \sum_{k=0}^{\infty} g_k z^k \frac{d^k \theta_1}{dz^k} = \frac{d^2 \theta_1}{dz^2} + z^{-r/s}\sigma_{Bi}, \qquad \bar{q}_{w2} = \bar{q}_{w1} - z^{-r/s} \qquad (3.27)$$

In Figure 3.8, the computed results are plotted for the case of the equal thermal resistances of both flows ($\sigma_{Bi} = 1/2$), $\theta_0 = 1$ and $\bar{q}_0 = -2$. The same pattern is observed. On the side on which the temperature head increases (its absolute value, dashed curve 2), the nonisothermicity coefficient χ_t is a little greater than unity, whereas on the other side where there is a section with decreasing temperature head, the value of χ_t sharply falls, becomes zero and then goes to $\pm\infty$ resulting in discontinuous curve $\chi_t(z)$ (Exer. 3.33 and 3.34).

■ **Example 3.8: A Plate with Inner Heat Source [355]**

The problem with sources is an inhomogeneous problem of which a solution is presented as a sum of general homogeneous solution and particular solution of inhomogeneous equations (S. 9.2.1). The former is found similar to other solutions using function ϑ_1 and ϑ_2. For the case of uniform or linear distributed sources, the particular solution in the form (3.14) can be obtained using functions ϑ_3 and ϑ_4 (Figs. 3.3 and 3.4). Figure 3.9 shows the results for turbulent flow over a plate with conditions at leading edge $\theta_0 = 1$, $q_0 = 0$ and inner linear heat source defined by relations (3.14) with $\bar{A} = 1$ and $\bar{B} = 2$.

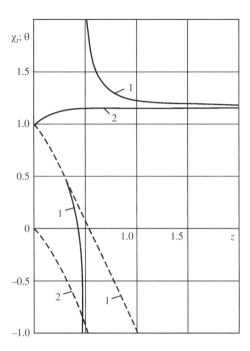

Figure 3.8 Variation of temperature head and nonisothermicity coefficient along the plate streamlined on both sides by turbulent flow, $\sigma_{Bi} = 0.5$, $\theta_0 = 1$, $\bar{q}_0 = -2$, ------χ_t, - - - -θ, 1, 2-different sides of a plate

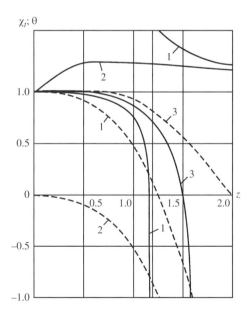

Figure 3.9 Heat transfer characteristics for the plate with inner heat sources streamlined by turbulent flow, ------- χ_t, - - - - θ, 1, 2- different sides of a plate, $\sigma_{Bi} = 0.5$, 3-one side streamlined plate ($\sigma_{Bi} = 0$)

Again we see that the values and variation of heat transfer coefficients strongly depend on temperature head distribution. For the plate streamlined from one side (curves 3) and for the side with temperature θ_1 (curves 1) of the plate's past two sides, for which curves $\theta(z)$ have decreasing sections, the heat transfer coefficient distributions $h(z)$ essentially differs from variations $h_*(z)$ for isothermal coefficients, leading to discontinued curve $\chi_t(z)$. At the same time, for another side of two sides streamlined plate where the absolute value of temperature head increases in flow direction, the ratio $\chi_t = h/h_*$ of heat transfer coefficients differs not much from unity (curve 2) (Exer. 3.35).

3.1.1.4 Applicability of Charts and Thermally Thin Assumption

Refining the Chart Data

In developing charts, the universal function was used with only three first terms of series. Although above (S. 2.1.1.1) it was shown that the other terms are small, the error arising by such simplification can be estimated by refining the data obtained using the charts. This is achieved applying the integral universal function (1.26), which accounts for the terms of equation (1.38) neglecting in charts creation. A first derivative $d\theta_w/dx$ required for integral (1.26) is computed via known chart data, which is considered as a first approximation. This procedure gives both a second approximation and the accuracy of the chart results by comparing these with the second approximation. The next approximation may be found by estimating the difference $\varepsilon(x) = (1/h_*)(q_w^{int} - q_w^{diff})$ between the second q_w^{int} and the first q_w^{diff} results obtained using the integral and restricted differential universal function, respectively. Incorporating this error in the restricted universal function leads to equation (2.37) or (2.38), which solution yields the third approximation q_w^{diff}. Then, the integral formula (1.26) provides the next approximation q_w^{int} and so on. This approach is developed in Section 2.2.2 and realized in examples 4.12 and 4.14.

Accuracy of Thin and Thermally Thin Assumptions

Another error occurs as the body is considered as a thermally thin object. We estimate the inaccuracy of such simplifications by integral method (S. 7.6) using polynomial $T/T_{av} = a_0 + a_1\eta + a_2\eta^2$ for describing the temperature distributions across the body thickness, where T_{av} is a cross-section average temperature and $\eta = y/\Delta$. To find the coefficients a_n, it is assumed that the heat fluxes on the body surfaces are known: q_{w1} at $y = 0$ and q_{w2} at $y = \Delta$. These data give two conditions. The third one is obtained knowing that integral of ratio T/T_{av} across a body thickness equals unity. From the first two condition and the Fourier law, one gets that at $\eta = 0$, $q_{w1} = -\lambda_w a_0$ and at $\eta = 1$, $q_{w2} = -\lambda_w(a_0 + a_1 + a_2)$. The last two equations together with the third relation $a_0 + a_1/2 + a_2/3 = 1$, which follows from the integral of temperature across the boy thickness, built up a system determining the coefficients a_n and as a result, the temperature distribution across a plate

$$T/T_{av} = 1 - (\omega_1/3) + (\omega_2/6) + \omega_1\eta - (\omega_1 + \omega_2)(\eta^2/2), \quad \omega = \frac{q_w\Delta}{\lambda_w T_{av}} = \frac{Bi\theta_w}{T_{av}} \qquad (3.28)$$

Setting $\eta = 0$ and $\eta = 1$, we obtain both surfaces temperatures

$$1 - \omega_1/3 + \omega_2/6, \qquad 1 + \omega_1/6 - \omega_2/3 \qquad (3.29)$$

for the plate top and bottom, respectively. The last two terms in these relations show how much the ratio T/T_{av} at the surfaces differs from unity, and hence, these terms can be used for estimating the accuracy of replacing some object by a thin or thermally thin body. It follows from the last relation (3.28) that the thin (Bi $\geq$ 1) and thermally thin (Bi $<$ 1) bodies correspond to $\omega > 1$ and $\omega < 1$, respectively. Therefore, in the case of a thin body, when ω is large, the two terms with ω in relations (3.29) identify how much these two terms, defining the major part of the surfaces temperatures, are greater than average body value of unity, giving an understanding whether the unity may be neglected comparing to that part of surface temperatures. In the case of thermally thin body, when ω is small, these two terms with ω in relations (3.29) identify how less these two terms, defining the small part of surface temperatures, are lesser than average body value of unity, showing whether a neglecting of that small part of two terms in comparison with unity is acceptable. Mathematically, these conditions are expressed by double inequalities

$$1 \ll \left| \frac{\text{Bi}\theta_{w2}}{3T_{av}} - \frac{\text{Bi}\theta_{w1}}{6T_{av}} \right| \ll 1 \qquad 1 \ll \frac{\text{Bi}\theta_w}{6T_{av}} \ll 1 \qquad (3.30)$$

where the left- and right-hand symbols pertain to thin and thermally thin assumptions, respectively, and the last inequality is related to a symmetrical streamlined body.

Physically, the right-hand symbol provides conditions at which a thin plate or body thermal resistance is small and due to that the cross-section temperatures are so close to each other that practically may be considered as constant. The conditions provided by the left-hand symbol physically mean that a large thermal resistance of a thin plate yields a linear temperature distribution across the plate thickness. The first physical interpretation of the constant cross-section temperature simple follows from definition of thermally thin plate (Com. 1.1). Although the linear temperature distribution across a thin plate or body could not be seen as simple as in the case of thermally thin body, it also follows from thin body definition given in Comment 1.1. To see that, compute the second derivative of the ratio T/T_{av} defined by (3.28), that according to the first equation (1.3) should be zero for the thin body in the case of $q_v = 0$. Such simple procedure gives: $\omega_1 = -\omega_2$ or $q_{w1} = -q_{w2}$, which shows that the temperature distribution is linear (Exer. 3.36 and 337).

Inequalities (3.30) differ from usual more simple relations Bi $\ll$ 1 and Bi $\gg$ 1 in three ways: (i) they are less strong, which results in widening of admissible parameters, (ii) according to these the result depends not only on Biot number, but rather on product of the Biot and temperature head ratio θ_w/T_{av} so that the greater is the Bi the smaller is possible temperature head, and (iii) for the case of asymmetric flow, the result depends on difference of parameters Biθ_w/T_{av} for both surfaces of a plate.

Comment 3.5 For pairs of thin metallic plate/nonmetallic coolant and of thin nonmetallic plate/liquid metal coolant, the inequalities (3.30) usually are satisfied with right (thermally thin assumption) and left (thin assumption) symbols, respectively, due to high and low plate conductivities in the first and second cases.

Comment 3.6 We approximate the temperature distribution by polynomial. Such simple procedure is applicable only to cases when the heat transfer characteristics are regular without specific features as, for example, for systems with near constant or increasing in flow direction or in time temperature heads. However, the parameter distributions in flows with decreasing temperature heads are usually complicated and could not be satisfactory described by polynomials. We discuss this question more detailed in Section 7.6, analyzing applicability of the Karman-Pohlhausen method. Nevertheless, the even less accurate course estimations by conditions (3.30) provide the order of errors and give an understanding whether or not the thin or thermally thin assumption is applicable.

3.1.1.5 Validation of Quasi-Steady Approximation

The methods outlined above for steady heat transfer processes are applicable for some classes of unsteady thermal processes. This holds in cases when unsteadiness is caused by variable thermal regime under the unchanged steady-state hydrodynamics. In particular, the hydrodynamic characteristics of the incompressible flows in the forced heat transfer systems are not affected by thermal unsteadiness. Consequently, in such cases, the methods of solving steady heat transfer problems are applicable to the similar unsteady problems if the quasi-steady state situation takes place. Physically, this means that the unsteady thermal effects in fluid are negligible small in comparison with that for a body, so that transport processes in fluid may be considered using steady state relations.

In the early work [311], Pomeranzev showed by qualitatively analysis that an unsteady conjugate convective problem may be considered as a quasi-state problem if the thermal capacity of a fluid is small in comparison with that for a wall as, for example, for a pair nonmetallic fluid/metal wall. Later, Perelman at al. [307] suggested a parameter that follows from the same reasoning by comparison between the times of a propagation of a heat impulse in the wall and in a fluid. Here, we present analysis of the quasi-steady approximation [117] based on exact solution of the unsteady thermal boundary layer equation outlined in Section 1.11.

Let the thermally thin plate with variable surface temperature $T_w(t)$. is flowing around symmetrically by incompressible flow. As we know from former discussion, the average across thickness temperature of such plate is given by the equation (1.3) as

$$\frac{1}{\alpha_w}\frac{\partial T_{av}}{\partial t} - \frac{\partial^2 T_{av}}{\partial x^2} + \frac{q_w}{\lambda_w \Delta} = 0 \qquad (3.31)$$

In the first chapter (S. 1.11) is shown that the thermal unsteadiness effect is basically determined by the term with the first time derivative-containing coefficient g_{01}, which is four times greater than coefficient g_1 of the term with the space derivative. Taking this into account and substituting in equation (3.31) for heat flux the series (1.55) restrained to only this term with the time derivative, we arrive at the following equation

$$\frac{1}{\alpha_w}\frac{\partial T_{av}}{\partial t} - \frac{\partial^2 T_{av}}{\partial x^2} + \frac{h_*}{\lambda_w \Delta}\left(T_w - T_\infty + g_{01}\frac{x}{U}\frac{\partial T_w}{\partial t} + \ldots\right) = 0 \qquad (3.32)$$

The quasi-steady approximation is applicable when the term with the derivative of fluid temperature $\partial T_w/\partial t$ is negligible in comparison with the term containing a derivative of solid temperature $\partial T_{av}/\partial t$. Thus, the quasi-steady regime existence requires satisfaction of the following inequality (Exer. 3.38)

$$\frac{1}{\alpha_w}\frac{\partial T_{av}}{\partial t} \gg g_{01}\frac{h_* x}{\lambda_w U \Delta}\frac{\partial T_w}{\partial t} \quad \text{or} \quad \frac{\partial T_{av}}{\partial t} \gg \frac{L}{\Delta}\frac{\mathrm{Nu}_{x*}}{\mathrm{Pe}}\frac{\mathrm{Lu}}{\mathrm{Pe}}\frac{\partial T_w}{\partial t} \qquad (3.33)$$

where Lu $= \rho c_p / \rho_w c_{pw}$ is Luikov number suggested by author in his book [119] in honor of Russian scholar A. V. Luikov who was a head of a group of scientists coined the term conjugate heat transfer and obtained among some others researchers' first solutions of conjugate problems.

For the boundary layer heat transfer problems, the inequality (3.33) usually is satisfied due to high Peclet numbers, especially if the body is thin and the Luikov number is not less that one. However, in an opposite case, when the body is thick ($\Delta/L \approx 1$) and Peclet number is relatively small, the result largely depends on the value of the Luikov number defined as a ratio of fluid/body capacities $\rho c_p / \rho_w c_{pw}$. In this case, the validity of quasi-steady regime is not obvious, and hence, the satisfaction of inequality (3.33) should be checked. The importance of capacity ratio (Luikov number) follows, in particular, from solved unsteady problems in which this parameter is often used as a criterion of the problem conjugation rate (see Exam. 3.17, 3.19, and 3.23).

Exercises

3.1 What is the reason to use relation for heat transfer coefficient in form (3.1)? What is the basic difference between two parts of this product? Hint: draw graphs taking, for example, $r/s = 1/2$ to see how these parts change along the plate.

3.2 Obtain equation (3.2) using the second equation (1.3) for thermally thin plate and universal function (1.38) together with equation (3.1) defining the heat fluxes for both streams. Follow directions from the text.

3.3 Obtain equation (3.4) by substitution of series (3.3) in equation (3.2). Perform required differentiating of both equations (3.3) and use the results as described in the text.

3.4 Recall or study in *Advanced Engineering Mathematics* the solution of differential equations in power series and method of undetermined coefficients to understand the procedure of collection terms with equal exponents. Repeat the change of indices in sums of equation (3.4) and perform the described in-text analysis leading to equations (3.5) for series coefficients. Explain why coefficient a_0 and a_s are free, where they come from, and how those two affect the other coefficients,

3.5 Derive equation (3.6) using the second equation (1.3) for thermally thin plate and universal function (1.38) together with equation (3.1) defining the heat fluxes for both streams. Follow the similar development of equation (3.2) in the exercise 3.2.

3.6* Perform the same analysis as in exercise (3.4) for the case of asymmetric flow to obtain equation (3.7) determining series coefficients in this case. Hint: change the exponent in the second equation (3.3) from $1/s$ to $1/s_1 s_2$ and substitute this equation into equation (3.6). Then, change the indices in the obtained equation and collect terms with equal power of variable ζ as it was done in exercise (3.4).

3.7 Repeat the first and third examples of estimation of structure of solutions (3.3) for laminar and turbulent flows to understand in details this procedure requires for the next exercises.

3.8   Show that in the case of non-Newtonian fluid considered in text, the two indicated terms $a_{13i}\zeta^{(13/8)i}$ and $a_{13i+8}\zeta^{(13/8)i+1}$ actually construct the series solution.

3.9   Obtain the structure of series (3.3) for the case of asymmetric laminar-turbulent flow to show that two terms s $a_{15i}\zeta^{(3/2)i}$ and $a_{15i+10}\zeta^{(3/2)i+1}$ designated in the text composed the solution in that case.

3.10   Determine the series structure for flow with exponent (-0.4) in relation (3.1) for an isothermal heat transfer coefficient.

3.11   Obtain the series structure for non-Newtonian fluid with $n = 0.25$. Hint: consider example 3.8.

3.12   Consider similarly to example 3.9 the asymmetric flow with values $r_1/s_1 = 1/3$ and $r_2/s_2 = 1/2$.

3.13*   Study the property of hypergeometric differential equation and corresponding Taylor series, for example, on Wikipedia, in order: (i) to show that a new variable $z = D_2 x^{2-r/s}$ transforms homogeneous equation (3.2) with first two derivatives to hypergeometric differential equation of the form $z(z-1)\theta''_z + (b_1 z + b_2)\theta'_z + b_3\theta = 0$, where b are constants and index z indicates the variable of differentiating, and (ii) to understand that the solution of the transformed equation is presentable as a sum of two hypergeometric functions in the form of relation (3.8).

3.14   Compare the Taylor series produced by two hypergeometric functions (3.8) with the second series (3.3) for turbulent flow, like is done for laminar flow in the text. Consider also other cases.

3.15   Obtain equations (3.10) from equation (3.2) for the case when coefficients g_k on both plate sides are the same. Explain why this form of equation is preferable for charts creation.

3.16   Obtain relation (3.14) for linear heat source $\bar{q}_v$ presented in equation (3.10). Why is such form of source presentation convenient for charts?

3.17*   Derive expressions (3.16) for dimensionless heat fluxes from and along the plate. Think: why the second formula consists of minus. Hint: follow the way described in the text, and note that for symmetrical flow the temperature head is presented as $(T_w - T_\infty)/(T_{w0} - T_\infty) = \theta/\theta_0$. To answer the second question, compare the direction of heat flow along a plate and sign of derivative of temperature in respect to coordinate.

3.18   Solve the problem from example 3.1 for the case of the given leading edge temperature T_{w0} and an isolated right edge. Compare results obtained in both cases. What are physical reasons of such results? Hint: think about effect of flow direction.

3.19   Solve the problem from example 3.2 for the case when the leading edge temperature T_{w0} and the value of heat flux $\bar{q}_x$ at the trailing end are given. Hint: use the second expression (3.16) and the values of C_1 and z_L, which can be determined due to the knowing temperature of leading edge and the flow regime (turbulent).

3.20   Explain why under assumption that specific capacity c_p does not depend on temperature, the stagnation point enthalpy can be substituted by adiabatic temperature.

3.21* Show that in the case of compressible flow, the Biot number in the formula (3.6) should be multiplied by ratio $C_x/\sqrt{C}$. Hint: write an equation using the steady-state equation (1.3) for a thermally thin plate and equation (1.49) for heat flux in the case of compressible flow to obtain equation (3.19), and compare this equation with similar equation (3.6) for incompressible flow.

3.22 Solve the problem from example 3.3 for the case of a given right edge temperature T_{wL} and heat flux at the leading edge $\overline{q}_x$. Hint: use the second formula (3.16) $\overline{q}_x = \theta'/\theta_0$ and two relations (3.15) for θ_0 and θ'_0. Compare the way of solution to that of problem from Exercise 3.19.

3.23 Repeat the analysis showing that additional radiation terms in series (3.3) have the form $d_{i-2s}\zeta^{(i+r)/s-2}$. Explain why the terms $Nd_{i-2s}\zeta^{(i+r)/s-2}$ should be added to the last equation (3.5), but not to the first one. Hint: compare indices of terms in equations (3.5) with those of additional radiation terms. Compare equations (3.21) with equations (3.7) to seize why equations (3.21) differ from equations (3.5) not only by radiating terms, and why they are similar to equations (3.7).

3.24* Obtain several first coefficients d_i of expansion of function $(\theta^4 - 1)$ in series using the chine rule for differentiation with respect to of variable $\zeta^{1/s}$. Simplify obtained relations applying coefficients a_i of series (3.3) for laminar flow.

3.25 Obtain equation (3.22) from equation (3.20) and solve it graphically or by trial and error approach. Hint: (i) for graphical solution find interaction point of two functions: $f_1 = \mathrm{Bi}_{*L}(\theta - 1)$ and $f_2 = \overline{q}_v - N(\theta^4 - 1)$, (ii) for trial and error, substitute some guessing values of θ in equation (3.22) and analyze the results obtained for the right-hand part of this equation; choose the next value of θ, which is between the more and less than zero and see how close to zero is the right-hand part. Continue this process to try to satisfy this equation.

3.26 Recall or study in *Advanced Engineering Mathematics* what initial and boundary value conditions for solving ordinary differential equations are in order to understand a numerical solution of equation (3.20) described in the text.

3.27 Repeat the estimation of the first coefficients a_i performed in example 3.4 and compute the two last coefficients of series (3.23). This will improve your expertise required for a solution the next two problems. Solve the problem from example 3.4 using first 6 and 7 coefficients. Compare results with that obtained in text employing 8 coefficients. Estimate yourself the accuracy of data by comparing results gained with different numbers of terms.

3.28 Solve the same problem 3.4 for radiating plate past one side by turbulent flow taking $\mathrm{Re} = 2 \cdot 10^6$. Compare the solution by series with numerical integration of integro-differential equation from [360] to see the relatively simplicity of series approach.

3.29 Derive equation (3.25) for total heat flux by integrating the first equation (3.16), and obtain the second equation (3.25) determining the ratio of total fluxes from plate in two cases. Hint: note that derivatives in expressions (3.16) are taken with respect to variable z defined by equation 3.11.

3.30   Think of and discuss with colleague the physical analysis of heat transfer character of two different flows in example 3.5 to understand the important role of temperature head variation in conjugate problems.

3.31   Show that two last relations (3.26) give proper values for $\overline{q}_0$ if the temperature head or heat flux is given at the trailing edge. Hint: find solutions of two problems similar to problem 3.6 with boundary conditions in the form of the second and third relation (3.26), respectively, and compare the results with the first expressions (3.26), knowing that all three solutions should be the same.

3.32   Derive expressions (3.27) from equation (3.10) following directions from the text. Hint: (i) take into account the dimensionless heat flux definition (3.26) and universal function (1.38), (ii) to transform equation (3.6) to variable z, present it as $z = z_L \zeta$, where z_L is a constant (see (3.11)).

3.33   Recall about function continuity to explain why the curve $\chi_t(z)$ for nonisothrmicity coefficient in Figure 3.7 becomes discontinuous. Hint: see Sec.2.4

3.34   Solve the problem from example 3.7 for laminar flow and parameters indicated on Figure 3.8. Compare both results. Hint: first solve the problem for turbulent flow to size the procedure of solution.

3.35   Solve the problem from example 3.8 for laminar flow and the same parameters as indicated in the text for turbulent flow. Compare both results. Hint: follow directions described for the turbulent flow case.

3.36   Obtain equation (3.28) and inequalities (3.30) following directions from the text.

3.37   Show that temperature distribution across the thin plate with relatively large thermal resistance is linear, and explain why from equality $q_{w1} = -q_{w2}$ it follows that the temperature distribution across the plate thickness is linear. Hint: use the Fourier law for the heat fluxes q_{w1} and q_{w2} at the plate top and bottom. See also text and Comment 1.1.

3.38   Repeat the deriving of inequality (3.33) and explain what quasi-steady approach means physically.

3.1.2   Conjugate Heat Transfer in Flows Past Bodies

As it indicated at the beginning of this chapter, the analysis of examples in this and next chapters includes problem formulation in its original form, model as basics equations, short description of solution, and most important results. Unlike the previous text, here, the discussion goes forward without detailed comments (offering only the most necessary short notices) and without exercises, referring a reader for further information to the cited original publications. To provide better reader orientation, examples are marketed by letters **a** (analytical) or **n** (numerical) showing at once what type of solution is employed.

■ **Example 3.9* n/a: Flow Past Rectangular Slab [412]**

A slab of finite dimensions with given bottom temperature T_{bt} is streamlined by incompressible flow. Because the slab is of finite length and height, the problem is governed by Navier-Stokes and energy equations for fluid, Laplace equation for a body (index s denotes solid) and following boundary conditions (S. 1.1)

$$\nabla^2 \psi = -\omega, \quad u\frac{\partial \omega}{\partial x} + v\frac{\partial \omega}{\partial y} = \frac{1}{Re}\nabla^2 \omega, \quad u\frac{\partial \theta}{\partial x} + v\frac{\partial \theta}{\partial y} = \frac{1}{Re}\nabla^2 \theta, \quad \nabla^2 \theta_s = 0$$

$$y = 0, \quad |x| > \frac{1}{2}, \quad \frac{\partial \theta}{\partial y} = 0, \quad x = \pm\frac{1}{2}, \quad -H \le y \le 0, \quad \frac{\partial \theta_s}{\partial x} = 0$$

$$y = -H, \quad |x| \le \frac{1}{2}, \quad \theta_{bt} = 1, \quad y \to \infty, \quad |x| \to \infty, \quad \omega \to 0 \quad (3.34)$$

Here, ψ and ω are the stream function and vorticity (S. 7.1.2.3), all variables are dimensionless and scaled by: a slab length L for x, a slab aspect $H = \hat{H}/L$ for y($\hat{H}$ is a slab height) U_∞ for u and v, LU_∞ for ψ, U_∞/L for ω and $\theta = (T - T_\infty)/(T_{bt} - T_\infty)$. The boundary conditions (3.34) stand for the first for uniform temperature (similar condition for uniform velocity $\partial^2 \psi / \partial y^2 = 0$) before and behind the slab, the second for insulated surfaces of the slab, the third for the given temperature θ_{bt} of the bottom slab surface, and the fourth for the flow far from a body, specifying that this flow is irrotational (S. 7.1. 2.5). Other conditions (not given here) are usual boundary and conjugate conditions on slab-fluid interface and far away from the body for a fluid (S. 7.2, Exam. 7.5).

The problem is solved by finite-difference method (S. 9.6). Because all equations (3.34) are elliptic, the boundary conditions at infinity before and behind the slab should be specified. The estimations conditions for ψ and ω are adopted from [325], whereas for θ these are derived employing the balance between the heat loss at the slab surface and that transported downstream by the flow. As a result, the following conditions are used:

for infinity before a slab: $\quad x = -x_\infty, 0 \le y \le y_\infty \quad$ and $\quad y = y_\infty, \quad -x_\infty \le x \le 0$

$$\theta = \omega = 0, \quad \psi \sim y + C_d[-1 + 1/\pi \tan(y/x)]$$

for infinity behind a slab: $\quad x = x_\infty, 0 \le y \le y_\infty \quad$ and $\quad y = y_\infty, 0 \le x \le x_\infty$

$$\theta \sim Cx^{-1/2} \exp[-Pe(y^2/4x)], \quad \psi \sim y + C_d[1/\pi \tan(y/x) - erf\,(yRe^{1/2}/2x^{1/2})$$

$$\omega \sim -C_d[yRe^{3/2}/2(\pi x)^{1/2}] \exp(-Rey^2/4x)$$

$$C_d = -\frac{1}{Re}\int_{-1/2}^{1/2} \omega(x,0)dx, \quad C = -\frac{1}{(\pi Pe)^{1/2}}\int_{-1/2}^{1/2} \frac{\partial \theta}{\partial y}(x,0)dx \quad (3.35)$$

Computations are performed for $10^2 \le Re \le 10^4$, $Pr = 10^{-2}, 10^2$, $\lambda_w/\lambda = 1, 2, 5, 20$, and two values of aspect ratio $H = 0.25$ and 1. The basic results are:

- The isotherms show that at lower Prandtl number $Pr = 10^{-2}$ and lower ratio λ_w/λ, the temperature drop across the slab is greater than that in the case of higher λ_w/λ. The isotherms indicate also that there is a kink at the boundary between the fluid and slab, which occurs due to a high slab conductivity. In the case of high Prandtl 10^2, the temperature drop across slab exists for both extreme values of ratio $\lambda_w/\lambda, = 1$ and 20.
- The solutions of Navier-Stokes equation for low Prandtl and Reynolds numbers $Pr = 10^{-2}$, $Re = 5 \cdot 10^2, 10^4$, both aspect ratio $H = 0.25, 1$, and all values of λ_w/λ show that the high value of local Nusselt number at the left-hand end of the boundary domain, first decreases

monotonically along most of the slab length and than increases remarkably reaching the other end of the slab. At the large Prandtl number 10^2 and the same values of the other parameters, the computations indicate that Nusselt number is approximately constant along the slab surface and increases as it approaches the right-hand end of a slab.

- It is seen from the results obtained for the low and large Prandtl numbers that increasing in Peclet number $Pe = Re\,Pr$ or in aspect ratio H decreases the boundary temperature, whereas an increase in conductivity ratio λ_w/λ leads to increase of boundary temperature.
- The dependences $\overline{Nu}(Re)$ indicate that average Nusselt number $(\overline{Nu})$ increases as usual with increasing Reynolds and Prandtl numbers and increases also as the aspect ratio and thermal conductivity ratio grow.
- The numerical results are compared for $Re \gg 1$ with data of two analytical approximate methods, and the range of applicability of both approaches is determined.

■ **Example 3.10* a: A Flush-Mounted Source on an Infinite Slab [377]**

The source is located at some distance from the leading edge of a slab. Therefore, the thermal boundary layer develops inside the dynamic boundary layer starting from the front edge of a heated strip. It is known (S. 7.7) that in such a case, the velocity distribution across the thermal boundary layer is close to linear $u = Cy$, where C is a constant. Because the axial conduction in fluid and in body is taken into account, the governing system consists of two elliptic equations and relevant boundary conditions

$$Cy\frac{\partial T}{\partial x} = \alpha\left(\frac{\partial^2 T}{\partial x^2} + \frac{\partial^2 T}{\partial y^2}\right), \qquad \frac{\partial^2 T_s}{\partial x^2} + \frac{\partial^2 T_s}{\partial y^2} = 0$$

$$T_s \to 0,\ |x_s| \to \infty, \quad T \to 0,\ x^2 + y^2 \to \infty, \quad \lambda\frac{\partial T}{\partial y}\bigg|_{y=0} = \lambda_w\frac{\partial T_s}{\partial y}\bigg|_{y=0},\ |x| > \frac{L}{2}, \quad \frac{\partial T_s}{\partial y}\bigg|_{y=-\Delta}$$

(3.36)

Here, L is the source length, the first condition relates to slab, the second pertains to fluid, the third is a conjugate condition for heated strip, and the last one denotes that the bottom slab is insulated.

The Fourier integral (S. 9.3.1) transforms the problem in the upper half-plane resulting in following subsidiary equations and boundary conditions

$$\frac{d^2\hat{T}_s}{dy^2} - \omega^2\hat{T}_s = 0, \qquad i\omega\,Cy\hat{T} = \alpha\left(\frac{d^2\hat{T}}{dy^2} - \omega^2\hat{T}\right)$$

$$\frac{d\hat{T}_s}{dy}(\omega, -\Delta) = 0, \quad \hat{T}_s(\omega,\,0) = \hat{T}(\omega,\,0) = \hat{T}_w(\omega) \tag{3.37}$$

The solution of the first equation is a simple task (Exam. 9. 9). The second equation (3.37) is transformed to Airy equation $\partial^2\hat{T}/\partial\zeta^2 - \zeta^2\hat{T} = 0$ by introducing a new variable $\zeta = (iP\omega)^{1/3}y + (iP)^{-2/3}\omega^{4/3}$ instead of y, where $P = Pe/L^2$ and $Pe = CL^2/\alpha$. Then, solutions of the first and the second equations (3.37) are

$$\hat{T}_s(\omega, y) = \hat{T}_w(\omega)\frac{\cosh[\omega(y+\Delta)]}{\cosh(\omega\Delta)}, \qquad \hat{T}(\omega, \zeta) = C_1(\omega)Ai(\zeta) + C_2(\xi)Bi(\zeta) \tag{3.38}$$

where $Ai(\zeta)$ and $Bi(\zeta)$ are Airy functions. Analysis [377] shows that $C_2(\omega) = 0$ leading after returning back to variable y and applying the last conditions (3.37) to solution

$$\hat{T}(\omega, y) = \hat{T}_w \frac{Ai[(iP\omega)^{1/3}y + (iP)^{-2/3}\omega^{4/3}]}{Ai[(iP)^{-2/3}\omega^{4/3}]} \qquad (3.39)$$

Both solutions for a body $\hat{T}_s(\omega, y)$ and a fluid $\hat{T}(\omega, y)$ contain unknown interface temperature $\hat{T}_w(\omega)$, which is defined by conjugate condition (3.36). Using this condition, authors obtain an equation for heat source distribution and straightforward transforms it in Fourier field as the first relation (3.40)

$$-\hat{q}(\omega) = \lambda \left.\frac{d\hat{T}}{dy}\right|_{y=0} - \lambda_w \left.\frac{d\hat{T}_s}{dy}\right|_{y=0}, \qquad \hat{T}_w = \frac{-\lambda(i\omega LPe)^{-1/3}\hat{q}(\omega)LAi}{\dfrac{Ai'[(iPe)^{-2/3}(\omega L)^{4/3}]}{Ai[(iPe)^{-2/3}(\omega L)^{4/3}]} - \omega L\lambda_w \tanh(\omega\Delta)}$$

$$(3.40)$$

Substituting equations (3.38) for body temperature $\hat{T}_s(\omega, y)$ and (3.39) for fluid temperature $\hat{T}_s(\omega, y)$ in that transformed condition for heat source, and solving the resulting equation for $\hat{T}_w(\omega)$ yields expression (3.40) for the interface temperature (Ai' is a derivative of the Airy function).

Comment 3.7 Airy functions (see Com. 3.13) are defined as follows:

$$Ai(x) = \frac{1}{\pi} \int_0^\infty \cos\left(\frac{\xi^3}{3} + x\xi\right) d\xi, \qquad Bi(x) = \frac{1}{\pi} \int_0^\infty \cos\left(-\frac{\xi^3}{3} + x\xi\right) + \sin\left(\frac{\xi^3}{3} + x\xi\right) d\xi$$

$$(3.41)$$

Because there is no known technique to inverse transformed expression (3.40), authors used the asymptotic series for large x (see a review of asymptotic solutions after example 3.19) and found the leading term of solution in physical field in the form

$$T_w(x) \cong \frac{3^{1/6}\Gamma(1/3)}{2\pi(x\lambda)^{2/3}(C\rho c_p)^{1/3}} \left(Q - \frac{2}{3}\frac{Q_{m1}}{x}\right) + O\left(x^{-2/3}\right) \qquad (3.42)$$

where Q and Q_{m1} are the total and the first moment amounts of heat released by the heat source. It is shown that this leading term can be physically classified into contribution from pure convection, from the interaction of convection and the conduction in solid, and from the interaction of convection and conduction in the fluid. It is also found that downstream of the heat source, the two leading terms come from pure convection. Some other characteristics of asymptotic solution are investigated. The asymptotic results are in good agreement with the author's numerical solution.

Comment 3.8 Signs $O(\varepsilon)$ and $o(\varepsilon)$ with letter "oh" are used to indicate the order of magnitude. For example, in series $\exp(-x) = 1 - x + x^2/2 - x^3/6 + \dots$ the order of magnitude of omitted terms is $o(x^3)$ or $O(x^4)$ since the next series term is $x^4/24$. The small "oh" tell us that omitting terms are not greater that the last one of retained terms, whereas the capital "oh" indicate that they are of the order of the next term of series.

■ Example 3.11a/n: Free Convection on Vertical and Horizontal Thin Plates [438]

The problem is governed by the following set of equations

$$u\frac{\partial u}{\partial x} + v\frac{\partial u}{\partial y} = -\frac{1}{\rho}\frac{\partial p}{\partial x} + \frac{\partial^2 u}{\partial y^2} + g\beta(T - T_\infty)\sin\phi$$

$$-\frac{\partial p}{\partial y} + g\beta(T - T_\infty)\cos\phi = 0, \qquad u\frac{\partial T}{\partial x} + v\frac{\partial T}{\partial y} = \frac{\partial^2 T}{\partial y^2} \qquad (3.43)$$

supplemented by continuity equation (1.9), conjugate conditions (1.18), and usual conditions for flow far away from the plate. The second equation (3.43) is the second boundary layer equation (S. 7.4.4.1), which as well as the last term in the first equation takes into account the natural convective effects (S. 7.8). For the horizontal plate $\phi = 0$, whereas for vertical plate $\phi = \pi/2$, and hence, in this case, the pressure gradient in both directions is zero. From conjugate conditions at a linear temperature distribution across a thin plate (S. 3.1.1.4.2), the Fourier law and scale analysis (7.4.4.1) show that the criterion ζ of a rate of conjugation in the case of free convection is given by first equation (3.44)

$$-\lambda\left(\frac{\partial T}{\partial y}\right)_{y=0} = \frac{\lambda_w(T_0 - T_w)}{\Delta}, \qquad \frac{T_0 - T_w}{T_w - T_\infty} \sim \frac{\lambda\Delta}{\lambda_w\delta} = \zeta \qquad (3.44)$$

where T_0 is the temperature of another plate surface and δ is the boundary layer thickness. Since the heat transfer coefficient $h \sim \lambda/\delta$, it is seen from the last equation that this criterion, like the others, in fact is the Biot number as it stated in Section 2.1.7.

It is assumed that the conjugate heat transfer may be investigated employing a combination of two well-known limiting cases of constant wall temperature and constant heat flux along the wall. Applying this idea and using formulae for boundary layer thickness in limiting cases on a vertical plate $\delta_T \sim x(\sigma \mathrm{Ra}_T)^{-1/4}$ and $\delta_q \sim x(\sigma \mathrm{Ra}_q)^{-1/5}$, where $\sigma = \mathrm{Pr}/(1 + \mathrm{Pr})$, the thickness for the conjugate free convection case is defined as

$$\delta \sim [\delta_T^4 + \delta_q^4] \sim x/\gamma, \qquad \gamma = [(\sigma\mathrm{Ra}_T)^{-1} + (\sigma\mathrm{Ra}_q)^{-4/5}]^{-1/4}$$

$$\mathrm{Ra}_T = g\beta(T_q - T_\infty)x^3/\alpha v, \qquad \mathrm{Ra}_T/\mathrm{Ra}_q = \lambda\Delta/\lambda_w x \qquad (3.45)$$

Introducing the new dependent f, θ and independent ξ, η similarity variables (S. 7.5.2)

$$f = \psi/\alpha\gamma, \quad \theta = [(T - T_\infty)/(T_0 - T_\infty)]\xi^{-1}, \quad \xi = [1 + \sigma\mathrm{Ra}_T/(\sigma\mathrm{Ra}_q)^{4/5}]^{-1}, \quad \eta = (y/x)\gamma$$
$$(3.46)$$

transforms the system of governing equations into set of ordinary differential equation

$$\mathrm{Pr}f''' + \frac{16 - \xi}{20}ff'' - \frac{6 - \xi}{10}f'^2 + (1 + \mathrm{Pr})\theta = \frac{1}{5}\xi(1 - \xi)\left[f'\frac{\partial f'}{\partial\xi} - f''\frac{\partial f}{\partial\xi}\right]$$

$$\theta'' + \frac{16 - \xi}{20}f\theta' - \frac{1 - \xi}{5}f'\theta = \frac{1}{5}\xi(1 - \xi)\left[f'\frac{\partial\theta}{\partial\xi} - \theta'\frac{\partial f}{\partial\xi}\right]$$

$$f(\xi,0) = f'(\xi,0) = f(\xi,\infty) = 0, \quad \xi\theta(\xi,0) - (1 - \xi)^{5/4}\theta'(\xi,0) = 1, \quad \theta(\xi,\infty) = 0$$
$$(3.47)$$

A similar consideration yields for horizontal plate an analogous system of equations with additional dimensionless terms with pressure scaled by $\rho a \nu \gamma^4 / \sigma x^2$

$$\mathrm{Pr}\, f''' + \frac{10-\xi}{15} ff'' - \frac{5-2\xi}{15} f'^2 + \frac{1}{15}(1+\mathrm{Pr})[(5+\xi)\eta p' - (10-4\xi)p] = \frac{1}{3}\xi(1-\xi)\left[f'\frac{\partial f'}{\partial \xi}\right.$$

$$\left. - f''\frac{\partial f}{\partial \xi} + (1+\mathrm{Pr})\frac{\partial p}{\partial \xi}\right], \qquad \theta'' + \frac{10-\xi}{15} f\theta' - \frac{1-\xi}{3} f'\theta = \frac{1}{3}\xi(1-\xi)\left[f'\frac{\partial \theta}{\partial \xi} - \theta'\frac{\partial f}{\partial \xi}\right]$$

$$\tag{3.48}$$

Here, $\theta = p'$ and the following correction in variables and boundary conditions are apt

$$\gamma = [(\sigma \mathrm{Ra}_T)^{-1} + (\sigma \mathrm{Ra}_q)^{-5/6}]^{-1/5}, \quad \xi = [1 + \sigma \mathrm{Ra}_T/(\sigma \mathrm{Ra}_q)^{5/6}]^{-1}$$

$$\xi\theta(\xi,0) - (1-\xi)^{6/5}\theta'(\xi,0) = 1, \qquad p(\xi,\infty) = 0 \tag{3.49}$$

These two sets of ordinary equations are solved by Keller's finite-difference method.

Comment 3.9 Keller's finite-difference method is an implicit numerical method in which initial second order differential equations are transformed by using new variables in the system of the first order differential equations. This method is of second order accuracy and is stable. More details may be found in [438] and [60].

The following results are formulated:

- The dimensionless velocity and interface temperature profiles for both vertical and horizontal plates are typical and as usually develop from the profiles of the constant wall heat flux at $\xi = 0$ to those of the constant wall temperature at $\xi = 1$.
- The interface temperature and friction increases as the conjugation criterion (3.44) ζ (local Biot number) decreases. Therewith, the interface temperature increases from T_∞ at $\zeta \to \infty$ $(x = 0)$ to T_0 at $\zeta = 0$ $(x/\Delta \to \infty)$.
- The heat transfer ratio q_w/q_{wq} decreases from this asymptotical value of the constant wall heat flux to another asymptotical value ratio q_w/q_{wT} of the constant wall temperature. This occurs due to the growing thermal boundary layer thickness.
- Comparison of the numerical data with well-known results for constant wall temperature or heat flux shows that the ranges of conjugation criterion in which the problem should be treated as conjugate depend on the Prandtl number. The analysis of such data reveals that this dependence is slight. Thus, these ranges varies from $\zeta = 0.087 - 25.5$ to $\zeta = 0.102 - 30.8$ for vertical plate and from $\zeta = 0.146 - 23.5$ to $\zeta = 0.159 - 30.8$ for horizontal plate for entire diapason of $\mathrm{Pr} = 0.001 - \infty$. The problems characterized by other values of criterion ζ can be solved by usual simple approach with error less than 5%.
- The local Nusselt number on the interface in the form Nu/γ decreases almost linearly with increasing ξ. This yields correlations for vertical $\mathrm{Nu}/\gamma = [1 - \xi\theta]/(1 - \xi)^{5/4}\theta$ and horizontal $\mathrm{Nu}/\gamma = [1 - \xi\theta]/(1 - \xi)^{6/5}\theta$ plates from which local temperature of the interface is simply obtained, and then, from relation $q_w = \mathrm{Nu}(\lambda/x)(T_0 - T_\infty)\xi\theta$, the local heat flux distribution on the interface can be estimated.

■ **Example 3.12a: Elliptical Cylinder in Laminar Flow [108]**

This problem is solved using the method of reducing a conjugate problem to a heat conduction problem. As explained in Section 2.2.2 in this case, the only conduction equation for the body is needed to be solved because the solution for fluid is given by the universal function in differential or integral form. Because of that, the governing system consists only of the two-dimensional Laplace equation (1.2) for cylinder with heat source q_v and two boundary conditions: the symmetry condition and the conjugate conditions (1.18) where q_w is the heat flux from the flow defined by universal function (1.40) and n is a normal to the cylinder surface in Fourier law

$$\frac{\partial^2 T}{\partial x^2} + \frac{\partial^2 T}{\partial x^2} + \frac{q_v}{\lambda_w} = 0, \qquad \frac{\partial T}{\partial y}\bigg|_{y=0} = 0, \qquad -\lambda\frac{\partial T}{\partial n}\bigg|_w = q_w \qquad (3.50)$$

This problem is convenient to consider in elliptic coordinates (u, v)

$$x = c\,chu\cos v \qquad y = c\,shu\,\sin v \qquad c = \sqrt{a^2 + b^2} \qquad (3.51)$$

were a and b are major and minor semi-axis. In coordinates (3.51), a half ellipse is mapped into a rectangle, one side of which corresponds to the surface of the semi-ellipse and three others pertain to the axes of symmetry. Then, relations (3.50) become

$$\frac{\partial^2 T}{\partial u^2} + \frac{\partial^2 T}{\partial v^2} = Q_v\left(sn^2u + \sin^2v\right), \qquad \frac{\partial\theta}{\partial v}\bigg|_{u=l} = \sqrt{1 - \frac{c^2}{a^2}\cos^2v}\,\frac{q_w a}{\lambda_w T_\infty}$$

$$\frac{\partial\theta}{\partial u}\bigg|_{u=0} = \frac{\partial\theta}{\partial v}\bigg|_{v=0} = \frac{\partial\theta}{\partial v}\bigg|_{u=\pi} = 0 \qquad (3.52)$$

Here, $\theta = (T - T_\infty)/T_\infty$, $Q_v = -q_v c^2/\lambda_w T_\infty$, and $(\partial\theta/\partial v)_{u-l}$ at $l = \ln[(a + b)/(a - b)]/2$ is the derivative $(\partial T/\partial n)_w$ from the last equation (3.50) in elliptical coordinates.

The solution of equation (3.52) subjected to the last three boundary conditions is found by separation of variables (S. 9.2)

$$\theta = \frac{Q_v}{8}(ch2u + \cos 2v) + N_0 + \sum_{k=1}^{\infty} N_k chk u\,\cos k v \qquad (3.53)$$

The constants N must be determined from the first boundary condition (3.52). Because the pressure gradient in flow around cylinder is variable, the integral universal function (1.40) depending on Görtler variable Φ (scaled by $U_\infty a$) in elliptical coordinates is used

$$q_w = \frac{\lambda Nu_* T_\infty}{a}\left\{\int_0^v f\left[\frac{\xi(\varepsilon)}{\Phi(v)}\right]\frac{d\theta_w}{d\varepsilon}d\varepsilon + \theta_w(0)\right\}$$

$$\Phi = \left(1 + \frac{b}{a}\right)(1 - \cos v), \qquad \xi = \left(1 + \frac{b}{a}\right)(1 - \cos\varepsilon) \qquad (3.54)$$

Here, ξ and ε are two dummy variables for Φ and v, respectively, (Exer. 1.17). The temperature head derivative with respect to variable v (or ε) containing in the relation (3.54) is determined by differentiating equation (3.53) and following setting $u = 1$ in order to eliminate the variable *u*. Substituting this result into equation (3.54) and denoting the integrals appear in the final outcome as J_k yield the expression for heat flux

$$\frac{d\theta_w}{d\varepsilon} = -\frac{Q_v}{4}\sin 2\varepsilon - \sum_{k=1}^{\infty} kN_k ch\, kl\, \sin k\varepsilon \qquad J_k = \int_0^v \left[1 - \left(\frac{1-\cos\varepsilon}{1-\cos v}\right)^{C_1}\right]^{-C_2} \sin k\varepsilon\, d\varepsilon$$

$$q_w = \frac{\lambda \mathrm{Nu}_* T_\infty}{a}\left[\frac{Q_v}{4}\left(ch^2 l - J_2\right) + N_0 + \sum_{k=1}^{\infty} N_k ch\, kl\,(1 - kJ_k)\right] \qquad (3.55)$$

Since exponent C_2 does not exceed 1, the integrals J_k, having the singularity at $\varepsilon = v$, converge. The equation for q_w determines the right-hand part of the first condition (3.52). The left-hand part of this condition is found from relation (3.53) by setting $u = l$

$$\left.\frac{\partial\theta}{\partial v}\right|_{u=l} = \frac{Q_v}{4}sh\, 2l + \sum_{k=1}^{\infty} kN_k shkl\, \cos kv \qquad (3.56)$$

Substituting this equation and equation (3.55) for q_w into the first condition (3.52), and combining the terms containing the same unknown coefficients N_k give an expression

$$\frac{sh\, 2l}{4} + \frac{\mathrm{Nu}_{**}\mathrm{Bi}}{4\sqrt{\mathrm{Re}}}\left(ch^2 l - J_2\right) = -\frac{\mathrm{Nu}_{**}\mathrm{Bi}}{\sqrt{\mathrm{Re}}}n_0$$

$$-\sum_{k=1}^{\infty} n_k\left[\frac{\mathrm{Nu}_{**}\mathrm{Bi}}{\sqrt{\mathrm{Re}}}ch\, kl\,\left(1 - kJ_k\right) + k\, sh\, kl\, \cos kv\right]$$

$$\mathrm{Bi} = \frac{\lambda\sqrt{\mathrm{Re}}}{\lambda_w} \qquad n_k = \frac{N_k}{Q_v} \qquad \mathrm{Nu}_{**} = \mathrm{Nu}_*\sqrt{1 - \frac{c^2}{a^2}\cos^2 v} \qquad (3.57)$$

which is used for the unknown coefficients n_k and N_k estimation. The Nusselt number Nu_* for isothermal surface containing in equation (3.57) was calculated by integral method (S. 7.6). Writing series (3.57) with the first $(k + 1)$ terms for $(k + 1)$ points in the interval $(0, \pi)$, one gets a system of linear algebraic equations defining the coefficients n_k and N_k required to calculate temperature head and heat flux distributions by equations (3.53) and (3.55), respectively. Calculations are performed with $k = 20$. Details may be found also in [119]. The following basic results are deduced:

- The Biot number mainly effects the distribution of the heat flux along the cylinder. Thus, at $\mathrm{Bi} = 1$, the heat flux is maximal in the region of the stagnation point, whereas at $\mathrm{Bi} = 10$, the maximum shifts to the central section of cylinder. This occurs because the disposition of minimal total thermal resistance depends on the Biot number.
- The temperature distributions along the cylinder obtained with and without conjugate effects differ significantly, especially at large Biot numbers when in some points the disagreement reaches 50%. At the same time, the heat flux distribution is less affected by the effect of conjugation which also increases as the Biot number grows.
- For $\mathrm{Bi} < 0.1$, the nonisothermicity coefficient is close to unity. For larger Biot numbers, the distribution of nonisothermicity coefficient along the cylinder becomes more complicated. Near the stagnation point, the values of χ_t is also close to unity. Then, when the distance

from stagnation point becomes larger, the nonisothrmicity coefficient grows, reaches the maximum and then falls steadily.

- Over the major part of the surface, with the exception of the end region, the larger Biot numbers correspond to the bigger nonisothermicity coefficients. In the end region, the situation is reversed: the values of χ_t decrease more rapidly for larger Biot numbers than for small Bi and finally becomes less than unity.

- The temperature head increases over the most of the surface, and because of that the effect of the nonisotermicity distribution is regular without features typical for decreasing temperature heads. Nevertheless, the maximum of the heat flux is 40–45% greater than that for isothermal surface. This maximum value takes place not as usually for Bi close to unity, but about Bi $= 10$. The reason of that is a special form of Biot number (3.57) used in this study.

■ **Example 3.13n: A Translating Fluid Sphere at Moderate Reynolds number [290]**

A fluid sphere of radius a and initial temperature T_0 falls at constant velocity U_∞ in infinite surrounding of another immiscible fluid at the same temperature. At the moment $t = 0$, the surrounding temperature changes stepwise from T_0 to T_∞.

In this study, the transient heat transfer between such sphere and its ambient fluid is considered under the following basic assumptions: (i) The sphere shape and size are invariable, and both flows outside and inside are steady and fully developed; (ii) The physical properties are constant and dissipation and buoyancy effects are neglected; (iii) There are no surface active forces and no rotation and oscillation of the sphere; (iv) The Reynolds numbers are moderate no higher than 50.

In contrast to boundary layer case at high (Re $\gg 1$) Reynolds numbers, when the inertia terms are the most important, and viscous terms may be neglected, at low (Re $\ll 1$) and moderate Reynolds numbers, the viscous terms play the dominant role, whereas inertia terms can be or omitted for the case of low Reynolds number or can be taken partly into account at moderate Reynolds numbers (S. 7.4). The mathematical model, which is formulated in spherical coordinates, consists of restricted equation similar to equation for creeping flow (S. 7.4.1) in terms of stream function

$$\nabla^4 \psi = \frac{\mathrm{Re}\sin\phi}{2} \left[\frac{\partial\psi}{\partial r} \frac{\partial}{\partial\phi} \left(\frac{E^2\psi}{r^2\sin^2\phi} \right) - \frac{\partial\psi}{\partial\phi} \frac{\partial}{\partial r} \left(\frac{E^2\psi}{r^2\sin^2\phi} \right) \right] = 0$$

$$\nabla^2 \psi = \frac{\partial^2\psi}{\partial r^2} + \frac{\sin\phi}{r^2} \frac{\partial}{\partial\phi} \left(\frac{1}{\sin\phi} \frac{\partial\psi}{\partial\phi} \right) \tag{3.58}$$

Equations (3.58) are used for both flows inside and outside of sphere. The velocities components and boundary conditions are defined as

$$u = \frac{1}{r^2\sin\phi} \frac{\partial\psi}{\partial\phi}, \qquad v = \frac{1}{r^2\sin\phi} \frac{\partial\psi}{\partial r}, \qquad \frac{\partial\psi}{\partial\phi} = 0,$$

$$\frac{\partial\psi_1}{\partial r} = \frac{\partial\psi_2}{\partial r}, \qquad \tau_{r\phi_1} = \tau_{r\phi_2}, \qquad \psi_2 = \frac{r^2\sin^2\phi}{2} \tag{3.59}$$

These conditions should be satisfied: the first at the sphere axis symmetry ($\phi = 0, \pi$) along with condition $\psi_1 = 0$, the second and the third together with condition $\psi_1 = \psi_2 = 0$, which are conjugate conditions at the sphere /ambience interface ($r = 1$), and the last one far away from sphere ($r \to \infty$). The variables are scaled by $U_\infty a$ for ψ, a for r, and Reynolds numbers are defined for sphere and surrounding as $Re_1 = 2\rho_\infty U_\infty a / \mu_1$ and $Re_2 = 2\rho_\infty U_\infty a / \mu_\infty$, respectively.

Equation (3.58) is solved using the series-truncation method [405]. The basic idea of his method is to determine the stream function and then the velocity components as a series of Legendre polynomials P_n in combination with so called radial functions $F_n(r)$

$$\psi = \sum_{n=1}^{\infty} F_n(r) \int_{\cos\phi}^{1} P_n(t)dt, \quad u = \sum_{n=1}^{\infty} \frac{F_n(r)}{r^2} P_n(\cos\phi), \quad v = -\sum_{n=1}^{\infty} \frac{F'_n}{r} \frac{P'_n(\cos\phi)}{n(n+1)} \quad (3.60)$$

These relations transform equation (3.58) into an infinite series of ordinary differential equations, which are truncated. The remaining ordinary equations are solved by the cubic finite-element method (S. 9.6). Details for $0.5 < Re_2 < 50$ can be found in paper [289].

The energy equations for the sphere and for the surrounding with the initial, boundary, and conjugate conditions are used as well in a special form introducing new variables $\varphi = \theta r$ for sphere and $\eta = 1/r$ for surrounding instead of θ and r, respectively

$$\frac{\alpha_2}{\alpha_1} \left[\frac{\partial\varphi}{\partial t_2} + \frac{Pe_2}{2} \left(u\frac{\partial\varphi}{\partial r} + \frac{\varphi}{r} \right) \right] = \frac{\partial^2\varphi}{\partial r^2} + \frac{\cos\phi}{r^2}\frac{\partial\varphi}{\partial\phi} + \frac{1}{r^2}\frac{\partial^2\varphi}{\partial\phi^2}$$

$$t = 0, \quad 0 < r < 1, \quad \theta = 1, \quad r = 0, \quad \varphi = 0, \quad \phi = 0, \quad \phi = \pi, \quad \partial\varphi/\partial\phi = 0 \quad (3.61)$$

$$\frac{\partial\theta}{\partial t_2} + \frac{Pe_2}{2} \left(-\eta^2 u\frac{\partial\theta}{\partial r} + \eta v\frac{\partial\theta}{\partial\phi} \right) = \eta^2 \left(\eta^2\frac{\partial^2\theta}{\partial\eta^2} + \cos\phi\frac{\partial\theta}{\partial\phi} + \frac{\partial^2\theta}{\partial\phi^2} \right)$$

$$t = 0, \quad 0 < \eta < 1, \quad \theta = 0, \quad r \to \infty, \quad \theta = 0 \quad (3.62)$$

$$r = 1, \quad \theta_1 = \theta_2, \quad \frac{\lambda_1}{\lambda_2} \left(\frac{\partial\varphi}{\partial r} - \frac{\varphi}{r} \right) + \frac{\partial\theta}{\partial\eta} = 0 \quad (3.63)$$

In equation (3.61), the variable φ is used because θ is undefined at the droplet center, whereas it is clear that $\varphi = 0$ when $r = 0$, $\theta = (T - T_\infty)/(T_{1,0} - T_\infty)$ is the temperature scaled by it initial value $T_{1,0}$, in contrast to time in equation (3.60), t_2 is time scaled by a^2/α_2, and $Pe_2 = 2aU_\infty/\alpha_2$ is the Peclet number. The second new variable η applied in relation (3.62) provides high density of nodes for numerical grid near the interface (S. 9.5). It is also seen that both types of variables the initial θ, r and the new ϕ, η are applied to formulate the conjugate conditions (3.63).

The system of equations (3.61)–(3.63) was solved numerically by alternating direction implicit method (ADI) [322].

Comment 3.10 Alternating direction method is a finite-difference approach for solving partial differential equations. The idea of ADI method is to split the finite differential equation into two others, with the x-derivative and y-derivative only, respectively. These simple equations may be solved, for example, using tridiagonal matrix algorithm (Com. 3.15).

The results listed below are obtained numerically for a range of Reynolds number $Re_2 = 2aU_\infty/\nu_\infty$ from 0 to 50, for both the viscosity μ_1/μ_2 and conductivity λ_1/λ_2 ratios from 0.333 to 3.0, the thermal diffusivity ratio $\alpha_1/\alpha_2 = 1$, and for $Pe_2 = 300$.

- The Nusselt number dependence from time at fixed other parameters shows that: (i) As Reynolds number increases the intensity of heat transfer considerable grows; this increasing is a result of strengthened circulation in both flows inside and outside of the sphere; (ii) The Nusselt number increases, and simultaneously, the fluctuating amplitudes decrease when the thermal conductivity ratio grows; both the increasing Nusselt number and decreasing the fluctuating amplitudes inside the droplet occur due to increased heat transfer from the surface to the center of the sphere; (iii) The increase of viscosity ratio μ_1/μ_2 leads to the lower Nusselt numbers since a higher viscosity causes the weaker internal circulation and yields a longer cycle of oscillation.
- In the limiting case of solid sphere ($\mu_1 \to \infty$), the dependence of heat transfer on Reynolds number is less strong than that in the case of a fluid sphere. The reasons of that are the zero velocity at the surface and absence of internal circulation for sphere. For another limiting case of a gas bubble ($\mu_1 \approx 0$), the internal circulation is maximal at the given conditions; in this case, the steady-state Nusselt number is almost independent on the Reynolds number.
- The results show that formula $1/Nu = \lfloor(\lambda_2/\lambda_1)(1/Nu_{in}) + (1/Nu_{ext})\rfloor$ obtained in [289] and in [360] reasonable predicts the asymptotic steady-state conjugate Nusselt number for a solid sphere at moderate Reynolds number in the case of $\alpha_1 = \alpha_2$.

Two other similar studies of unsteady heat transfer from solid and liquid spheres for low Reynolds number are published. The case of identical thermal properties in both inside and outside flows of sphere is investigated in [5]. The effect of variable ratio of volumetric heat capacities at equal thermal diffusivities is studied in [289].

■ **Example 3.14a: Radiating Plate with Heat Source in Laminar or Turbulent Flow [360]**

The heat flux on thin isolate at the bottom plate is determined by a sum of contributions of conduction (first term), heat source (second), and radiation (third)

$$q_w = \left(\lambda \frac{d^2 T_w}{dx^2} + q_v\right)\Delta - \varepsilon\sigma\left(T_w^4 - T_\infty^4\right) \tag{3.64}$$

Applying the Duhamel integral in form (1.23) and influence function (1.24), the authors derived a formula for temperature head–like known relation (1.69), which we used instead of their formula. Substituting heat flux q_w into relation (1.69) and applying dimensionless variables, leads to an expression defining the temperature head

$$\theta = 1 + C\frac{\lambda_w}{\lambda}\overline{x}^{-r}\int_0^x \left[1 - \left(\frac{\xi}{x}\right)^{C_1}\right]^{1-C_2}\left(\overline{\Delta}\frac{d^2\theta}{d\xi^2} - \frac{\theta^4}{\Lambda} + \frac{1+\overline{q}}{\Lambda}\right)d\xi$$

$$C = K\,Pr^{-n}\,Re^{-m}, \quad \overline{x} = \frac{x}{L}, \quad \overline{\Delta} = \frac{\Delta}{L}, \quad \theta = \frac{T}{T_\infty}, \quad \Lambda = \frac{\lambda_w}{\varepsilon\sigma T_\infty^3 L}, \quad \overline{q} = \frac{q_v\Delta}{\varepsilon\sigma T_\infty^4}$$

$$\tag{3.65}$$

Expending the influence function in binominal series and using corresponding constants: $K = 0.623, 3.32$, $n = 1/3, 3/5$, $m = 1/2, 4/5$, $r = 1/2, 4/5$, $C_1 = 3/4, 9/10$, and $C_2 = 1/3, 1/9$ for laminar and turbulent flows, respectively, gives computational expressions

$$\theta = 1 + C\frac{\lambda_w}{\lambda}\bar{x}^{-1/2}\left\{3.53\left(\frac{1+\bar{q}}{\Lambda}\right) + \sum_{i=1}^{j}\left(\frac{-d^2\theta_i}{\Delta \ d\xi^2} - \frac{\theta_i^4}{\Lambda}\right)\left[\frac{1}{j-1} + \frac{8}{21}\frac{i^{7/4} - (i-1)^{7/4}}{(j-1)^{7/4}}\right.\right.$$

$$\left.\left. + \frac{2i^{5/2} - i^{5/2}}{9(j-1)^{5/2}} + \cdots\right]\right\} \tag{3.66}$$

$$\theta = 1 + C\frac{\lambda_w}{\lambda}\bar{x}^{1/5}\left\{9.83\left(\frac{1+\bar{q}}{\Lambda}\right) + \sum_{i=1}^{j}\left(\frac{-d^2\theta_i}{\Delta \ d\xi^2} - \frac{\theta_i^4}{\Lambda}\right)\left[\frac{1}{j-1} + \frac{80}{171}\frac{i^{19/10} - (i-1)^{19/10}}{(j-1)^{19/10}}\right.\right.$$

$$\left.\left. + \frac{160i^{14/3} - i^{14/3}}{567(j-1)^{5/2}} + \cdots\right]\right\} \tag{3.67}$$

Solution of these equations is obtained by broken the interval of independent variable in small subintervals where the temperature head is considered as practically constant (see S. 1.3.1). Such procedure leads to system of linear algebraic equations for which there are standard methods of solution (Com. 3.15). Some basic conclusions are:

- The high value of plate conductivity (parameter Λ) results in an appreciable decrease of the plate temperature at small $\bar{x}$, but with increasing of $\bar{x}$ this effect reduces.
- Increasing internal heat generation or the plate thickness enlarging at low conductivity as well as reducing of the radiation part of heat (parameter $\bar{q}$) yields increasing of the plate temperature.
- Neglecting radiation in the case of turbulent flow does not cause as severe errors. The maximum error in plate temperature caused by neglecting radiation is about 45.5 % for laminar flow, whereas the corresponding error for turbulent flow is only 5.5%.

A similar problem was considered in [122] using linearization of integro-diffrential equation obtained from relation (3.64) after defining heat flux by Duhamel integral.

■ **Example 3.15n: Radiative Thin Plate in Laminar Compressible Boundary Layer [330]**

Unsteady heat transfer between a thin plate with insulated outer surface and back end and flowing over it a gray, absorbing, and scattering airflow is investigated. The plate is considered as thermally thin using equation (1.3), and the heat transport in airflow is treated as quasi-steady (S. 3.1.1.5) applying relevant energy equation. Thus, the model contains a system of equation for a radiative thin plate and a steady-state energy equation

$$\frac{\partial T_w}{\partial t} = \frac{\partial^2 T_w}{\partial x^2} - \frac{1}{\lambda_w \Delta}\left[-\lambda\frac{\partial T}{\partial y} + E\right]_{y=0},$$

$$\rho c_p\left(u\frac{\partial T}{\partial x} + v\frac{\partial T}{\partial y}\right) = \frac{\partial}{\partial y}\left(\lambda\frac{\partial T}{\partial y}\right) - \frac{\partial E}{\partial y} + \mu\left(\frac{\partial u}{\partial y}\right)^2 \tag{3.68}$$

This system is solved under usual no-slip on the plate and the asymptotic far away from the plate conditions. The conjugate conditions and radiation flux are defined as

$$
-\lambda_w \frac{\partial T_s}{\partial y}\Big|_{y=0} = \lambda \frac{\partial T}{\partial y}\Big|_{y=0} + E, \qquad E = 2\pi \int_{-1}^{1} I(y,\gamma)\gamma\,d\gamma \tag{3.69}
$$

where I is integral radiation intensity. The energy equations (3.68) for compressible flow is transformed to the incompressible form employing Dorodnizin's type variable η (S. 1.8), and then, the Blasius velocity profile (S.7.5.1.1) for incompressible fluid in terms of this variable $f(\eta)$ is used. The further simplification is achieved by applying the dimensionless variables presenting the governing equations and boundary conditions as:

$$
\frac{\partial^2 \theta}{\partial \eta^2} + \frac{f}{2}\frac{\partial \theta}{\partial \eta} - \Pr\xi\frac{\partial f}{\partial \eta}\frac{\partial \theta}{\partial \xi} - \frac{\text{Sk}}{\text{Re}}\tau_L\xi(1-\omega)(\theta^4 - \phi_I) + \Pr\text{Ec}\left(\frac{\partial^2 f}{\partial \eta^2}\right)^2 = 0
$$

$$
\frac{\partial \theta_w}{\partial t} = \frac{\partial^2 \theta_w}{\partial \xi^2} - \frac{\lambda L}{\lambda_w \Delta}\text{Sk}\, Q_w, \quad \phi_I(\tau) = 2\pi\int_{-1}^{1}\frac{I(\tau,\gamma)}{4\sigma T_\infty^4}\,d\gamma, \quad \phi(\tau) = \frac{E(\tau)}{4\sigma T_\infty^4} = 2\pi\int_{0}^{1}\frac{I(\tau,\gamma)}{4\sigma T_\infty^4}\gamma\,d\gamma
$$

$$
\eta = \left(\frac{\rho_\infty U_\infty}{\mu_\infty x}\right)^{1/2}\int_{0}^{y}\frac{\rho}{\rho_\infty}\,dy, \quad \text{Sk} = \frac{4\sigma T_\infty^4 L}{\lambda_\infty}, \quad \text{Ec} = \frac{U_\infty^2}{c_p T_\infty}, \quad Q_w = \frac{\text{Re}^{1/2}}{\text{Sk}\,\xi^{1/2}}\frac{\partial \theta}{\partial \eta}\Big|_{\eta=0} + \phi(0)
$$

$$
t = 0, \quad \xi = \xi_0, \quad \theta_w = \theta_{w0}, \quad \xi = \xi_L, \quad \partial\theta/\partial\xi = 0 \tag{3.70}
$$

Here, t is time scaled by L^2/α_w (Fourier number), $\theta = T/T_\infty$, $\xi = x/L$, Q_w is the total heat flux at the wall, Ec, Sk are Eckert and Starks numbers, $\phi(\tau)$ and $\phi_I(\tau)$ are the radiation flux and radiant energy density, $\tau = \tau_L(\xi/\text{Re})^{1/2}\eta$ and $\tau_L = bL$ are the optical depth of boundary layer in section ξ_L and the characteristic optical thickness, where b is the extinction coefficient and index L denotes the end of the plate section. The other notation are: ω is the single scattering albedo, σ is the Stefan-Boltzmann constant, $\theta_0(\eta)$ is the self-similar temperature profile for radiation free heat transfer case, and ξ_0 is the origin coordinate. To avoid the singularity at $\xi = 0$, a small part of a heated plate is assumed to be at ambient temperature.

Radiation flux $\phi(\tau)$ and radiant energy density $\varphi_I(\tau)$ are determined employing the equation of radiative transfer for emitting, absorbed, and scattering medium. The modified mean flux method [329] is applied to reduce the integro-differential equation for radiative heat transfer to a system of two ordinary nonlinear differential equations

$$
\frac{d(\varphi^+ - \varphi^-)}{d\tau} + (1-\omega)(m^+\varphi^+ - m^-\varphi^-) = (1-\omega)\,\theta^4,
$$

$$
\frac{d(m^+\delta^+\varphi^+ - m^-\delta^-\varphi^-)}{d\tau} + (1-\omega\hat\xi)(\varphi^+ - \varphi^-) = 0
$$

$$
\tau = 0, \quad \varphi^+ = \varepsilon\theta_w^4/4\theta + r\varphi^- \qquad \tau = \tau_\infty, \quad \varphi^- = \theta_\infty^4/4
$$

$$
m(\tau) = \int_{0}^{1} I(\tau,\gamma)\,d\gamma \Big/ \int_{0}^{1} I(\tau,\gamma)\gamma\,d\gamma, \quad \delta(\tau) = \int_{0}^{1} I(\tau,\gamma)\gamma^2\,d\gamma \Big/ \int_{0}^{1} I(\tau,\gamma)\,d\gamma,
$$

$$\hat{\zeta} = \frac{1}{2} \int_{-1}^{1} z(\zeta) \zeta d\zeta \qquad (3.71)$$

Here, $\tau_\infty = \tau_L (\xi/\text{Re})^{1/2} \eta_\infty$, η_∞ is the value of the external boundary layer edge, r and $\varepsilon = 1 - r$ are the reflectivity and the emissivity of the plate surface, $\hat{\zeta}$ is the mean cosine of the scattering angle, $z(\zeta)$ is the scattering indicatrix, and ζ is the cosine of the angle between incident and scattering beams. Expressions for ϕ, m and δ contained in relations (3.70) and (3.71) are denoted as ϕ^+, m^+ and δ^+, whereas the similar expressions for ϕ^-, m^- and δ^- differ from those only by integral limits, which are (-1 and 0) instead of (0 and 1). The radiation flux and radiant energy density contained in (3.70) are determined as $\phi = \phi^+ - \phi^-$ and $\phi_l = m^+ \phi^+ - m^- \phi^-$, where m and δ are the transfer coefficients.

Comment 3.11 Albedo is a ratio of reflected radiation from the surface to incident radiation upon it. Scattering indicatrix is an imaginary ellipsoidal surface representing the spatial distribution of scattering.

The system of equations (3.70) and (3.71) is solved iteratively applying the finite-difference approach. The energy equation (3.70) for fluid is solved jointly with system (3.71) assuming the value of θ_w to use the boundary conditions. Then, the total heat flux Q_w is calculated to solve the energy equation (3.71) for plate and to obtain the new value of θ_w. Usually three to five iterations are required to achieve the converging results. Details of numerical scheme and authors' mean flux method are given in [330] and more completely in the original article [329]. The calculations are performed for plate of length $L = 1\ m$ and thickness $\Delta = 0.01\ m$. The conjugate parameter $K = \lambda_\infty L / \lambda_w \Delta$ is varied from 0.01 to 10. The following basic results are formulated:

- In the case of transparent medium ($\tau_L = 0$), the spatial-temporal plate temperature distribution considerable depends on the conjugate parameter. The greater K, the faster the steady-state is achieved. Thermal radiation yields in higher level of temperature than that for a nonradiated plate, and as a result, the steady-state situation is achieved faster.
- The conjugate effects are more noticeable at small values of conjugation parameter. The results for different values of Reynolds and Starks numbers compare the intensity of the convective and radiative heating. In particular, it follows from these data that when $\text{Re}^{1/2}/.Sk < 0.1$, the convection heating component may be neglected.

Comment 3.12 The parameter K does not taken into account the flow thermal resistance (except λ_∞). Therefore, the lesser is that parameter, the greater is the influence of fluid resistance, which results in more noticeable conjugate effects. The more proper parameter is the Biot number, which takes into account both body and fluid thermal resistances (S. 2.1.7).

- The results for absorbing, emitting, and scattering medium for three values of optical thickness $\tau_L = 1$, 10 and 100 indicate that the time required for achieving the steady state increases as the optical thickness grows. The reason of this is the radiation flux diminution in the boundary layer. The effect of scattering is also more considerable at the small values of K. In the case of anisotropic scattering and a black wall, the maximum increase in plate temperature is noticed, whereas the minimum is obtained for a reflective surface and an isotropic scattering

- The viscous dissipation significantly affects the temperature distribution on a plate and influences the behavior of radiation component in the total heat flux. In the case of Ec = 0 when the viscous dissipation is ignored, the radiation flux monotonically decreases along the plate. In contrast, at the significant dissipation (Ec > 1), the behavior of radiation flux is complicated. When the Eckert numbers are appreciable, there are two cases with the maximum in radiation flux distribution along the plate, namely, at moderate thickness ($\tau_\infty \approx 1$) and thick ($\tau_\infty > 4$) optical regions. On the other hand, at the small optical thickness, the local radiation flux is almost constant. The described behavior at $\tau_\infty \approx 1$ should be most significant in the absorbing media ($\omega \approx 0$), whereas in a scattering media ($\omega \to 1$), this phenomenon seems to be only moderate.

- The data of this study indicate that the conjugate approach should be used in the radiative-convective heat transfer studies, especially in the cases of complex temperature distribution.

OTHER WORKS. A comprehensive review of studies of radiative-convective heat transfer from early works in the 1960s of the last century to current results, but basically without conduction effects is given in [418]. The review of early works considered coupling of free convection with conduction on the plate until the 1990s of the last century may be found in [438], analysis of results of interaction between a natural convection, radiation and conduction in enclosures obtained in the last 20–25 years is presented in [347], and a survey of forced heat transfer from flat plate starting from early studies to the end of the last century is given in [412].

Problems similar to those considered above are solved in [78, 156, 281]. In the first article, the problem of a heated small strip similar to the one reviewed in example 3.10 is solved using Green's function method (S. 9.4), in paper [281], the translating liquid drop (similar to that considered in Example 3.13) under electric field influence at low Peclet number is investigated analytically. More complicated problem for a plate of finite thickness, taking into account the temperature-dependent conductivity is solved semi-analytically and numerically in the paper [156] published in the last year. This study shows that the variable temperature distribution on the interface significantly affects the final results. The opposite situation is considered in another recently published article [235] that presents examples where the conjugate effects are negligible, and a common simple approach with isothermal heat transfer coefficient leads to satisfactory accurate solutions. Both results are in line with the general idea formulated in Section 2.2.1 and that is frequently underlined in this text that not each heat transfer problem should be considered as the conjugate one. Moreover, it is shown, as suggested in Section 2.2.1, that the approach of approximate estimation of error caused by a common simple method helps to understand whether or not the conjugate solution is required. Finally, the results of discussion of this topic including some recommendations are summarized at the end of Part I.

We close the review of other works by mentioning several studies of conjugate problems of the same type, which take into account additional effects or other specific conditions such as the rising of bubbles [211], the heat transfer between a jet and a slab [190], the unsteady heat transfer from blunt bodies in supersonic flow [320], an analytical analysis of heat transfer from rods by Galerkin's method (S. 9.6) [168], the heat transfer from cylinder at low Reynolds numbers [381], the heat transfer from a sphere heated by X-ray or laser [263], the turbulent free convection in rectangular enclosure [347], and unsteady free convection in vertical cylinder filled with a porous medium [349].

3.2 Internal Flows-Conjugate Heat Transfer in Pipes and Channels Flows

■ **Example 3.16n: Fully Developed Laminar Flow in a Pipe Heated at the Outer Surface** [31]

The governing system includes energy equations for the fluid, conduction equation for body, boundary, and conjugate conditions in cylindrical coordinates

$$u\frac{\partial \theta}{\partial x} = 4\left(\frac{\partial^2 \theta}{\partial r^2} + \frac{1}{r}\frac{\partial \theta}{\partial r}\right), \qquad \theta(0, r) = \frac{\partial \theta}{\partial r}(x, 0) = 0$$

$$4\left(\frac{\partial^2 \theta_s}{\partial r^2} + \frac{1}{r}\frac{\partial^2 \theta_s}{\partial r^2}\right) = \frac{1}{\mathrm{Pe}^2}\frac{\partial^2 \theta_s}{\partial x^2}, \qquad \frac{\partial \theta_s}{\partial x}(0, r) = \frac{\partial \theta_s}{\partial x}(L, r) = 0$$

$$\frac{\partial \theta_s}{\partial r}(x + 2\Delta) = \frac{\lambda}{\lambda_w(1 + 2\Delta)}, \qquad \theta_s(x, 1) = \theta(x, 1), \qquad q_w = \frac{\lambda_w}{\lambda}\frac{\partial \theta}{\partial r}(x, 1) \qquad (3.72)$$

Dimensionless variables are scaled using the pipe radius R, mean velocity U, initial Temperature T_e, and outer uniform hat flux q_0. Peclet number and temperatures are defined as: $\mathrm{Pe} = 2RU\rho c/\lambda$, $\theta = (T - T_e)\lambda/q_0 R$, $\theta_s = (T_s - T_e)\lambda_w/q_0 R$. The energy equation for fluid is solved employing Duhamel integral in the form similar to (1.69)

$$\theta_w - \theta_{bl} = \frac{2q_w(0)}{\mathrm{Nu}_{xq}(x)} + 2\int_0^x \frac{dq_w(\xi)}{d\xi}\frac{d\xi}{\mathrm{Nu}_{xq}(x - \xi)}$$

$$\theta_{bl} = 8\int_0^x q_w(\xi)d\xi, \qquad 2\frac{\partial \theta_w}{\partial r}(x, 1) = \frac{\lambda}{\lambda_w}(\theta_w - \theta_{bl})\mathrm{Nu}_x, \qquad (3.73)$$

whereas the conduction equation for the body is solved using finite-element method (S. 9.6). In relations (3.73), θ_{bl} is the bulk temperature, and Nu_{xq} is the Nusselt numbers for the case $q_w = \mathrm{const}$ for the flows in the ducts given in the book [343].

The system (3.72) is solved by iterations starting from guessing the temperature distribution and computing the left part of last equation (3.73), which is an auxiliary equation using to improve the iterations convergence. This gives the derivative $(\partial \theta_w / \partial r)_{r=1}$ on the fluid-body interface. Then, the procedure proceeds by: (i) solving the second equation (3.72) for body applying the finite-element approach and knowing the boundary condition for this equation to get the body temperature, (ii) computing the heat flux and the bulk temperature using the last equation (3.72) and the second equation (3.73), respectively, (iii) solving equation (3.73) for fluid to get the temperature on the interface in the second approximation. These results give data for the next iteration using again the last equation (3.72). The following basic results are formulated:

- An isothermal region is observed on the interface at the nearness of the inlet. Here, the wall-to-fluid temperature difference is such that the heat is transferred from the fluid to the wall. This is similar to what is indicated in other studies for the case of uniform $q_w =$ const. heating. An almost isothermal temperature exists also along the heated section in the case of higher conductivities, wall thicknesses, and for the lower lengths of the heated section and Peclet numbers. The plots for the same heated section length have a common point of intersection in the second part of this section that is shifted to the end of the pipe with increasing pipe length.

- In the initial part of the heated section, the heat flux decreases from very high values to nominal magnitude $q_w = 1$. Then, it varies in two ways. The first is when it reaches the minimum and then goes up back to the nominal value but dose not reach it. In this case, the wall temperature distribution is close to the one-dimensional, and the wall may be considered as thermally thin. In the other case, the heat flux decreases along the heated section monotonically so that the axial component of the temperature gradient decreases sharply from the outer to the inner wall surface. In that case, the wall should be seen as thermally thick because disregarding the wall conduction leads to large errors.
- The dependence of difference $\mathrm{Nu}_x - \mathrm{Nu}_{x*} = f(x)$ starts from zero near the inlet and close to the end decreases sharply, reaching the curve $\mathrm{Nu}_{xq} = f(x)$ in such a way that Nu_x always remains lower than Nu_{xq}
- The increasing Peclet number reduces the effect of axial conduction much more than the corresponding decreasing wall-to-fluid conductivity ratio or the wall thickness. The other way to reduce the effect of axial wall conductivity is to increase a pipe length.

■ Example 3.17a: Turbulent Flow in a Parallel Plates Duct at Periodical Inlet Temperature

The thermal development of hydrodynamically developed turbulent flow inside a channel at periodically variable inlet temperature is investigated. The original complex system of equations and boundary conditions is transformed using quasi-steady approach (S. 3.1.1.5) and assumption of thermally thin walls to simplify this periodic problem [155]

$$\bar{u}(\bar{r}) \frac{\partial \theta(\bar{r}, \bar{x})}{\partial \bar{x}} = \frac{\partial}{\partial \bar{r}} \left[\varepsilon(\bar{r}) \frac{\partial \theta(\bar{r}, \bar{x})}{\partial \bar{r}} \right] - i\bar{\omega}\theta(\bar{r}, \bar{x})$$

$$T(\bar{r}, \bar{x}, \bar{t}) - T_\infty / \Delta T = \theta(\bar{r}, \bar{x}) \exp(i\bar{\omega}\bar{t}), \quad \theta(\bar{r}, 0) = 1, \ 0 \leq \bar{r} \leq 1, \quad \frac{\partial \theta}{\partial \bar{r}}(0, \bar{x}) = 0, \ \bar{x} > 0$$

$$\frac{\partial \theta}{\partial \bar{r}}(1, \bar{x}) + \left(\mathrm{Bi} + \frac{\overline{\Delta i\bar{\omega}}}{\mathrm{Lu}} \right) \theta(1, \bar{x}) = b \frac{\partial^2 \theta}{\partial \bar{x}^2}(1, \bar{x})$$

$$\bar{x} = \frac{x}{r_1 \mathrm{Pe}}, \quad \bar{r} = \frac{r}{r_1}, \quad \bar{t} = \frac{\alpha t}{r_1^2}, \quad \bar{\Delta} = (r_2 - r_1)/r_1, \quad \bar{u} = \frac{u}{u_{\max}}, \quad \bar{\omega} = \frac{\omega r_1^2}{\alpha},$$

$$\mathrm{Pe} = \frac{u_{av} D_e}{\alpha}, \quad \mathrm{Bi} = \frac{h_\infty r_1}{\lambda}, \quad \mathrm{Lu} = \frac{\rho c}{\rho_w c_w}, \quad b = \frac{\bar{\Delta}\,\bar{u}^2 \lambda_w}{\mathrm{Pe}^2 \lambda}, \quad \varepsilon = 1 + \frac{v_{tb}}{\alpha}, \tag{3.74}$$

Here, r_1 and r_2 are distances of the inner and outer wall surfaces from the centerline, $D_e = 4r_1$ is equivalent diameter, h_∞ is ambient heat transfer coefficient, u_{av} is the average channel velocity, v_{tb} is the eddy viscosity for turbulent flow (S. 8.3.3), Lu is Luikov number (S. 3.1.1.5), and ΔT is amplitude of inlet temperature.

In deriving the system (3.74) some additional simplifications are used: (i) It is assumed that the solution has the same periodic exponential form as the variable inlet temperature and is presented by the second equation (3.74); (ii) The amplitude $\theta(\bar{r}, \bar{x})$ of this periodic solution is determined by quasi-steady simplified initial energy equation (3.74) for fluid subjected to usual boundary conditions given by the third and fourth equations (3.74); (iii) The conduction equation for walls from initial system is replaced by the last equation (3.74) derived under thermally thin walls assumption, which is used as the boundary condition on the outer channel surface for the simplified energy equation (3.74). As the result of these assumptions, the initial system of unsteady energy equation for fluid and conduction equation for channel walls with periodic inlet temperature is substituted by system (3.74) consisting of simplified energy equation for fluid with the usual boundary conditions for walls and a boundary condition on the outer surface of the channel.

The solution of this simplified problem strongly depends on the evaluation of the complex eigenvalues and eigenfunctions of the corresponding nonclassical Sturm-Liouville problem in the complex domain (S. 9.2.3). For such a problem, no known solution is available. To find an approximate solution, the generalized integral transform (Com. 9.2) is used by considering a subsidiary problem, which is related to the classical steady Graetz problem (Com. 3.14)

$$\frac{d}{d\bar{r}}\left[\varepsilon\left(\bar{r}\right)\frac{d\psi(\mu_i, \bar{r})}{d\bar{r}}\right] + \mu_i^2 \bar{u}(\bar{r})\, \psi(\mu_i, \bar{r}) = 0$$

$$\frac{d\psi(\mu_i, \bar{r})}{d\bar{r}} = 0, \quad \bar{r} = 0, \qquad \frac{d\psi(\mu_i, \bar{r})}{d\bar{r}} + \text{Bi}\psi(\mu_i, \bar{r}) = 0, \quad \bar{r} = 1 \qquad (3.75)$$

The eigenfunctions set of this system yields the pair of integral transform and inversion

$$\tilde{\theta}(\bar{x}) = \int_0^1 \bar{u}(\bar{r})\frac{\psi(\mu_i, \bar{r})}{N_i^{1/2}}\theta(\bar{r}, \bar{x})d\bar{r}, \quad \theta(\bar{r}, \bar{x}) = \sum_{i=1}^\infty \frac{\psi(\mu_i, \bar{r})}{N_i^{1/2}}\tilde{\theta}(\bar{r}), \quad N_i = \int_0^1 \bar{u}(\bar{r})\psi^2(\mu_i, \bar{r})d\bar{r}$$

$$(3.76)$$

where $\tilde{\theta}$ is transformed function (S. 9. 3). Transforming system (3.74) using equations (3.76) and performing tedious mathematical manipulations given in [155], authors present final results in the complex form in terms of amplitude $A_w(\bar{x})$ and phase $\varphi_w(\bar{x})$

$$\Theta_w(\bar{x}) = \frac{T(\bar{x}, \bar{r}, \bar{t}) - T_\infty}{\Delta T} = A_w(\bar{x})\exp[-i\phi_w(\bar{x})],$$

$$A_w(\bar{x}) = \{[\text{Re}\Theta_w(\bar{x})]^2 + [\text{Im}\Theta_w(\bar{x})]^2\}^{1/2}, \qquad \phi_w(\bar{x}) = \tan^{-1}\frac{\text{Im}\Theta_w(\bar{x})}{\text{Re}\Theta_w(\bar{x})} \qquad (3.77)$$

Numerical results show the effects of Biot and Luikov numbers and the wall conduction parameter b defined by notation (3.74):

- For the smallest values of Luikov number $\text{Lu} \approx 5 \cdot 10^{-4}$, the oscillations in the fluid temperature are dampened within a short distance from the duct inlet because of the larger thermal capacitance of the walls in comparison with that of the fluid. For the larger values of $\text{Lu} \approx 5 \cdot 10^{-3}$, the thermal wave penetrates further downstream due to the relatively smaller thermal capacitance of the walls and, consequently, a longer length required for storing the heat.

- The amplitudes of the wall temperature flattened when parameter b increases. This occurs due to improved heat diffusion along the wall, which becomes more intensive close to the inlet where the thermal gradients are larger. The amplitudes decay slower when the Reynolds number increases from 10^4 to 10^5. In the case of $b = 0$, the wall temperature amplitudes decay faster as the value of Luikov number Lu decreases. The similar effect of Lu on the amplitude of the bulk fluid temperature is observed, whereas the effect of axial wall conduction on bulk fluid temperature turns out to be little.
- Similar trends are found with increasing the Biot number, which yields decreases of the wall amplitudes as well because the external thermal resistance becomes smaller at larger Biot number. In this case, the smaller values of Lu lead to the larger heat flux amplitudes. This occurs due to significant attenuation in the wall temperature. On the other hand, at the larger Lu, the axial wall conduction yields increasing heat flux amplitudes. This effect is more pronounced at the inlet and is negligible for smaller Lu.
- These results are obtained in the case of fixed dimensionless frequency $\bar{\omega} = 0.1$ and indicates, in particular, that for systems of gases flowing inside metal walls, the effect of conjugation cannot be neglected in the regions close to inlet.

■ Example 3.18n: Laminar Flow in a Thick-Walled Parallel-Plate Channel with Moving Wall

The model contains of the system of equations: (i) the steady boundary layer energy equation (1.11) with the fully developed Poiseuille-Couette velocity profile u (S. 7.3.2) and the boundary conditions at the channel inlet ($x = 0$) and outlet ($x = L$), (ii) the two-dimensional Laplace equation (1.2) for the walls (here not shown), (iii) the thermal conditions at the walls as the constant temperature and heat flux at the upper ($y = H$) and lower ($y = -\Delta$) walls, respectively, and (iv) usual conjugate conditions (1.18) as equalities of temperatures and heat fluxes at the fluid/walls interfaces [164]

$$\rho c_p u \frac{\partial T}{\partial x} = \lambda \frac{\partial^2 T}{\partial y^2} + \mu \left(\frac{\partial u}{\partial y} \right)^2, \quad x = 0, \ T = T_e, \ \frac{\partial T_s}{\partial x} = 0,$$

$$x = L, \ \frac{\partial}{\partial x} \left(\frac{T_w - T}{T_w - T_m} \right) = 0, \ \frac{\partial T_s}{\partial x} = 0$$

$$u = [6U_m - 2U - (6U_m - 3U)(y/H)](y/H) \qquad y = H, \ T = T_e, \quad y = -\Delta, \ q_s = q_0 \quad (3.78)$$

Here, T_e is the entrance temperature, H is the channel height, T_m and U_m are the fluid mean temperature and velocity and T and U are the temperature and velocity of the moving surface. The fluid temperature field is computed using finite-difference method, whereas for the walls the Laplace equation is solved applying boundary-element approach. The advantages of such technique are discussed in Section 9.6. The energy equation for fluid is modified so as to cluster the grid points close to the both moving and fixed walls.

To conjugate both solutions, the iterative procedure is used. The calculation is started by guessing of the heat flux distribution on the interface and solving the Laplace equation to find the temperature distribution on the interface. Then, using these data, the energy equation for fluid is solved. The new heat flux distribution is estimated using the approximate equation $q^{n+1} = (\omega_s q_s^n + \omega q^n)/(\omega_s + \omega)$, where ω is a weighting functions (S. 9.6). By numerical experiments it is found that for relatively rapid convergence of interactions, the weighting functions ratio should satisfy the inequality $\omega/\omega_s < 0.03$.

The numerical calculations give the following results:

- As the conductivity ratio of the body/fluid increases from 80.6 to 8060, the interfacial temperature at the leading edge becomes higher and changes of the interfacial temperature become smaller. As this ratio becomes small, the wall temperature at the entrance approaches the entrance flow temperature and results in a greater change in the interfacial temperature. At the same time, the conductivity ratio very little affects the Nusselt number distribution.

- At the changes of dimensionless wall thickness from 0.0042 to 0.21, the observed effect is similar. When the wall thickness increases, the interfacial temperature becomes closer to the entrance fluid temperature, similar to that in the case of smaller conductivity ratio. In contrast, when the wall thickness decreases, the conjugate problem get closer to the case of constant heat flux with only difference in the temperature at the entrance region. The thin wall case shows a higher dimensionless temperature, which is caused by the heat storage in the wall despite the fact that it is thin.

- Both the interfacial temperature and heat flux are strongly influenced by the channel aspect ratio, which increases from 0.001 to 0.127. When this ratio is small, the fluid flow rate reduces, and the temperature increases more rapidly. In the case of large aspect ratio, the temperature changes slightly, but the Nusselt number varies greatly.

- The effect of the ratio of the free surface to the fluid mean velocities was studied for the values from 0.075 to 2.03. When the mean fluid velocity increases with unchanged free surface velocity, the changes in interfacial temperature are small. This is because in this case the free surface temperature is low. When the mean fluid velocity is low, the convective heat transfer in the cross-channel direction is high, but in the opposite case, this part of heat transfer is greatly reduced, and this causes the temperature to increase down the channel. This is confirmed by Nusselt number distribution because at the high velocity ratio, the low Nusselt number is observed.

■ Example 3.19* a: Hydrodynamically and Thermally Developed Flow in a Thick-Walled Pipe

This problem is solved using asymptotic series in eigenfunctions. This method is employed for solution of many problems that we have briefly reviewed below. In this case, the unsteady problem is considered for the beginning time period $t_{max} < x/2u_m$. This is very short period during which the first portion of flow passes only small entrance part of the channel or tube. Nevertheless, transient processes at this short time are important for startup, shutdown, or other off-normal surge of the thermal systems operations. Because in this case the longitudinal derivative can be neglected, the unsteady energy equation for fluid and conduction equation for a tube wall consist of only radials derivatives [288]

$$\frac{\partial T}{\partial t} = \alpha \frac{1}{r}\frac{\partial}{\partial r}\left(r\frac{\partial T}{\partial r}\right) - \mu\left(\frac{\partial v}{\partial r}\right)^2, \quad 0 < t < t_{max}, \quad 0 < r < R$$

$$\frac{\partial T_s}{\partial t} = \alpha_w \frac{1}{r}\frac{\partial}{\partial r}\left(r\frac{\partial T_s}{\partial r}\right), \quad R < r < R_0$$

$$t = 0,\ T = T_s = T_i, \qquad t > 0,\ r = 0,\ \frac{\partial T}{\partial r} = 0, \qquad r = R,\ T = T_s,\ \lambda \frac{\partial T}{\partial r} = \lambda_w \frac{\partial T_s}{\partial r}$$

$$r = R_0,\ q = -\lambda_w \frac{\partial T_s}{\partial r} \quad \text{or} \quad T_s = T_w \qquad (3.79)$$

In this system, R and R_0 are internal and external pipe radiuses, T_i is initial temperature defining the initial conditions, the next two conditions at $t > 0$ specified the boundary condition at the symmetry axis ($r = 0$) and the conjugate condition ($r = R$), and the last two conditions determine the heat flux or temperature imposed on the tube at $r = R_0$.

Using dimensionless variables, this system of equations is transformed into one equation with a single domain with discontinuous thermophisical properties and sources

$$\frac{\partial \theta}{\partial \bar{t}} = f(\bar{r}) \frac{1}{\bar{r}} \frac{\partial}{\partial \bar{r}} \left(\bar{r} \frac{\partial \theta}{\partial \bar{r}} \right) + \sigma(\bar{r}), \qquad f(\bar{r}) = \left(\frac{\alpha}{\alpha_w} \right), \qquad \sigma(\bar{r}) = \left(\begin{matrix} c\bar{r}^2 \\ 0 \end{matrix} \right) \quad \begin{matrix} 0 \le \bar{r} \le 1 \\ 1 \le \bar{r} \le R_0 \end{matrix}$$

$$\bar{t} = 0,\quad \theta = 0,\quad \bar{t} > 0,\ \bar{r} = 0,\ \frac{\partial \theta}{\partial \bar{r}} = 0,\quad \bar{r} = 1,\ \theta^- = \theta^+,\ \left(\frac{\partial \theta}{\partial \bar{r}} \right)^- = \frac{\lambda_w}{\lambda} \left(\frac{\partial \theta}{\partial \bar{r}} \right)^+$$

$$\bar{r} = R_0,\quad \frac{\partial \theta}{\partial \bar{r}} = -\frac{\lambda}{\lambda_w}, \quad (a)\ \theta = \frac{T - T_i}{T_w - T_R},\ c = \frac{16 \mu u_m}{\lambda(T_w - T_R)},$$

$$(b)\ \theta = \frac{\lambda(T - T_i)}{qR},\ c = \frac{16 \mu u_m}{qR}$$

$$(3.80)$$

In equations (3.80), $\bar{r} = r/R$, $\bar{t} = \alpha t/R^2$ (Fourier number), and T_R is a reference temperature. The third equation (3.80) determines the heat source $\sigma(\bar{r})$, which takes into account dissipation energy equals $c\bar{r}^2$ in fluid ($0 \le \bar{r} \le 1$) and zero in wall ($1 \le \bar{r} \le R_0$). The constant c is defined by velocity of developed flow $u = 2u_m(1 - r^2/R^2)$, and the different values of θ related to the last boundary conditions (3.79) on the outer wall when is prescribed a) temperature or b) heat flux. The second line in (3.80) presents in dimensionless variables conditions (3.79) given by third line of this system.

The eigenvalue problem (S. 9.2.3) associated with equation (3.80) is expressed by Bessel equation, which solutions G_1 for fluid ($0 \le \bar{r} \le 1$) and G_2 for wall ($1 \le \bar{r} \le R_0$) are given using the Bessel functions J_0 and Y_0:

$$f(\bar{r}) \frac{1}{\bar{r}} \frac{\partial}{\partial \bar{r}} \left(\bar{r} \frac{\partial G}{\partial \bar{r}} \right) = -g^2 G, \qquad \bar{r} = 0,\ \frac{\partial G}{\partial \bar{r}} = 0$$

$$\bar{r} = 1,\quad G^- = G^+, \quad \left(\frac{\partial G}{\partial \bar{r}} \right)^- = \frac{\lambda_w}{\lambda} \left(\frac{\partial G}{\partial \bar{r}} \right)^+, \qquad \bar{r} = R_0,\ \frac{\partial G}{\partial \bar{r}} = 0 \quad \text{or} \quad G = 0$$

$$G_1 = C_1 J_0(g\bar{r}) + C_2 Y_0(g\bar{r}), \qquad G_2 = C_3 J_0(g\bar{r}/\sqrt{\alpha_w/\alpha}) + C_4 Y_0(g\bar{r}/\sqrt{\alpha_w/\alpha}) \qquad (3.81)$$

where G is eigenfunction and $-g^2$ is a separation constant (like $-\mu^2$ in equations (9.7)).

Comment 3.13 Special functions that give solutions of particular differential equations usually are called by scholars first consider these equations. We encountered with Airy functions in Example 3.10. Bessel functions J_0 and Y_0 using in this example satisfied the considered here Bessel equation (3.81). Special functions are determined by series or integrals because they could not be expressed through elementary functions.

To satisfy the condition (3.81) for $\bar{r} = 0$, it must be $C_2 = 0$ in solution G_1, and the remaining three conditions (3.81) for $\bar{r} = 1$ and $\bar{r} = \bar{R}_0$ should be met to specify three other constants. To achieve this, the Sturm-Liouville problem (9.3.2) is investigated. The details of this complex procedure may be found in the paper [288]. As the result, the two temperature distribution $\theta(\bar{r}, \bar{t})$ along the interface are obtained for two boundary conditions (3.80) imposed at the outer wall (a) and (b)

$$\theta = \theta_w + \sum_{n=1}^{\infty} G_n(\bar{r}) \left[c_n \exp\left(-g_n^2 \bar{t}\right) + \frac{V_n}{g_n^2} \right]$$

$$\theta = W_0 \bar{t} - \frac{\lambda}{\lambda_w \bar{R}_0} \left(\frac{\bar{r}^2}{2} - \frac{1}{4} \right) + \sum_{n=1}^{\infty} G_n(\bar{r}) \left[c_n \exp\left(-g_n^2 \bar{t}\right) + \frac{W_n}{g_n^2} \right]$$

$$V_n = \frac{c \displaystyle\int_0^1 G_{1n}(\bar{r}) \bar{r}^3 d\bar{r}}{\displaystyle\int_0^1 G_{1n}(\bar{r}) \bar{r} d\bar{r} + \frac{1}{\mathrm{Lu}} \int_0^{\bar{R}_0} G_{2n}^2(\bar{r}) \bar{r} d\bar{r}},$$

$$W_n = \frac{\displaystyle\int_0^1 \left(c\bar{r}^2 + \frac{2\lambda}{\lambda_w \bar{R}_0} \right) G_{1n}(\bar{r}) \bar{r} d\bar{r} + \frac{1}{\mathrm{Lu}} \int_0^{\bar{R}_0} \frac{2\mathrm{Lu}}{\bar{R}_0} G_{2n}(\bar{r}) \bar{r} d\bar{r}}{\displaystyle\int_0^1 G_{1n}^2(\bar{r}) \bar{r} d\bar{r} + \frac{1}{\mathrm{Lu}} \int_0^{\bar{R}_0} G_{2n}^2(\bar{r}) \bar{r} d\bar{r}} \qquad (3.82)$$

In two equations for temperature, c_n are coefficients of generalized Fourier series (S. 9.2. 3) defined by formulae resulting from Sturm-Liouville analysis (like equation (9.24)).

The numerical examples are obtained using 50 terms of asymptotic series in eigenfunctions, which was enough to get data with four accurate digits.

The following basic results are obtained:

- The temperature decreases with increasing values of the conjugation parameter $\sqrt{\alpha_w/\alpha}/(\lambda_w/\lambda) = \sqrt{\mathrm{Lu}}$, where Lu is Luikov number. In the case (a), the temperature reaches the steady state; in the case (b), the temperature decreases with increasing the conjugation parameter and time as well, but does not reach the steady state.
- The greater the conjugation parameter is, the greater the effect of fluid properties, so that at a given time, the interface temperature becomes to be closer to the initial condition $\theta = 0$.
- The fluid temperature increases along the radial coordinate and with increasing the viscous dissipation.
- Despite the dissipation, in the case of heat extraction from the outer wall and of high value of conjugation parameter, the heat flux reversal occurs at the interface.

OTHER SOLUTION IN ASYMPTOTIC SERIES: Asymptotic solutions are obtained for small and large axial coordinates in various conjugate heat transfer problems. Luikov et al. in their early work [247] gave the solution of heat transfer problems for a plate in a compressible fluid and for flows in the channel and tube. They used the Fourier sine transform, reducing the system of governing equations to an integral equation of which a solution was presented by asymptotic series. Stein et al. [377] presented an asymptotic analysis of conjugate heat transfer from a flush-mounted source in an incompressible fluid flow with linear velocity distribution across the boundary layer (Example 3.10). Wang et al. [416] obtained an asymptotic solution of conjugate heat transfer between a laminar impinging jet and a disk with given temperature or heat flux distribution on its surface. The asymptotic conjugate analysis of drying process on a continuously moving porous plate was given by Grechannyy at al. [151, 152] who studied the effect of the temperature and concentration heads distribution on heat and mass transfer between the porous plate and flow entrained due to it moving (S.1.10). Several authors published asymptotic solutions of conjugate heat transfer in the channels and tubes. David and Gill [86] considered heat transfer between the Poiseuille-Couette flow and walls of parallel plates channel analyzing the effects of axial conduction by eigenfunctions series. The method of eigenfunction series is also used by Hickman [167] who obtained an asymptotic solution for the Nusselt-Graetz problem taken into account the effect of the wall but neglecting the axial conduction effects. Lee and Ju [222] solved a conjugate problem for a duct in the case of high Prandtl number, and Papoutsakis and Ramkrishna [302] presented a solution of a more complicated case of a low Prandtl number when the axial conduction in a fluid should be taken into account (see review in Exam. 4.9).

Comment 3.14 The solutions in asymptotic eigenfunctions series based on the expansion of Graetz method are efficient for high abscissas but converge slowly for small values of coordinate x. (Leo Graetz first considered such problem). In the next example, we present a solution of a conjugate heat transfer problem for a duct given by Pozzi and Lupo [313], which is efficient for small values of x.

■ **Example 3.20a: Fully Developed Laminar Flow in the Entrance of a Plane Duct [313]**

The goal of considering this article is to show a way of construction a solution that requires only few terms of the eigenfunctions series to get a satisfactory accuracy for small abscissas. The model contains two simplified equations for temperatures of thin walls of a duct and of fluid for fully developed flow with velocity profile $u = 1 - y^2$

$$T_s = T_w + (T_0 - T_w)\frac{y-1}{\overline{\Delta}}, \qquad (1 - y^2)\frac{\partial\theta}{\partial x} = \frac{\partial^2\theta}{\partial y^2}$$

$$\theta(0, y) = \left.\frac{\partial\theta}{\partial y}\right|_{y=0} = 0, \qquad \theta_w - 1 = -\Lambda\left.\frac{\partial\theta}{\partial y}\right|_{y=1} \qquad (3.83)$$

where $\theta = (T - T_i)/(T_0 - T_i)$, T_0 and T_i are outer wall and fluid inlet temperatures, R is the half-height of the duct, the reference lengths for dimensionless x and y are RPe and R, respectively, $Pe = u_{max}R/\alpha$, and $\Lambda = \lambda\Delta/\lambda_w R$ is a conjugate parameter. The first relation is written assuming the linear temperature distribution across the walls, and the second one, the energy equation, does not consist of the second inertia term because the flow is parallel. The first boundary condition (3.83) specifies the uniform temperature at the entrance, and the second formula presents the conjugate condition.

The Laplace transform (S. 9.3.2) of the set of equation (3.83) yields

$$\frac{\partial^2 \tilde{\theta}}{\partial y^2} - s(1 - y^2)\tilde{\theta} = 0, \quad \left.\frac{\partial \tilde{\theta}}{\partial y}\right|_{y=0} = 0, \quad \tilde{\theta}(s, 1) - \frac{1}{s} = -\Lambda \left.\frac{\partial \tilde{\theta}}{\partial y}\right|_{y=1} \tag{3.84}$$

The solution in Laplace field for temperature of fluid and then for interface at $y = 1$ are:

$$\tilde{\theta}(s, y) = \exp\left[\frac{(1 - y^2) i\sqrt{s}}{2}\right] \frac{F_1(s, y)}{F_1(s, 1)} \tilde{\theta}(s, 1)$$

$$\tilde{\theta}(s, 1) = \frac{F_1(s, 1)}{s}\left[\left(1 - i\sqrt{s}\Lambda\right) F_1(s, 1) + \Lambda\left(s + i\sqrt{s}\right) F_2(s, 1)\right] \tag{3.85}$$

where the two confluent hypergeometric functions $F(\gamma, \delta, x)$ are defined as (Com. 3.13)

$$F_1(s, y) = F\left(\frac{1 - i\sqrt{s}}{4}, \frac{1}{2}, i\sqrt{s}y^2\right), \quad F_2(s, y) = F\left(\frac{5 - i\sqrt{s}}{4}, \frac{3}{2}, i\sqrt{s}y^2\right) \tag{3.86}$$

After series of mathematical transformations and simplifications described in the paper [313], authors present the final interface temperature in physical space by two relations

$$\theta_w = \frac{a_1}{b_1}\left[\frac{1 - \Lambda x^{-1/3}}{a_1\Gamma(2/3)} + \frac{\Lambda^2 x^{-2/3}}{a_1^2\Gamma(1/3)}\right], \quad \theta_w = \frac{1}{\Lambda}\sum_{n=0}^{\infty} c_n \frac{x^{(1+n)/3}}{\Gamma[(4 + n)/3]} \tag{3.87}$$

The first relation is obtained using the asymptotic formula for hypergeometric function, and the second is a result of the inversion of series given solution in Laplace space in physical variables. Expressions (3.87) are valid for $\Lambda \leq 0.01$ and $\Lambda \geq 1$, respectively, with coefficients $a., b., c_n$ given in [313]. The fluid bulk temperature and the local Nusselt number are defined as:

$$\theta_{bl}(x) = \frac{3}{2\Lambda}\int_0^x [1 - \theta_w(\zeta)]d\zeta, \quad \text{Nu} = \frac{4(\partial\theta/\partial y)_{y=1}}{\theta_w - \theta_{bl}} \tag{3.88}$$

The accuracy of equations (3.87) for small and large Λ in a range of small dimensionless abscissa $0 \leq x \leq 0.001$ is shown in [313] by comparison of Nusselt numbers with the data adopted from [343], which was calculated using 120 terms of the asymptotic series.

■ Example 3.21n: Fully Developed Laminar Flow in a Horizontal Channel of Finite Length Heated From Below by Constant Heat Flux [71]

Because the channel is of finite length, the mathematical model consists of full two-dimensional continuity (1.4), Navier-Stokes (1.5), (1.6), and energy (1.8) (not simplified) equations. To take into account effects of pressure gradient and buoyancy forces (S. 7.8) arising due to heating from below, the additional terms in the form of sources are used: $S(x, y) = -dP/dx$ and $S(x, y) = -dP/dy + (Gr/Re^2)\theta$ for equations (1.5) and (1.6), respectively. Here, $\theta = (T - T_\infty)\lambda/q_w H$ and H is a channel height. For the walls, the two-dimensional Laplace equation (1.2) is considered.

The conjugate boundary conditions are formulated in conformity with the model, which authors used for experimental study. This setup has been insulated by the fiberglass and plexiglass on the top wall and two sections at the inlet and outlet of the bottom wall, whereas the heated section is located between these two insulated sections. The heat loss through insulated channel parts to the ambience is included in conjugate conditions by Biot number so that these conditions for top wall and bottom isolated and heated sections are:

$$(\partial \theta / \partial y)_{tp} = -\text{Bi}_{tp}\theta \quad (\partial \theta / \partial y)_{is} = \text{Bi}_{is}\theta / (\lambda_{is}/\lambda) \quad (\partial \theta / \partial y)_{ht} = -\lambda / \lambda_w \tag{3.89}$$

where $\text{Bi}_{tp} = 1.27$ and $\text{Bi}_{is} = 1.22$ are Biot numbers estimated by taken into account the total resistance of insulations and of natural convection from the walls to the surrounding.

The velocity and temperature conditions at the entrance and exit sections should be defined also since the channel length is finite (S. 7.2). It is assumed that the flow enters with parabolic velocity profile (S. 7.3.2) and with uniform ambient temperature. These assumptions are verified by measurements. The flow at the exit is quite complex due to recirculation and entrainment of ambient air. According to some known data, it is accepted that $\partial u / \partial x = v = 0, \theta = 0$ if $u \leq 0$ (inflow) and $\partial \theta / \partial x = 0$ if $u \geq 0$ (outflow).

A finite-volume technique with a uniform staggered grids (S. 9.7.1.1) and different mesh size for body and flow was employed. The successive over-relaxation code was used for computing the pressure (S. 9.7.1.1) and Tridiagonal matrix algorithm was applied for the solution of nonlinear coupled system of momentum, energy, and continuity equations. The grid density was increased until the two successive solutions of flow and thermal fields becoming different by less than 1%.

Comment 3.15 Tridiagonal matrix algorithm (TDMA) and successive over-relaxation method are approaches for solving linear systems of algebraic equations. The first is based on a tridiagonal matrix that has only three nonzero elements in main diagonal and in two first diagonals above and below it. A second approach is a variant of Gauss-Seidel method with relaxation factor improving the convergence of iterations (Com. 4.12).

The numerical results are in agreement with author's experiments. The basic conclusions are as follows:

- In the case of low Reynolds number (Re = 9.48), the buoyancy causes two rolls with axis of rotation perpendicular to the flow direction. The upstream roll produces a recirculation zone, whereas the downstream roll entrains flow from outside. As the Reynolds number increases to Re = 29.7, these rolls become smaller, and flow entrainment is reduced. It is observed that the intensity of these and transverse rolls as well as oscillatory motion and turbulence depend on the channel cross-section aspect ratio, flow, and heating rates. The study shows that the effect of the wall conductivity on the rolls location is small.
- The horizontal asymmetry of temperature profiles is caused by bulk flow through the channel. The heated region temperature is higher than that of insulation, which results in generation of thermal plume above the surface. In the case of aluminum heated region, the temperature distribution is highly uniform, which differs significantly from the case of ceramic heated region when the temperature uniformity is reduced due to poor thermal material diffusivity and to an increased heat transfer to the insulation.
- The comparison of the results obtained for conjugate and nonconjugante approaches demonstrates a significance of the conjugate modeling. This comparison is made for different heated region conductivities in the range corresponding to the materials such as plexiglass,

ceramics, stainless steel, and aluminum. In the case of uniform heated surface, the nonconjugate model predicts a highly nonuniform temperature profile. In contrast, the conjugate model gives the highly uniform temperature profile. This occurs not only due to the redistribution of the thermal energy by itself, but also to the increase of the thermal energy loss to the insulation. These results are confirmed by analysis of the effect of the conductivity of the heated and insulation regions on their average temperature predicted by nonconjugate and conjugate models.

- The numerical and experimental results show that the conjugate effects are significant and usually should be taken into account, except two cases: (i) for the thin walls, low insulation or high thermal conductivity of the heated regions, the surface can be considered as an isothermal one, and (ii) at a low heated region thermal conductivity, the heat transfer may be modeled as $q_w = $ const. process.

■ Example 3.22: Unsteady Heat Transfer in a Duct with Laminar Flow and Outside Heating

A fluid flowing inside the duct with a steady laminar, fully developed velocity profile is at initial temperature T_i when suddenly the outside of the duct wall is exposed to an ambient fluid at temperature T_0 and constant surface coefficient h_0. Under usual assumption, the governing equation for fluid and for a thin wall are [378]

$$\frac{\partial \theta}{\partial t} + 3\left(y - \frac{y^3}{2}\right)\frac{\partial \theta}{\partial x} = \frac{\partial^2 \theta}{\partial y^2}, \qquad \frac{h_0 R}{\lambda_0}(1 - \theta) = -\frac{\partial \theta}{\partial y} + \frac{\Delta}{\mathrm{Lu}}\frac{\partial \theta}{\partial t}$$

$$t = 0, \quad x \geq 0, \quad \theta = 0, \qquad y = 1, \quad \frac{\partial \theta}{\partial y} = 0 \qquad (3.90)$$

All variables are dimensionless and are scaled by: half channel height R for y and Δ, $R^2 u_m/\alpha$ for x, R^2/α for t, $\theta = (T - T_i)/(T_0 - T_i)$, and $\mathrm{Lu} = \rho c_p/\rho_w c_w$ is the Luikov number. The first equation (3.90) is the unsteady boundary layer energy equation for flow in a parallel plate duct with fully developed parabolic velocity profile, and the second one is the usual equation for thin plate with an additional last term taking into account the unsteady effects. At the same time, the last equation is the conjugate conditions, which becomes clear from dimension form of this equation (we use the same notations for dimension and dimensionless Δ and t) $h_0(T_0 - T) = -\lambda(\partial T/\partial y)_{y=0} + \rho_w c_{pw}\Delta(\partial T/\partial t)$.

The boundary conditions (3.90) stand for initial temperature at $t = 0$ and for symmetry condition on the central channel line at $y = 1$, respectively.

The problem is solved by using the finite-difference method. Details of numerical scheme and proof of consistency of the corresponding set of finite-difference equations are present in [378]. The accuracy of numerical results is checked by comparisons with some analytical solutions of similar problems.

The following results are obtained:

- The comparison of the results obtained for the limiting case $\mathrm{Lu} \to \infty$ and for a large, but finite value of Luikov number shows that the agreement between these two results at fixed magnitude of Lu depends on the value of parameter $h_0 R/\lambda_0$, which determines the value of the minimum time when the approximation $\mathrm{Lu} \to \infty$ is acceptable. As the parameter $h_0 R/\lambda_0$ increases, the values of the minimum time become smaller. Thus, for $h_0 R/\lambda_0 = 10$, the approximation is acceptable for $t > 0.2$, whereas for $h_0 R/\lambda_0 = 2$, it becomes acceptable only for times $t > 0.5$.

- Both wall and fluid temperatures increase monotonically to their state distribution in x for all values of Lu and $h_0 R/\lambda_0$. The greater the Luikov number is, the shorter is the time required to reach the steady-state condition. For fixed value of Lu, this time is shorter for the larger values of parameter $h_0 R/\lambda_0$.

- The behavior of the surface heat flux is strongly dependent on the value of Lu and is significantly different for Lu $\rightarrow \infty$ and for other finite values of Luikov number. For Lu $\rightarrow \infty$, the surface heat flux increases stepwise to its maximum value right at $t = 0$ and then monotonically decreases to its steady-state distribution. In contrast, for the finite values of Lu, the heat flux is zero at the beginning and then exhibits a quite complicated behavior. At some values of Lu, $h_0 R/\lambda_0$ and x, the heat flux increases from initial zero value to its maximum and then decreases below the steady-state condition approaching it from below. Such behavior occurs in the second time domain after transient period of time $t_{max} < x/2u_m$ (see Example 3.19). Similar effect was observed in the earlier works, from which the results are compared with the data obtained in this study.

- Higher values of parameter $h_0 R/\lambda_0$ lead to higher values of both wall and bulk fluid temperatures for all Lu, t and x. The heat flux is greater at higher values of Luikov number and at lower values of time. For the larger values of time, heat flux is greater for lower x but is lower at the larger x.

- The comparison between data obtained for conjugate model for Lu $\rightarrow \infty$ and simple usual nonconjugate approach with isothermal heat transfer coefficient shows that the large errors could result in such simple calculations. The analysis indicates that the results of simple approach and conjugate solution data would get closer as the parameter $h_0 R/\lambda_0$ or the Luikov number become smaller, as time gets larger in the first time domain or as x gets larger in the second time domain.

■ **Example 3.23n: Transient Heat Transfer in a Pipe with Surface Constant Temperature [219]**

The unsteady conjugate heat transfer in a circular tube with fully developed flow and imposed from outside constant temperature is studied. The heated section is located on the wall between two insulated parts at the entrance and exit of the tube, similar to that in Example 3.21. Both full two-dimensional equations for fluid and wall are used in the cylindrical coordinates with zero circumferential components (axisymmetric flow) as:

$$\frac{\partial \theta}{\partial t} + \text{Pe}(1 - y^2)\frac{\partial \theta}{\partial x} = \frac{1}{y}\frac{\partial}{\partial y}\left(y\frac{\partial \theta}{\partial y}\right) + \frac{\partial^2 \theta}{\partial x^2}, \qquad \frac{\alpha}{\alpha_w}\frac{\partial \theta_s}{\partial t} = \frac{1}{y}\frac{\partial}{\partial y}\left(y\frac{\partial \theta_s}{\partial y}\right) + \frac{\partial^2 \theta_s}{\partial x^2}$$

$$\theta_s(x, R_0, t) = 1, \qquad \frac{\partial \theta_s}{\partial y}(x, R_0, t) = 0 \qquad (3.91)$$

Here, the first boundary condition specifies the heated section temperature, whereas the second one determines the condition on the insulated sections. The dimensionless variables are scaled as: R_i for x and y, R_i^2/α for t, and $\theta = (T - T_e)/(T_0 - T_e)$, where R_i and R_0 are inside and outside radii, T_e is initial and inlet flow temperature, and T_0 is the external surface temperature. The other conditions are common: the conjugate conditions, uniform velocity and temperature inlet profiles, and the symmetry condition.

Equations (3.91) are solved using the finite-difference scheme of Patankar [306] computing procedure through one domain including both subdomains for fluid and wall, conjugate conditions on interface, and outer boundary conditions (S. 1.2.1 and 9.7.1.1). Detailed description of mesh discredization, and employing grids are given in [219].

The analysis of results yields the following conclusions:

- The basic characteristics: the bulk temperature, the interfacial heat flux, and the Nusselt number are calculated by following expressions

$$\theta_{bl} = \int_0^1 (y - y^3)\theta\, dy / \int_0^1 (y - y^3)\, dy, \qquad q_w = -\left.\frac{\partial\theta}{\partial y}\right|_{y=1}, \qquad \text{Nu} = \frac{2q_w}{\theta_w - \theta_{bl}} \qquad (3.92)$$

These characteristics depend on four dimensionless parameters: the ratio of the wall-fluid conductivities λ_w/λ, the ratio of radii R_0/R_i, Peclet number $\text{Pe} = 2R_i u_e/\alpha$ and ratio of the wall-fluid diffusivity α_w/α. During the initial transient time ($t < 0.016$), the interfacial heat flux increases stepwise from zero to the maximum value and then monotonically decreases to its steady-state value. Since at this period the heat is basically transported by radial conduction in the wall, the fluid temperature increases less than interfacial temperature. Therefore, the large wall-fluid temperature difference forms that result in a rapid increase in fluid temperature. For $t < 0.016$, q_w is relatively uniform across the heated section, but near the entrance and the exit of the heated section, the interfacial heat flux peaks are observed. After reaching the maximum, q_w decreases until it becomes steady-state conditions. Analysis shows that steady state is reached more quickly near the exit end than at other positions of the heated section. The negative values of q_w appear in the downstream region close the heated section and in the upstream region also close to the heated section but for the smaller value of α_w/α.
- A decrease in thermal resistance in the wall that corresponds to large values of λ_w/λ leads to higher interfacial temperatures. In the upstream region of the heated section, both the interfacial heat and preheating increase as the conductivity ratio grows.
- Thinner wall thicknesses correspond to smaller resistances and energy storage of the walls, and due to that the energy from the outside surface to the fluid becomes easier to transport. This causes a rapid response of the interfacial heat flux and shorter time for reaching the steady state. The time to reach the steady state becomes shorter also for decreasing the conductivity ratio. For the small thickness after early period, the convective effect becomes dominant, whereas for a thicker wall, the conduction remains dominant. The reverse heat transfer is observed downstream of the heated section.
- A lower thermal diffusivity ratio corresponds to a greater wall capacity and, hence, delays the increase of interfacial temperature, causing the increase of the time required to reach the steady state. In the early time period, the response of q_w is faster for the larger diffusivity ratio. Later, the convective heat transfer becomes dominant, and its effect increases with decreasing α_w/α.
- At larger Pe, the greater convection results in faster heat transportation from the wall to the fluid, and a decrease in Pe decreases the interfacial heat flux. According to the computation results, the preheating length in the upstream region decreases with an increase of the Peclet number. Conversely, the postheated length downstream of the directly heated portion is increased with Pe. The results show also that reverse heat transfer takes place earlier for

larger values of Pe. The time required to reach the steady state is shorter for increasing Peclet numbers.

OTHER WORKS: The same as in the last example conjugate problem with different conditions at the outer surface of the pipe or channel is considered in three other papers. In [234] the uniform heat flux is applied to the outer pipe surface instead of the uniform temperature imposed on the surface in [428]. In [429], the two cases of stepwise changes of temperature are considered: of the imposed temperature on the outer pipe surface in the first case and of ambient temperature at the outer channel wall in the second one. The mathematical models, solution approaches and basic conclusions in these papers are almost the same as in [219]. The reviews of heat transfer studies in pipes and channels including the solutions of conjugate problems one may find in [31] for relatively early results, in paper [429] for studies published until the end of the last century, and in [96] for more recent publications. Some conjugate heat transfer problems different from those discussed above may be mentioned. The conjugate heat transfer in compressible flows is investigated in [95], in thick-walled pipe of arbitrary cross-section with the fully developed flow is considered in [73] applying boundary-element approach, and the conjugate heat transfer in a rectangular channel at simultaneously developing velocity and temperature fields is analyzed in [12]. The conjugate heat transfer from radiating fluid in a rectangular channel is considered in [271]. The natural convective conjugate heat transfer in enclosures with openings and in square cavity is studied in [426] and [417] respectively, and the experimental results of conjugate heat transfer of water flow boiling in a channel are presented in [46]. Finally, we mention some examples of the latest publications. The finite element modeling is used in article [77] to investigate the heat transfer in flexible elastic tubes. The other recently appeared articles present the following research results: the interaction of surface radiation with mixed convection from vertical channel with heat sources [240] and with natural convection in enclosure with energy source [253], the influence of natural convection on the conjugate heat transfer characteristics in the minichannel of thermal storage system during the liquid material melting [252], the effect of axially varying or periodic heat transfer subjected at the outer wall of a duct on conjugate heat transfer characteristics [22], and the conjugate heat transfer from a sudden expansion investigated using nanofluid in [192].

4

Specific Applications of Conjugate Heat Transfer Models

In this chapter, the examples are considered shortly as well as the conjugate solutions are discussed in the second part of third chapter, beginning from Section 3.2. Some of the examples might demand additional physical and mathematical information that one could get from the third part of the textbook. The author hopes that this knowledge is enough to understand the majority of presented solutions, otherwise, any course of *Advanced Engineering Mathematics* that consists of extra theory and exercises will do. For additional information of fluid flow and heat transfer, a reader is referred to Schlichting's book "Boundary Layer Theory" [338] and for more specific knowledge of turbulent flow to Wilcox's monograph "Turbulence Modeling for CFD" [422].

4.1 Heat Exchangers and Finned Surfaces

4.1.1 Heat Exchange Between Two Fluids Separated by a Wall (Overall Heat Transfer Coefficient)

We begin from simple classical model of heat transfer between two fluids separated by thin plate. Recall that we analyzed this model in Example 2.11 when the errors arising by using the third kind boundary conditions are estimated. Here, we consider six conjugate solutions of this classical problem that simulate different real situations. All six models analyzed heat transfer between two fluids but differ from each other by separated wall-applying horizontal or vertical plates or a vertical tube with inside and outside flows. They differ also by regime and type (con- or counter-) of flows, mathematical models, and methods of solutions. The data obtained show how much the conjugate solutions are more reliable and physically grounded than a simple formula for overall heat transfer coefficient. In the next section, we compare data obtained for thin separated walls with the exact solution of the same problem with a thick separated wall to understand where the approximate results are acceptable and in what cases the correction should be done.

Applications of Mathematical Heat Transfer and Fluid Flow Models in Engineering and Medicine,
First Edition. Abram S. Dorfman.
© 2017 John Wiley & Sons Ltd. Published 2017 by John Wiley & Sons Ltd.

■**Example 4.1a: Two Concurrent Laminar Flows Separated by a Thin Plate [100, 103]**

For a thin plate ($\Delta/L \ll 1$) with thermal resistance greater than that of both liquids, the longitudinal conductivity of the plate is negligible (S. 3.1.1.4). In this case, the temperature distribution across the plate thickness is practically linear, and hence, the heat fluxes on both sides of the plate are equal resulting in two simple equations for heat fluxes (Fig. 4.1). These equations after using for heat fluxes the universal function (1.38) with two first derivatives only (S. 2.1.1.1) gives the following system of equations

$$-q_{w1} = q_{w2} = (\lambda_w/\Delta)(T_{w1} - T_{w2})$$

$$T_{w2} - T_{w1} = \frac{h_{*1}\Delta}{\lambda_w}\left(T_{w1} - T_{\infty 1} + g_{11}x\frac{dT_{w1}}{dx} + g_{21}x^2\frac{d^2T_{w1}}{dx^2}\right) \qquad (4.1)$$

$$-h_{*1}\left(T_{w1} - T_{\infty 1} + g_{11}x\frac{dT_{w1}}{dx} + g_{21}x^2\frac{d^2T_{w1}}{dx^2}\right)$$

$$= h_{*2}\left(T_{w2} - T_{2\infty} + g_{12}x\frac{dT_{w2}}{dx} + g_{22}x^2\frac{d^2T_{w2}}{dx^2}\right)$$

The second and the third equations (4.1) are found after substitution (1.38) into equations $-q_{w1} = (\lambda_w/\Delta)(T_{w1} - T_{w2})$ and $-q_{w1} = q_{w2}$, respectively, which follow from the first one.

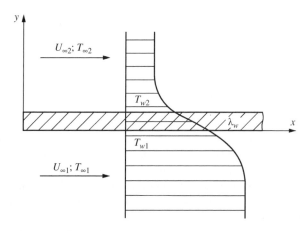

Figure 4.1   Scheme of heat transfer between two flows through a thin plate

To complete the model, the boundary conditions should be specified. Since equations (4.1) are of the second order, two conditions are needed. The first we get knowing that at the leading edge, the temperature of each side of the wall is equal to corresponding initial fluid temperature. The second condition is obtained taking into account that as the distance from the leading edge increases, the heat flux between fluids decreases due to growing boundary layer so that at $x \rightarrow \infty$, the heat flux reaches asymptotically zero. This leads to asymptotically decreasing temperature derivatives finally resulting in a constant temperature $T_{w\infty}$ of both fluids and the plate.

Setting temperature derivatives to zero in the second equation (4.1) yields the limiting value of this temperature, which we use to form dimensionless variables and boundary conditions

$$T_{w\infty} = \frac{\text{Bi}_{*1}T_{\infty 1} + \text{Bi}_{*2}T_{\infty 2}}{\text{Bi}_{*1} + \text{Bi}_{*2}}, \quad \text{Bi}_* = \frac{h_*\Delta}{\lambda_w}, \quad \theta = \frac{T_w - T_\infty}{T_{w\infty} - T_\infty} \tag{4.2}$$

$$x = 0, \quad \text{Bi}_* \to \infty, \quad \theta_1 = \theta_2 = 0$$

$$x \to \infty, \quad \text{Bi}_* \to 0, \quad \theta_1 = \theta_2 = 1, \quad \theta_1' = \theta_2' = \theta_1'' = \theta_2'' = 0$$

Analysis shows [103] that the system (4.1) in variables (4.2) for the case of equal coefficients for both fluids $g_{k1} = g_{k2}$ can be transformed into one equation containing only one parameter Bi_Σ that is a combination of Biot numbers Bi_{*1} and Bi_{*2}.

$$\theta(1 + \text{Bi}_\Sigma) + \hat{g}_1 \text{Bi}_\Sigma^2 \frac{d\theta}{d\text{Bi}_\Sigma} + \hat{g}_2 \text{Bi}_\Sigma^3 \frac{d^2\theta}{d\text{Bi}_\Sigma^2} = 1 \tag{4.3}$$

$$\text{Bi}_\Sigma = \frac{1}{1/\text{Bi}_{*1} + 1/\text{Bi}_{*2}}, \quad \text{Bi} = \frac{q_w\Delta}{\lambda_w(T_{\infty 2} - T_{\infty 1})} = 1 - \theta$$

In this case, the dimensionless fluid temperatures are equal $\theta_1 = \theta_2 = \theta$, and the total heat flux across the plate Bi (the overall heat transfer coefficient) is simple determined through the dimensionless temperature by the last equation (4.3).

The coefficients $\hat{g}_k$ appear in this equation depend on coefficient g_k of universal function (1.38) given in Figures 1.3 and 1.4. Calculations indicate that first coefficient $\hat{g}_1 \approx -0.53$ is practically the same for $\text{Pr} > 0.5$ and changes almost linearly to $\hat{g}_1 \approx -0.7$ in the range of Prandtl number $0.5 > \text{Pr} > 0.001$. The second one is $\hat{g}_2 = -0.11$ for whole diapason of Prandtl numbers [100, 119]. Therefore, equations (4.3) derived for equal coefficients $\hat{g}_k$ are valid if both fluids are the same, or their Prandtl numbers both are in the range $\text{Pr} > 0.5$. When both streams or one of them have $\text{Pr} < 0.5$, equations (4.3) and the results presented below are valid with small inaccuracy caused due to substitution coefficients $\hat{g}_1$ in the range $0.53 \div 0.7$ by their average value. More accurate solutions (4.3) for different coefficients $\hat{g}_k$ one may found in [100] or [103].

The ordinary differential equation (4.3) was solved in three approximations: the first one with only the first term in the left-hand side (boundary condition of the third kind), the second with the two first terms, and the third using equation with both derivatives. Equation (4.3) in the first approximation is a simple algebraic equation. The second approximation of that equation is an ordinary differential equation of the first order solved by standard method. These both solutions are as follows:

$$\theta = \frac{1}{1 + \text{Bi}_\Sigma}, \quad \theta = \left(-\frac{1}{\hat{g}_1\text{Bi}_\Sigma}\right)^{1/\hat{g}_1} \exp\left(\frac{1}{\hat{g}_1\text{Bi}_\Sigma}\right) \int\limits_0^{-1/\hat{g}_1\text{Bi}_\Sigma} \xi^{-1/\hat{g}_1} \exp(\xi)d\xi \tag{4.4}$$

For the limiting cases $\text{Pr} \to 0$ ($\hat{g}_1 = -1$) and $\text{Pr} \to \infty$ ($\hat{g}_1 = -1/2$) relation (4.4) simplifies to $\theta = 1 - \text{Bi}_\Sigma[1 - \exp(-1/\text{Bi}_\Sigma)]$ and $\theta = 1 - 0.5\,\text{Bi}_\Sigma^2[1 - \exp(-2/\text{Bi}_\Sigma)]$. The solution of a full equation (4.3) reveals that the second approximation (4.4) is satisfactory accurate [103].

As it should be, the first equation (4.4) along with last relation (4.3) for total heat flux yields the formula for overall heat transfer coefficient $1/\text{Bi} = 1 + 1/\text{Bi}_{*1} + 1/\text{Bi}_{*2}$, whereas the second conjugate solution (4.4) presents the refined data. These conjugate results are plotted in

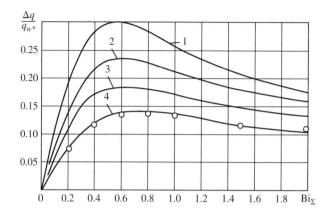

Figure 4.2   Effect of conjugation in heat exchange between two fluids through thin wall in laminar flow, 1 − Pr = 0, 2 − 0.01, 3 − 0.1, 4 − Pr → ∞, points-the third approximation

Figure 4.2 as function $\Delta q_w / q_{w*} = f(\mathrm{Bi}_\Sigma)$, where Δq_w is the difference between heat fluxes q_w and q_{w*} calculated with and without effects of conjugation, and where Bi_Σ is defined by the second formula (4.3).

It is seen that the maximal effect of conjugation: (i) is 15–25% for the medium to large (gases and liquids) and small (liquid metals) Prandtl numbers, respectively, (ii) takes place close to the case of equal thermal resistances of both fluids, at the range $\mathrm{Bi}_\Sigma = 0.5 - 0.7$, and (iii) is in line with approximate estimations of error caused by boundary conditions of the third kind obtained in Example 2.11 where it was noted that the effect of conjugation is moderate because the temperature head increases in flow direction on both sides of a plate. It was also estimated that the error of overall heat transfer coefficient for turbulent flow is about 10% for low Prandtl and Reynolds numbers and becomes smaller as Pr or Re increases.

■ Example 4.2a: Two Concurrent Laminar and Turbulent Flows Separated by a Thin Plate

The same problem was solved in [408] using Duhamel integral (1.23). The approach is similar to that applied in solving the heat transfer problem for radiation plate (Example 3.14) when the limits of integrals are broken up in small intervals assuming the constant temperature derivatives within each small interval (S.1.3.1). In this case, the substitution of equation (1.23) for heat fluxes into temperature distribution (4.1) across a plate as in Example 4.1 and using relation for an isothermal heat transfer coefficient $h_* = K\mathrm{Re}_x^m \mathrm{Pr}^n$ leads to equations similar to two relations (4.1) in the form

$$-\int_0^{\overline{x}_1} f\left(\frac{\xi}{\overline{x}_1}\right) \frac{d\theta_1}{d\xi} d\xi = H \frac{\overline{x}_1^{1-m_1}}{\overline{x}_2^{1-m_2}} \int_0^{\overline{x}_2} f\left(\frac{\xi}{\overline{x}_2}\right) \frac{d\theta_2}{d\xi} d\xi, \quad \theta_2(\overline{x}_2) - \theta_1(\overline{x}_1) = \frac{G}{\overline{x}_1^{1-m_1}} \int_0^{\overline{x}_1} f\left(\frac{\xi}{\overline{x}_1}\right) \frac{d\theta_1}{d\xi} d\xi$$

$$\theta = \frac{T - T_{\infty 2}}{T_{\infty 1} - T_{\infty 2}}, \quad \overline{x} = \frac{Ux}{\nu} = \mathrm{Re}_x, \quad H = \frac{K_2 \lambda_2 \mathrm{Pr}_2^{n_2} (U_2 \nu_1)^{m_2}}{K_1 \lambda_1 \mathrm{Pr}_1^{n_1} (U_1 \nu_2)^{m_1}}, \quad G = \frac{K_1 \lambda_1 U_1^{m_1} \mathrm{Pr}_1^{n_1}}{\lambda_w \nu_1^{m_1}}$$

$$(4.5)$$

In these equations, the coordinates on both sides of a plate are considered as $\bar{x}_2 = \bar{x}_1$ for concurrent flows and $\bar{x}_2 = \mathrm{Re}_{L1} - \bar{x}_1$ for countercurrent flows, where Re_{L1} is the Reynolds number at the end of a plate on the side one. The influence function in integrals (4.5) is defined in standard form (1.24) with exponents 3/4 and $(-1/3)$ and 9/10 and $(-1/9)$ for laminar and turbulent flows, respectively.

If the interval $0 < \bar{x} < \mathrm{Re}_{L1}$ is divided into N small subintervals $\Delta_{i,j}$ assuming constant values of $d\theta/dx = s_{i,k}$ on each of k subintervals, the temperature on this space is defined as $\Delta\theta_k = s_{i,k}\Delta_k$, and the sums of these products yield the temperatures θ_1 and θ_2 of both streams. Then, the equations (4.5) for the concurrent flows become

$$-\sum_{k=1}^{j} s_{1,k} \int_{\bar{x}_{1,k-\Delta_k}}^{\bar{x}_{1,k}} f\left(\frac{\xi}{\bar{x}_{1,j+1}}\right) d\xi = H \frac{\bar{x}_{1,j}^{-1-m_1}}{\bar{x}_{2,j+1}^{-1-m_2}} \sum_{1}^{j+1} s_{2,k} \int_{\bar{x}_{2,k-\Delta_k}}^{\bar{x}_{2,k}} f\left(\frac{\xi}{\bar{x}_{2,j+1}}\right) d\xi \quad \theta_1(\bar{x}_{1,j}) = 1 + \sum_{k=1}^{j} s_{1,k}\Delta_k$$

$$-\sum_{k=1}^{j+1} s_{2,k}\Delta_k - \sum_{1}^{j} s_{1,k}\Delta_k - 1 = \frac{G}{\bar{x}_1^{-1-m_1}} \int_{\bar{x}_{2,k-\Delta_k}}^{\bar{x}_{2,k}} f\left(\frac{\xi}{\bar{x}_{1,j}}\right) d\xi, \quad \theta_2(\bar{x}_{2,j+1}) = \sum_{k=1}^{j+1} s_{2,k}\Delta_k \quad (4.6)$$

The first and third equations (4.6) present a system of $2N$ linear algebraic equations specifying values of $s_{1,k}$ and $s_{2,k}$ required to calculate the flow temperatures by second equations (4.6). Calculations using standard approaches for solving systems of algebraic equations (Com. 3.14) are performed for laminar flow resulting in the following conclusions:

- The wall temperature distribution depends on the dimensionless parameters H (a measure of the relative heat transfer conductance of the streams), G (a measure of the conductance of the plate relative to that of the one of the streams), and coordinate $\bar{x}$.
- For any values of H, a decrease in G and/or an increase in $\bar{x} = \mathrm{Re}_x$ results in a limit of wall temperature $\theta_w = 1/(1 + H)$.
- For fixed values of H and G, the heat flux decreases as Re_{x1} increases because the boundary layer becomes thicker. As $\mathrm{Re}_x \to 0$, the heat flux approaches unity, which is a result of neglecting conduction in fluids.
- For given values of G and Re_x, an increase in H results in an increased heat flux. However, for large values of H this effect becomes significantly smaller.
- The neglecting of the actual wall temperature variation (conjugation effect) leads to 20–25% errors of order in comparison to that obtained for uniform wall temperature or uniform heat flux, which authors count as a serious inaccuracy.
- The slug approximation velocity profile (S. 7.7) may be used with reasonable accuracy when the dynamic boundary layer is relatively thin, that is, in the vicinity of the leading edge and for the large Biot numbers.

Comment 4.1 The heat transfer characteristics in this study are determined by three parameters $(H, G, \bar{x}_1)$, which in fact are defined by only two Biot numbers, as follows: $H = \mathrm{Bi}_2/\mathrm{Bi}_1$ and $\bar{x}_1 G = \mathrm{Bi}_1$. That might be expected because the Biot number is a general criteria for estimating resistances ratio of thermally interacting objects (S. 2.1.7).

■ **Example 4.3n: Two Countercurrent Laminar Flows Separated by a Thin Plate [354]**

This problem is considered assuming that the thermal resistance of a plate is negligible in comparison with that of both fluids. This assumption simplifies the problem that is more complicated than similar concurrent conjugate problems. At the same time, such a simplified problem retains the basic qualities of the same problem for the plate with finite resistance. If $\zeta = x/L$ is the dimensionless coordinate, the equality of heat fluxes at the positions ζ and $1 - \zeta$ for one and for another counter-flowing streams, respectively, leads to equation determining the interface temperature. In the case of laminar flow, when the universal function (1.38) with only two first derivatives is employed for heat fluxes, such equation is present in the following form [354]

$$[g_{22}\zeta^{3/2} + g_{12}\sigma(1-\zeta)^{3/2}]\theta'' + [g_{12}\zeta^{1/2} - g_{11}\sigma(1-\zeta)^{1/2}]\theta'$$
$$+ [\zeta^{-1/2} + \sigma(1-\zeta)^{-1/2}]\,\theta - \zeta^{-1/2} = 0 \qquad (4.7)$$

Here, $\theta = (T_w - T_{\infty 1})/(T_{\infty 2} - T_{\infty 1})$ and $\sigma = h_{*1}/h_{*2}$ is a ratio of the streams resistances, which may be viewed as a ratio of Biot numbers characterizing the conjugate effects. The indices 1 and 2 denote the streams for which the ending ($x = L$) and starting ($x = 0$) sections of a plate are, respectively, the initial in the countercurrent flows. Boundary conditions $\theta(0) = 1$ ($T_w = T_{\infty 2}$) and $\theta(1) = 0$ ($T_w = T_{\infty 1}$) one gets, knowing that at each end the plate temperature is equal to temperature of the fluid for which this end is an initial.

A linear differential equation (4.7) was integrated numerically for different values of σ starting from some small value $\zeta_i > 0$ by using two solutions satisfying the simple boundary conditions: $\vartheta_1(0) = \vartheta_1'(0) = \vartheta_2(0) = 1$ and $\vartheta_2'(0) = 0$ (similar to procedure in (S. 3.1.1.2) employed for chart creation) and presenting the final result satisfying the boundary condition for θ in terms of functions ϑ as:

$$\theta = \frac{\vartheta_2(1)\vartheta_1(\zeta) - \vartheta_1(1)\vartheta_2(\zeta)}{\vartheta_2(1) - \vartheta_1(1)} \qquad (4.8)$$

The initial values of θ for $\zeta \leq \zeta_i$ are calculated by series (3.3) using first six coefficients: $a_0 = 1$, $a_1 = a_3 = 0$, $a_2 = \vartheta'(0)$, $a_4 = (a_2 g_{11} - 1)/2g_{21}$, and $a_5 = -4a_2(1 + g_{12})/g_{21}\sigma$.

Calculation data obtained for Pr > 0.5, for which coefficients g_k are almost independent of Prandtl number, are plotted in Figure 4.3. These indicate that in the case of countercurrent flows:

• The temperature along the interface changes significantly (Fig. 4.3a). In the case of equal thermal resistances ($\sigma = 1$) when the effect of conjugation is maximal, the heat flux is about 30% bigger than that calculated by ignoring the temperature variation along the interface obtained in conjugate solution.
• The temperature head grows in flow direction, and therefore, the distributions of heat transfer coefficient does not have singularities and differs not as much from an isothermal one (Fig. 4.3b) as in the case of decreasing temperature head.
• Nevertheless, the effect of conjugation in this case is more significant than that in the case of concurrent flows when a zero resistance plate is isothermal (see Equation (4.3) at

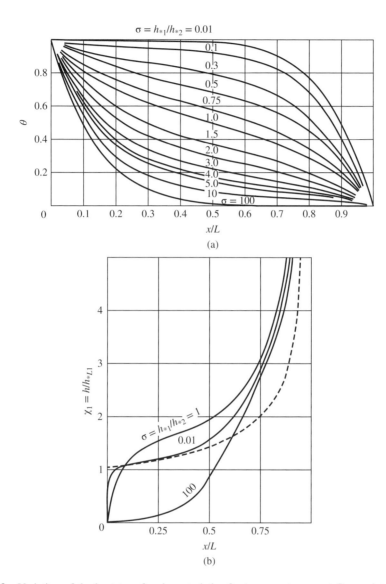

Figure 4.3 Variation of the heat transfer characteristics for two countercurrent flows: a) temperature head, b) heat transfer coefficient, - - - - for the case of isothermal plate

$Bi_* = 0$). Therefore, the effect of conjugation in the countercurrent flows should be taken into account.

- The results at the ends of streams interaction obtained in the conjugate and in traditional approaches differ in essence because (Fig. 4.4): (i) In the traditional analysis, the interface is isothermal, and due to that, the resistance of the stream that starts at each end is zero ($h_*|_{x=0} \rightarrow \infty$). Therefore, the heat transfer coefficient at the beginning point is equal to that

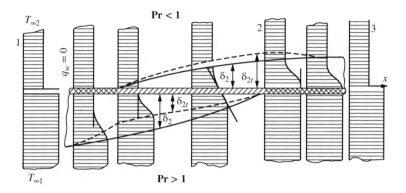

Figure 4.4   Scheme of the temperature profiles deformation in countercurrent heat transfer

of another interacting stream, and hence, the heat fluxes at the ends in this analysis are finite. (ii) In contrast to that, the conjugate procedure shows that the heat fluxes at the ends of both contacting streams due to conjugation equals each other and both are zero because the temperature head of starting stream according to the first series coefficient ($a_1 = 1$) goes to zero as x, whereas an isothermal coefficient at this point tends to infinity $x^{1/2}$ resulting in heat flux $q_w = \theta h_*$ which goes to zero as $x^{1/2}$ as $x \to 0$.

- The stream on the other side of interface reaches at the end the insulated plate or inactive flows interface, and its temperature profile deforms in the same way as in the case when flow impinges on an adiabatic wall. The scheme of such deformation is plotted in Figure 4.4 where one sees the gradually changes of streams temperature profiles. We considered this phenomenon in Section 2.1.4.3 showing that physically this process occurs because the heat flux drops abruptly to zero as flow reaches an insulated surface, whereas the temperature head decreases gradually, becoming practically zero only at certain distance from the entrance point.

Comment 4.2   In reality, the heat fluxes at the ends are not zero, due to finite boundary layer thickness. In consequence, the described deformation process starts not at the edges rather on some small distance from them.

■ Example 4.4a/n: Two Quiescent Fluids Separated by a Vertical Thin Plate [409]

It is considered a reservoir with warm and cold water separated by a thin vertical wall. Because in such system the density gradients are directed down and up at both sides of a wall, the two different oriented natural convection flows arose (like in countercurrent streams). This model simulates processes in the real systems including heat exchange equipment, cooling of electronic devises, and others containing the fluid-filled containers.

Although the heat transfer between such two boundary layers streams is similar to others considered above, the analysis in the case of natural convection is more complex due to non-linearity associated with coupled dynamic and energy equations (S. 7.8). In particular, the Duhamel integral is not applicable for nonlinear systems (S. 1.3.1).

In the reviewed study, the heat fluxes are estimated by formula derived in [316]

$$\overline{q}_w = \frac{q_w \lambda_w}{(T_{\infty 1} - T_{\infty 2})\Delta} = C(\text{Pr}) \frac{\lambda \Delta}{\lambda_w L} \text{Ra}^{1/4} \theta_w^{5/3}(\overline{x}) \left[\int_0^{\overline{x}} \theta_w^{5/3}(\xi) \, d\xi \right]^{-1/4} \tag{4.9}$$

where $\theta_w = (T_w - T_\infty)/(T_{\infty 1} - T_{\infty 2})$ and $C(\text{Pr})$ is a tabulated function. Using this formula and the first equation (4.1) for the temperature distribution across thin plate as in former examples yields a system of two relations similar to second and third equations (4.1)

$$C_1 \frac{\lambda_1}{\lambda_w} \text{Ra}_1^{1/4} \theta_{w1}^{5/3}(\overline{x}_1) \left[\int_0^{\overline{x}_1} \theta_{w1}^{5/3}(\xi) \, d\xi \right]^{-1/4} = C_2 \frac{\lambda_2}{\lambda_w} \text{Ra}_2^{1/4} \theta_{w2}^{5/3}(\overline{x}_2) \left[\int_0^{\overline{x}_2} \theta_{w2}^{5/3}(\xi) \, d\xi \right]^{-1/4}$$

$$1 - \theta_{w1}(x_1) + \theta_{w2}(x_2) = C(\text{Pr}_1) \frac{\lambda_1 \Delta}{\lambda_w L} \text{Ra}_1^{1/4} \theta_{w1}^{5/3}(\overline{x}_1) \left[\int_0^{\overline{x}_1} \theta_{w1}^{5/3}(\xi) \, d\xi \right]^{-1/4} \tag{4.10}$$

Here, $\overline{x} = x/L$, $\overline{x}_2 = 1 - \overline{x}_1$ (like in countercurrent flows).

Comment 4.3 The formula (4.9) is a result gained in [316] as approximation of a typical for natural convection temperature distributions curve by two straight lines presenting a linear and constant temperature variations at its initial and terminal parts, respectively.

Equations (4.10) are solved numerically by iterations using 40 intervals for $\overline{x}$ between 0 and 1. Results computed are in reasonable agreement with author's data obtained experimentally and show that:

- The plate temperatures depend on two parameters $H = (\lambda_1/\lambda_w)(\Delta/L)\text{Ra}_1^{1/4}$ and $G = (\lambda_1/\lambda_2)(\Delta/L)(\text{Ra}_1/\text{Ra}_2)^{1/4}$ determining ratios of thermal resistances of one fluid and wall and of both fluids, respectively.
- When the thermal resistances of both fluids are equal, the sharp temperature variations are observed near the top and bottom of the plate. In this case the heat flux is symmetrical about the midpoint of a plate and practically constant over about 80% of it.
- The more the parameter G departs from unity, the greater is the asymmetry of the heat fluxes due to increasing difference of both fluid thermal resistances.
- The effect of conjugation of two natural streams is moderate. The heat transfer coefficients are about 12% higher than obtained using a constant wall temperature.

■ Example 4.5a/n: Two Flows Separated by a Vertical Thin Plate [239]

This problem is similar to the previous one, but the flowing fluids along the plate with zero resistance are considered instead of quiescent fluids in former example. The governing system contains: continuity (1.9), dynamic (1.10), and thermal (1.11) boundary layer steady-state equations without pressure gradient and dissipation terms and with additional term taking into account the natural convection effects (S. 7.8). The problem is solved using so called local similarity approach when the system of equations is modified using similar variables (S. 7.5.2) to the following form

$$\Pr F''' + FF'' - \frac{2}{3}F'^2 + (1 + \Pr)\Theta = \frac{4}{3}\xi\left(F'\frac{\partial F}{\partial \xi} - F''\frac{\partial F}{\partial \xi}\right), \quad F(\xi, \eta) = \frac{\psi(1 + 1/\Pr)^{1/4}}{(4\xi/3)^{3/4}}$$

$$\Theta'' + F\Theta' = \frac{4}{3}\xi\left(F'\frac{\partial \Theta}{\partial \xi} - \Theta'\frac{\partial F}{\partial \xi}\right), \quad \Theta(\xi, \zeta) = \frac{|T - T_\infty|}{T_{\infty 1} - T_{\infty 2}} \tag{4.11}$$

These equations in similar variables $\xi = x$, $\zeta = y/(4x/3)^{1/4}$, $\eta = \zeta(1 + 1/\Pr)^{1/4}$ refer to both fluids, for which $T_{\infty 1} > T_{\infty 2}$ and $x_2 = 1 - x_1$. The boundary conditions are simple: $\eta = 0$, $F = F' = 0$, $\eta = \infty$, $F' = \Theta = 0$, whereas conjugate conditions are more complicated

$$\frac{\Theta'_1(\xi_1, 0)}{(4\xi_1/3)^{1/4}} = \frac{\lambda_2}{\lambda_1}\left[\frac{\mathrm{Ra}_2\Pr_2\left(1 + \Pr_1\right)}{\mathrm{Ra}_1\Pr_1(1 + \Pr_2)}\right]^{1/4}\frac{\Theta'_2(\xi_2, 0)}{(4\xi_2/3)^{1/4}}$$

$$\Theta_1(\xi_1, 0) + \Theta_2(\xi_2, 0) - 1 = \frac{\Theta'_2(\xi_2, 0)}{(4\xi_2/3)}\frac{\lambda_2\Delta}{\lambda_1 L}\left(\frac{\mathrm{Ra}_2\Pr_2}{1 + \Pr_2}\right)^{1/4} \tag{4.12}$$

Comment 4.4 The variable ξ, which equals x, in fact, is a dummy variable. Such a variable is applied instead of x in order to avoid confusing situations (as in the case of integration discussed in Comment 1.3). In particular, such a problem occurs when we use the chain rule defining derivatives with respect to new variables. In this case, without introducing variable ξ, we arrive at situation when the confusing derivative of x with respect to x is necessary to compute. Introducing the variable $\xi = x$ solves this problem leading to usual derivative $\partial \xi/\partial x$. instead of a puzzle $\partial x./\partial x$.

Comment 4.5 The idea of local similarity approach is based on presenting governing equations in the form (4.11) so that the left-hand parts consist of only ordinary derivatives as it should be for similarity equations, and the right-hand parts include other terms that do not satisfy these conditions and depends on partial derivatives. When the defining functions F and Θ are independent of coordinate ξ, the right-hand parts becomes zero, and these equations take a self-similar form. Therefore, the right-hand parts of equations in such form may be viewed as a measure of the local deviation from similarity.

Equations (4.11) are solved numerically using consecutive approximations. The first approximation is obtained assuming that the right-hand parts of equations (4.11) are zero. Each of the following approximations is performed calculating these parts and satisfying conjugate conditions using the data of previous step. The details may be found in [163] and [239]. The one merit of local similarity approach is that the equations remain ordinary in any approximation. However, the convergence of iterations depends on the right hand parts of equations and is satisfactory only when those are relatively small.

The conclusions obtained in this study basically agree with those formulated in article [409] reviewed in the previous example, except the following:

- The temperatures near the ends of the plate change more gradually, and the temperatures at the top of the cool sides and at the bottom of the warm sides are higher.

- The temperature distributions have sigmoidal form showing that the greatest variations of temperatures occur on about 20% of the surface length at each end so that even neglecting small differences over central region results in overall departures from isothermal conditions of about 40% of the plate.

■ **Example 4.6n: A Vertical Pipe with Forced Flow Inside and Natural Flow Outside [371]**

It is assumed that the flow enters the pipe with the zero thermal resistance wall at uniform temperature $T_e > T_\infty$ and a fully developed velocity profile. The governing system in cylindrical coordinates consist of the continuity, momentum, and energy equations with additional buoyancy term for external natural convection flow and only energy equation (the last one) for internal forced flow since the velocity profile for inside flow is assumed to be known and fully developed

$$\frac{\partial(ru)}{\partial x} + \frac{\partial(rv)}{\partial r} = 0, \quad u\frac{\partial u}{\partial x} + v\frac{\partial u}{\partial r} = \frac{(\text{Ra}/\text{Pe})\theta}{8\text{Pr}_o} + \frac{1}{r}\frac{\partial}{\partial r}\left(r\frac{\partial u}{\partial r}\right)$$

$$u\frac{\partial\theta}{\partial x} + v\frac{\partial\theta}{\partial r} = \frac{1}{r}\frac{\partial}{\partial r}\left(r\frac{\partial\theta}{\partial r}\right), \quad (1-r^2)\frac{\partial\theta}{\partial x} = \frac{1}{r}\frac{\partial}{\partial r}\left(r\frac{\partial\theta}{\partial r}\right) \qquad (4.13)$$

$$r \to \infty, x \to 0, \quad \theta, u \to 0 \quad x = 0, \theta = 1, r = 0 \quad x = 1, \theta_i = \theta_o, \quad \left(\frac{\partial\theta}{\partial r}\right)_i = \frac{\lambda_o}{\lambda_i}\left(\frac{\partial\theta}{\partial r}\right)_o,$$

The system (4.13) is written in dimensionless variables scaled using R for r, R Pe for x, R Pe$/v_o$ for u, R/v_o for v, $\theta = (T - T_\infty)/(T_e - T_\infty)$, Ra $= g\beta_o(T_e - T_\infty)D^3 \text{Pr}_o/v_o^2$, Pe $= u_{av}D/\alpha_i$, and subscripts o and i denote outer and inner flows. Because the wall is considered as a thin with no thermal resistance, the dimensionless variable r is less than one ($r \leq 1$) and more than one ($r \geq 1$) for inside and outside of pipe, respectively. The boundary conditions (4.13) are zero velocity and temperature for external flow far away from a tube ($r \to \infty$) and at the entrance ($x = 0$), initial temperature at $x = 0$ for internal flow and conjugate relations at the wall ($x = 1$).

Comment 4.6 Compare energy equations (4.13) and (3.72), which both are used for fully developed flow in a tube. Although they look different, in fact, both equations are alike because the second equation may be obtained from the first one by differentiating the right-hand part and taking into account different definition of dimensionless coordinates.

Numerical solutions of both the energy equation inside flow and the full system of equations for outside flow are performed in [371] using the Patankar-Spalding approach [305], which is similar to the SIMPLE described in [306] and discussed in Section 9.7.1. The iterative procedure is applied to conjugate both the inside and outside solutions. The procedure starts from solving external problem using the boundary condition of uniform wall temperature. To improve the convergence of iterations, the local heat transfer coefficient distribution on the interface is used. The authors claim that due to using the heat transfer coefficient distribution (instead of usual applying the temperature or heat flux distribution), the convergence of iterations is achieved during three to five iterations.

The following results are obtained

- The Nusselt number of the inside flow is insensitive to the thermal boundary conditions. Values of conjugate Nusselt number for different parameters λ_o/λ_i, Pr_o and Ra/Pe are bounded between those for uniform wall temperature and uniform heat flux. Near the inlet, the Nusselt number values are closer to uniform wall temperature, whereas for the larger distance from the inlet, Nusselt values become closer to uniform heat flux.

- The Nusselt number for the external flow is compared with the results for isothermal vertical cylinder Nu_{cy}. Analyzing the ratio Nu_o/Nu_{cy} shows that the values obtained in conjugate problem are always lower than those in the case of uniform temperature. Increasing both ratios λ_o/λ_i and Ra/Pe leads to rapid decreasing of the temperature head ($T_w - T_\infty$) along the pipe, and results in a decreasing of the Nusselt number. At the fixed values of these parameters, the variations of Pr_o from 0.7 (air) to 5 (water) do not practically affect the values of the Nusselt number,

- The wall temperature decreases along the pipe. This effect intensifies at large values of λ_o/λ_i and Ra/Pe. It is clear that this occurs due to the small resistance of external air flow in comparison with that of internal water flow. The results obtained for $Pr_o = 0.7$ and $Pr_o = 5$ indicate that this parameter does not play a significant role.

- The dimensionless bulk temperature ratio $(T_{bl} - T_\infty)/(T_e - T_\infty)$ determines the heat transfer effectiveness because it compares the heat transfer rate for the length of pipe from entrance to certain location and that for infinitely long pipe at the same mass flow. It is found that bulk temperature decreases with x as heat is transferred from the inside flow to external flow. Higher values of λ_o/λ_i and Ra/Pe increase the heat transfer rate, and results in more rapid decrease in the bulk temperature. The small effect of Pr_o is also confirmed by variation of the bulk temperature along the pipe length.

4.1.2   Applicability of One-Dimensional Models and Two-Dimensional Effects

We considered six conjugate solutions of heat transfer between two fluids separated by a thin wall obtained by different mathematical models for diverse wall configuration. Two studies are used universal function with only the first two derivatives and horizontal plates (Exam. 4.1 and 4.3), two others applied the Duhamel integral for horizontal (4.2) and approximate formula for natural flow on vertical (4.4) plates, and the local similarity approach for vertical plate (4.5) and numerical solution for both out and inside flows in vertical tube (4.6) are employed in the two last examples. We have seen that conjugate results differ by 15–25% from data of common overall heat transfer coefficient gained under assumption of isothermal walls. Effects of various conditions are investigated indicating that differences are larger for small Prandtl numbers, close each to other thermal resistances of fluids, countercurrently flows, flowing natural flows in comparison with quiescent fluids, and in some other studied specific situations.

   In this section we show when the conjugate heat transfer data obtained for thin walls are applicable and where and how those results should be corrected. To realize this, we compare the solution for thin plate from Example 4.1 with exact two-dimensional conjugate solution of the same problem solved in [111] and translated in [115].

■ **Example 4.7* a/n: Two Streams Separated by a Thick Plate [115]**

   Because the plate is not thin, the solution of Laplace equation (1.2) is needed instead of linear temperature distribution (4.1). For further study, it is convenient to present such solution in the form of sum and difference of variables ϑ_1 and ϑ_2

$$\vartheta_1 + \vartheta_2 = \frac{1}{\pi}\int_0^\infty (\mathrm{Bi}_1 + \mathrm{Bi}_2)\ \ln[4sh|z - \zeta|\pi\ \ sh(z + \zeta)\pi]dz \quad \vartheta = \frac{T_w - T_{w\infty}}{T_{\infty 1} - T_{\infty 2}}$$

(4.14)

$$\vartheta_1 - \vartheta_2 = \frac{1}{\pi}\int_0^\infty (\mathrm{Bi}_2 - \mathrm{Bi}_1)\ \ln\left[cth\frac{|z - \zeta|}{2}\pi\ \ cth\frac{z + \zeta}{2}\pi\right]dz \quad \mathrm{Bi} = \frac{q_w\Delta}{\lambda_w(T_{\infty 1} - T_{\infty 2})},$$

where $\zeta = x/\Delta$ and asymptotic temperature $T_{w\infty}$ is determined by the first equation (4.2). The heat fluxes from fluids on both sides of a plate are obtained as well as in Example 4.1 by universal function (1.38), which in variables (4.14) become

$$\mathrm{Bi}_1 = \mathrm{Bi}_{\Delta*1}\ \zeta^{-1/2}\left(\vartheta_1 - \frac{1}{\sigma + 1} + g_{11}\zeta\vartheta_1' + g_{21}\zeta^2\vartheta_1'' + \cdots\right)$$

$$\mathrm{Bi}_2 = \mathrm{Bi}_{\Delta*2}\ \zeta^{-1/2}\left(\vartheta_2 + \frac{\sigma}{\sigma + 1} + g_{12}\zeta\vartheta_2' + g_{22}\zeta^2\vartheta_2'' + \cdots\right)$$

(4.15)

Here, $\mathrm{Bi}_{\Delta*} = h_{\Delta*}\Delta/\lambda_w$, $\sigma = \mathrm{Bi}_{\Delta*1}/\mathrm{Bi}_{\Delta*2} = h_{\Delta*1}/h_{\Delta*2}$ and $h_{\Delta*}$ is an isothermal heat transfer coefficient defined through the plate thickness (which means using $\mathrm{Re} = u\Delta/v$).

The set of equations (4.14) and (4.15) presents an exact solution of the conjugate problem of heat transfer between two streams separated by a plate. In [115], this system is solved by successive approximations. As a first one, the solution for a thin plate from Example 4.1 is used. Equations (4.3) indicate that in this case the dimensionless heat fluxes and temperatures for both streams are equal. Substitution of first approximation results (4.3) into relations (4.14) gives the second approximation for temperatures

$$-\mathrm{Bi}_1^{(1)} = \mathrm{Bi}_2^{(1)} = 1 - \theta^{(1)}, \quad \vartheta_1^{(2)} = -\vartheta_2^{(2)} = \frac{1}{\pi}\int_0^\infty \mathrm{Bi}^{(1)}\ \ln\left[cth\frac{|z - \zeta|}{2}\pi\ cth\frac{z + \zeta}{2}\pi\right]dz$$

(4.16)

The corresponding heat fluxes $\mathrm{Bi}_1^{(2)}$ and $\mathrm{Bi}_2^{(2)}$ are found by substituting $\vartheta_1^{(2)}$ and $\vartheta_2^{(2)}$ into equations (4.15). Then, returning to relations (4.14) yields $\vartheta_1^{(3)}$ and $\vartheta_2^{(3)}$, and so on.

Details of solution and analysis of results of this complex problem one may find in [111] and [115]. Here, we present the basic conclusions obtained in this study:

- The temperature distributions along the plate in the first and the second approximations computed for the case $\sigma = 1$ and $\mathrm{Pr} \to 0$ by second equation (4.4) and relation (4.16), respectively, show that starting from a small length ζ_0, the second and the first approximations virtually coincide (Fig. 4.5). This means that for $\zeta > \zeta_0$ a semi-infinite plate can be considered as a thin one, and heat transfer for this part may be defined as for thin plate. This result obtained for the case of maximal conjugation effect ($\sigma = 1$, $\mathrm{Pr} \to 0$) is qualitatively valid in general.
- Analogous calculations are performed for turbulent boundary layer. In this case, the calculations are made for $\mathrm{Pr} = 0.01$ and $\mathrm{Re}_{\delta_1} = 10^3$ under which the influence of conjugation is also close to the maximal.

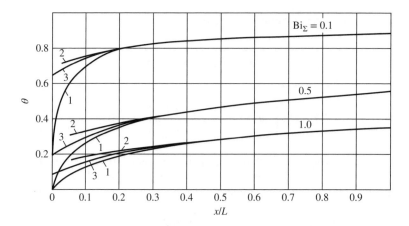

Figure 4.5 Comparison of one- and two-dimensional solutions for the heat exchange between two fluids separated by a plate: 1-thin plate, 2-second approximation for thick plate according to equation (4.16), 3-final two-dimensional results for thick plate according to equation (4.21)

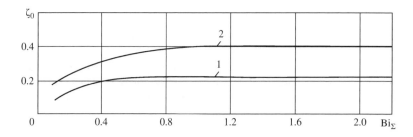

Figure 4.6 Initial length of the plate as a function of a ratio of thermal resistances of the plate and fluids: 1-turbulent flow and 2-laminar flow

- Figure 4.6 shows the values of initial length ζ_0 as a function of the ratio Bi_Σ of resistances of fluids and isothermal plate (see relation (4.3)). It is seen that in turbulent flows the initial length is smaller than in laminar flow.
- At the domain $\zeta < \zeta_0$, the plate could not be considered as a thermally thin or as a thin because all sizes of this domain are comparable so that neither the longitudinal nor transverse thermal resistance may be neglected in comparison with others (Com. 1.1).
- It follows from the solution of the Laplace equation considered above that in the case of equal fluids resistances ($\sigma = 1$), the heat transfer behavior in the vicinity of the leading edge is governed by two equations

$$\theta = 1 - \frac{2}{\pi} \int\limits_0^\infty \mathrm{Bi} \ln \left[\mathrm{cth} \frac{z - \zeta}{2} \pi \, \mathrm{cth} \frac{z + \zeta}{2} \pi \right] dz,$$

$$\mathrm{Bi} = \frac{1}{2} \mathrm{Bi}_{\Delta*} \zeta^{-1/2} (\theta + g_1 \zeta \theta' + g_2 \zeta^2 \theta'' + \dots) \qquad (4.17)$$

These equations are obtained from system (4.14)–(4.16) for the case of $\sigma = 1$ by taking into account the following chain of conclusions: (i) equalities $\theta_1 = \theta_2$ and $-\text{Bi}_1 = \text{Bi}_2$ are accurate irrespectively of whether the plate is considered as a thick or as a thin because these equalities follow from facts that are valid for both semi-infinite configurations (thick and thin) plate: the correctness of the first equality follows from the definition of dimension-less temperatures (which is easy to check by substitution of asymptotic temperature $T_{w\infty}$ into formula (4.2) for θ), and then, the correctness of the second equality of Biot numbers becomes evident from universal function (1.38) since $\theta_1 = \theta_2$ and $\sigma = 1$ ($h_{\bullet 1} = h_{\bullet 2}$), (ii) it is seen from the first equation (4.14) that the condition $-\text{Bi}_1 = \text{Bi}_2$ leads to another equality $\vartheta_1 = -\vartheta_2$ of dimensionless temperatures defined by (4.14), (iii) under condition $\sigma = 1$, both dimensionless temperatures are coupled by formulae $\theta_1 = 1 - 2\vartheta_1$ and $\theta_2 = 1 + 2\vartheta_2$, which gave the relation $\pm 2\vartheta = (1 - \theta)$, and (iv) substitution of this result in the second equations (4.16) and both equations (4.15) yield relations (4.17).

- It is shown in [115] that in the vicinity of $\zeta = 0$, the solution of system (4.17) can be pre-sented by the series in variable $\zeta^{1/2}$ similar to (3.3) for a thermally thin plate

$$\theta = \sum_{n=0}^{\infty} a_n \zeta^{n/2} \qquad \text{Bi} = \sum_{n=-1}^{\infty} a_{n+1} d_n \zeta^{n/2} \qquad (4.18)$$

The coefficients a_n and d_n are found by substitution of series (4.18) in equations (4.17). This complicated procedure performed in [115] reveals that the first and the second approx-imations computed using four and eight series coefficients differ insignificantly. Therefore, to determine heat transfer characteristics at initial domain, it is possible to use series (4.18) with first four coefficients, three of which are $a_1 = 4a_0 d_{(-1)}$, $a_2 = a_3 = 0$, where coefficient a_1 is specified by first one a_0 and coefficient $d_{(-1)}$ defined as follows

$$a_0 = 1 - \int_0^{\infty} \text{Bi} \ln cth \frac{z\pi}{2} dz, \quad d_n = \frac{\text{Bi}_{\Delta *}}{2}\left[1 + \sum_{k=1}^{\infty} g_k \frac{(n+1)(n-1) \quad \dots \quad (n-2k+3)}{2^k}\right]$$
$$(4.19)$$

It is seen that coefficient a_0 depends on the integral with semi-infinite limits. That occurs because the considered two-dimensional problem is of elliptic type, for which the solution requires the information of whole computation domain (Com. 1.5 and 1.6). To perform such integration, we divide the interval $(0, \infty)$ into two parts: the first part for limits $(0, \varepsilon)$ in which the series (4.18) are valid, and the second one within the limits (ε, ∞) in which the results for thin plate can be used. Then, the first coefficient (4.19) becomes

$$a_0 = \frac{\pi - 4I_\infty}{\pi + 4d_{(-1)}[I_{(-1)} + 4d_0 I_0]}, \quad I_n = \int_0^{\varepsilon} z^{n/2} \ln cth \frac{z\pi}{2} dz, \quad I_\infty = \int_\varepsilon^{\infty} \text{Bi} \ln cth \frac{z\pi}{2} dz$$
$$(4.20)$$

- Thus, in the case of equal thermal resistances of both streams, the temperature and heat flux on the initial domain can be computed by the series (418) with four terms

$$\theta = a_0(1 + 2\text{Bi}_{\Delta *}\zeta^{1/2}), \quad \text{Bi} = (1/2)a_0\text{Bi}_{\Delta *}(\zeta^{-1/2} + 4d_0), \qquad (4.21)$$

where θ and Bi are defined by formulae (4.2) and (4.14), respectively, $\text{Bi}_{\Delta *} = h_{\Delta *}\Delta/\lambda_w$, and $\zeta = x/\Delta$. The values of a_0 computed using equations (4.20) are given in Figure 4.7.

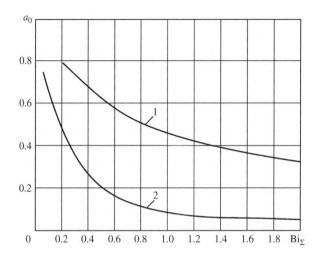

Figure 4.7 Dependence of the coefficient a_0 in equations (4.21) on the ratio of thermal resistances of the plate and fluids (Biot number): 1-turbulent flow, 2-laminar flow

- The heat fluxes on the initial domain defined according to (4.21) differ also significantly from that on a thin plate given for the same case of equal thermal resistances ($\sigma = 1$) by the last simple formula (4.3) and formula for θ in the limiting case $\mathrm{Pr} \to 0$ (see relations after (4.4)) resulting in expression $\mathrm{Bi} = (\mathrm{Bi}_*/2)[1 - \exp(-2/\mathrm{Bi}_*)]$. However, at the leading edge ($x = 0$), the usual singularity $h \to \infty$ associated with the use of boundary layer model for fluids persists in both cases.
- Due to the small length of the initial segment in a turbulent flow (Fig. 4.6), the only first terms of the formulae (4.21) may be used: $\theta = a_0$, $\mathrm{Bi} = 1/2a_0\mathrm{Bi}_{\Delta*}\zeta^{-1/5}$, where the values of a_0 are defined by Figure 4.7.

In the next several sections we present examples of more specific models of conjugate heat transfer including the most recent publications.

4.1.3   Heat Exchanger Models

■ **Example 4.8a/n: Double-Pipe Heat Exchanger Model with Laminar Flow [297]**

Two flows in steady state with the fully developed velocity profiles inside a double pipe are considered. The double pipe consists of inner central tube and outer annual channel. The thermal conduction in the fluids and viscose dissipation are neglected. The governing system in cylindrical coordinates includes energy equations (4.22) (compare with equations (4.13)) for the inner and outer streams ($T_{01} < T_{02}$), the Laplace equation (4.23) for separating wall, boundary, and conjugate (4.24) conditions

$$u_1 \frac{\partial \theta_1}{\partial x} = 4\left(\frac{1}{r}\frac{\partial \theta_1}{\partial r} + \frac{\partial^2 \theta_1}{\partial r^2}\right), \quad \theta_1(0, r) = \theta_{01}, \quad \frac{\partial \theta_1}{\partial r}(x, 0) = 0$$

$$u_2 \frac{\partial \theta_2}{\partial x} = 4\frac{\lambda_2}{(mc_p)_2}[R_i^2 - (1 + \Delta)^2]\left(\frac{1}{r}\frac{\partial \theta_2}{\partial r} + \frac{\partial^2 \theta_2}{\partial r^2}\right)$$

(4.22)

$$\theta_2(0, r) = \theta_{20}, \qquad \theta_2(L, r) = \theta_{20}, \qquad \frac{\partial \theta_2}{\partial r}(x, R_i) = 0$$

$$\mathrm{Pe}_1^2 \left(\frac{1}{r} \frac{\partial \theta_s}{\partial r} + \frac{\partial^2 \theta_s}{\partial r^2} \right) + \frac{\partial^2 \theta_s}{\partial x^2} = 0 \qquad \frac{\partial \theta_s}{\partial x}(0, r) = \frac{\partial \theta_s}{\partial r}(L, r) = 0 \tag{4.23}$$

$$\theta_1(x, 1) = \theta_w(x, 1), \qquad \frac{\partial \theta_1}{\partial r}(x, 1) = \lambda_w \frac{\partial \theta_w}{\partial r}(x, 1)$$

$$\theta_2(x, 1 + \Delta) = \theta_w(x, 1 + \Delta), \qquad \frac{\partial \theta_2}{\partial r}(x, 1 + \Delta) = \frac{\lambda_w}{\lambda} \frac{\partial \theta_w}{\partial r}(x, 1 + \Delta) \tag{4.24}$$

The dimensionless variables are scaled as follows: linear sizes by radius R_i of the inner duct, the velocities by mean axial velocity U, and temperatures by inlet temperature T_{01} of the inner fluid, thermal conductivity by λ_2, and capacity $(mc_p)_2$ by corresponding values of the inner flow. The first and the second boundary conditions (4.22) for outer fluid pertain to con-and countercurrent cases, respectively (see Example 4.3).

The concurrent and countercurrent cases have been studied, however, the only concurrent case is presented in detail. The Duhamel's integral (1.23) in the form

$$\theta_w - \theta_{b1} = \frac{2q_1(0)}{\mathrm{Nu}_q} + \int_0^x \frac{2}{\mathrm{Nu}_q(x - \xi)} \frac{dq_1}{d\xi} d\xi \tag{4.25}$$

is used to solve the energy equations (4.22). In equation (4.25), Nu_q and θ_{bl} are Nusselt number for the uniform heat flux and bulk temperature. The equation (4.23) for the wall is solved numerically applying the finite element method (S. 9.6). The iterative procedure starts with guessing of the distributions of the bulk temperature and Nusselt number, which are used as boundary condition for the wall equation. As a result, new distributions of the wall temperature and Nusselt number are obtained. Then, the updated values of the Nusselt numbers and the bulk temperatures are calculated from conjugate conditions (4.24) and the refined wall temperature distribution is obtained, again using equation (4.25). The convergence is achieved in less than 14 iterations. The numerical results are obtained for: $L = 10$ and 100, $\Delta = 5$ and 2, $R_i = 3$ and 6, $\lambda_w = 1, 10, 100, 1000$ and $10, 000$, $\lambda_2 = 0.1, 1$ and 5, $\mathrm{Pe}_1 = 500, 1000$ and $10,000$, $(mc_p)_2 = 0.5, 1$ and 2. The following results are formulated:

- The conjugate effect is studied by comparing results obtained with and without [296] the thermal wall conduction. In contrast to the latter case, two isothermal zones are created at the interface, and the wall temperatures do not coincide with temperature of inlet fluid. The length of these zones, as well as the wall-to-fluid temperature difference, increases due to axial wall conduction. With increasing wall conductivity, the wall temperature becomes more uniform, and the outlet temperature of the internal stream decreases, whereas the outlet temperature of the external stream increases correspondingly. For the relatively small wall conductivities up to $\lambda_w = 100$, the two streams are uncoupled and a central zone with uniform and equal heat fluxes for both sides exist. This is similar to the case without wall conduction effect. For high wall conductivities, the situation completely changes, and the heat fluxes monotonically decreases from the inlet to the outlet with one crossing point. Near the inlet, the Nusselt number values coincide with those for the isothermal condition. Downstream, both the Nusselt number and heat flux reach the asymptotic values corresponding to isothermal or to uniform heat flux conditions.

- Distribution of the local entropy production in the wall and streams calculated as suggested in [39] are sensitive to wall conductivity. For $\lambda_w = 10,000$, the entropy production monotonically decreases downstream from the maximum at the inlet. For reducing wall conductivities, the entropy production decreases progressively at the inlet, increasing in the outlet regions. Analyzing the distribution of entropy generation rate in the fluids indicates that the maxima in entropy production correspond to high values of heat flux or wall-to-fluid temperature difference. Distribution of entropy production in the wall shows that the wall radial thermal resistance is dominant when the conductivity of the wall is low. For $\lambda_w = 1$, the entropy production distribution is similar to that for the wall with zero conductivity. For the high conductivities, the maximum is found at the middle of the wall instead of minimum existence here in the case of low conductivities.

- The dependence of the heat exchanger effectiveness on the wall conductivity changes according to the wall conduction effects. Instead of monotonically increased effectiveness with growing wall conductivity, for the intermediate λ_w, the maximum effectiveness exists. For low wall conductivities, the entropy production is concentrated in the wall. In the short device, the increasing λ_w leads to monotonically decreasing the wall contribution. In a long exchanger, the minimum of the entropy production is observed. The reduction of effectiveness due to wall conduction effect increases for increasing wall thickness, but it is slightly affected by variation of the pipe diameter ratio. The effectiveness reduces also with increasing Peclet number. Increasing the fluids conductivity ratio has strong and positive effect on effectiveness.

- The proper choice of the wall material is needed for optimization. For example, in the case of countercurrent water streams separated by copper wall, the conductivity of the wall yields a small reduction in effectiveness. The optimum can be achieved by using a steel wall with $\lambda_w \approx 100$. For the glass wall, the order of magnitude drops to 1, and the effectiveness decreases due to the high radial resistance. For gaseous fluids, the wall conduction effect is more pronounced indicting, for instance, that corresponding value of λ_w becomes higher than 10,000 for two streams of air separated by copper wall.

■ Example 4.9a: Double Pipe with Concurrent or Countercurrent Flow (Conjugate Graetz Problem)

A similar model of two fully developed streams one (1) in a tubular space of the double pipe, and the other (2) in annual space is used in [302]. The axial conduction in fluids and in a wall is taken into account. The one energy equation for both fluids, the conjugate boundary conditions, the symmetry condition, the developed velocity profiles for flows, and two boundary conditions of the third kind for walls build up the model:

$$-\frac{1}{r}\frac{\partial}{\partial r}\left[\lambda r\frac{\partial T}{\partial r}\right] - \lambda\frac{\partial^2 T}{\partial r^2} + \rho c_p u_x(r)\frac{\partial T}{\partial x} = 0, \quad -\lambda_1\frac{\partial T}{\partial r}\bigg|_{r_1} = -\lambda_2\frac{r_1 + \Delta}{r_1}\frac{\partial T}{\partial r}\bigg|_{r_1+\Delta} = 0,$$

$$\frac{\partial T}{\partial r}\bigg|_{r=0}, \quad u_{x1} = U_1\left(1 - \frac{r^2}{r_1^2}\right), \quad u_{x2} = \pm U_2\left[1 - \left(\frac{r}{R}\right)^2 + (\overline{\Delta}^2 - 1)\frac{\ln(r/R)}{\ln\overline{\Delta}}\right] \quad (4.26)$$

$$-\lambda_2\frac{\partial T}{\partial r}\bigg|_{r=R} = h_e[T(x,R) - T_0], \quad -\lambda_1\frac{\partial T}{\partial r}\bigg|_{r_1} = h[T(x,r_1) - T(x,r_1 + \Delta)]$$

This model is more complicated then the similar previous one due to taken into account conduction in fluids and in the walls. To simplify the problem, the two conditions of the third kind (4.26) are included in the model specifying the thermal resistances in the walls. Actually, such simplified problem is intended to investigate the conjugate heat transfer between two fluids, which may be considered as conjugate Craetz problem (Com. 4.7).

In relations (4.26), U is the characteristic velocity, h and h_e are heat transfer coefficients for the resistance of separated and outer walls, respectively, Δ, r_1, r_2, R are the width of the annulus, the inner and outer radii of inner tube, the radius of outer tube and $\overline{\Delta} = (r_1 + \Delta)/R$. The plus and minus at u_{x2}, are for the concurrent and for countercurrent problems, respectively. To decompose energy equation (to get separated equations for each fluid) two functions $\hat{S}_1$ ($0 < r < r_1$) and $\hat{S}_2$ ($r_1 + \Delta < r < R$) are used resulting in solution of the same problem for both fluids as follows

$$\hat{S}_1 = \int_0^r \left(-\lambda_1 \frac{\partial T}{\partial x} + \rho_1 c_{p1} u_{x1} T\right) 2\pi \zeta d\zeta, \quad \hat{S}_2 = \hat{S}_1 + \int_{r_1+\Delta}^r \left(-\lambda_2 \frac{\partial T}{\partial r} + \rho c_p u_{x2} T\right) 2\pi \zeta d\zeta$$

$$\frac{\partial \hat{S}}{\partial x} = -2\pi r \left(-\lambda \frac{\partial T}{\partial r}\right), \quad \frac{\partial \hat{S}}{\partial r} = 2\pi r \left(-\lambda \frac{\partial T}{\partial x} + \rho c_p u_x T\right) \tag{4.27}$$

The numerical results are obtained for the flow in an annual space around a solid cylinder with uniform heat source. Two versions are considered: with heat transfer through the outer wall and with insulated outer wall. Solutions are presented in the usual Graetz form of asymptotic series of the eigenfunctions. It is known that such series are efficient for high abscissas but converge slowly for small values of x (Com. 3.14). Authors claimed that in their examples the series converge unusually fast at Pe = 5, but did not point out the reason of this. They found that unlike the single-stream Graetz problem the effect of axial heat conduction in the fluid cannot be ignored, even for Peclet numbers larger than 40–50, and only for Peclet numbers considerable higher than 100, the axial conduction in the fluid becomes insignificant.

Comment 4.7 The conjugate Graetz heat transfer problem in the complete formulation including the conduction effects is a complex task. The analysis of difficulties arising in solution of such problem is given in [302] and in earlier publications of these authors. As it mentioned in Chapter 3 (see "other solutions in asymptotic series"), some asymptotic solutions of simplified (without effects of conduction) conjugate Graetz problem were published. The previous Example 4.8 actually presents the numerical solution of the conjugate Graets problem of this type, whereas the Example 4.9 is an attempt to take into account the conduction effects in conjugate Graetz problem.

■ Example 4.10n: Microchannel Heat Sink as an Element of Heat Exchanger [135]

The microchannel heat sink model applied in this work is an element of a modern microchannel heat exchanger. The heat transfer characteristics of such exchanger should be significantly better due to great reduction of the thickness of thermal boundary layer and overall notable capacity based on large surface/volume ratio. The studied model is a rectangular silicon channel with hydraulic diameter $D_h \approx 100\,\mu m$ that is a basic element of experimentally investigated model in [200].

Comment 4.8 The surface/volume ratio is an important characteristic for small objects because it determines the value of surface per unit volume. This ratio increases as the body size decreases since the surface is proportional to second power of an object size, whereas the volume changes as third power of it. Therefore, for sphere, this parameter is inversely proportional to radius. In many cases, the value of surface/volume ratio helps to understand the basic features of phenomenon. In particular, it is clear that heat exchanger efficiency should significantly grow with the size decreasing.

The following assumptions in the reviewed study are used: (i) a range of Knudsen number $Kn = l/D_h$ (Com. 7.5) lies in continuum flow regime, and hence, the Navier-Stokes equations are appropriate, (ii) the flow is incompressible, laminar, and steady state; the thermophysical properties are temperature dependent, (iii) the largest temperature gradients and thermal stresses are expected to occur at the inlet of the channel; therefore, the development of the flow and temperature at the inlet should be carefully resolved, (iv) thermal radiation is negligible since the typical operation temperature should be below $100°C$.

The tree-dimensional continuity, Navier-Stokes, and energy equations for fluid plus the Laplace equation (1.1) for walls with depending on temperature properties are used in the vector form (S. 7.1.2.1)

$$\nabla(\rho \mathbf{V}) = 0, \quad \mathbf{V} \cdot \nabla(\rho \mathbf{V}) = -\nabla p + \nabla \cdot (\mu \nabla \mathbf{V})$$

$$\mathbf{V} \cdot \nabla(\rho c_p T) = \nabla \cdot (\lambda \nabla T), \quad \nabla(\lambda_w \nabla T_S) = 0 \tag{4.28}$$

A uniform heat flux is imposed at one of the channel walls, whereas the others are isolated. The entering flow velocity and temperature are given, and gradients of velocity and temperature at the exit are taken to be zero. The no-slip condition at the walls and conjugate conditions at the fluid-solid interface should be satisfied. Patankar's technique of discretization [306] and SIMPLER algorithm (S. 9.7) are used to solve the system of governing equations (4.28). The predicted here calculations are in agreement with the experimental data from [200]. The following results and conclusions are deduced:

- The local temperature distribution shows that the walls are isothermal, but the temperature field in fluid is essentially nonuniform. Initially, the high temperature gradient zone forms at the inlet of the channel and then increasing fluid core temperature is observed. Three basic conclusions can be stated: (i) the maximum heat fluxes occur at the inlet of channel, (ii) the heat flux imposed at the wall is spread out by conduction within walls and finally is transferred to fluid, and (iii) due to effect of conjugation, the thermal development occupies the entire channel; the temperature distributions at the inner and outer wall surfaces show a very complex pattern which occurs because of convective heat transfer and three-dimensional conduction.
- The distribution of the local heat fluxes on the inner walls, which are the fluid-walls interface, confirms the observation deduced by the temperature distribution analysis in the first conclusion. In particular, the local heat fluxes are the greatest at the inlet of the channel where the temperature gradients are high. The local heat flux inside the channel is distributed high nonuniformly so that magnitude variation in the heat fluxes reaches several orders. The reason of this is the difference in spacing between channel walls. The channel cross-section is a stretched rectangle such that the distance between two walls in one direction is about three times smaller than that between two other walls. As a result, the boundary layer

between small spacing walls is much thinner, and therefore, the convective heat transfer rate is much larger. In the corners, the complicated heat flow structure is observed. Here, the negative heat fluxes directed from the fluid to walls exist, which occurs due to interaction of both boundary layers developed along adjacent walls. In such a case, the heat transfer coefficients are also negative (S. 2.4) when the traditional methods are not applicable, and only conjugate solutions give the realistic results.

- The average characteristics in general conform to local distribution quantities. The average wall temperature increases significantly in the entering portion due to high local temperatures in this area. In contrast to this, the fluid bulk temperature grows gradually along the channel approaching the wall temperature at the exit. Large temperature gradients in the inlet channel portion may result in significant thermal stresses, which is important to take into account during design. The average heat fluxes and average heat transfer coefficient gradually decrease along the channel. The average heat flux of all walls is smaller than initially imposed heat flux everywhere except inlet portion, where the average heat flux is greater than imposed one. This occurs because the area where the heat flux imposed is mach smaller than that of the inside wall surfaces.

4.1.4 Finned Surfaces

■ **Example 4.11* n: Fin Array on a Horizontal Base [319]**

The configuration studied consists of two adjacent long vertical fins setting up on a horizontal base. The temperature T_{w0} is higher than ambient air temperature T_∞, and hence, heat is transferred from the fins to ambient air by convection and radiation. The problem is formulated as for a closure composed by fins and base, and is governed by the momentum equation in vorticity-stream ($\omega - \psi$) variables (4.29) (S. 7.1.2.3 and 7.1.2.4), energy equation (4.30) and two one-dimensional equations (4.31) for fins

$$\frac{\partial \omega}{\partial t} + u\frac{\partial \omega}{\partial x} + v\frac{\partial \omega}{\partial y} = \frac{\partial T}{\partial y} + \frac{1}{Gr^{1/2}}\frac{\partial^2 \omega}{\partial x^2} + \frac{\partial^2 \omega}{\partial y^2}, \quad \omega = \frac{1}{Gr^{1/2}}\frac{\partial^2 \psi}{\partial x^2} + \frac{\partial^2 \psi}{\partial y^2} \tag{4.29}$$

$$\frac{\partial T}{\partial t} + u\frac{\partial T}{\partial x} + v\frac{\partial T}{\partial y} = \frac{1}{Pr}\left(\frac{1}{Gr^{1/2}}\frac{\partial^2 T}{\partial x^2} + \frac{\partial^2 T}{\partial y^2}\right) + \frac{1}{Pr\,Gr^{1/4}}\left(2q_{R1} + \frac{q_{R3}}{A_R}\right) \tag{4.30}$$

$$\frac{Pr\,Gr^{1/2}}{\alpha}\frac{\partial T_w}{\partial t} = \frac{\partial^2 T_w}{\partial x^2} + \Lambda\frac{\partial T}{\partial y}\bigg|_{y=0}, \quad \frac{Pr\,Gr^{1/2}}{\alpha}\frac{\partial T_w}{\partial t} = \frac{\partial^2 T_w}{\partial x^2} - \Lambda\frac{\partial T}{\partial y}\bigg|_{y=Gr^{1/4}} \tag{4.31}$$

The radiation heat fluxes from the fins and base are included in equation (4.30) as sources

$$q_{R1} = \frac{\varepsilon_1}{1-\varepsilon_1}[S_{13}(J_1 - J_3) + (S_{14} + 2S_{15})(J_1 - E_{b4})], \quad J_1 = \frac{a_{22}b_1 + S_{11}b_2}{a_{11}a_{22} - 2S_{13}S_{34}} \tag{4.32}$$

$$q_{R3} = \frac{\varepsilon_3}{1-\varepsilon_3}[2S_{31}(J_3 - J_1) + (S_{14} + 2S_{15})(J_3 - E_{b4})], \quad J_3 = \frac{2S_{31}b_1 + a_{11}b_2}{a_{11}a_{22} - 2S_{13}S_{31}} \tag{4.33}$$

These expressions take into account that the radiation exchange occurs between the left (subscript 1) and right (2) fins surfaces, the base (3), and the walls of the room through the open top (4), front side (5), and rear side (6) of closure. The open top and sides are considered as imaginary surfaces. The black body irradiations of the fins and base are $E_{b1} = E_{b2} = \sigma T_w^4$ and $E_{b3} = \sigma T_{w0}^4$, $E_{b4} = E_{b5} = E_{b6} = J_4 = J_5 = J_6 = \sigma T_\infty^4$, whereas $E_{b4} = E_{b5} = E_{b6} = J_4 = J_5 = J_6 = \sigma T_\infty^4$, $S_{ij} = F_{ij_i}(1 - \varepsilon_i)/\varepsilon_i$, $a_{11} = 1 + S_{13} + S_{14}$, $a_{22} = 1 + 2S_{31} + S_{34}$, $b_1 = E_{b1} + S_{14}E_{b4}$, $b_2 = E_{b3} + S_{34}E_{b4}$, where J is radiosity, ε is emissivity, S is a spacing between fins, $i = 1, 3$ and $j = 1 - 6$. The shape factors F_{ij} are taken from [172]. The following scales are used to form the dimensionless variables in equations (4.29)–(4.31): S, $S/Gr^{1/4}$, $S^2/\nu Gr^{1/2}$, $\nu Gr^{1/2}/S$, $\nu Gr^{1/4}/S$, $T_{w0} - T_\infty$, $\nu Gr^{1/4}$, $\nu Gr^{3/4}/S^2$, α, $\lambda(T_{w,0} - T_\infty)Gr^{1/4}/S$ and σT_∞^4 for x, y, t, u, v, $T - T_\infty$, ψ, ω, α_w, q_R, E_b, respectively, and Grashof number is defined as $Gr = g\beta(T_{w,0} - T_\infty)S^3/\nu$.

The boundary conditions include: $T = T_{w0}$, no slip condition for the fins and the base and $\partial T/\partial x = 0$ for the fin tips. For open top, the following boundary conditions are adopted: $v = \partial u/\partial x = \partial \psi/\partial x = \omega = 0$ and $\partial T/\partial x = 0$. The results for various fin arrays are presented applying the conduction-convection Λ, radiation N_R, parameters, and the aspect A_R, and temperature γ ratios

$$\Lambda = \frac{\lambda PS}{\lambda_w A_w}Gr^{1/4}, \quad N_R = \frac{S\sigma T_\infty^4}{\lambda Gr^{1/4}(T_{w0} - T_\infty)}, \quad A_R = \frac{H}{S}, \quad \gamma = \frac{T_{w0} - T_\infty}{T_\infty} \qquad (4.34)$$

Here, $P = \Delta + W$, $A_w = W\Delta/2$, Δ, H and W are the half perimeter, the half section area, the thickness, height, and width of the fin.

The alternating direction method (Com. 3.10) is used to solve numerically system (4.29)–(4.31). The temperature distributions are obtained in the fluid from the first and in the fins from the fourth and fifth equations, respectively. The vorticity and stream function are calculated using the first and the second equations (4.29), and the components of velocity are obtained knowing the stream function. The procedure is continued until steady state fields for all variables are obtained. The following results are formulated:

- Average Nusselt numbers for a four-fin array for low and high emissivities agree with known experimental data. Analysis shows that the contributions of the fins, base and end fins to total heat transfer are 36, 13.5 and 50.5%, which agrees with data from [318]. The effect of fins spacing on heat fluxes is studied for arrays with different number of fins over a fixed base. As the number of fins increases from 4 to 16, and the value of spacing S decreases from 20 to 2.8 mm, the heat fluxes from fin and from base decrease from 149 to 44 W/m^2 and from 379 to 148 W/m^2. Despite increased numbers of fins, the heat transfer rate and effectiveness remain almost the same, but the average heat transfer coefficient decreases remarkable from 5.29 to 1.48 W/m^2 K. The effect of the base temperature is studied for the case investigated experimentally in [376]. Both results are in reasonable agreement and indicate that the total heat transfer rate increases as the base temperature grows for any studied values of spacing and heights. The effectiveness increases as well for all heights, but for small values of S, effectiveness decreases as the base temperature grows. The results for different fin thicknesses indicate that in the case of low heights and high thermal conductivities, the heat flux from fin practically does not depend on thickness. It is found that the conductivity decreasing leads to reduction of the fin heat flux, and increasing in emissivity yields growing heat flux due to increasing radiation.
- The temperature profiles obtained for two different spacing show that the temperature far away from the fins is lower for higher spacing than that for smaller spacing. At the same

time, the velocity profiles indicate that there is a recirculation zone at larger S, resulting in higher velocities near the wall and lower velocities at the distance $S/2$. The isotherms and streamlines for the same two enclosures indicate that the air temperature is high in the middle of the enclosure with smaller spacing, whereas in the enclosure with the larger spacing, the heating is confined to the air near the fins and the base. The streamline distribution shows that the streamlines travel upward along the fins, where the temperature is high compared to that of the air in the enclosure.

■ Example 4.12* a/n: Transverse Flow Over Finned Surface [149]

A fluid flows along a finned surface transversely to the fins. Since the flow is normal to the fins, the eddy forms in the each interfin space (Fig. 4.8). On the side 1 of the fin, the temperature gradient increases, whereas on the side 2 it decreases in the flow direction. On the base 3, the temperature gradient may be considered as constant. It is assumed that at intersections, the surfaces are rounded, owing to which stagnant zones or secondary eddy flows do not form in the corners. The boundary layer develops on the finned surface from the end face to the base on the front (in the direction of the flow) fin surface 1 and increases from the base to the end on the back fin surface 2 (Fig. 4.8). The pattern presented here of the dynamic boundary layer development on the walls of the cavern with eddy flow inside it was proposed by Batchelor [63].

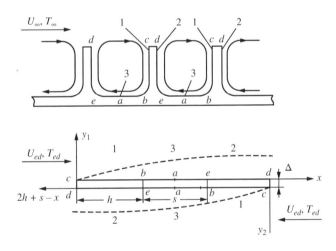

Figure 4.8 Scheme of a transverse flow over a finned surface

Assuming that the conditions in all cells are identical, the model of the problem of heat transfer in fins is formulated as bilateral flow over the body schematically shown below in Figure 4.8. Here, the body surface represents the surface of the fin and of two adjacent cells. The upper body surface represent the right cell surface, the ends of the model body corresponds to the ends of the two adjacent fins, and the lower body surface represent the surface of the left cell. The numbers in the model correspond to these in the scheme of the finned surface. Thus, the numbers 1, 2, and 3 correspond to sections with increasing, decreasing and constant temperature gradients, respectively, and the model represents the case of countercurrent flows with complicated velocity and temperature distributions. The distances h and s on the model correspond to the fins height H and to the length between the fins H_B.

For thermally thin fins, the governing equation is the steady-state equation (1.3). The boundary conditions are the zero temperature gradient at the entrance $(x = 0)$ and prescribed base temperature T_0 at the point $x = H$ (or at $x = h$)

$$\lambda_w \Delta \frac{d^2 T_w}{dx^2} = q_1(x) + q_2(x), \quad \frac{dT}{dx}\bigg|_{x=0} = 0, \quad T|_{x=H} = T_0 \tag{4.35}$$

The heat fluxes $q_1(x)$ and $q_2(x)$ from surfaces of fin are calculate by equation (1.26)

$$q_1(x) = h_*(x)\left[T_E - T_D + \int_0^x f\left(\frac{\xi}{x}\right) \frac{dT_w}{d\xi} d\xi \right], \quad q_2(x) = h_*(x)(2H + H_B - x)\left[T_E - T_D \right.$$

$$\left. + \int_{2H+H_B}^{H+H_B} f\left(\frac{2H + H_B - \xi}{2H + H_B - x}\right) \frac{dT_w}{dx} d\xi + \int_{H_B}^{x} f\left(\frac{2H + H_B - \xi}{2H + H_B - x}\right) \frac{dT_w}{d\xi} d\xi \right] \tag{4.36}$$

Here, T_E and T_D are temperatures of the fin end and the eddy flow inside interfin space that plays the role of the temperature of the external flow for the boundary layer on the fin. The expression for heat flux on the back surface $q_2(x)$ takes into account that the boundary layer in this case develops starting from the end point c (Fig. 4.8) of the front surface. Therefore, the heat transfer on the back surface depends on the temperature distribution on the front surface (section bc) and the length H_B of the interfin section. This fact takes into account the first integral in the equation for $q_2(x)$, whereas the second integral in this relation determines the effect of the section (ed) on the back surface.

Substituting expressions (4.36) into equations (4.35) yields the intego-differential equation and boundary conditions determining the temperature over the height of the fin

$$\frac{d^2\theta_w}{d\eta^2} = N^2\left\{ \phi(\eta)\left[\theta_E + \int_0^\eta f\left(\frac{\zeta}{\eta}\right) \frac{d\theta_w}{d\zeta} d\zeta \right] + \phi(\eta_0 - \eta)\left[\theta_E + \int_0^1 f\left(\frac{\zeta}{\eta_0 - \eta}\right) \frac{d\theta_w}{d\zeta} d\zeta \right. \right.$$

$$\left. \left. - \int_\eta^1 f\left(\frac{\zeta}{\eta_0 - \eta}\right) \frac{d\theta_w}{d\zeta} d\zeta \right] \right\} \quad \frac{d\theta}{d\eta}\bigg|_{\eta=0} = 0, \quad \theta|_{\eta=1} = 1 \tag{4.37}$$

$$\theta_w = \frac{T_w - T_D}{T_0 - T_D}, \quad N^2 = \frac{\overline{h}_* H^2}{\lambda_w \Delta}, \quad \phi(\eta) = \frac{h_*(\eta)}{\overline{h}_*}, \quad \overline{h}_* = \frac{1}{H}\int_0^{H_B} [h_*(\zeta) + h_*(2H + H_B - \zeta)] d\zeta$$

Here, $\eta = x/H$, $\eta_0 = 2 + H_B/H$, $\overline{h}_*$ is the average heat transfer coefficient of an isothermal fin and the parameter N^2 is the Biot number (note that H^2/Δ is a linear characteristic) determining the conjugation effect. Intego-differential equation (4.37) is solved by reduction of conjugate problem to an equivalent conduction problem (S. 2.2. 2). According to this approach, equation (4.37) is reduced to the system of ordinary differential equation and to the equation for error

$$\frac{d^2\theta_w}{d\eta^2} = N^2 \left\{ \phi(\eta) \left[\theta_w + g_1\eta\frac{d\theta_w}{d\eta} + g_2\eta^2\frac{d^2\theta_w}{d\eta^2} \right] + \phi(\eta_0 - \eta) \left[\theta_w - g_1(\eta_0 - \eta)\frac{d\theta_w}{d\eta} + \right. \right.$$

$$\left. \left. g_2(\eta_0 - \eta)^2\frac{d^2\theta_w}{d\eta^2} \right] \right\} + \varepsilon(\eta) \quad \varepsilon(\eta) = \theta_w^{\text{int}} - \theta_w^{\text{diff}} \tag{4.38}$$

A first approximation is obtained assuming $\varepsilon(\eta) = 0$ and solving the ordinary differential equation. Using the first approximation results in $\theta_w(\eta)$, the next approximation is obtained by integrating the right hand part of intego-differential equation (4.37), which gives θ_w^{int} and calculating $\varepsilon(\eta)$ by applying the first approximation $\theta_w(\eta)$ as θ_w^{diff}. Then, including $\varepsilon(\eta)$ in differential equation (4.38) gives the next approximation for $\theta_w(\eta)$. The procedure is continued until a desired accuracy is achieved.

Numerical results are obtained for laminar flow: $C_1 = 3/4$, $C_2 = 1/3$, $g_1 = 0.63$, $g_2 = -0.14$ (S. 1.6). The following conclusions are formulated:

- The solutions of the conjugate problem are compared with the results obtained by approximate calculations using local and average isothermal heat transfer coefficients. It is found that in the range $0 \leq N^2 \leq 2$, the results of the conjugate and both simplified methods are in agreement. For $N^2 > 2$, the deference between results attained in the conjugate and simplified problems becomes more significant. The simplified values of the fin efficiency and the total heat flux removed by the fin are too low, and the errors grow as the conjugate parameter N^2 increases, reaching for large N^2 the values of 60–70%. The distributions of the heat transfer coefficient and heat flux over the fin height obtained by conjugate approach are substantially nonuniform with maximum values at the fin end.
- On the back side of the fin, for $N^2 \geq 1.9$, the heat flux inversion is observed when the heat flux is directed toward the fin despite the temperature head is still positive (S. 2.4). As the value of N^2 increases, the absolute magnitude of the inversed heat flux and the length of the heated end section (Fig. 4.8) increase. Since the inversion effect cannot be explained by the simplified approach based on heat transfer coefficient, neglecting the conjugation of the problem in this case not only yields quantitative errors, but also leads to qualitative incorrect results. The reason for this is that on the back side of the fin, the temperature head decreases in flow direction. In such a case, the conjugate effect should always be taken into account (S. 3.1.1.3, Com. 3.4).
- It is found that for large values of conjugate parameter $N^2 > 2$, the heat flux removed by fin is maximal for fins with $1 < H_B/H < 1.5$. This result is a consequence of nonisothermicity of the finned surface and also cannot be obtained by simple methods.

OTHER WORKS: The similar double-pipe heat exchanger model considered in example 4.8 is investigated in [363] for countercurrent flows. The system of governing equations is solved numerically using Galerkin's method (S. 9.6). The heat exchanger model in the form of tube-fin is considered in [80]. Heat transfer in a microchannel similar to the one analyzed in [135] (Exam. 4.10) is studied in papers [188]. Other examples of fins are considered in articles [407] and [237]. Both works investigate the fins embedded in porous medium. In the first one, the vertical fin with rounded tip is considered using finite-difference solution, and in the second paper, a mixed convection along a cylindrical fin is analyzed employing local similarity method (Com. 4.5). In the recent study [171], the effect of micropolar fluid on the heat exchange is examined by considering the flow in concentric annulus. The construction and the results of the experimental verification of a new compact heat exchanger are presented in the other recently published article [176].

4.2   Thermal Treatment and Cooling Systems

4.2.1   Treatment of Continuous Materials

In Section 1.10 we obtained the universal functions for the continuously moving sheet model schematically showing in Figure 1.12. It was explained that such model simulates the systems of the production of different materials in which a tape or a thread extruded from a slot is cooled by pulling though a surrounding. There, we consider the features of the boundary layer growing on continuously moving materials, which differs from usual boundary layers formed on streamlined or flying bodies. Here, some solutions of conjugate heat transfer problems for continuously moving plates are reviewed.

■ **Example 4.13a: Continuous Plate Moving Through Surrounding [69]**

An infinite flat plate or circular cylinder is moving out of a slot with constant velocity U_w into a viscous medium at temperature T_0 and is cooled pulling through it. The governing system for fluid consists of the boundary layer equations: continuity (1.9), momentum (1.10) for zero pressure gradient and energy (1.11) without dissipation term. The equation for the moving plate and relevant boundary conditions are as follows

$$U_w \frac{\partial T}{\partial x} = \alpha_w \frac{\partial^2 T}{\partial y^2}, \quad y = 0, \Delta \quad u = U_w, T = T_w(x), \quad y \to \infty, u \to 0, T \to T_\infty \quad (4.39)$$

Comment 4.9 The first boundary condition specifies velocity and temperature on both surfaces of a plate, and the second one indicates that far from the plate, the surrounding at temperature T_∞ is unmoving. The problem for continuously moving plate with such boundary conditions in the fixed coordinates is described by usual boundary layer equations (1.9)–(1.11), but as it is seen from relations (4.39), the boundary conditions are opposite to those for streamlined plate. In contrast to that, in coordinate system attached to a moving surface, the boundary conditions are identical with usual ones, but boundary layer equations are unsteady, and thus, they differ from usual system for a plate (S. 1.10).

Dimensionless variables transform system (4.39) to the form similar to well investigated classical Blasius-Pohlhausen's equations (S. 7.5.1)

$$\frac{d^3 f}{d\eta^3} + \frac{f}{2}\frac{d^2 f}{d\eta^2} = 0, \quad \frac{\partial^2 \theta}{\partial \eta^2} + \Pr\frac{f}{2}\frac{\partial \theta}{\partial \eta} = \Pr\frac{df}{d\eta}\xi\frac{\partial \theta}{\partial \xi}, \quad \gamma\frac{\partial \theta}{\partial \xi} = \frac{\partial^2 \theta}{\partial \zeta^2} \quad (4.40)$$

$$\eta = 0, f = 0, \frac{df}{d\eta} = 1, \quad \eta \to \infty, \frac{df}{d\eta} = 0, \quad \xi = 0, \theta_w = 1, \theta = \theta_0$$

$$\eta = \xi = 0, \quad \theta = \theta_w \quad -\frac{\partial \theta}{\partial \eta} = \frac{(\Pr \xi)^{1/2}}{\gamma} \frac{\partial \theta}{\partial \zeta} \quad \gamma = \frac{\lambda}{\lambda_w} \mathrm{Lu}$$

$$f = \frac{\psi}{(\nu x U_w)^{1/2}}, \quad \xi = \frac{x \lambda \mathrm{Lu}}{U_w \Delta^2 (\rho c)_w}, \quad \zeta = \frac{y}{\Delta}, \quad \theta = T - T_\infty, \quad \eta = y\sqrt{U_w/\nu x}$$

Because in this case, the conjugate problem is considered, the conjugate condition (the second and third equations in the third line) is included in a set of boundary conditions. The first equation (4.39) has been solved numerically by Sakiadis [333]. In the system (4.40), this equation is presented as third equation. Other equations are solved using analogous technique in reviewed paper [69]. The details may be found in this work.

The experiments are performed by pulling steel and plastic endless belts through the air and water leading to the following basic conclusion: the calculation data for the plate temperature distributions well agree with the experimental results and both show that the conjugate parameter $\gamma = (\lambda/\lambda_w) \mathrm{Lu}$ and dimensionless coordinate ξ (4.40) are the proper variables governed the conjugate heat transfer in the case of moving plate.

Comment 4.10 It was shown that Biot number is a measure of conjugate rate of steady-state problems (S. 2.1.7) and Luikov number plays similar role in the case of unsteady heat transfer (S. 3.1.1.5). The conjugate parameter for the moving plate $\gamma = (\lambda/\lambda_w) \mathrm{Lu}$, which follows from third equation of system (4.40), in fact is a combination of Biot and Luikov numbers. This becomes clear if one takes into account that dimensionalities of λ and $h\Delta$ are equal ($\lambda \sim h\Delta$), so that $\gamma = (\lambda/\lambda_w) \mathrm{Lu} \sim \mathrm{Bi} \cdot \mathrm{Lu}$.

■ Example 4.14a/n: Continuous Plate Moving Through Surrounding (Two Parts Solution)

Here, the same problem for moving through surrounding continuous plate is solved using the reduction of conjugate problem to the equivalent conduction problem (S. 2.2.2) [110]. The initial governing system is the same as in previous example. The solution of this problem consists of two parts. The first part presents the solution for initial section of a plate from $x = 0$ to $x = x_{cr}$ where the thermal boundary layers growing along the plate joined (Fig. 1.12). It is shown [110] that for this part equation (4.39) for continuously moving plate in variables x/L and $\varphi = y\sqrt{U_w/2\nu x}$ coincides with equation defining coefficient g_k for universal function (1.38) in the case $\Pr \to 0$. Thus, the heat on the moving plate is determined by same series (1.38) with coefficients (1.42)

$$q_w = \frac{\lambda_w \sqrt{\mathrm{Re}}}{\theta_w} \sqrt{\frac{\Pr}{x\pi}} \left(T_w - T_0 + \sum_{k=1}^{\infty} g_k x^k \frac{d^k T_w}{dx^k} \right), \quad g_k = \frac{(-1)^{k+1}}{k!(2k+1)} \tag{4.41}$$

At the same time, the heat flux from fluid for moving plate is given as well as for fixed plate by series (1.38) with coefficient g_k obtained in Section 1.6. Substituting both series in conjugate condition $q_{w1} = q_{w2}$ yields the equation

$$K(T_w - T_\infty) + (T_w - T_0) + \sum_{k=1}^{\infty} (Kg_{k1} + g_{k2})x^k \frac{\partial^k T_w}{\partial x^k} = 0, \quad \frac{T_w - T_\infty}{T_0 - T_\infty} = \frac{1}{1 + K}, \tag{4.42}$$

from whence it follows that the plate temperature at the initial section at a die is constant and is defined by the second equation (4.42). That is because the first equation may be satisfied only assuming that each term is zero $\partial^k T_w / \partial x^k = 0$ resulting in dimensionless temperature (4.42) with $K = g_0 \mathrm{Pr}^{1/2} \sqrt{\pi/\gamma}$ and the same $\gamma = (\lambda/\lambda_w) \mathrm{Lu}$(Exam. (4.13)).

Comment 4.11 The physical reason of the plate temperature constancy is the similarity of temperature distributions in the fluid and in the solid leading to the proportional grows of the thermal boundary layer in the moving plate and in flow entrained by it.

For the second part of the plate for $x > x_{cr}$ where the boundary layers on the inner plate surfaces interact, the conjugate problem is solved by reduction to the conduction problem similar to solution of equation (4.37) in example 4.12. In this case, the system of boundary layer equations (1.9)–(1.11) and equation (4.39) for the moving plate is solved numerically using two last relations (4.39) as boundary conditions. Three iterations provide acceptable accurate solution. Details are given in [110]. The results for the symmetric flow around plate for various Pr, γ and $\phi = U_\infty / U_w$ are plotted in Figure 4.9. The curve 7 for $\phi = 0$ and $g_1 = 0$ represents solution obtained with boundary condition of the third kind. The dashed curve pertains to the

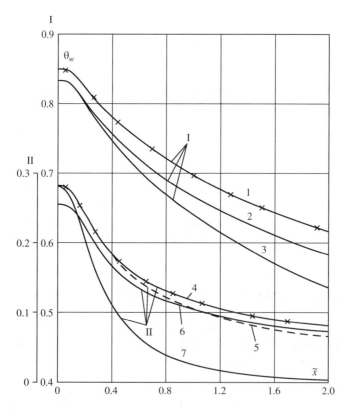

Figure 4.9 Temperature of the moving plate surface in symmetrical flow: I)$\gamma = 0.042$, Pr $= 5.5$; 1) $\phi = 0.2$) $\phi = 0.8$, 3) $\phi = 0$, $g_1 = 0$, $\varepsilon = 0$, II) $\gamma = 8.51$, Pr $= 6.1$;4) $\phi = 0$, 5) $\phi = 0$, $\varepsilon = 0$,6) $\phi = 0.8$, 7) $\phi = 0$, $\varepsilon = 0$, $g_1 = 0$

first iteration when only the first derivative is taken into account and $\varepsilon = 0$. It is seen that the result obtained after correction (curves 1 and 4) well agree with experimental data from [69] showing by crosses, whereas the significant deviation of the temperature distribution calculated using boundary condition of the third kind (curve 7) implies that the problem of heat transfer between a moving plate and the surrounding must be solved as a conjugate problem.

■ Example 4.15n: Horizontal and Vertical Moving Continuous Plates [189]

The same problem as in two previous examples is considered for both horizontal and vertical moving plates with additional effects of unsteady natural convection and two-dimensional aspects near the slot. Solution of such problem requires more complex governing equations, which are used in the following vector form (S.7.1.2.1)

$$\nabla \cdot \mathbf{v} = 0, \quad \frac{\partial \mathbf{v}}{\partial t} + \mathbf{v} \cdot \nabla \mathbf{v} = -\frac{\nabla p}{\rho} + \nu \nabla^2 \mathbf{v} - g\beta(T - T_\infty), \quad \mathbf{v} = \mathbf{i}u + \mathbf{j}v$$

$$\frac{\partial T}{\partial t} + \mathbf{v} \cdot \nabla T = \alpha \nabla^2 T, \quad (\rho c_p)_w \frac{\partial T}{\partial t} + (\rho c_p U)_w \frac{\partial T}{\partial x} = \lambda_w \nabla^2 T \tag{4.43}$$

Here, the first and second equations are continuity and Navier-Stokes equations, the fourth one is the energy equation and the last is unsteady equation for solid. The boundary condition are as usual no-slip at the moving plate and on the slot walls, zero velocity far away from the plate, conjugate conditions on the interface, and the symmetry condition on the x axis. The last condition is not valid if the buoyancy effect is taken into account, except the cases of vertical plate or when the surface with x axis is insulated. The special condition $-x_{in} < x < 0$, $T = T_0$ is used to simulate the proper initial condition implying that at time $t = 0$, the very long plate at temperature T_∞ starts also moving with constant velocity, when the heat input upstream at $x = 0$ is turned on.

Equations (4.43) are transformed in the stream function-vorticity $(\psi - \omega)$ form (S. 7.1.2.3 and 7.1.2.4) using dimensionless variables scaled by: $\Delta/2$ for x and y, U_w for u and v, $\Delta/2U_w$ for t, $2U_w/\Delta$ for ψ and ω. Re= $U_w\Delta/2\nu$, Gr = $g\beta(T_0 - T_\infty)\Delta^3/8\nu^2$, Pe = $U_w\Delta/2\alpha_w$ and $\theta = (T - T_\infty)/(T_0 - T_\infty)$. For the plate moving vertically upward, the transformed equations (4.43) are

$$\nabla^2\psi = -\omega, \quad \frac{\partial \omega}{\partial t} + \mathbf{v} \cdot \nabla \cdot \omega = \frac{\nabla^2\omega}{\mathrm{Re}} - \frac{\mathrm{Gr}}{\mathrm{Re}^2}\frac{\partial \theta}{\partial y}, \quad \frac{\partial \theta}{\partial t} + \mathbf{v} \cdot \nabla\theta = \frac{\nabla^2\theta}{\mathrm{Re}\,\mathrm{Pr}}, \quad \frac{\partial \theta}{\partial t} + \frac{\partial \theta}{\partial x} = \frac{\nabla^2\theta}{\mathrm{Pe}}$$
$$\tag{4.44}$$

For a plate moving vertically downward, the sign of the buoyancy term is reversed. For a horizontal plate, the buoyancy gives rise to a pressure gradient normal to plate. Because of that, the last negative derivative with respect to y in the second equation should be changed to the positive derivative with respect to x.

The computation domain is divided in two regions. For the first region, starting from the slot, the Navier-Stokes and full energy equations are solved because here the nonboundary layer effects are important. These two equations of system (4.44) and the third one, the conduction equation for the plate (the system (4.44) except first equation), are solved applying alternate direction implicit (ADI) method (Com. 3.10). The first equation of this system, the Poisson's equation (S. 1.1) for stream function, is solved using overrelaxation method (Com. 4.12). For the second region downstream from the slot, the boundary layer equations are employed. The boundary conditions at the interface between two regions obtained numerically are used to join both solutions. The procedure of iterations and other details are described in the reviewed paper [189].

Comment 4.12 The relaxation means are used to improve the convergence of Gauss-Seidel method by choosing the proper value of relaxation factor η, which relates the successive iterations; the method is known as underrelaxation and overrelaxtion depending on using factors from ranges $0 < \eta < 1$ and $1 < \eta < 2$, respectively.

The basic results for different Pe, Gr, Pr, λ/λ_w and Re are listed below:

- Streamlines and isotherms for an aluminum plate moving vertically upward show that flow starts close to the moving plate. Then, a recirculating flow appears near the slot and moves away downstream. At first, heat is conducted along the plate and afterwards is going to ambient fluid. Finally, the transient process approaches the steady state. When the plate is started heating after the fully flow is developed, no recirculating flow arises, and the entrained fluid comes from ambience. Although the transient process is different from a steady state, the velocity and temperature distributions are identical.
- The buoyancy affects the corresponding transient velocity and temperature distribution only in time after beginning of the transient process, and the velocity maxima occurs during this process but not at the steady state. The Nusselt number decreases with the distance and after approaching minimum value increases. The minimum value of the Nusselt occurs in the recirculation zone mentioned above.
- As the same aluminum plate moves horizontally, the steady state is reached later than in the case of a vertically moving plate. The reason of this is the different directions of buoyancy forces, which aid the flow in the case of vertical plate but are normal to flow direction for the horizontal plate. The temperatures of a horizontal plate are higher than those for vertical plate. This occurs due to higher velocities near the vertical plate, which results in a higher amount of removal heat from the plate.
- In a transient process, the maximum of relatively velocity u/U_w monotonically increases near the slot. As the distance from the slot increases, an overshoot phenomenon arises, and due to buoyancy forces increasing, it becomes larger with distance.
- Although in the steady state the flow patterns are similar for different values of parameters, the flow in the case of air (Pr = 0.7) is stronger than that in the case of water (Pr = 7) because the buoyancy effects are larger in the air. The temperature on the aluminum plate is almost uniform, which is because of a high thermal conductivity. Near the slot, the strong cooling is observed. This effect is not always desirable because it may result in residual thermal stresses. For the case of air and Teflon with low thermal conductivity, the plate is not cooling much.
- As Grashof number increases, the flow becomes stronger, and the heat transfer rate from the plate to the fluid as well as the Nusselt number grows. The plate velocity affects the induced flow and finally changes the plate temperatures and Nusselt number. Therefore, when the plate velocity is higher, these quantities are also higher. However, a certain region is observed where despite the higher Nusselt number in the case of higher plate speed, the greater distance is required to cool the plate to a given level.
- The data obtained for different ratio λ/λ_w show that heat transfer coefficient near the slot basically is determined by plate thermal conductivity so that the aluminum plate has the

highest Nu in this region. The comparison of these data with results for stationary isothermal or with uniform heat flux plate indicates that the temperature distributions in the transverse direction in both cases are similar. At the same time, the velocity distributions are different, and the Nusselt number is higher on the moving plate.

4.2.2 Cooling Systems

4.2.2.1 Electronic Packages

Miniaturization of contemporary electronic systems leads to increasing the amount of heat flux per unit volume, which must be removed to provide stable and reliable operation conditions. This requires high accuracy predictions of the heat transfer characteristics and appropriate efficient cooling systems.

■ **Example 4.16n: Protruding Heat Sources on the Wall of the Horizontal Channel [314]**

Four volumetric sources are mounted on the bottom wall of a horizontal channel. The heat transfer occurs by mixed convection and radiation. The air is considered as cooling agent. The flow is laminar and hydrodynamically and thermally developed. The system of two-dimensional continuity (1.4), Navier-Stokes (1.5) and (1.6) and energy (1.8) (without dissipation) equations is used for fluid. To take into account the natural convection, the term $(Gr/Re)\theta$ for buoyancy force (S. 7.8) is added in equation (1.6). The following two-dimensional Laplace equations are employed for determining the temperatures of the walls and streamlined sources

$$\frac{Pe}{Lu_w} \frac{\partial \theta}{\partial t} = \frac{\lambda_w}{\lambda} \left(\frac{\partial^2 \theta}{\partial x^2} + \frac{\partial^2 \theta}{\partial y^2} \right), \qquad \frac{Pe}{Lu_h} \frac{\partial \theta}{\partial t} = \frac{\lambda_h}{\lambda} \left(\frac{\partial^2 \theta}{\partial x^2} + \frac{\partial^2 \theta}{\partial y^2} \right) + \frac{R^2}{L\Delta} \qquad (4.45)$$

The scales for dimensionless variables are: R for x and y, U for u and v and ρU^2 for p, R is the channel height, Δ and L are height and width of sources, $Lu_w = \rho c/(\rho c)_w$, $Lu_h = \rho c/(\rho c)_h$ are Luikov numbers for channel walls and for sources, respectively, subscript h denotes the heat source, $Re = UR/\nu$, and $Pe = UR/\alpha$. The temperature and the Grashof number are defined in terms of the volumetric heat generation q_v as follows: $\theta = (T - T_\infty)/\Delta T_{ref}$ and $Gr = g\beta q_v L\Delta R^3/\lambda\nu^2$ where $\Delta T_{ref} = q_v L\Delta/\lambda$. The source term q_v/λ in energy equation (4.45) (the last one) defined in terms of these scales is presented only through geometric parameters.

The uniform inlet conditions for fluid and no-slip boundary conditions at the walls are applied. The outer wall surfaces are assumed to be adiabatic. At the outlet of the channel an extended domain is used to avoid the influence of large recirculation occurs at the last source. It is assumed that the boundary conditions: $\partial u/\partial y = v = \partial\theta/\partial y = 0$ and $\partial^2 u/\partial x^2 = \partial^2 v/\partial x^2 = \partial^2 \theta/\partial x^2 = 0$ are appropriate for top and bottom of extended domain and of it outlet, respectively.

Comment 4.13 The extended domain and boundary conditions are adopted to provide uniform velocity profile at the channel outlet for Dirichlet problem formulation (S. 1.1).

Radiative heat transfer is calculated using the radiocity-irradiation approach [172] as in Example 4.11. The surfaces are considered as opaque, diffuse, and gray, and the inlet and outlet of the channel are treated as black surfaces at ambient temperature. A radiative heat flux q_{Ri} from a discrete surface and radiocity $J_i = \sigma T_\infty^4$ are determined as

$$q_{Ri} = \frac{\varepsilon_i}{1 - \varepsilon_i} \left[\left(\frac{T_i}{T_\infty} \right)^4 - \frac{J_i}{\sigma T_\infty^4} \right], \quad \frac{J_i}{\sigma T_\infty^4} = \varepsilon_i \left(\frac{T_i}{T_\infty} \right)^4 + (1 - \varepsilon_i) \sum_{j=1}^{n} \frac{J_i}{\sigma T_\infty^4} F_{ij} \quad (4.46)$$

where ε and F_{ij} are emissivity and shape factor. Temperatures at the channel walls/fluid interface and at source/fluid interface are determined from energy balance equations

$$-\frac{\lambda_w}{\lambda} \left(\frac{\partial \theta}{\partial y} \right)_w = -\frac{\partial \theta}{\partial y} + q_R N_{RC} \quad N_{RC} = \frac{\sigma T_\infty^4}{\lambda} \frac{R}{\Delta T_{ref}} \quad (4.47)$$

$$\frac{\lambda_h}{\lambda} \left(\frac{\partial \theta}{\partial x} \right)_w \Delta y - \left[\frac{\lambda_h}{\lambda} \left(\frac{\partial \theta}{\partial y} \right)_w + \left(\frac{\partial \theta}{\partial y} \right)_n + \frac{\lambda_h}{\lambda} \left(\frac{\partial \theta}{\partial x} \right)_e + \frac{R^2}{L\Delta} \Delta y - q_R N_{RC} \right] \Delta x = Pe \frac{\partial \theta}{\partial t} \Delta x \Delta y$$

where Δx and Δy are dimensionless width and height (scaled by R) of an element chosen for the energy balance and subscripts e and n refer to extended domain and nodal points.

The SIMPLE algorithm (S. 9.7) and the point-by-point Gauss-Siedel iteration method are used to solve the governing equations for velocity components, pressure, temperature, and radiosity. The following basic results and conclusions are obtained:

- The temperature of the first chip is lower than others due to contact with fresh air. The maximum temperature of the last chip is lower than that of others due to high recirculation at the last chip. The radiation heat transfer from the first and the last chips is higher than from the others, because these are exposed to the atmosphere. The radiation effect and the maximum temperature become smaller as the Reynolds number grows.
- The buoyancy effect was studied using different Grashof numbers obtained by changing the value of the volumetric heat generation q_v. The results show that with increasing Grashof number, the dimensionless temperature decreases linearly. It follows from this data that effect of buoyancy is negligible for the range of parameters studied. Despite the dimensionless temperature decreases, the actual dimension temperature increases as the Grashof number increases, as is expected.
- The dimensionless temperature decreases as the emissivity of heat sources and of the walls increase if all Reynolds numbers are analyzed. The effect of the wall emissivity is more significant than that of the heat sources. The temperature decreases also when the emissivity of the substrate grows. As the emissivity of substrate changes from 0.02 to 0.85, the maximum temperature decreases by $11°C$, whereas the same change of emissivity of heat sources gives only $4°C$ drops in maximum temperature.
- The dimensionless temperature decreases as the thermal conductivities of the heat sources and substrate increase. At $\lambda_h/\lambda_w = 500$, heat sources become isothermal. When the thermal conductivity ratio of heat sources and fluid λ_h/λ changes from 50 to 500, the calculation shows a 20% drop in the maximum temperature. Similar behavior of dimensionless temperature is observed when the wall/fluid conductivity ratio is varied.
- As the emissivity of substrate and top wall increases, the contribution by radiation decreases, whereas the convective contribution increases. Increasing of the heat sources emissivity leads to increasing radiation contribution. At $Re = 250$, the radiation contribution increases from 10.5% at emisivity of heat sources at $\varepsilon_h = 0.02$ to 19% at $\varepsilon_h = 0.85$. As the Reynolds number increases, the radiation fraction becomes less.

■ **Example 4.17n: Closed Unit with Electronic Heat-Emitting Elements [210]**

Unsteady heat transfer in a small hermetically sealed unit of an electronic system with a heat source is considered. The heat sources have a core of constant power. It is assumed that the housing walls are thermally insulated. A computing domain consists of four walls formed an air filled cavity with a heat-emitting element (e.g., a transistor), which generates a constant heat flux.

The mathematical model similar to that in previous example consists of two-dimensional unsteady Navier-Stokes equation in $\omega - \psi$ variables with additional term for natural convection, energy equation for temperature, and conduction equation with source of generation for walls. Relations $\psi = \omega = 0$ and $T = T_0$ are initial conditions; no-slip and thermal insulation are conditions for walls. The conjugate conditions are used in the form of equalities of temperatures and heat fluxes for gas and walls and for heat sources and gas. The problem is solved by finite-difference method. The calculations are carry out for square steel and fiberglass boxes with 0.1 m of length and 0.005 m walls thickness containing the heat source of 0.02 m length and 0.015 m height with volumetric heat generation in interval $10^5 - 10^7$ W$/m^3$. Other model characteristics are given in [210].

The following conclusions are formulated:

- For heat generation $q_v = 10^5$W$/m^3$, two convective cells with smaller secondary flows in the corners are formed in the cavity. Immediately above the heat source, ascending flow occupies a domain, whereas the descending flows exist near the walls. The energy is transmitted from the source to the gas-giving rise to a thermal plume, whose position is determined by the ascending flows. In this case, the conductivity heat transfer plays a dominant role, whereas the convective heat transfer contribution is small.
- As the power of volumetric generation increases, the flow velocities in convective cells grow and the size of these vortices decreases. In this case, the size of secondary vortices and the gas temperature increase as well leading to more clear shape of thermal plume. Inside the heat source, the temperature inhomogeneity is observed.
- Similar investigations for different wall material and conditions of operation show that: (i) increase in the thermal diffusivity of the wall material leads to significant decreasing in the temperature of the air inside the cavity and of the temperature of heat source itself, (ii) increase in the heat emission capacity results in a growing the cavity gas temperature close to the heat source, (iii) as it was expected, the temperature of the fiberglass wall is higher than that of the steel wall, (iv) the extension of time of unsteady operation significantly modifies the velocity and temperature fields leading, in particular, to much smaller secondary flows of different form moving to the upper part of the cavity, (v) the power of heat source generation plays critical role in formation the thermo-hydrodynamic regimes, and (vi) using the heat transfer coefficient, especially for unsteady regimes is unreasonable.

4.2.2.2   Turbine Blades and Vanes Cooling

■ **Example 4.18n: Film Cooled System For Blades and Vanes [197]**

Film cooling is used to protect turbine blades and vanes of the first and second rows from direct contact with streams of hot gas. Injected cold air covers the surface of the blade or vane and produces a layer of cold air between protected object and hot gas. The reviewed article presents the temperature field of the model of the real engine with film cooled vanes [165] obtained by numerical solution of relevant conjugate heat transfer problem. The NASA explicit finite volume code Glenn-HT is used to solve Navier-Stokes equations for fluid domain, and boundary element method (BEM) is employed for solving the Laplace equation for the body. Such combination approach as applied to conjugate problem provides two advantages arising from the fact that BEM requires only body surfaces discretization (S. 9.6). The first benefit of that is the reducing of computing time and storage saving, since there is no need to involve in calculation the whole body domain. This feature of BEM is especially important for the conjugate problems, whose solution is associated with iterations when the temperature distribution on the interface obtained by Laplace equation is used as a boundary condition for Naver-Stokes equation, and Navier-Stokes solution data gives the boundary condition for the Laplace equation. Because in such iteration procedure, only the surface temperature under Laplace equation solution is required, using BEM results in eliminating the rest part of solid descretization, which in this case is not needed. Another merit of using BEM in conjugate problem follows from the fact that in BEM the temperature and heat fluxes on the interface are directly obtained satisfying the conjugate conditions, and thus there is no need in the solid temperature differentiating usually employed to get heat fluxes on the interface.

The kinetic energy turbulence and the specific dissipation rate are determined by two-equations $k - \omega$ turbulence model (S. 8.4.3.1). For turbulence viscosity and turbulence thermal conductivity, the Boussinesq formulae with $Pr_{tb} \approx 0.9$ are used. The Kirchhoff transform is employed to convert the equation for body with conductivity dependent on temperature $\lambda_s(T)$ to usual form of the Laplace equation.

Comment 4.14 The Kirchhoff transform is defined as integral of conductivity $\lambda_s(T)$ with λ_0 and T_0 as reference parameters in the form of the first equation

$$\lambda_{kf}(T) = \frac{1}{\lambda_0} \int_{T_0}^{T} \lambda_s(\xi)d\xi, \quad \frac{\partial}{\partial x}\left[\lambda_s(T)\frac{\partial T}{\partial x}\right] = \frac{\partial}{\partial x}\left(\frac{\partial \lambda_{kf}}{\partial T}\frac{\partial T}{\partial x}\right) = \frac{\partial^2 \lambda_{kf}}{\partial x^2}, \quad \nabla \cdot [\lambda_s(T)\nabla T] = 0$$

(4.48)

It follows from the first equation that $\lambda_s = \partial \lambda_{kf}/\partial T$. Then, from the second equation one may see that using this relation and chain rule transform the Laplace equation (1.1) with variable conductivity in usual form of a second derivative but of Kirchhoff variable λ_{kf} instead of temperature. The third relation (4.48) is the Laplace equation with dependent on temperature conductivity in the general vector form.

A system of three-dimensional governing equations in integral form and other details of solution are given in [197]. The numerical calculations are performed under conditions that match the planning experiment at NASA Glenn Research Centre. The geometry of the model vane is based on the scaled factor 2.943 to mach the exit Mach number M = 0.876 and Reynolds number Re = $2.9 \cdot 10^6$ computed on real chord 0.206 m. The vane has two plena that feed 12 rows of film-cooling holes. A special program is adopted to model this complex geometry. The basic results are as follows:

- The comparison of the temperature distribution along the vane obtained from conjugate problem with that calculated by standard two-temperature method shows a significant difference between those results indicating that conjugate prediction for the minimum temperature is lower and for the maximum temperature is higher.
- These two temperature distributions differ not only in separate values or at discrete points, but are entirely different. In particular, according to conjugate data the trailing part of the blade is much hotter; whereas the forward blade section as it follows from the conjugate solution is considerable cooler.
- It is evident that such temperature changes are important for stresses analysis and design of the highly loaded turbine blades.

■ Example 4.19n: System with Radial Channels For Blades Cooling [82]

The other type of cooling system for turbine blades is performed using radial channels with flowing cold air. Such channels are made through whole blade in radial direction perpendicular to a blade cross-section. The set of governing equations is similar to that employed in the previous example. The two-dimensional Navier-Stokes equations are used for the fluid domain as well as the Laplace equation is applied to compute the heat conduction in solid. In this model, the Navier-Stokes equations are written in curvilinear coordinates. The finite-volume method (S. 9.6) is applied for numerical solution of both the Navier-Stokes and Laplace equations. The turbulence is taken into account by Baldwin-Lomax model (S. 8.3.5). To calculate the heat fluxes from the blade to cooling air in the radial channels, the heat transfer coefficient is prescribed.

To conjugate solutions of different domains, the iteration procedure is employed. As usual, the Dirichlet problem is considered for fluid, and the Neumann problem formulation is used for heat conduction equation (S. 1.1). It is noted that such approach gives the stable solutions, but in the case of vice versa, when the heat fluxes are used as a boundary conditions for the Navier-Stokes equations, and the surface temperature distribution is employed to solve the Laplace equation, unstable results may be obtained.

Three blade configurations with different size and location of cooling channels at two exit Mach numbers $M = 0.59$ and $M = .0.95$ have been investigated. Distribution of parameters around blades with different cooling systems are calculated showing that:

- Cooling duct configurations have a little effect on pressure distribution around the blade, but the temperature distribution strongly depends on size and location of the cooling ducts.
- The blades with small channels and relatively uniform positions show better cooling effects and smaller mass flow rate. This effect is more pronounced on the pressure blade side.
- On both blade sides, the temperature decreases approaching the minimum close to stagnation point and then rapidly increases to its maximum.
- Mach number gradually increases on the suction side of the blade, but on the pressure side, Mach number increases almost until the exit and then goes down. For higher Mach number at the exit ($M = 0.95$), this behavior is more pronounced and leads to the supersonic values at the maximum.

- Accordingly, the gauge pressure on the suction side starting from stagnation point is almost constant except small area close to the exit where it drops to zero. At the same time, on the other side in conformity with Mach number behavior, the pressure decreases, approaches the negative values of gauge pressure at the minimum, and then grows to zero at the exit.

4.2.2.3 Thermal System Protection in Rockets and Nuclear Reactors

■ **Example 4.20* n: Charring Protection of Rocket and Reentry Vehicle [24]**

A charring materials exposed to high temperature is a process of decomposition and loss of surface material by ablating to absorbing the heat. Such processes are used, for instance, for internal thermal protection of rocket combustion chambers or for thermal shield of reentry vehicles. This article studies the charring material process using the three-dimensional model composed of three zones: the virgin zone, the decomposition zone, and the char zone. These zones are disposed one over another so that on the top, the chair zone appears along with the working fluid flows. In the first zone, the material changes are negligible; in the second zone, the material undergoes chemical and physical changes and energy is absorbed by decomposition; and in the third zone, composed mainly of char, the heat is transferred by conduction and convection. The changes in the material proceed by two ways: by the free material surface recession and by its decomposition, when the surface does not move, but the material properties are changed. During this process due to heating, the material releases pyrolysis gas, which passes through the solid into fluid that flows along the upper chair zone.

To simplify the mathematical model, two basic assumptions are used: (i) the pyrolysis gas velocity is approximately orthogonal to receding surface, and (ii) the surface regression is locally uniform and occurs along the normal direction to the surface. The governing equations for such simplified model for the fluid consist of conservation laws of mass, momentum, and energy equations in the vector form

$$\frac{D\rho}{Dt} + \rho \nabla \cdot \mathbf{v} = 0, \quad \rho \frac{D\mathbf{v}}{Dt} = \rho g - \nabla p + \mu \nabla^2 \mathbf{v} + \frac{\mu}{3} \nabla(\nabla \cdot \mathbf{v})$$

$$\rho c_p \frac{DT}{Dt} = \frac{Dp}{Dt} + \nabla \cdot (\lambda \nabla T) + \mu S \tag{4.49}$$

The energy equation for the decomposing charring material and it derivation on the base of the Arrhenius decomposition law are given in [24] as follows

$$\rho c_p \left(\frac{\partial T}{\partial t} \right)_\xi = \nabla \cdot (\lambda \nabla T) + \rho c_p \mathbf{v} \cdot \nabla T + (J_g - \hat{J}) \left(\frac{\partial \rho}{\partial t} \right)_x + \rho_g \mathbf{v}_g \cdot \nabla J_g \tag{4.50}$$

$$\hat{J} = \frac{\rho_v J_v - \rho_c J_c}{\rho_v - \rho_c}, \quad J_c(T) = J_c(T_r) + \int_{T_r}^{T} c_{pc}(t)dt, \quad J_v(T) = J_v(T_r) + \int_{T_r}^{T} c_{pv}(t)dt$$

$$\lambda = x\lambda_v + (1-x)\lambda_c, \quad c_p = xc_{pv} + (1-x)c_{pc}, \quad x = \frac{\rho_v}{\rho_v - \rho_c}\left(1 - \frac{\rho_c}{\rho}\right)$$

Here, **v** is the thermal protection recession velocity vector, S is a dissipation function in energy equation, J is solid enthalpy, $\hat{J}$ is weighted solid enthalpy, T_r is the reference temperature for enthalpy estimation, x is a fraction of virgin and subscripts g, c, v denote: pyrolysis gas, charred, and virgin values, whereas subscripts x and ξ refer to derivatives at constant x in a fixed frame and at constant ξ in a moving frame.

Comment 4.15 The Arrhenius equation determines the chemical reaction constant k as a function of the temperature and activating energy E in the form $k = A \exp(-E/RT)$, where A is so called pre-factor (the experimentally determined constant) and R is gas constant. It is shown that equation (4.50) follows from the conservation equations (4.49) and two other basic relations determining the density and its variation inside the decomposing material

$$\rho = \Gamma(\rho_1 + \rho_2) + (1 - \Gamma)\rho_3, \quad \left(\frac{\partial \rho}{\partial t}\right)_i = -A \exp(-E/RT)\rho_{oi}\left(\frac{\rho_i - \rho_{ri}}{\rho_{oi}}\right)^{m_i} i = 1, 2, 3$$

(4.51)

The first equation (4.51) estimates the density and corresponding derivative with respect to time (after differentiation) as a combination of three component densities, with $i = 1$ and $i = 2$ components of resin, the reinforcing material ($i = 3$), and weighting coefficient Γ(S. 9. 6). The second equation (4.51) defines the derivatives of those three components as a function of temperature according to Arrhenius equation with apt correction, where ρ_i, ρ_{0i}, ρ_{ri} are the density of component i, the original and residual (S.9.6) densities of component i.

In equation (4.50) the term on the left-hand side is sensible energy accumulation, the first term on the right-hand side is the conduction term, the second term is the energy removed by the motion of reference frame, the third term is the difference between the energy removed away by pyrolysis gas and the chemical energy, and the fourth one is the energy removed by pyrolysis gases passing through the solid. The turbulence is taken into account using the $k - \omega$ turbulence model (S. 8.4.3.1). The governing equations are solved numerically applying three-dimensional code Phoenics. This software uses a finite volume element approach and, as well as other similar programs, like SIMPLE or SIMPLER, can be used for solving the energy equation of both the fluid and solid by accounting for corresponding boundary and velocity conditions (S. 9.7). The details of numerical performance are given in [24]. The predicted results for some problems are compared with known experimental data or with analytical or numerical solutions:

- Results of simulation of heat transfer in a blast tube with thermal protection are compared with data obtained by CMA program [8]. The initial temperature is 300 K. The laminar flow of combustion gas in the tube is at temperature 3600 K. The density and temperature profiles in the fluid and solid obtained in both studies are in agreement.
- The process in Material Test Motor (MTM) for testing new ablative materials with protection is simulated for the charring material ES59A with a low-density thermal protection ESA-ESTEC developed for space rockets [47]. The dimensions of the model cross-section have been extrapolated from test section in MTM and the curvilinear mesh is used. The velocity and pressure as well as turbulent viscosity and conductivity show low numerical errors, which indicates that these are computed correctly. The results indicate that assumptions of negligible propellant reduction and constant pressure in chamber during burning are possible. The temperature profiles obtained in this case are similar to these in the blast tube with a typical sudden derivative variation arising due to passage through the different computation domains. The density changes suddenly between charred and virgin zones at

approximately one half of the solid thickness. The comparison of Material Affected Depths (MAD) obtained experimentally and predicted by three different approaches shows a reasonable agreement, indicating that after 25.3 seconds burning time, the MAD is about 4 mm. The mass flow rate and heat fluxes predicted by conjugate heat transfer approach are slightly underestimated, whereas the MAD predictions are basically overestimated.

- The heat exchange analysis of igniter of solid propellant rocket during the turbulent combustion is performed. Some approximations are employed to simplify the problem model. The pressure at rocket exit is assumed to be constant, the external thermal protection of the igniter is considered as nondecomposing, and the igniter switching time is assumed to be negligible. Two thermal protections are considered: one from aramidic fiber and another by using silica phenolic with reduction erosion rate. The adiabatic combustion gas temperature 3424 K is uniformly kept everywhere except for the solid parts which are at 300 K. The velocities in the chamber are slow after a short transient period, and therefore, the buoyancy convection is taken into account, whereas the radiation is assumed to be negligible. The MAD and the thermal fields in gas and solid are calculated. The results indicate that the steel interface temperature is between 350 and 400 K under the initial temperature of 300 K. The temperature profiles in the internal thermal protection of the igniter are calculated as well as are the profiles of the solid density in the thermal protection.

■ **Example 4.21a: Emergency Cooling in Nuclear Reactor [118]**

One of the ways of reducing the fuel temperature during emergency situation in nuclear water reactors consists of using the rewetting process for cooling the hot surface by adjacent liquid film. The model here reviewed, in contrast to other knowing studies of heat transfer between a hot surface and liquid moving film, takes into account a transient character of the real rewetting process. The essential specialty of this approach is that instead of usual assumption of infinite long surface, in this case, a semi-infinite object should be considered. As a result, the length of the wet portion covered by film and the surface temperature at the moving film front become unknown functions of time leading to considerable complicated problem. The following assumptions are used to simplify the model: (i) the lower plate surface is adiabatic, (ii) the heat transfer coefficient between the plate and the film is constant and known, (iii) the film is supplied at constant velocity U, and (iv) the heat losses to surrounding are negligible.

Comment 4.16 This problem is an intrinsic conjugate even at a known heat transfer coefficient between a plate and film because the rewetting process divides the plate in two parts with time variable lengths: the wet (covered by film) and dry portions, which should be conjugated.

The one-dimensional conduction equation in the moving frame with the origin at the film front and initial and boundary conditions for wet and dry plate portions are:

$$\lambda_w \frac{\partial^2 T}{\partial x^2} - \frac{h}{\Delta}(T - T_f) = \rho_w c_w \frac{\partial T}{\partial t} \tag{4.52}$$

$$wet: \quad T(x, \ t = 0) = T_i, \quad T(x = 0, \ t > 0) = T_w(t), \quad \frac{\partial T}{\partial x}(x = -U_w t, \ t > 0) = 0$$

$$dry: \quad T(x, t = 0) = T_i, \quad T(x = 0, t > 0) = T_w(t), \quad T(x \to \infty, t > 0) = T_i$$

Here T, T_i, T_f, T_w are temperatures, subscripts i, f, w refer to the initial values, to the film properties and to the plate characteristics at the moving front, the product $U_w t$ in the last condition of the second line is the length of the wet portion at time t, Δ is the plate thickness, and U_w is the front velocity, which is less than constant supplied velocity U due to the evaporation.

The dimensionless variables

$$\eta = x/U_w t, \quad z = th/c_w \rho_w \Delta, \quad \theta = (T - T_f)/(T_i - T_f), \quad \theta_w = (T_w - T_f)/(T_i - T_f) \quad (4.53)$$

transform system (4.52) with the time-dependent wet portion of length $x = -U_w t$ to the system with an independent of time length of wet portion, which is equal $\eta = -1$

$$\text{Ls}\frac{\partial^2 \theta}{\partial \eta^2} + z\eta\frac{\partial \theta}{\partial \eta} - z^2\frac{\partial \theta}{\partial z} - z^2\theta = 0 \quad \text{Ls} = \frac{\lambda_w h}{\rho_w^2 c_w^2 U_w^2 \Delta} = \frac{\text{Bi}}{\text{Pe}^2}, \quad (4.54)$$

$$\theta(\eta, z = 0) = 1, \quad \theta(\eta = 0, z > 0) = \theta_w(z), \quad \frac{\partial \theta}{\partial \eta}(\eta = -1, z > 0) = 0$$

where $\text{Pe} = U_w \Delta/\alpha_w$ and $\text{Bi} = h\Delta/\lambda_w$. System (4.54), unlike the initial set of equations (4.52), depends on only one parameter Bi/Pe^2. As shown below, this ratio controls the rate of the transient cooling process. The greater this ratio, the shorter is a dimensionless time z required to cool the plate to a given dimensionless temperature θ. Because cooling by a thin film proceeds in boiling transitional state, it is suggested in [118] to name a ratio Bi/Pe^2 the Leidehfrost number, Ls, similar to the Leidenfrost point.

For the dry region we use different variables

$$\frac{\partial^2 \vartheta}{\partial \xi^2} - \frac{\partial \vartheta}{\partial z} = 0, \quad \xi = x\sqrt{\frac{h}{\lambda \Delta}}, \quad \vartheta = \frac{T - T_i}{T_f - T_i} \quad \vartheta_w = \frac{T_w - T_i}{T_f - T_i} \quad (4.55)$$

$$\vartheta(\xi, z = 0) = 0, \quad \vartheta(\xi = 0, z > 0) = \vartheta_w(z), \quad \vartheta(\xi \to \infty, z > 0) = 0$$

Here, h is the same heat transfer coefficient for wet portion, whereas the dry portion is considered under zero heat transfer.

As it mentioned above, the plate temperature at the moving film front is unknown. This is a typical situation when the temperature of the interface of two domains of some conjugate problem is not known a priori. Therefore, the conjugation procedure requires special numerical or analytic methods as it explained in Section 1.2. In this case, to conjugate solutions of equations (4.54) and (4.55) for the wet and dry portions, we use series of derivatives similar to universal function (1.38)

$$\theta = \sum_{n=0}^{\infty} G_n(\eta, z)\frac{\partial^n \theta_w}{\partial z^n}, \quad \vartheta = \sum_{n=0}^{\infty} H_n(\eta, z)\frac{\partial^n \vartheta_w}{\partial z^n} \quad (4.56)$$

Substitution of these series into equations (4.54) and (4.55) leads to two infinite systems of equations with constant initial and boundary conditions

$$\text{Ls}\frac{\partial^2 G}{\partial \eta^2} + \eta z\frac{\partial G_n}{\partial \eta} - z^2\frac{\partial G_n}{\partial z} - z^2 G_n - z^2 G_{n-1} = 0, \quad \frac{\partial^2 H_n}{\partial \xi^2} - \frac{\partial H_n}{\partial z} - H_{n-1} = 0 \quad (4.57)$$

$$G_{-1} = 0, \quad z = 0, G_0 = 1, G_n = 0, \quad \eta = 0, G_0 = 1, G_n = 0, \quad \eta = -1, \frac{\partial G_n}{\partial \eta} = 0$$

$$H_{-1} = 0, \quad z = 0, H_n = 0, \quad \xi = 0, H_0 = 1, H_n = 0, \quad \xi \to \infty, H_n = 0$$

The exact solutions of equation (4.57) with constant coefficients for functions H_n are given by the error functions (S. 9.1). The first two functions are

$$H_0(\xi, z) = 1 - erf\frac{\xi}{2\sqrt{z}}, \quad H_1(\xi, z) = \frac{\xi^2}{2}erfc\frac{\xi}{2\sqrt{z}} - \xi\sqrt{\frac{z}{\pi}}\exp\left(-\frac{\xi^2}{4z}\right) \qquad (4.58)$$

Equations (4.57) for G_n with variable coefficients are solved by approximate method of moments (S. 9.6, Com. 8.2). The results are obtained in the following form

$$G_n(\eta, z) = \left(\frac{\eta}{2} + \eta\right)A_n(z) + \left(\frac{\eta^3}{3} + \eta\right)B_n(z) + C_n \quad C_0 = 1, \ C_n = 0 \text{ for } n = 1, 2, 3 \ldots \qquad (4.59)$$

The functions A_n and B_n are given by ordinary differential equations, which for $n = 0$ are

$$\frac{dA_0}{dz} = A_0(z)\left(-\frac{28Ls}{z^2} + \frac{14}{3z} - 1\right) + B_0(z)\frac{16}{3}\left(\frac{Ls}{z^2} - \frac{1}{z}\right) + 28 \quad A_0(0) = 0$$

$$\frac{dB_0}{dz} = A_0(z)10\left(-\frac{2Ls}{z^2} + \frac{1}{3z}\right) + B_0(z)\left(\frac{20Ls}{z^2} + \frac{11}{3z} + 1\right) + 20 \quad B_0(0) = 0 \qquad (4.60)$$

Details of solutions and similar equations for A_n and $B_n (n > 0)$ may be found in [118].

The temperature fields (4.56) in the wet and dry portions just obtained depend on the plate temperature T_w at the moving film front, which is still undetermined. To find this temperature, which is in fact the temperature of the wet/dry plate portions interface, the energy balance at the moving film front is required. Physically, it is clear that the heat $q^+(t)$ conducted from the hot dry region is slightly absorbed by evaporation and sputtering $q_w(t)$ at the moving film front, whereas the majority of the heat $q^-(t)$ is transferred to the wet region. Consequently, the balance equation is as follows

$$x = \eta = 0 \quad q^+(t) = q^-(t) + q_w(t,), \quad q_w(t) = h_w(T_w - T_{wet}) \qquad (4.61)$$

where h_w and T_{wet} are the heat transfer coefficient of evaporation and sputtering at moving film front and the rewetting temperature.

Comment 4.17 This study simulates the initial transient part of the rewetting process when the body temperature at the moving film front is higher than rewetting temperature. The developed model does not describe the second part of the rewetting process with practically constant rewetting temperature. Therefore, formula (4.61) is applicable until $T_w > T_{wet}$. As the Leidenfrost number increases, the temperature at the film front decreases and finally reaches the minimum temperature T_{min} at time z_{min}. Therefore, formula (4.61) and the model are applicable until $T_{min} > T_{wet}$.

Heat fluxes at the moving front consisting in equation (4.61) are calculated according to Fourier law by computing the derivatives of expressions (4.56) for the temperatures of both plate parts at $\eta = \xi = 0$. This procedure gives two relations

$$q^-(t) = -\frac{\lambda(T_i - T_f)}{Ut}\sum_{n=0}^{\infty}\left(\frac{\partial G}{\partial \eta}\right)_{\eta=0}\frac{\partial^n\theta_w}{\partial z^n}, \quad q^+(t) = (T_i - T_f)\sqrt{\frac{\lambda h}{\Delta}}\sum_{n=0}^{\infty}\left(\frac{\partial H}{\partial \eta}\right)_{\xi=0}\frac{\partial^n\vartheta_w}{\partial z^n} \qquad (4.62)$$

Substituting these heat flux components into balance equation (4.61) leads to ordinary differential equation for the plate temperature at the moving film front

$$[z + (a_1 - a_0)\sqrt{z\pi Ls}]\frac{d\theta_w}{dz} + \theta_w\left[1 + a_0\sqrt{\frac{\pi Ls}{z}} + \frac{h_w}{h}\sqrt{z\pi Bi}\right] - \left[1 + \frac{h_w}{h}\sqrt{z\pi Bi}\theta_{wet}\right] = 0$$

(4.63)

and initial condition $\theta_w(z = 0) = 1$. Functions a in this equation depend on functions (4.60): $a_0(z) = A_0(z) - B_0(z)$ and $a_1(z) = [A_1(z) - B_1(z)]/z$. If the first term in equation (4.63) is assumed to be relatively small, the first approximation result for θ_w can be easily obtained solving the algebraic equation (4.63) without the first term

$$\theta_w = \frac{1 + (h_w/h)\sqrt{z\pi Bi}\theta_{wet}}{1 + a_0(z)\sqrt{\pi Ls/z} + (h_w/h)\sqrt{z\pi Bi}}, \quad \theta_{min} = \frac{1}{1 + a_0(z_{min})\sqrt{\pi Ls/z_{min}}}$$

(4.64)

Then, taken in account that in the limiting case we have $\theta_w = \theta_{wet} = \theta_{min}$ (Com. 4.17), we substitute in the first expression (4.64) θ_{min} for both θ_w and θ_{wet} and get the relation between the minimal temperature and time z_{min} required to reach this temperature at the moving film front.

The analysis of final equations and calculation result in following conclusions:

- The first formula (4.64) indicates that at fixed dimensionless time z and parameters Bi, θ_{wet}, and h_w/h, the plate temperature changes inversely to square root of Leidenfrost number
- In the case of negligible heat absorbed at the moving front ($h_w = 0$), both relations (4.64) coincide leading to the same conclusion of Leidenfrost number inversely proportionality for the moving front temperature and it minimal value.
- From these results, it follows that the Leidenfrost number controls the cooling process so that, in system with a greater Leidenfrost number, the plate temperature is lesser for the same dimensionless time z, or vice versa, as it stated above, the greater Leidenfrost numbers, the shorter is the dimensionless time z required to cool the plate to a given dimensionless temperature.
- The calculation indicates that the plate temperature at the moving front sharply decreases at the beginning of the cooling process. Then, the rate of cooling decreases and finally becomes zero at the point of minimal moving front temperature.
- Correlation formulae for minimal temperature and time for reaching it are found

$$\theta_{min} = 0.036\, z_{min} - 0.03 \quad z > 2 \quad z_{min} = 5 - 6\log(Ls) + 3\log^2(Ls)$$

(4.65)

These relations are applicable whether the heat absorbed by evaporation and sputtering is taken into account or not because, as it follows from comparison of two equations (4.64), the onset time does not depend, on heat absorbed by evaporation at the moving front (at least in the first approximation).

An example of second approximation found in [118] by solving the differential equation (4.63) shows that the differences of both results are small for small and large times and reaches 15% to 20% for middle values of time.

OTHER WORKS: A survey of results obtained for heat transfer from moving continuous materials was published by Jaluria [179]. Three types of approaches are considered: the problems with a given heat transfer coefficient, the problems with isothermal or uniform heat flux moving surfaces, and the conjugate heat transfer problems. The thermal treatment of wood is studied numerically and experimentally in [437]. Heat transfer in electronic packages is investigated in articles [415, 436], and [439] considering flows with discrete heating elements. In the first two articles, the channels with electronic packages are studied. In the third paper, an analysis of heated packages mounted in-line on a printed circuit board is presented. The backward-facing step and cube on the wall cooling by jet are investigated in [191] and [312], respectively. Heat transfer in flow past sphere-blunted cone with a blowing-in of a gas is considered in [447]. Unsteady heat transfer around an ablating plate is investigated in [331]. The cooling process in a nuclear reactor considered in the last example is also investigated in [317] by numerical analyzing the dissipation fission heat of a nuclear fuel element into its surrounding. Two recently published works present modern results of the improved cooling systems of turbine blades. The utilizing a finely dispersed water-in-air mixture producing very high cooling rates is suggested in [183]. Authors of a second paper [427] developed improved method of design cooling systems for high performance turbine blades combining the network fluid analysis with conjugate heat transfer approach.

4.3   Simulation of Industrial Processes

Comment 4.18 It is usually difficult to simulate complicated industrial processes in their entirety. To simplify such problems, the real computation domain is divided in several subdomains with relatively simple processes, which could be easier to model. Solutions obtained for the subdomains are then coupled using properly conjugate conditions to get the model of the whole process. Thus, those problems are of conjugate type even if some other interaction effects, for example, wall conduction influence are not taken into account.

■ **Example 4.22* n: Twin-Screw Extruder [334]**

   The actual complicated flow domain as a circular region is divided into two subdomins named translation and intermeshing regions. The translation domain is modeled similar to single screw extruder as a channel with cross section $H/W \ll 1$, where H and W are height and width of the channel. For steady developing, the creeping type approximation (S.7.4.1) is used for two-dimensional flow at low Reynolds number. In contrast to the boundary layer approximation, in this case the inertia terms are neglected, whereas the viscose terms are important. The momentum conservation equations are considered in $y - z$ cross section, where z is directed along the screw helix, y is normal to this direction, and x is normal to $y - z$ cross-section. The temperature dependence of physical properties and dissipation effect are taken into account because the treatment materials are strongly viscous. Thus, the governing system for the translating domain is

$$\frac{\partial p}{\partial x} = \frac{\partial}{\partial y}\left(\mu \frac{\partial u}{\partial y}\right), \quad \frac{\partial p}{\partial z} = \frac{\partial}{\partial y}\left(\mu \frac{\partial w}{\partial y}\right), \quad \frac{\partial p}{\partial y} = 0$$

$$\rho c_p w \frac{\partial T}{\partial z} = \frac{\partial}{\partial y}\left(\lambda \frac{\partial T}{\partial y}\right) + \mu\left(\frac{\partial u}{\partial y}\right)^2 + \mu\left(\frac{\partial w}{\partial y}\right)^2$$

(4.66)

The power law nonNewtonian fluid (S.1.9) is used for temperature dependent viscosity

$$\mu = \mu_0 \left(\frac{\dot{\gamma}}{\dot{\gamma}_0}\right)^{n-1} \exp[-m(T - T_0)], \quad \dot{\gamma} = \left[\left(\frac{\partial u}{\partial y}\right)^2 + \left(\frac{\partial w}{\partial y}\right)^2\right] \quad (4.67)$$

where $\dot{\gamma}$ is the strain rate and m is the temperature coefficient. Equations (4.66) for adiabatic screw under no-slip boundary conditions and specified barrel temperature are solved using finite difference method (S. 9.6).

The flow in intermeshing domain is also simplified using steady creeping flow equations for the case of completely filled with fluid screw channel. Because the axial length of most extruders is much larger than others, the z velocity component, which is directed along the extruder axis, is taken as a small compared with x and y components. The system of governing equations consist of three equations

$$\frac{\partial p}{\partial x} = \frac{\partial \tau_{xx}}{\partial x} + \frac{\partial \tau_{xy}}{\partial y}, \quad \frac{\partial p}{\partial y} = \frac{\partial \tau_{xy}}{\partial x} + \frac{\partial \tau_{yy}}{\partial y}, \quad \frac{\partial p}{\partial z} = \frac{\partial \tau_{xz}}{\partial x} + \frac{\partial \tau_{yz}}{\partial y} \quad (4.68)$$

supplemented by continuity equation. In relations (4.68) $\tau_{i,j} = \mu(u_{i,j} + u_{j,i})$ is the stress tensor (Com. 1.9) presented in Einstein notations (S. 7.1.2.2). The energy equation and expression for strain rate are similar to equations (4.66) and (4.67), but more complicated due to taken into account dissipation (S. 1.14) using two-dimensional function S (1.8)

$$\frac{\partial}{\partial x}\left(\lambda \frac{\partial T}{\partial x}\right) + \frac{\partial}{\partial y}\left(\lambda \frac{\partial T}{\partial y}\right) + \rho c_p \left(u \frac{\partial T}{\partial x} + v \frac{\partial T}{\partial y} + w \frac{\partial T}{\partial z}\right) + \mu S = 0, \quad \dot{\gamma} = S^{1/2} \quad (4.69)$$

As in the previous domain, the no-slip boundary conditions, assumption of adiabatic screw, and specified barrel temperature are applied. The first two equations (4.68) are solved by finite element method, and the last one is solved by Galerkin method (S. 9.6).

It is found numerically by varying the location of the interface of two domains that intermeshing domain should occupy two-thirds of the total circular region. To modeling the flow in screw extruder, two known approaches are used: by the screw moving and by moving the barrel. The flow in the translating domain is simulated by the barrel approach, and the flow in the intermeshing domain is modeled by the screw formulation. To change the type of flow modeling, the profiles of temperature and velocity components are matched at the interface. An iterative procedure is used. Starting with a guessed profile at the inlet, the iterations are continued until the profile at the outlet becomes close to the one obtained from the outlet of another domain. The calculations are performed for a concrete screw with diameter of 23.3 mm, channel depth 4.77 mm, and barrel diameter 30.84 mm, resulting in the following conclusions:

- The flow temperature in for translating region increases above the barrel temperature, which is 275°C. The reason of this is the viscous dissipation. As a result, the heat goes from the fluid to barrel. The temperature variations slightly affect the velocity field, which is defined mainly by flow rate. Knowing the velocity and temperature distributions, one may calculate the stress, strain, and other characteristics.

- The flow and pressure are screw-symmetric in the intermeshing domain. At the centre of this region, a significant pressure change is seen. The linear velocity profile at the outlet yields a nonlinear pressure change along the annulus. The temperature at the root of the screw is lower than that at the barrel, and the temperature rise in this domain is small in comparison with that for translating domain. A portion of total flow in the screw channel goes into the other channel, whereas the remaining flow is retained in the same channel. The ratio of flow portion leaving the channel to the total flow decreases as the index n in the power law (4.67) increases, but it increases when the throughput increases or the depth of screw channel decreases.
- The velocity, temperature, shear, and pressure fields indicate that the region where the both domain flows interact is relatively small. The pressure patterns show the different regions with low, high, and uniform pressure. The pressure rise in the intermeshing region is small as it is compared with that in the translating region, and the same conclusion is valid for the bulk temperature. At the smaller die openings when the throughputs are smaller also, the bulk temperature is higher, because in this case the velocity profiles are much steeper, giving higher viscous dissipation.

■ Example 4.23* n: Optical Fiber Coating [433]

In the typical optical fiber coating system, a bare fiber is drown from a furnace where it riches over $1,600°C$. After cooling to a proper wetting temperature, fiber moves through a coating applicator with a coating fluid, laser micrometer to control a diameter, and finally gets through ultra-violet curing oven to a take-up a spool.

In this study, an axisymmetric two-dimensional process is considered in cylindrical coordinates with the radial distance r measured from a centre of the fiber and the axial distance z measured upward from a die exit. The coating material is a UV-curable acrylate with viscosity that highly depends on the temperature. This yields the coupled momentum and energy equations with nonlinear diffusion terms. The system of governing equations is used in the general variable (S. 7.1.1)

$$\frac{1}{r}\frac{\partial(\rho r u f)}{\partial r} + \frac{\partial(\rho r u f)}{\partial z} = \frac{1}{r}\frac{\partial}{\partial r}\left(r\mu\frac{\partial f}{\partial r}\right) + \frac{\partial}{\partial z}\left(\mu\frac{\partial f}{\partial z}\right) + I_f$$

$$f = \begin{pmatrix} 1 \\ u \\ v \end{pmatrix}, \quad I_f = \begin{pmatrix} 0 \\ -\dfrac{\partial p}{\partial r} + \dfrac{1}{r}\dfrac{\partial}{\partial r}\left(r\mu\dfrac{\partial u}{\partial r}\right) + \dfrac{\partial}{\partial r}\left(\mu\dfrac{\partial v}{\partial r}\right) - 2\mu\dfrac{u}{r} \\ -\dfrac{\partial p}{\partial z} + \dfrac{1}{r}\dfrac{\partial}{\partial r}\left(r\mu\dfrac{\partial u}{\partial z}\right) + \dfrac{\partial}{\partial r}\left(\mu\dfrac{\partial v}{\partial z}\right) + \rho g_z \end{pmatrix} \qquad (4.70)$$

$$\rho_w c_w v_w \frac{\partial T}{\partial z} = \frac{2\lambda_{in}}{r_w}\frac{\partial T}{\partial r}, \quad \frac{1}{r}\frac{\partial(\rho r u T)}{\partial r} + \frac{\partial(\rho r u T)}{\partial z} = \frac{1}{r}\frac{\partial}{\partial r}(r\lambda) + \frac{\partial}{\partial z}\left(\lambda\frac{\partial T}{\partial z}\right) + \mu S$$

$$S = 2\left[\left(\frac{\partial u}{\partial r}\right)^2 + \left(\frac{u}{r}\right)^2 + \left(\frac{\partial v}{\partial z}\right)^2\right] + \left(\frac{\partial u}{\partial z} + \frac{\partial v}{\partial r}\right)^2$$

The additional source term I_f in the first equation as well as the last term with function (1.8) S in cylindrical coordinates in energy equation arise due to variable viscosity.

The first equation (4.70) is valid for continuity equation at $f = 1$ and $I_f = 0$ as indicates the first line of the second equation and for momentum equation at $f = u, f = v$ and I_f defined according to the second and third lines of second equation, respectively. The energy equation for fiber (the third equation (4.70)) is simplified, taken into account that: (i) the Biot number $Bi = 2hr_w/\lambda_w$ is small due to small fiber diameter $2r_w = 125\mu m$, and hence, the radial temperature component can be averaged, (ii) the axial conduction may be neglected because the energy carried from fiber by convection is much higher than axial conduction, and (iii) the buoyancy effects are negligible since $Gr/Re^2 \ll 1$. In this equation λ_{in} is a harmonic thermal conductivity at the fiber/fluid interface. The last equation (4.70) is the energy equation for coating fluid.

Comment 4.19 The term harmonic mean is referred to the reciprocal of the arithmetic mean of the reciprocals of the items. For example, the harmonic mean of 1, 2 and 4 is: a reciprocal of arithmetic mean of thee reciprocal numbers $1/[(1/1) + (1/2) + (1/4)]/3 = 7/12$ or in general $1/[(1/x_1) + (1/x_2) + (1/x_3) + \ldots]/n$. This is one of the average types that is used in conformity with physical situation. As an example, consider a harmonic mean when the averaging of observation data is performed under mitigating the influence of large and increasing the effect of small values.

The boundary conditions are: (i) a shear free and adiabatic conditions at the top surface of the pressurized applicator, (ii) at the fiber surface no slip radial condition, a given axial speed, and conjugate thermal conditions, (iii) no slip condition at all walls, isothermal condition at die wall, an adiabatic condition or a given heat transfer coefficient at the applicator wall, (iv) uniform speed and temperature 298.15 K of coating material at the feed inlet, and (v) at the die exit, a fully developed condition for the flow and thermal fields and the specified meniscus. Details of free surface modeling are given in [434].

Comment 4.20 The meniscus is formed across the interface of two fluids due to surface tension. It creation is described by Young-Laplace equation that relates the pressure difference across the fluids interface to the curvature of the surface or wall.

The new variables $\xi = (r - r_i)/(r_o - r_i)$ and $\eta = z/L$, where r_i and r_o are inner and outer radii, are used to solve governing equations for velocity, pressure, and temperature. Highly clustered grids are employed in the regions with large velocity and temperature gradients. The second-order upwind scheme and an algorithm like SIMPLE (S. 9.7) are applied. For validation, the final coating thicknesses at variable fiber speed in laminar flow in a circular duct are calculated to compare with known data.

The basic results are as follows:

- The moving fiber creates the thermal field in which the cooler fluid removes the energy from the fiber. Nevertheless, the fiber temperature increases with speed. The reason of this is the extremely high viscous dissipation, which occurs with the growing fiber speed. As the moving fiber first meets the fluid, it loses energy to the cooler fluid due to the high Nusselt numbers. As the fluid heats up, the temperature gradient becomes smaller. When the speed increases, the Nusselt number becomes negative, and the fluid heats the fiber. It is shown that the smallest gap between die wall and fiber is responsible for the rise of Nusselt number and the fall of the fiber temperature near the die exit.

- The temperature gradient at the fiber is very sharp and temperature increase is the highest in the die. This very large temperature at the high speed should be avoided since the polymer may start to cross-link and degrade. As the fiber moves away from the dynamic contact point to the die exit, the fiber temperature decreases due to removal heat at the fiber surface. Then, after the speed increases, the temperature grows, and the Nusselt number reaches an asymptotic value.

- The die exit diameter is one of the critical variables determining the quality of the coating process. The numerical data show that the coating thickness increases linearly with the die exit diameter at the different fixed speed values. The ratio of thermal conductivities of the fiber and coating fluid affect the process significantly. At the fixed material properties, the increase in coating fluid conductivity leads to an increasing in coating thickness. This effect becomes unnoticeable when the fiber speed grows. If the fluid thermal conductivity is high, the Nusselt number variation is strong as well. This and increasing the coating thickness with the growing fluid conductivity is due to enhanced thermal diffusion in coating fluid.

- The effect of entrance temperature is high. As entrance temperature increases, the coating thickness decreases linearly. The fiber speed affects this dependence only slightly. At the lower entrance temperature, the viscous heating becomes a major factor. The condition on the outer wall of applicator does not significantly affect the resulting coating thickness so that practically, the conditions at the applicator wall can be flexible.

■ Example 4.24* n: Continuous Wires Casting [45]

The wires casting production directly from the melt has special benefits: (i) the near-net shape formation of metallic materials in which plastic deformation or too great reduction in cross-section area is unfeasible, accompanied by reduction of energy, time, and labour, (ii) the development of new functional properties caused by structural modification, such as the formation of non-equilibrium phases, (iii) an increased control over the process through automation, (iv) an improvement of mechanical properties caused by decreased segregation and the refinement of grain size.

The simulated wire casting system involves a casting channel formed by a static component and a rotating copper wheel. Liquid metal is fed into the cavity at the top of a shoe and drawn into a gap of decreasing cross-section as the wheel rotates. To simplify the modeling, the following assumptions are adapted: (i) steady state is achieved, (ii) fluid motion within the melt is Newtonian, incompressible and laminar, (iii) equilibrium solidification with negligible undercooling applies, and (iv) an instantaneous and complete filling of channel cavity.

Conjugate heat transfer is modeled through coupling of solid/fluid and solid/solid regions. Mixed convection/radiation boundary conditions are assigned to the external walls. The Al-4.5% Cu alloy is used as a model material, which detailed properties and other characteristics are presented in [45]. The mathematical model consists of the equation governing the phase changing process, and the energy equation for solid region

$$\frac{\partial(\rho J)}{\partial t} + \frac{\partial(\rho \Delta I)}{\partial t} + \frac{\partial(\rho u_i J)}{\partial x_i} = \frac{\partial(\lambda \nabla T)}{\partial x_i} + Q, \qquad \frac{\partial(\rho J)}{\partial t} + \frac{\partial(u_i \rho J)}{\partial x_i} = \frac{\partial}{\partial x_i}\left(\lambda \frac{\partial T}{\partial x_i}\right) + q_v \quad (4.71)$$

Here, J is enthalpy, ΔI is the latent heat content that during cooling varies between zero, ∇ is the Hamilton operator (S. 7.1.2.1), Q and q_v are heat sources, x_i and u_i are coordinates and velocity components. The second term on the left-hand side of the second equation represents heat transfer due to rotational or translational motion of solid. The heat flux through the external wall is defined as a sum of convection and radiation parts $q_a = h_a(T_a - T_w) + \varepsilon_a \sigma(T_a{}^4 - T_w{}^4)$, where ε is emissivity and index a means ambience.

The phase change in the melt is modeled using software FLUENT based on the enthalpy-porosity approach in which the mushy zone is represented by liquid fraction defined through liquidus T_l and solidus T_s temperatures as follows:

$$f_i = \frac{\Delta I}{\Lambda} = \begin{cases} 1 & T > T_l \\ (T - T_s)/(T_l - T_s) & T_s \le T \le T_l \\ 0 & T < T_s \end{cases} \qquad (4.72)$$

where, Λ is the latent heat of solidification and ΔI is calculated applying the lever rule.

Comment 4.21 The enthalpy method is an approach applicable to problem with phase changes, when it is necessary to take into account the latent heat, like, for example, in process of solidification. The adventure of such approach is that in this case, the energy balance is satisfied at the phase front explicitly tracking the required interface position.

Comment 4.22 Liquidus is the temperature above which a substance is as liquid and solidus specifies the temperature below which a substance is solid.

Comment 4.23 The lever rule or binary phase diagram is used to determine the percent weight of liquid and solid phases in binary composition at the temperature that is between the liquidus and solidus.

The solution of three-dimensional system of equations (4.71) is performed using FLUENT in several steps: (i) the temperature and fluid fraction are obtained by iterations applying auxiliary relation $\Delta I^{n+1} = \Delta I^n + 0.7 c_p [T - (T_l - T_s)T_l - T_s]$, where 0.7 is the underrelaxaction factor (Com. 4.12), (ii) to satisfy the continuity equation when the velocity components u_i are computed, the special correction equation is derived (the difficulties arising in this process are discussed in S. 9.7.1.1), (iii) heat flux through an external wall and heat transfer to the wall from the solid fraction are estimated as a sum of convection and radiation heat similar to formula for q_a given above, (iv) the thickness of air gap arising between the melt and rotating drum is estimated applying simple ratio $d = \lambda/h$. These iterations are continued until the convergence criteria is met. Other details of computing procedure are given in the original article.

The basic results and conclusions are as follows:

- Three parameters are most influential on temperature, liquid fraction and velocity profiles in the casting channel: the initial inlet melt velocity, the melt temperature at the inlet, and the rotation speed of the wheel.
- The cast material must be sufficiently solidified on reaching the exit of the casting channel because at too low temperature near the exit, the wire will fail due to a high deformation resistance.

- The relationship between the wheel (drum) speed and melt depth is the main factor determining strip thickness in metal melt spinning. The maximum achievable mass flow rate at the inlet is $23 \, \text{g s}^{-1}$ (corresponding to a line speed of $0.78 \, \text{m s}^{-1}$). The inlet melt velocities below this value cause excessive solidification throughout the channel so that the corresponding wheel speed is fixed at $9.85 \, \text{m s}^{-1}$.
- The only thermal contact between shoe and the wheel should be through the melt. Simultaneous run with the thermal contact between the shoe and wheel caused complete solidification of the melt within the channel.
- The liquid fraction profiles versus distance and in normal direction to melt flow as well as temperature profiles parallel to flow show distribution of heat transfer characteristics throughout the casting assembly.
- The velocity profiles within the melt as it travels through the casting channel give the understanding of the effects of solidification, casting geometry, and wheel speed on the behavior of the melt within the channel.
- The model here presented is only an approximate because it does not take into account the non-equilibrium solidification effects. The model also assumes that solidified melt is a very viscous fluid and neglect the effects of velocity and 'turbulence' on the melt due to solidification. Nevertheless, the heat transfer model is reasonably accurate.

OTHER WORKS: Several studies similar to what is considered in Example 4.22 for twin-screw extruder were published by Jaluria with co-authors during the last decade of last century. In papers [70, 196, 232], and [233], the processes in different types of extruders and flows of melts through dies are investigated numerically. Extrusion flows at special slip boundary conditions on the wall is investigated in [217]. In the recent paper [309], the fluid flow and heat transfer of advanced U-M/Al and U-M/Mg fuels of research reactor fuels are numerically investigated. A three-dimensional numerical study of the thermal behavior of the building envelope is presented in [30], and the simulation of the process in solar energy storage system is analyzed in [127], considering flow in channels with special phase-change material. The erosion (often associated with heat transfer in industrial processes) in the hearth of a blast furnace is studied in [62].

4.4   Technology Processes

4.4.1   Heat and Mass Transfer in Multiphase Processes

Problems of modeling multiphase processes are challenges associated with complex procedures of taking into account moving boundaries, the exchange of the thermal energy between phases, their different thermophysical properties, and latent heat of melting or solidification. Modeling processes of this type have practical interest in metallurgy, purification of metals, crystal grows, material production, etc.

■ **Example 4.25a: Model of Wetted-Wall Absorber [88]**

   The model consists of a tube with gas flow and the tubular walls filled by flowing liquid. Such a model simulates many chemical, metallurgical, and other systems involving absorption process. To simplify the problem, it is assumed that: (i) the gas flow rate is high, so that influence of solute on the flow and mass transfer is negligible, (ii) the velocity profiles in both phases are fully developed, and (iii) the liquid film thickness is small versus the tube radius. Under those assumptions, the problem is governed by two diffusion equations written by lower letters and capitals for liquid and gas, respectively

$$u\frac{\partial c}{\partial z} = D_L\frac{\partial^2 c}{\partial r^2} + k(c - c_{eq}),\ R \leq r \leq R_e,\quad U\frac{\partial C}{\partial z} = D_G\frac{1}{r}\frac{\partial}{\partial r}\left(r\frac{\partial c}{\partial r}\right),\ 0 \leq r \leq R \quad (4.73)$$

Here, c and C are concentrations, R and R_e are internal and external radii, k is the first order reaction constant, and subscripts mean: L-liquid, G-gas, and eq- equilibrium value. For the case of $\Delta/R_e \ll 1$ (Δ is the thickness of tabular wall), the velocities containing in equations (4.73) are obtained from Navier-Stokes equation and presented in the form

$$u = a - \kappa\frac{\mu_G}{\mu_L}\frac{yR_e}{2} - u_c\frac{y^2}{R_e^2},\quad U = a - \kappa\phi_G\frac{yR_e}{2} - U_c\frac{y^2}{R_e^2}$$

$$u_c = \frac{R_e^2}{4}\left[\phi_L + \frac{\kappa}{(2-\kappa)}\left(\phi_L - \phi_G\frac{\mu_G}{\mu_L}\right)\right],\quad U_c = \frac{\phi_L R_e^2}{4}\quad \kappa = 1 - \frac{\Delta}{R_e} \quad (4.74)$$

$$\phi = \frac{1}{\mu}\left(\rho g - \frac{\Delta p}{L}\right)\quad a = \frac{R_e^2}{4}\left[\phi_L\left(1-\kappa^2\right) - \frac{\kappa(\kappa-1)(\kappa-3)}{(2-\kappa)}\left(\phi_L - \frac{\mu_G\phi_G}{\mu_L}\right)\right]$$

where $y = r - R$ is the transverse coordinate with origin at gas-liquid interface. For inlet are prescribed: concentrations $c(y, 0) = c_0$, $C(y, 0) = C_0$, no fluxes conditions $(\partial c/\partial y)_{y=\Delta} = (\partial C/\partial y)_{y=0} = 0$ and conjugate conditions $C(0.z) = K_{eq}c(0, z)$ and $D_G(\partial C/\partial y)_{y=0} = D_L(\partial c/\partial y)_{y=0}$ (K_{eq} is equilibrium constant).

A Green's functions (S. 9.4) associated with equation (4.73) are:

$$G_L(\eta_L, \zeta_L - \zeta_L', \eta_L') = \sum_{n=1}^{\infty} \frac{Y_{nL}(\eta_L')Y_{nL}(\eta_L)}{\|Y_{nL}\|^2}\exp[-\gamma_{nL}^2(\zeta_L - \zeta_L')]$$

$$G_G(\eta_G, \zeta_G - \zeta_G', \eta_G') = \sum_{n=1}^{\infty} \frac{Y_{nG}(\eta_G')Y_{nG}(\eta_G)}{\|Y_{nG}\|^2}\exp[-\gamma_{nG}^2(\zeta_G - \zeta_G')] \quad (4.75)$$

$$\eta_L = \frac{y}{\Delta},\quad \zeta_L = \frac{z}{Pe_L\Delta},\quad \eta_G = \frac{R+y}{R},\quad \zeta_G = \frac{z}{RPe_G},\quad Pe_L = \frac{u_c\Delta}{D_L},\quad Pe_G = \frac{RU_c}{D_G}$$

Comment 4.24 A sign $\|\ \ \|$ is used for norm, which is a quantity employed to perform expressions to standard structure (for normalization). In this case, the norm $\|Y_{nG}\|^2$ is used to transform the Green's functions to the orthogonal expressions (S. 9.2.3).

The solution of corresponding Sturm-Liouville equations (S. 9.2.3) presented in [87] shows that eigenfunctions can be expressed in terms of the confluent hypergeometric function $F(A, B, X)$ (S. 3.1.1.2) with constants A, B, d_n and eigenvalues γ_n given in [88]

$$Y_{nL} = \exp(-X/2)[d_n F(A_0, B_0, X) + X^{1/2}F(A_1, B_1, X)]$$

$$Y_{nG} = \exp(-\gamma_{nG}\eta_G^2/2)\ F[(2 - \gamma_{nG}\hat{B})/4, 1, \gamma_{nG}\eta_G^2]\quad X = \gamma_{nL}(\eta_L - b/2)^2 \quad (4.76)$$

$$\hat{B} = \frac{R_e^2}{4U_c}[\phi_L + \kappa^2(\phi_G - \phi_L) + \phi_L(1 - \kappa^2)] - \frac{a}{U_c}\quad b = -\frac{\kappa R_e^2\phi_L}{2u_c}\frac{\mu_G}{\mu_L}$$

Applying these results yields solutions of equations (4.73) for concentrations

$$c(\eta_L, \zeta_L) = c_0 \int_0^1 w(\eta')G(\eta, \zeta, \eta')d\eta' + \int_0^\zeta \int_0^1 c_w(\zeta')w(\eta')\frac{\partial}{\partial \zeta'}G(\eta, \zeta - \zeta', \eta')d\eta'd\zeta' -$$

$$\sigma^2 \int_0^\zeta \int_0^1 c_w(\zeta')G(\eta, \zeta - \zeta', \eta')d\eta'd\zeta' + \sigma^2 c_{eq} \int_0^\zeta \int_0^1 G(\eta, \zeta - \zeta', \eta')d\eta'd\zeta' \qquad (4.77)$$

$$C(\eta_G, \zeta_G) = C_0 \int_0^1 (\hat{B} - \eta'^2)\eta' G(\eta, \zeta, \eta')d\eta' + \int_0^\zeta \int_0^1 C_w(\zeta')(\hat{B} - \eta'^2)\frac{\partial}{\partial \zeta'}G(\eta, \zeta - \zeta', \eta')d\eta'd\zeta'$$

Here $\sigma = (\kappa R_e^2/D_L)^{1/2}$. These final equations contain unknown concentrations on the interface c_w and C_w. The one of conjugate conditions is used to express C_w via c_w. The unknown concentration c_w is found from second conjugate condition employing an integral equation, which is solved by Laplace transform (S. 9.3.2) resulting in expression

$$c_w(\zeta_L) = \sum_{k=1}^\infty \frac{p(Z_k)}{q'(Z_k)}\exp(Z_k\zeta_L), \quad Z_k < 0 \qquad (4.78)$$

Detailed calculation and polynomials $p(Z_k)$ and $q(Z_k)$ are given in [88]. Numerical data are obtained for carbon dioxide absorption in water leading to following conclusions:

- Interfacial concentration c_w is determined by three parameters: σ, $\omega = \zeta_G/\zeta_L$, $\varepsilon = D_L k R_e/D_G\Delta$, the initial concentrations, and equilibrium constant K_{eq}. The solution for the case of absorption without chemical reaction is obtained by setting $\varepsilon = 0$. If the parameter ε is small, the gas phase resistance is negligible, and for large ε the gas phase basically determines the system behavior.
- Ten eigenvalues for each phase are calculated and listed. It is found that further eigenvalues did not affect the results. Series (4.78) rapidly converges so that only three or four terms are needed to get satisfactory accuracy. However, near $\zeta = 0$ this series as well as the eigenvalues expansions converge slowly as it usually does for Graetz series (Com. 3.14).
- Variations of mass fluxes of both phases as functions of Sherwood numbers

$$Sh_L = \frac{1}{c_w - c_{av}}\frac{\partial c}{\partial \eta_L}(0) \quad Sh_G = \frac{1}{C_w - C_{av}}\frac{\partial C}{\partial \eta_G}(1) \qquad (4.79)$$

show that the inlet liquid concentration does not affect the gas phase Sherwood number, but the liquid phase Sherwood number shows peculiar behavior. For values near zero, it decreases sharply to a minimum for $\zeta_L < 0.01$ and then increases to an asymptotic value of 2.89 for $\zeta_L > 1$. For the case of larger values of the inlet liquid concentration, the liquid phase Sherwood number decreases monotonically to its asymptotic value, which is almost independent of the Reynolds number Re_L.
- At fixed chemical reaction constant k, the dimensionless liquid phase concentration rises sharply along the channel from zero to a maximum value that depends on constant k. As the reaction proceeds, the liquid concentration decreases.

■ Example 4.26* n: Two Three-Dimensional Models of Concrete Structure [85, 387]

The solid skeleton of the hardened cement paste is porous hygroscopic material including various chemical compounds and pores filled with liquid and vapor water and dry air. The articles in question consider the heat and moisture transfer in concrete exposed to high temperature of fire. On the condition of high temperature, heat changes the chemical compounds and fluid content resulting in changes in physical structure that affects the mechanical and other properties of the concrete. Both works used three-dimensional models taken into account the basic phenomena, but additional effects of capillary pressure and adsorbed water are studied in [85].

Three basic assumptions are used: (i) the equilibrium exists between phases, (ii) vapor and air behave as ideal gases, and (iii) the temperature dependence of mass fluxes is negligible. The conservation of mass and energy equations for dry air and moisture are

$$\frac{\partial \varepsilon_G \tilde{\rho}_A}{\partial t} = \nabla \cdot \mathbf{J}_A, \qquad \frac{\partial \varepsilon_G \tilde{\rho}_V}{\partial t} + \frac{\partial \varepsilon_{FW} \rho_L}{\partial t} - \frac{\partial \varepsilon_D \rho_L}{\partial t} = -\nabla \cdot (\mathbf{J}_V + \mathbf{J}_{FW})$$

$$\rho c \frac{\partial T}{\partial t} - \Lambda_E \frac{\partial \varepsilon_{FW} \rho_L}{\partial t} + (\Lambda_E + \Lambda_D) \frac{\partial \varepsilon_D \rho_L}{\partial t} = \nabla \cdot (\lambda \nabla T) + \Lambda_E \nabla \cdot \mathbf{J}_{FW} - (\rho c \mathbf{v}) \cdot \nabla T$$

(4.80)

Here ε and $\mathbf{J}$ are volume fraction and mass flux of phase, subscripts A, D, G, L, S, V and FW denote: dry air, chemically bound water released by dehydration, gas, liquid, solid, vapor, and free water (combined liquid and adsorbed), Λ_E and Λ_D are specific heat of evaporation and dehydration. The mass fluxes of dry air, vapor, and free water are defined by Darcy's and Fick's laws (S. 7.1.1) ignoring the diffusion of the surface adsorbed water

$$\mathbf{J}_A = \varepsilon_G \tilde{\rho}_A [\mathbf{v}_G - D_{AV} \nabla(\tilde{\rho}_A/\tilde{\rho}_G)], \quad \mathbf{J}_V = \varepsilon_G \tilde{\rho}_A [\mathbf{v}_G - D_{VA} \nabla(\tilde{\rho}_F/\tilde{\rho}_G)], \quad \mathbf{J}_{FW} = \varepsilon_{FW} \rho_L \mathbf{v}_G$$

$$\mathbf{v}_G = (K K_G/\mu_G) \nabla p_G, \quad \mathbf{v}_L = (K K_L/\mu_L) \nabla p_G, \quad \varepsilon_{FW} = (\varepsilon_{CM}/\rho_{CM}) f[(p_V/p_{ST}), \ T]$$

(4.81)

Comment 4.25 Darcy's law describes the flow of fluid through a porous medium. It states that the flow flux is proportional to the permeability coefficient k and pressure gradient $q = -k \Delta p / \mu$. The Darcy number is defined as $\mathrm{Da} = k/L^2$.

In relations (4.81), $D_{AV} \approx D_{VA}$ are diffusion coefficients of dry air and water vapor, $\tilde{\rho}$ is mass of a phase per unit volume, K, K_G and K_L are the permeability of dry concrete and the relative permeabilities of the gas and the liquid. The pressure of air p_A and of vapor p_V are determined using the state equation of ideal gas $p = R\tilde{\rho}T$ (R is the gas constant) as well as the pressure in a gas and water $p_L \approx p_G = p_A + p_V$. The volume fraction of free water ε_{FW} is determined from the equation of sorption isotherms that relates the ratio of free water content to the cement content $\varepsilon_{CM}/\rho_{CM}$ to the temperature and relative humidity p_V/p_{ST}, where p_{ST} is the saturation pressure of the vapor. For temperatures above the critical point of water ($374.14°C$), the free water content is zero and the gas volume fraction can be defined from relation $\phi = \varepsilon_{FW} + \varepsilon_G$, where ϕ is the concrete porosity.

Comment 4.26 Desorption (opposite of sorption) is a phenomenon whereby a substance (usually water) is released from or through surface. Desorption (sorption) isotherm gives in a graphical or analytical form the equilibrium relation between water content in a material and relative humidity (the amount of water vapor in air) at constant temperature.

Using basic relation (4.80), the system of differential equations is formulated which consists of ten equations in [387] and an additional six equations in [85]. The boundary conditions that are the same in both models include the following expressions:

$$\frac{\partial T}{\partial n} = \frac{h_{GR}}{\lambda}(T_\infty - T), \quad \mathbf{J} \cdot \mathbf{n} = \gamma(\tilde{\rho}_V - \tilde{\rho}_{V\infty}), \quad \frac{\partial \tilde{\rho}_V}{\partial n} + \frac{K_{VT}h_{GR}}{K_{VV}\lambda}(T_\infty - T) + \frac{\gamma}{K_{VV}}(\tilde{\rho}_{V\infty} - \tilde{\rho}_V)$$
(4.82)

where h_{GR} and γ are coefficients of heat transfer and of vapor mass transfer on boundary. The first and the second equations are the energy and gaseous mixture mass balances, and the third expression determines the vapor content gradient on the boundary.

The following basic results are formulated:

- In both studies a steep drying front is observed. The vapor and liquid water content of the phase mixture changes from high and low on the hot side to low and high on the cold side. The water content increases ahead of drying front due to the recondensation in the cooler zone, which is named the "moisture clog zone". The maximum of peaks in a gas pressure and vapor content obtained in the modified model [85] are lower than for the initial model [387]. This may be significant in analyzing potential spalling, because the internal pore pressure is considered as cause of it. The modified model also predicts more extensive moisture clog zones in which the liquid pressure gradient drives water away from the face exposed to fire. These results show that ignoring the adsorbed water flux can significantly affect the predicted values of free water flux, vapor content, and gas pressure.
- The values of liquid pressure predicted by two models are considerably different. The capillary pressure is zero according the initial model, but it increases rapidly with a decrease in relative humidity according to modified model. However, the overall results given by both models are very similar, showing that the capillary pressure has a little or no effect on the transport in concrete under intense heating.
- If both capillary pressure and adsorbed water are taken into account as in [85], the gas pressure and vapor content are higher than shown by the initial or by the modified model, including only capillary pressure. In this case, the results of modified model are physically realistic in contrast to the unrealistic behavior of capillary pressure when the adsorbed water is ignored.

■ Example 4.27* n: Simulation of the Crystal Growth Process [285]

The Czochralski process is used for growing semiconductios (e.g., silicon) crystals. The process is performed in an apparatus with cylindrical crucible heated in a furnace above the melting temperature of the melt. The crystal and the crucible are rotated in opposite directions. The resulting crystal is vertically pulled from the crucible.

Since the temperature in this process is high, the radiative heat transfer in addition to conduction and convection should be taken into account. The three-dimensional heat transfer model is studied. Due to axial symmetry, only the cross-section containing schematic liquid, solid, and filling gas areas is considered. In addition to the usual assumptions, the following are used: (i) the diameter of the crystal pulled from the crucible is constant, (ii) the studied region with liquid, solid and gas is an enclosure, (iii) the effect of rotation is negligible, and (iv) the crystal and liquid have the same radiative properties. The steady state momentum with buoyancy $\rho(\mathbf{v} \cdot \nabla)\mathbf{v} = -\nabla p + \mu\nabla^2\mathbf{v} + \rho\mathbf{g}\beta(T - T_\infty)$, the energy with radiative term $\nabla T(\mathbf{r}) = (1/\alpha)\mathbf{v} \cdot \nabla + (1/\lambda)q(\mathbf{r})$, and continuity $\nabla \cdot \mathbf{v} = 0$ equations governed the problem [286]. The radiative heat source is calculated considering the liquid and solid phases as semi-transparent medium with constant absorption coefficients k_L and k_S, whereas the surrounding gas is considered as transparent medium with $k = 0$. The radiative source q^r is defined as a radiant energy absorbed by infinitesimal surface within infinitesimal time. In 3-D case, this amount of energy is determined by the system of complex coupled integral equations [286, 418]

$$q^r(\mathbf{p}) + \varepsilon(\mathbf{p})e_b[T(\mathbf{p})] = \varepsilon(\mathbf{p}) \int_S \left\{ e_b[T(\mathbf{r})] + \frac{1 - \varepsilon(\mathbf{r})}{\varepsilon(\mathbf{r})} q^r(\mathbf{r}) \right\} \tau(\mathbf{r}, \mathbf{p}) K(\mathbf{r}, \mathbf{p}) dS(\mathbf{r})$$

$$+ \varepsilon(\mathbf{p}) \int_S \left\{ \int_{L_{rp}} k(\mathbf{r}') e_b[T^m(\mathbf{r}')\tau(\mathbf{r}', \mathbf{p})] dL(\mathbf{r}') \right\} K(\mathbf{r} \cdot \mathbf{p}) dS(\mathbf{r})$$

$$q_v^r(\mathbf{p}) + 4k(\mathbf{p})e_b[T^m(\mathbf{p})] = k(\mathbf{p}) \int_S \left\{ e_b[T(\mathbf{r})] + \frac{1 - \varepsilon(\mathbf{r})}{\varepsilon(\mathbf{r})} q^r(\mathbf{r}) \right\} \tau(\mathbf{r}, \mathbf{p}) K(\mathbf{r}, \mathbf{p}) dS(\mathbf{r})$$

$$+ k(p) \int_S \left\{ \int_{L_{rp}} k(\mathbf{r}') e_b[T^m(\mathbf{r}')\tau(\mathbf{r}', \mathbf{p})] dL(\mathbf{r}') \right\} K_{\mathbf{r}}(\mathbf{r}' \cdot \mathbf{p}) dS(\mathbf{r}) \qquad (4.83)$$

Here, $\mathbf{r}$ and $\mathbf{p}$ refer to current and observed points, whereas point $\mathbf{r}'$ is located on the line L_{rp} connecting points $\mathbf{r}$ and $\mathbf{p}$, the integration is performed over the surface of the computational domain S and along the line L_{rp}, ε is the emissivity, $e_b(T)$ and $e_b(T^m)$ are the black-body emissivities of the computational domain surface and of the semitransparent medium, respectively. The kernel functions and transmissivity are

$$K(\mathbf{r}, \mathbf{p}) = \frac{\cos\phi_r \cos\phi_p}{\pi|\mathbf{r} - \mathbf{p}|^2}, \quad K_{\mathbf{r}}(\mathbf{r}, \mathbf{p}) = \frac{\cos\phi_p}{\pi|\mathbf{r} - \mathbf{p}|^2} \quad \tau(\mathbf{r}, \mathbf{p}) = \exp\left[-\int_{L_{rp}} k(\mathbf{r}') dL_{rp}(\mathbf{r}')\right]$$

$$(4.84)$$

Derivation of these equations and other details one may find in [418].

The usual boundary conditions should be satisfied on the external surfaces of studied enclosure. The conjugate condition on the phase-change front involves continuity of the melting temperature, a jump in heat flux, and no-slip condition for a melt velocity

$$T_L(\mathbf{r}) = T_S(\mathbf{r}) = T_{ph}, \quad -\lambda_L \frac{\partial T_L}{\partial n_{ph}} + \lambda_S \frac{\partial T_S}{\partial n_{ph}} = -\Lambda_{ph}\rho_S(\mathbf{v} \cdot \mathbf{n}_{ph}), \quad \mathbf{v}_L = \mathbf{v}_x \qquad (4.85)$$

On the solid-gas and liquid-gas interfaces, the equalities of temperatures and heat fluxes including the radiative component q_r should be satisfied:

$$T_L(\mathbf{r}) = T_G(\mathbf{r}), \qquad \lambda_L \frac{\partial T_L}{\partial n} = \lambda_G \frac{\partial T_G}{\partial n} + q_{rL}, \qquad T_S(r) = T_G(r), \qquad \lambda_S \frac{\partial T_S}{\partial n} = \lambda_G \frac{\partial T_G}{\partial n} + q_{rS}$$

(4.86)

In conditions (4.85) and (4.86), T_{ph}, $\mathbf{n}_{ph}$ and Λ_{ph} are the phase-change temperature, vector normal to phase-change surface, and the latent heat.

Numerical solution is performed using commercial package FLUENT, which is based on finite-volume approach (S. 9.6) and software based on boundary element method (BEM) (S. 9.6). The FLUENT is applied for determining velocity, pressure, and temperature fields under known distribution of the radiative heat source. The domain is divided into liquid, solid, and gas subdomains that are numerically analyzed separately. To calculate the radiative heat fluxes and source, the BEM code is used. The iterative procedure is employed to couple both numerical results. More details are given in [418]. As an example, the velocity and temperature profiles in liquid phase and radiative heat source in a typical 3-D Czochralski process are given in [286].

4.4.2 Drying and Food Processing

■ **Example 4.28a/n: Drying of a Pulled Continuous Material [99, 150]**

The present work is concerned with the conjugate problem of heat and mass exchange between a heat transfer agent (air) and a continuous material pulled through it. The model is schematically presented in Figure 1.12. In the flow domain, the problem is governed by boundary layer equations for velocity and temperature (steady-state equations (1.9)–(1.11) for zero pressure gradient and without dissipation) and additional equation for vapor density ρ_{10} similar to the energy equation (1.11)

$$u \frac{\partial \rho_{10}}{\partial x} + v \frac{\partial \rho_{10}}{\partial y} = D \frac{\partial^2 \rho_{10}}{\partial y^2}$$

(4.87)

This system of boundary layer equations is solved under following boundary conditions

$$y = \Delta/2, \quad u = U_x, \quad v = U_y = -\frac{D}{1 - \rho_{10,w}} \left. \frac{\partial \rho_{10}}{\partial y} \right|_w, \quad \rho_{10} = \rho_{10,w}, \quad T = T_w$$

$$y \to \infty, \quad u = v \to 0, \quad \rho_{10} \to \rho_{10,\infty}, \quad T \to T_\infty$$

(4.88)

The boundary layer equations are used for steady-state regime because the problem is considered as a quasi-steady for the coolant (S. 3.1.1.5). The heat and mass transfer in capillary-porous body is defined by equations of Luikov theory [244, 245]

$$c_M \rho_3 \frac{\partial T}{\partial t} = \frac{\partial}{\partial x}\left(\lambda_M \frac{\partial T}{\partial x}\right) + \frac{\partial}{\partial y}\left(\lambda_M \frac{\partial T}{\partial y}\right) + \rho_3 \varepsilon \Lambda \frac{\partial M}{\partial t}$$

$$\rho_3 \frac{\partial M}{\partial t} = \frac{\partial}{\partial x}\left[\rho_3 \alpha_M \left(\frac{\partial M}{\partial x} + \gamma \frac{\partial T}{\partial x}\right)\right] + \frac{\partial}{\partial y}\left[\rho_3 \alpha_M \left(\frac{\partial M}{\partial y} + \gamma \frac{\partial T}{\partial y}\right)\right]$$

(4.89)

In equations (4.87)–(4.89), the subscripts 0, 1, 2, 3 refer to air, vapor, liquid, and dry material, respectively, so that subscript 10 means "vapor in the air", D is a diffusion coefficient, subscript

M denotes the moist material, and hence, c_M, λ_M, and α_M are the specific heat, the thermal conductivity and diffusivity coefficients of the moist substance, U_x is the velocity of the pulled material, U_y is the transverse velocity on the surface, γ is the thermal diffusion coefficient, ε is a phase change coefficient defining the part of the vapor in the moisture content M, and Λ is a heat of evaporation.

The first and second equations (4.89) present the balance of heat and mass (moisture) transfer, respectively, for porous body. The left-hand side of these equations defines the quantity of heat or mass of the body element that changes during the unit of time. This change occurs due to conduction in x and y direction of heat introduced by two first terms in the right-hand side of the first equation and due to mass diffusion represented by the terms containing the derivatives $\partial M/\partial x$ and $\partial M/\partial y$ in the right-hand part of the second equation. The additional terms determine: the heat associated with evaporation caused by changing the moisture content (the last term in the first equation) and the mass change induced by thermal diffusion (two terms with derivatives $\partial T/\partial x$ and $\partial T/\partial y$ in the second equation). Estimation of the magnitude of terms in equations (4.89) shows that the terms determining the heat and mass transfer in the x direction are relatively small and these may be neglected. Then, the previous equations and the usual initial and symmetry conditions take the form

$$c_M \rho_3 \frac{\partial T}{\partial t} = \frac{\partial}{\partial y}\left(\lambda_M \frac{\partial T}{\partial y}\right) + \rho_3 \varepsilon \Lambda \frac{\partial M}{\partial t}, \quad \rho_3 \frac{\partial M}{\partial t} = \frac{\partial}{\partial y}\left[\rho_3 \alpha_M \left(\frac{\partial M}{\partial y} + \gamma \frac{\partial T}{\partial y}\right)\right] \quad (4.90)$$

$$t = 0, \quad T = T(0), \quad M = M(0), \quad y = 0, \; \partial T/\partial y = 0, \quad \partial M/\partial y = 0$$

The conjugate conditions consist of four equations. Three of these are equalities of temperatures, vapor densities and mass fluxes $I(x)$ defined on the interface from coolant (+) and body (−) sides. The forth condition is the balance on the interface: the difference between heats incoming from coolant and absorbing by material (two terms on the left-hand side) is used for evaporation (the right-hand side part)

$$T_w^+(x) = T_w^-(x), \quad \rho_{10,w}^+(x) = \rho_{10,w}^-(x), \quad I_w^+(x) = I_w^-(x), \quad -q_w^+ + q_w^- = (1 - \varepsilon_w)I_w^-\Lambda \quad (4.91)$$

The temperatures T_w^+ and T_w^- are determined from the boundary layer energy equation (1.11) and system (4.90), respectively. The vapor density at the material surface from coolant side $\rho_{1,w}^+$ is defined from equation for vapor concentration at the surface $\rho_{10,w}^+ = \rho_{1,w}^+/(\rho_{1,w}^+ + \rho_{0,w})$ and two relations $\rho_{0,w} = (p_\infty - p_{1,w}^+)/R_0 T_w$ and $p_{1,w}^+ = \rho_{1,w}^+ R_1 T_w$ gained by considering the air and vapor as ideal gases. Since these three equations consist of three unknown $\rho_{0,w}$, $\rho_{1,w}^+$, and $p_{1,w}^+$, simple algebra yields a relation for desired density

$$\rho_{1,w}^+ = \frac{p_\infty}{R_0 T_w[(1/\rho_{10,w}^+) - 1 + R_1/R_0]}, \quad \frac{1}{\phi(T_w, M_w)} = \frac{p_{st}(T_w)}{p_\infty}\left[1 + \frac{R_0}{R_1}\left(\frac{1}{\rho_{10,w}} - 1\right)\right] \quad (4.92)$$

To determine the vapor density at the surface from body side, the equation of desorption isotherm (Com. 4.26) $\rho_{1,w}^-/\rho_{st}(T_w) = \phi(T_w, M_w)$ should be used, where $\rho_{st}(T_w)$ is the saturated vapor density. Substituting $\rho_{1,w}^+$ from the first equation (4.92) and $\rho_{1,w}^-$ from the equation of desorption isotherm gives the conjugate condition in the form of the second expression (4.92) that relates the temperature T_w and moisture content M_w at the material surface to relative density of vapor at the surface $\rho_{10,w}$.

The heat flux at the surface from coolant side q_w^+ is defined as a difference between the incoming coolant heat (the first term) and the heat being carrying away by transverse vapor flux I_w^+ that is found as a sum of diffusion and convective $\rho_{1,w}U_y$ fluxes

$$q_w^+ = -\lambda \left.\frac{\partial T}{\partial y}\right|_w^+ - I_w^+ i_{1,w}, \quad I_w^+ = -\rho_\infty D \left.\frac{\partial \rho_{10}}{\partial y}\right|_w + \rho_{1,w}U_y = -\frac{\rho_\infty D}{1-\rho_{10,w}} \left.\frac{\partial \rho_{10}}{\partial y}\right|_w, \quad (4.93)$$

where $i_{1,w}$ is the vapor enthalpy at the surface. The last result (4.93) is obtained after substitution of expression (4.88) for the transverse velocity U_y and taken into account that $\rho_\infty \rho_{10,w} = \rho_{1,w}$. Similarly defined is the heat flux at the surface from a body side that is also found as a sum of the conductive heat and heat taken by flux I_w^- of vapor. The last one that comes across a body consists of the diffusion and thermal diffusion fluxes

$$q_w^- = -\lambda_M \left.\frac{\partial T}{\partial y}\right|_w^- - I_w^- i_{1,w}, \quad I_w^- = -\rho_3 \alpha_M \left.\frac{\partial M}{\partial y}\right|_w - \rho_3 \alpha_M \gamma \left.\frac{\partial T}{\partial y}\right|_w \quad (4.94)$$

Substituting vapor fluxes I_w^+ and I_w^- and heats q_w^+ and q_w^- into two last equations (4.91) yields two other conjugate conditions

$$\frac{\rho_\infty D}{1-\rho_{10,w}} \left.\frac{\partial \rho_{10}}{\partial y}\right|_w = \rho_3 \alpha_M \left[\left.\frac{\partial M}{\partial y}\right|_w + \gamma \left.\frac{\partial T}{\partial y}\right|_w^-\right]$$

$$\lambda \left.\frac{\partial T}{\partial y}\right|_w^+ - \lambda_M \left.\frac{\partial T}{\partial y}\right|_w^- = -(1-\varepsilon_w)\Lambda \alpha_M \rho_3 \left[\left.\frac{\partial M}{\partial y}\right|_w + \gamma \left.\frac{\partial T}{\partial y}\right|_w^-\right] \quad (4.95)$$

Thus, the conjugate problem under consideration is reduced to solving the system of: (i) boundary layer equations (1.9)–(1.11) supplemented by equation (4.87) under the boundary conditions (4.88) for coolant (ii) two equations (4.90) subjected to showing below them initial and symmetry conditions for a body and (iii) to conjugating the results using the second condition (4.92) and two conditions (4.95). Such a multiple conjugate problem usually is solved numerically. In this study, another way is employed when the set of boundary layer equations is solved applying the integral universal functions (1.40). Because the relation (1.40) is a solution of boundary layer equation at arbitrary surface temperature distribution, the energy equation (1.11) for temperature and equation (4.87) for vapor density are solved in two similar expressions

$$q_w = h_* \left\{ T_w(0) - T_\infty + \int_0^x \left[1 - \left(\frac{\xi}{x}\right)^{C_1}\right]^{-C_2} \frac{dT_w}{d\xi} d\xi \right\}$$

$$I_w = h_{*m} \left\{ \rho_{10,w}(0) - \rho_{10,\infty} + \int_0^x \left[1 - \left(\frac{\xi}{x}\right)^{C_1}\right]^{-C_2} \frac{d\rho_{10,w}}{d\xi} d\xi \right\} \quad (4.96)$$

The second equation (4.96) that relates mass flux I_w to vapor density at the surface $\rho_{10,w}$ has the same form as the first one for temperature because the energy and diffusion equations are similar and differs only by dimensionless numbers $Pe = UL/\alpha$ in the first case and $Re_m = ReSc = UL/D$ in the second case (S.7.1.1). The mass transfer coefficient h_{*m} in the second equation is connected with analogues heat transfer coefficient through Lewis number $h_{*m} = (h_*/c_p) \, Le^{1/2}$, which equals practically unit for gases. The exponents C_1 and C_2 depend on Schmidt number, and hence, in this case are the same as those in universal function (1.40) for temperature because for air $Sc \approx Pr$.

Comment 4.27 The drying process is usually considered as consisting of two parts: the initial period with practically constant rate of drying, and the second period with falling rate of drying, finally becoming zero when the moisture content of body reaches the equilibrium with the drying agent. During the first period the drying surface remains saturated, and hence, the partial pressure of the vapor and the surface temperature remain constant. This lasts until the material moisture exceeds a maximum value $M_{m,s}$ of sorptive isotherm.

For the initial period, when the partial pressure of the vapor above the surface and temperature equals the saturation values and are constant, the function in desorption isotherm in second equation (4.92) $\phi = 1$, and the phase coefficient $\varepsilon = 0$. In this case, Luikov equations (4.90) and the corresponding initial and symmetry conditions simplifies to the following system in dimensionless variables

$$\frac{c_M}{c_3} \frac{\partial \theta}{\partial Fo} = \frac{\partial}{\partial \eta} \left(\frac{\lambda_M}{\lambda_3} \frac{\partial \theta}{\partial \eta} \right), \qquad \frac{\partial \vartheta}{\partial Fo} = \frac{\partial}{\partial \eta} \left[\frac{\alpha_M}{\alpha_3} \left(\frac{\partial \vartheta}{\partial \eta} + \gamma \frac{T(0) - T_\infty}{M(0) - M_{m,s}} \frac{\partial \theta}{\partial \eta} \right) \right] \quad (4.97)$$

$$Fo = 0, \quad \theta(0) = 1, \quad \vartheta(0) = 1, \quad \eta = 0, \quad \frac{\partial \theta}{\partial \eta} = 0, \quad \frac{\partial \vartheta}{\partial \eta} = 0$$

$$Fo = \frac{x \alpha_3}{U_x \Delta^2} = \frac{t \alpha_3}{\Delta^2}, \quad \eta = \frac{y}{\Delta}, \quad \theta = \frac{T - T_\infty}{T(0) - T_\infty}, \quad \vartheta = \frac{M - M_{m,s}}{M(0) - M_{m,s}}$$

Expressions (4.96) determine the parameters on the interface from the coolant side. To simplify and convert the conjugate conditions (4.95) to new variables, the parameters on the interface from body side are expressed in terms of knowing data of coolant side. This is done in several steps: (i) the expression in the brackets on the right-hand side of second equation (4.95) is replaced by the term of the left-hand side of the first equation (4.95), (ii) the derivative $\partial T/\partial y|_w^-$ is found from the same first equation (4.95) and then, (iii) the derivative $\partial M/\partial y|_w$ is obtained from the second equation (4.95) knowing that $\varepsilon = 0$ and applying the result (ii), (iv) because in the first period the vapor density is equal to the saturation density and function $\phi = 1$, the derivative of vapor density from body side is expressed as $d\rho_{10,w}/dx = (d\rho_{st}/dT_w)(dT_w/dx)$, where the derivative $d\rho_{st}/dT_w$ of saturation vapor density is known.

The first three steps present the derivatives $\partial T/\partial y|_w^+$ and $\partial M/\partial y|_w$ from the body side as functions of the heat $q_w = \rho_\infty D(\partial \rho_{10}/\partial y)_w$ and mass $I_w = \lambda(\partial T/\partial y)_w^+$ fluxes from the coolant side (see (4.93)) that are defined by two integral universal functions (4.96). Taken this into account together with the note (iv), one obtains the conjugate conditions (4.95) as derivatives $\partial T/\partial y|_w^+$ and $\partial M/\partial y|_w$ that in variables (4.97) get the following form

$$-\frac{\partial \theta}{\partial \eta}\bigg|_w = \frac{g_0(Pr)\lambda_3}{\lambda_M} \left(\frac{\lambda Lu}{\lambda_3 \, Pr \, Fo} \right)^{1/2} \left\{ 1 + \int_0^{Fo} \left[1 - \left(\frac{\hat{Fo}}{Fo} \right)^{C_1} \right]^{-C_2} \frac{d\theta_w}{d\hat{Fo}} d\hat{Fo} \right\} + \frac{\Lambda}{(1 - \rho_{10,w})c_p}$$

$$\times \left\{ \frac{\rho_{10,w}(0) - \rho_{10,\infty}}{T(0) - T_\infty} + \int_0^{Fo} \left[1 - \left(\frac{\hat{Fo}}{Fo}\right)^{C_1} \right]^{-C_2} \frac{d\rho_{10,w}}{dT_w} \frac{d\theta_w}{d\hat{Fo}} d\hat{Fo} \right\} \qquad (4.98)$$

$$-\frac{\partial \vartheta}{\partial \eta}\bigg|_w = \frac{g_0(Pr)c_3}{(\alpha_M/\alpha_3)(1 - \rho_{10,w})c_p} \left(\frac{\lambda Lu}{\lambda_3 \, Pr \, Fo}\right)^{1/2} \left\{ \frac{\rho_{10,w}(0) - \rho_{10,\infty}}{M(0) - M_{m,s}} + \frac{T(0) - T_\infty}{M(0) - M_{m,s}} \right.$$

$$\left. \times \int_0^{Fo} \left[1 - \left(\frac{\hat{Fo}}{Fo}\right)^{C_1} \right]^{-C_2} \frac{d\rho_{10,w}}{dT_w} \frac{d\theta_w}{d\hat{Fo}} d\hat{Fo} \right\} + \gamma \frac{T(0) - T_\infty}{M(0) - M_{m,s}} \frac{\partial \theta}{\partial \eta}\bigg|_w \qquad (4.99)$$

Here, $\hat{Fo}$ is dummy variable, $g_0(Pr)$ is a factor in formula for heat transfer coefficient on the isothermal moving continuous plate (Fig. 1.15), $Lu = c_p\rho/(c\rho)_3$ is Luikov number, physical properties of moist air are calculated additively as, for example, specific heat $c_p = c_{p1}\rho_{10,w} + c_{p0}(1 - \rho_{10,w})$, and vapor concentration is defined from second equation (4.92) that for initial period at $\phi = 1$ is solved for $1/\rho_{10,w} = 1 + (R_1/R_0)[(p_\infty/p_{st}) - 1]$. System (4.97) under initial and symmetry conditions (4.90) and conjugate conditions (4.98) and (4.99) are solved numerically using the tridiagonal matrix algorithm (Com. 3.15) and implicit difference scheme (S. 9.6). The first equation (4.97) is solved with the condition (4.98), and thereafter, the second equation (4.97) with the condition (4.99) that contained derivative $(\partial\theta/\partial\eta)_w$ already obtained by solving the first equation (4.97). Iterations are applied to get coefficients depending on sought variables.

Calculations are performed for the following conditions: $T_\infty = 90°C$, $\rho_{10,\infty} = 0.125$ and $M(0) = 0.25$. Two cases with initial temperature of the material $T(0) = 70$ and $50°C$ are considered. The first of them is higher, and the second is lower than the dew point temperature corresponding to the assigned relative vapor concentration in the heat transfer agent $\rho_{10,\infty} = 0.125$. Therefore, in the first case, drying proceeds from the beginning, whereas in the second case, the material is first moistened, and drying begins after some time interval. The thermophysical characteristics of the paper-type material are: $\rho_3 = 800 \text{ kg/m}^3$, $c_3 = 1500 \text{ J/kg}$ K, $c_M = c_3 + c_{H_2O}M$, the maximum absorptive moisture content at the dew point temperature and the thermal conductivity of the moist material are taken as $M_{m,s} = 0.19$ and $\lambda_M = 0.4$ W/m K. The conjugate parameter and dimensionless coefficient of moisture diffusion gained using these data are: $(\lambda/\lambda_M)Lu = 6 \cdot 10^{-5}$ and $\alpha_M(c\rho)_3/\lambda_M = 0.125$.

Prediction leads to the following conclusions:

- The temperature and moisture content varies little across the material and their values on the surface do not actually differ from those mean integrals over the thickness.
- The rate of heat and mass transfer predicted with conjugation effects is lower than that resulting from the calculation with the third kind boundary conditions and the heat and mass transfer coefficients for constant temperature and concentration heads. In the case of drying, the material moisture content is higher, and in the case of moistening lower than the corresponding values predicted by the third kind of boundary conditions. The reason of this is that, as we seen above, in general, when the head grows, the heat and mass transfer

coefficients are higher, and in the reverse case they are lower than the corresponding coefficients for the constant heads.

- In the considered both cases, the concentration heads diminish, whereas the temperature head increases in the drying process and decreases in the moistening process. In conformity with this, the mass transfer coefficients in both cases are smaller then the isothermal coefficient $h_{m*}(\chi_p < 1)$, and the heat transfer coefficients are smaller than h_* for the moistening $(\chi_t < 1)$ and larger than h_* for the drying $(\chi_t > 1)$ processes. Despite in the considered drying process $h > h_*$, the resulting heat transfer rate decreases because the mass transfer coefficient decreases more than the heat transfer coefficient increases.

Comment 4.28 Here $\chi_p = h_m/h_{m*}$ is the coefficient of nonisobaricity showing how much the mass transfer coefficient in conjugate problem differs from that at constant concentration that is similar to nonisotrmicity coefficient $\chi_t = h/h_*$ (S. 2.1.1.1).

- Because the processes of the convective drying and moistening proceed under falling concentration head, the established here decrease in the rate will take place in any process of this type. The quantitative results and the degree of this decrease will be determined by the specific conditions, but qualitatively, the results will be the same.
- The analogy between heat and mass coefficients, frequently employed in predictions, is not observed. This is because for such analogy, the coincidence is required not only of differential equations describing the heat and mass transfer, but also of the appropriate boundary conditions determined the distribution of the temperature and concentration heads along the surface or in time.
- The data show that the distributions of the temperature and concentration heads substantially differ resulting in considerable different heat and mass transfer coefficients. While the heat transfer coefficients are close to the isothermal coefficient h_*, the mass transfer coefficients differ drastically from the isobaric one h_{m*}, especially in the moistening process when the mass transfer coefficient even reduces to zero.
- In the course of moistening, there occurs the mass flow inversion—the phenomenon similar to well-known heat flow inversion (Sec. 2.4). In this case, the mass flux reduces to zero much earlier than the concentration head does, which is in contrast to nonconjugated solution showing that both zero points coincide. In the conjugate problem, at the point where the mass flux reduces to zero, the concentration head is finite and, therefore, the mass transfer coefficient becomes zero. After this point, the mass flux changes its sign, and the drying process begins even though the direction of the concentration head remains the same $(\rho_{10,w} < \rho_{10,\infty})$. Thus, the mass transfer coefficient is negative in this section. Such a pattern remains up to the point at which the head of concentration vanishes. Since at this point the mass flux is finite, the zero concentration head results in the infinite mass transfer coefficient that virtually looses meaning. This inversion situation is analogous to that in heat transfer inversion detailed considered and physically explained in Section 2.4.

The same approach is applied in [150] for the second drying period investigation, when the moisture content of the material M_w is less than maximum sorptive moisture content $M_{m,s}$(Com. 4.27). In this case, the body is assumed to be thin, and due to that the parameters of the materials are averaged across the thickness. If in addition, the heat and mass fluxes are defined by integral relations (4.96), the problem is reduced to a system of two ordinary integro-differential equations

$$
[1 + (c_2/c_3)M_w]\frac{d\theta}{dz} + 1 + \left\{1 + \int_0^z \left[1 - \left(\frac{\xi}{z}\right)^{C_1}\right]^{-C_2}\frac{d\theta}{d\xi}d\xi\right\}
$$

$$
+ \frac{\Lambda\left\{\rho_{10}(0) - \rho_{10,\infty} + \int_0^z \left[1 - \left(\frac{\xi}{z}\right)^{C_1}\right]^{-C_2}\frac{d\rho_{10,w}}{d\xi}d\xi\right\}}{[T_\infty - T(0)][c_{p1}\rho_{10,w} + c_{p0}(1 - \rho_{10,w})](1 - \rho_{10,w})}
$$

$$
\left.\frac{\partial M_w}{\partial \phi}\right|_w\left[\frac{\phi_w^2 p_{st}(T_w) R_0}{\rho_{10,w}^2 P_\infty R_1}\frac{d\rho_{10,w}}{dz} - \frac{\phi_w}{p_{st}(T_w)}\left.\frac{\partial p_{st}}{\partial T}\right|_w\frac{dT_w}{dz}\right] + \left.\frac{\partial M_w}{\partial \phi}\right|_w\frac{dT_w}{dz}
$$

$$
+ \frac{c_3\left\{\rho_{10}(0) - \rho_{10,\infty} + \int_0^z \left[1 - \left(\frac{\xi}{z}\right)^{C_1}\right]^{-C_2}\frac{d\rho_{10,w}}{d\xi}d\xi\right\}}{[c_{p1}\rho_{10,w} + c_{p0}(1 - \rho_{10,w})](1 - \rho_{10,w})} \tag{4.100}
$$

The first equations is written in dimensionless temperature and longitudinal coordinate

$$
\theta = \frac{T_\infty - T_w}{T_\infty - T(0)}, \ z = \frac{2}{U_x\rho_3 c_3\Delta}\int_0^x h_* d\zeta \tag{4.101}
$$

The same dimensionless coordinate z is used in the second equation, whereas the sought function—the material moisture content M_w is considered as dimension variable. The desorption function ϕ in this equation is defined by second equation (4.92).

Comment 4.29 Despite the fact that relations (4.100) are ordinary equations, the second one contains the partial derivative $(\partial M_w/\partial \phi)_w$ which should not be present in ordinary equation. Indeed, this is not usual derivative as a function, rather that is a local value of a derivative on the surface, which is considered as a constant parameter at fixed surface point.

Ordinary integro-differential equations (4.100) are solved by standard Runge-Kutta numerical method using parameters at the end of the first drying period as initial conditions. The two basic conclusions are deduced:

- The rates of heat and mass transfer in drying predicted by conjugate and common approaches differ substantially. The heat and mass transfer coefficients obtained in conjugate solution are considerable less than corresponding coefficients for moving surface with constant temperature and concentration heads. This difference grows as the distance from die increases and for large distances reaches ten times. The reason of his is the significantly decreasing temperature and concentration heads along the surface.

- As well as in the first period, there is no analogy between heat and mass transfer coefficients. In the case considered, the first one is greater than the second, and the ratio h/h_m increases as the distance from a die grows becoming finally about 1.5.

We considered the Luikov drying theory in details because it is one of general conception among few others in this area that is often used in applications. Two such examples are presented next.

■ Example 4.29n: Wood Board Drying [291]

A drying of a suspended rectangular wood board placed in air stream is studied using the two-dimensional plane model. The basic assumption are: (i) equilibrium exist at each point and time, (ii) a board is unsaturated and homogeneous (uniform), (iii) gravity effects are negligible, (iv) the characteristic length of the drying medium is much smaller than that of the external fluid, (v) the thickness of the interfacial surface is negligible.

The governing system of equations as well as in former example consist of the boundary layer equations for a coolant, Luikov equations [244] for porous board, and the conjugate conditions. All equations are presented in the vector form

$$\nabla \cdot \mathbf{u} = 0, \quad \mathbf{u} \cdot \nabla \mathbf{u} = -\nabla p + \frac{1}{\text{Re}} \nabla^2 \mathbf{u}, \quad \mathbf{u} \cdot \nabla T = \frac{1}{\text{Pe}} \nabla^2 T, \quad \mathbf{u} \cdot \nabla \omega = \frac{1}{\text{Re Sc}} \nabla^2 \omega$$

$$\lambda_a \nabla T \cdot \mathbf{n} + \Lambda \rho_a D \nabla \omega \cdot \mathbf{n} = (\lambda_a + \varepsilon \Lambda \alpha_M \gamma) \nabla T \cdot \mathbf{n} + \varepsilon \Lambda \alpha_M \nabla M \cdot \mathbf{n} \qquad (4.102)$$

$$T = T_w, \quad \omega = 0.62198 \ \phi p_{st}/(p - \phi p_{st}), \ \rho_a D \nabla \omega \cdot \mathbf{n} = \alpha_M \gamma \nabla T \cdot \mathbf{n} + \alpha_M \nabla M \cdot \mathbf{n}$$

Here, $\phi = 1 - \exp(a M^b)$ is an empirical function for the desorption isotherm (Com. 4.26) with constant coefficients a and b adopted from [448], ω is the water mass fraction (used here instead of ρ_{10} in the former example). The variables in (4.102) are scaled by L, u_∞, ρu_∞^2, L/U_∞, T_∞ and ω_∞ for the coordinates, velocity, pressure, time, temperature, and vapor fraction, $\mathbf{n}$ is unit vector, and the index a denotes air. The other notations are the same as in the former example.

The finite-element method (S.9.6) is used to solve the set of governing equations. The initial conditions of the stream of air and solid are: $T_\infty = 60°C$, $\omega_\infty = 0.116 \text{ kg/kg}$, Re $= 200$, $Sc = 0.6$, and $T_0 = 25°C$, $M_0 = 222°M$ (40% moisture content), respectively. Two examples are considered. In the first example, the drying of only upper surface of the board is studied, whereas both vertical surfaces are adiabatic and impermeable. In the second example, drying of the entire board is investigated. The results for both cases are:

- In the first example, the solid temperature and moisture content distributions at 3 h time drying show that both the heating and the drying of the board are nonuniform. The leading part of the board heats much faster than the rest of it. As the air flows along surface, the temperature gradient between the flow and solid decreases. Due to this, the heat fluxes decrease in the flow direction. For the longer drying time due to evaporation, the heat fluxes increase in the flow direction. A dry zone appears close to the leading edge, while the remainder remains wet. In time, it becomes more saturated, and both the humidity gradients and moisture fluxes of the surface decrease in flow direction.

- In the second case, the results are close to those outlined for the first example. Only the temperature and moisture content distribution are different due to the fact that the vertical surfaces are also involved in the process. Both heat and mass transfer are less intensive on the vertical surfaces because the air velocities at the leading and aft edges are significantly reduced. The drying front near the aft edge is less penetrated than that in the area close to leading edge. The average moisture content variations show that board drying in the second example occurs faster than that in the first case. Despite the fact that the difference in final moisture content is only about 1%, the variation in moisture distribution is considerable, which can be important for the using the wood board.

■ **Example 4.30n: Porous Material in Drying Air Flow [254]**

The external air flow is flowing parallel to porous material that is assumed to be unsaturated, homogeneous, and nondeformable. This relatively simple model is used to simulate various situations studying the behavior of heat and mass transfer coefficients at the interface during the drying process. The system of governing equations is almost the same as in the previous example. For coolant, the set of boundary layer equations are used in the same form, and the Luikov equations similar to equations (4.102) as well as in the former example are employed. The same notations are also basically used for the porous medium and for conjugate conditions of fluxes. However, for another conjugate equation, that relates the water content M of a body and relative humidity c, expressions

$$c = \frac{p_v M}{M_a(p - p_v) + M p_v}, \quad p_v = p_{st} \exp \frac{gM\psi}{RT} \tag{4.103}$$

based on the Kelvin equation is used instead of the desorption isotherm (Com. 4.26). In relations (4.103), p_v is vapor partial pressure, p_{st} is the saturation pressure, and ψ is the capillary potential—the driving force that causes moisture to move in capillary.

Comment 4.30 The Kelvin equation describes the change in vapor pressure due to a curved liquid/vapor interface (meniscus, Com. 4.20), which is higher than that of non-curved surface. Here, this equation is used to estimate the vapor pressure in pores of porous material. The finite element method is used to solve the problem. Both the first and the second periods are considered. The main results are as follows:

- For the first period, the initial moisture content is assumed to be 8%. The results show the thermal and mass leading edge effect, which leads to high heat and mass fluxes at the interface. Intensive evaporation and mass transfer toward the interface occur.
- For the second period, the initial moisture content is assumed to be 3%. The obtained temperature and moisture fields show a dry or sorption zone linked to recession of an evaporation front. In the dry zone, the basic moisture transport occurs by vapor. In a sorption zone, the bound water exists. The leading edge effect is also seen in this case.
- The heat and mass transfer analogy is observed in the first drying period. This result can be expected since the specific humidity at the interface depends only on the temperature. However, depending on situation, this analogy may or may not be valid (see Example 4.28). In the second period, the specific humidity on the interface depends on the temperature and moisture content. Therefore, the boundary conditions at the interface for heat and mass transfer may be different. Thus, in this case, the analogy between heat and mass transfer may be not valid.

- Because the temperature and the specific humidity at the interface are not uniform, the obtained heat and mass transfer coefficients differ from the standard values corresponding to the case of plate with constant temperature and moisture content. At the same time, the variations from the reference values in this study are only about 10%. This result is in conflict with much larger variations from standard values obtained by many authors (for instance, [67] and [99]).
- Comparing two- and one-dimensional solutions of the same problem shows that one-dimensional approach cannot generally give realistic results because in this case, the effects of the boundary layer leading edge is ignored.

■ Example 4.31n: Freeze Drying of Slab-Shaped Food [280]

Unlike the conventional drying process, which is based on capillary motion and evaporation of water, the freeze-drying process uses sublimation of ice to dry the object. The low temperature and pressure below the triple point in freeze drying provide high- quality freeze-dried products. Despite the high quality of those products, the conventional drying methods are basically used in food production due to the long drying time and the high cost of freeze-drying process. Among many investigations of drying, most of them studied the alternative conventional drying methods.

The reviewed model consists of planar and slab-shaped products that are divided by a sublimation interface into the dried and frozen parts. The moisture or ice in the products sublimate under vacuum pressure and developed vapor m_V diffuses through pores to exit. The energy for sublimation comes from the bottom by conduction and by radiation from the upper heating plate. As a result, the uniform sublimation interface below the top forms in the planar product. In the slab-shaped product, the additional radiation energy comes through the lateral surface opened to the drying chamber. Due to this, the sublimation interfaces formed below the top and beside the lateral surface are nonuniform. Such a drying process in slab-shaped product in contract to that in planar product requires more complicated analysis.

The problem is governed by the mass and energy conservation equations

$$
\varepsilon \frac{(1-S)\rho_V - (1-S^0)\rho_V^0}{\Delta t} \Delta V + \sum_{j=E,\,W,N,\,S} (\mathbf{m}_V)_j \cdot \mathbf{n}_j = -\varepsilon\rho_I \frac{S-S^0}{\Delta t} \Delta V
$$

$$
(\rho c_p)^0 \frac{T-T^0}{\Delta t} \Delta V + \sum_{j=E,\,W,N,\,S} (-\lambda \nabla T)_j \cdot \mathbf{n}_j = -\varepsilon\rho_I \Lambda_S \frac{S-S^0}{\Delta t} \Delta V
$$

(4.104)

Here, ΔV is the volume of grid cell, $S = \varepsilon_I/\varepsilon$ is the ice saturation, ε and ε_I denote the porosity and the fraction of the ice volume to the total volume, Λ_S is heat of sublimation of ice, indices V and I indicate water and ice, $\mathbf{n}_j$ is outward normal vector, the subscript j denotes the control surface of a cell with East, West, North, and South faces, and superscript 0 refers to initial values. The vapor flux is determined by summing the rates of diffusion and of flow through porous products

$$
m_V = -(1-S)\frac{m_M}{RT}\left[D + p_V \frac{K}{\mu_V}\right]\nabla p_V
$$

(4.105)

where m_M, D and K are molecular mass, the effective diffusivity and permeability.

The first term of the mass conservation in the first equation (4.104) represents the change in the vapor containing in a particular cell, and the second term determines the vapor flow out of this cell. These changes yield the reduction of the rate of ice saturation S defined by right term of this mass conservation equation. Similarly, the first and second terms of the energy conservation equation (4.104) represent the change of energy inside the cell and heat flux through the control surface. The sum of these two terms is equal to the last term in this equation that defines the latent heat arising due to the ice sublimation.

The temperature boundary conditions for the top, bottom and lateral surface are

$$q_T = \sigma F_T(T_H^4 - T_T^4), \quad q_B = h(T_H - T_B), \quad q_S = \sigma F_S(T_H^4 - T_S^4) \qquad (4.106)$$

where subscripts T, B, S and H denote top, bottom, side surfaces of product and heating plates, respectively, F is the radiation shape factor, and h is the overall heat transfer coefficient for the bottom. The pressure boundary condition for surfaces opened to drying chamber (subscript C) is $p_V = p_{VC}$. As the initial condition at $t = 0$, the uniform temperature, pressure and rate of ice saturation are used. The numerical procedure starts from computing the temperature distribution by solving the second equation (4.104). Then, the distribution of vapor pressure is obtained from the first equation (4.104) only for dried cells with $S = 0$. To calculate the pressure in the frozen and sublimation cells, those are treated using Dirichlet formulation (S.1.1) assuming the local thermodynamic equilibrium for saturated vapor when its pressure is defined as $p_V = p_{st}(T)$. The evolution of ice saturation S in the frozen and sublimation cells is performed by solving the mass conservation equation (the first equation (4.104)). Iterative calculations are carried out. The numerical results are obtained using the beef as a product

- The average sublimation temperature of the slab-shaped product is 5–10% lower than that of planar product. The reason of this is that curved sublimation interfaces in slab-shaped product caused by literal surface opened to the drying chamber. As a result, the diffusion length is decreased and interfacial area is increased. These two effects enabled the shorter drying time with a lower sublimation temperature, and the primary direction of drying changes from vertical from top to bottom in planar product to radial in slab-shaped product from lateral surface to inner core.

- The distribution of ice saturation, temperature, and vapor pressure in planar product show the existence of the second sublimation interface near the bottom. The vapor from the secondary interface is transported out of a product by diffusion or is deposited in the frozen region as ice. The maximum temperature and vapor pressure increase in time. The ice saturation in the slab-shaped product in the frozen region remains relatively constant at about 0.7. The distribution of the temperature and vapor pressure in the slab-shaped product show that the heat and mass transfer during the frozen drying in this case is fully multi-dimensional process. Despite the fact that spatial temperature gradients in the dried area are larger than those in the frozen region, the heat transfer through this area is small due to low thermal conductivity. The spatial vapor pressure gradients in the frozen region caused by the temperature gradients according to dependence for saturated pressure $p_{st}(T)$ are also relatively small because of a small pore space in the frozen area.

- The main source of energy is the conduction from bottom, whereas the radiation from the top and lateral surface supplies also about 40% of total energy. Since a dried region is developed near the bottom, the heat flow from the bottom decreases from the initial 2.86 to the final 0.93 J/s. The initial vapor flow through the lateral surfaces is about two times larger than that through the top surface. At the end of drying process, almost 80% of vapor flows through the lateral surface.

- The results obtained for product of heights 5, 10, 15, and 20 mm indicate the relatively constant drying rate for planar product. The slab-shape product shows more nonlinear behavior. The high drying rate at the beginning changes to lower drying rate in the latter parts of the drying process. The reason for this is that the insulating dried region surrounded the frozen area. For the height 5 mm both planar and slab-shaped products exhibit almost the same drying time, but the difference increases with time because the drying time of the slab-shaped product is less sensitive to the product height than that of the planar product. The configuration of the sublimation interfaces of the slab-shaped product with different heights indicates that the primary drying direction is the radial one from the lateral surface to the inner core.
- The lateral surface of the slab-shaped product is favorable for the reduction of both the drying time and the sublimation temperature by increasing the vapor diffusion and the interfacial area for sublimation.

OTHER WORKS: Solutions of two other conjugate problems using similar technique are given in the work [88] reviewed in example 4.25. The heat transfer from a thick-walled tube is considered in the first one, and the falling film reactor is simulated in the second problem. Simulation of solidification processes in enclosed regions by studying the systems with moving boundaries between different phases is presented in articles [410] and [411]. Solidification process during the continuous casting (we considered continuous wires casting in example 4.24) is analyzed in [15]. A special kind of problem with moving boundaries, known as Stefan problem, is investigated in article [72]. Melting and solidification of paraffin is considered in paper [393]. Drying and food processing reviewed in the last paragraph are also simulated in several other works including some results recently published. In four papers published in the last five years, the drying of porous bodies is investigated. In the first article [212], the drying of a rectangular body streamlined by parallel flow confined in the channel is studied using Luikov theory [244], the heat and mass transfer coefficients of a porous plate at low Reynolds numbers are analyzed using conjugate modeling in the second work [93], the drying of porous materials of building envelope is considered in the third one [401], and an original approach of biosubstrates drying by providing an exposure to air jet impingement of a blunt body is outlined in the fourth study [89]. The results of food processing studies are presented in articles [53] and [356]. The three-dimensional turbulent model for freezing food is used for predictions and comparing the results with experimental data in the first article, and crystallization from the food solutions is studied in another. The microwave freeze-drying of cylindrical porous media with dielectric cores is numerically simulated in [385].

Summary of Part I

Effect of Conjugation

The general physical analysis and a number of examples considered in this part show that the basic characteristics and intensity of conjugate heat transfer depend on:

Table 4.1 Conjugation effect of different factors

Decreasing temperature head		Increasing temperature head
Comparable thermal resistances		Incomparable thermal resistances
Counter-current flows		Concurrent flows
Unfavorable pressure gradients	Zero gradient flows	Favorable pressure gradients
Laminar flows	Transition flows	Turbulent flows
Small Reynolds numbers	Mean Reynolds numbers	High Reynolds numbers
Unsteady heat transfer		Steady heat transfer
Non-Newtonian fluids with $n > 1$	Newtonian fluids	Non-Newtonian fluids with $n < 1$
Small surface curvature		Large surface curvature
Porous surface with injection	Nonporous surface	Porous surface with suction
Continuously moving surface		Streamlined surface

- Temperature head distribution on the body/fluid interface that mainly specifies the effect of a problem conjugation so that an increasing in time or in flow direction temperature head leads to moderate growing heat transfer coefficient, whereas the decreasing temperature head results in a dramatic falling in heat transfer intensity in comparison with the case of isothermal surface.

- At the given temperature head distribution, the Biot number, which is a ratio of body/fluid thermal resistances, is the second parameter basically determining the effect of conjugation showing that the maximum effect of conjugation is observed in the systems with comparable thermal resistances, and the opposite case of small conjugation effects is typical for structures where one of thermal resistances is negligible in comparison with another.

- Different factors affecting more or less a problem conjugation are listed in the Table 4.1 where they are arranged so that next to the right issue, each represents a subject with a lower effect of conjugation. For instance, because the turbulent flow is located to the right of laminar flow, this means that conjugation effect in the problems with turbulent flows is less than that in corresponding problems with laminar flows.

- The data of conjugation effects except of interest of itself is useful in making a decision whether the conjugate solution is needed in a particular problem, or the common simple approach is enough to solve it with satisfactory accuracy. Although such decision depends on the aim of specific problem, on desired accuracy of solution, and on some other conditions, the summarized comparative effects of conjugation are handy, at least for preliminary analysis. For example, some problem is likely to be solved by common method without conjugation if the temperature head increases in flow direction and Biot number for isothermal conditions is small or large, but the accurate solution of a problem with opposite characteristics with decreasing temperature head and Biot number close to unity may be obtained only applying the conjugate approach.

- The more reliable way to answer this key question of a proper method of solution of a particular problem is described in Section 2.2.1. In this approach, one starts with the common solution using boundary conditions of the third kind. Then, the error arising by using this traditional approach may be approximately estimated by computing the second term of universal function applying the knowing data of traditional solution. Examples show that such estimations are in reasonable agreement with corresponding conjugate solutions giving the understanding of necessity of conjugate solution.

- Besides approximate estimations of conjugation effects, some general results are obtained indicating that in two cases the role of conjugation is known a priori:

  Statement 1 Convective heat transfer problems containing temperature head decreasing in flow direction or in time should be considered as a conjugate because in this case, the effect of conjugation is usually significant.

  Statement 2 For the turbulent flow of the fluids with high (say higher than 100) Prandtl numbers, the convective heat transfer problems may be solved using traditional approach with boundary condition of the third kind, because for such fluids, the conjugation effect is negligible.

Part II

Applications in Fluid Flow

Here, the term "fluid flow" is used instead of incompressible fluid flow that is a relatively simple type of flows in fluid mechanics. At the same time, as it will be clear soon, the new methods and corresponding applications in fluid flow considered in this part are much complicated than the modern conjugate heat transfer problems, including the applications outlined in the previous part of the textbook. That is one of the reasons why the Part II is somewhat shorter than Part I despite the fact that both problems—the peristaltic flows and modern simulation of turbulence—comprising this part are highly important. The other cause of that shortness is the relatively late beginning of the mathematical modeling in these specific portions of hydro-dynamics. Nevertheless, the theoretical principles and basic features of peristaltic flows and of direct turbulence simulation together with examples of application presented in Chapters 5 and 6, respectively, show the contemporary situation in two involved areas of knowledge.

Applications of Mathematical Heat Transfer and Fluid Flow Models in Engineering and Medicine,
First Edition. Abram S. Dorfman.
© 2017 John Wiley & Sons Ltd. Published 2017 by John Wiley & Sons Ltd.

5

Two Advanced Methods

The both advanced methods considering in this chapter are developed and intensively used during the last fifty years. The two phenomena—the turbulent flow and flow in flexible channels underlying the physical grounds of those methods—came from nature. The first one is most often flow regime in different occurrences in the world, and the second is the basic mechanism of human and primate organs transporting physiological fluids such as blood or urine. Due to that, the development of both methods was practically important, especially in a view of the possible wide applications. However, in the 1960s of the last century when the advent of the computer made possible the development of new methods, the initial scientific situation in two considered areas was completely different. At that time, the status of turbulence theory was close to the contemporary situation consisting of the Reynolds average equations (RANS) and even $k - \varepsilon$ and $k - \omega$ turbulence models (S. 8.4.3). In contrast to that, the knowledge of flows in flexible channels of human organs at the same time correspond to the situation in biology that in the middle of the last century was basically a descriptive science consisting only verbal and illustrative information. This decisive distinction in starting conditions gives an understanding why the considered below new methods in turbulence are the top of the contemporary resources, whereas the modern means and results in the other topic studied here, the peristaltic motion in human organs, are only at the beginning level of present-day possibilities.

This part as well as previous text consists of theory with examples, detailed explanations, comments, and exercises (Chapter 5) and the applications (Chapter 6) presented more shortly, similar to applications description in Chapters 3 and 4, containing problem formulation, mathematical model as the basic equations, brief presentation of solution, and the most important results.

5.1 Conjugate Models of Peristaltic Flow

5.1.1 Model Formulation

We consider the peristaltic phenomenon as a conjugate problem because any peristaltic system is inherently a conjugate. This follows from the fact that the peristaltic flow occurs due

Applications of Mathematical Heat Transfer and Fluid Flow Models in Engineering and Medicine,
First Edition. Abram S. Dorfman.
© 2017 John Wiley & Sons Ltd. Published 2017 by John Wiley & Sons Ltd.

to close interaction between a flexible wall and fluid containing inside the channel. The word peristalsis came from the Greek *peristaltikos* and means clasping and compressing. Peristalsis or peristaltic motion arises when the progressive wave moving along the flexible tube results in contraction and expansion of the channel wall, inducing the motion of fluid inside the channel in the wave direction. Although there are a lot of engineering applications of peristaltic motion, originally this idea was adapted from creation. Muscular walls of human or primate organs in the form of a tube provide consecutive narrowing and relaxing of a wall portion, which travels in longitudinal direction resulting in a conjugation effect in movement of fluid inside a tube. Some examples of such organs are: (i) the gastrointestinal tract, in which muscles provide a swallowing and movement of the chyme from the mouth, through esophagus, stomach, small and large intestines, to rectum and anus, resulting in the digestive process, (ii) the ureter transporting due to its muscles the urine from the kidney to the bladder against the pressure gradient, (iii) the urethra that discharges the urine outside from the body, (iv) a male reproductive tract supplying the spermatic flows ejection, (v) the ovum motion inside the fallopian tube and the transport of embryo in uterus, (vi) small blood vessels, like venues and arterioles, as well as lymphatic channels (Exers. 5.1 and 5.2).

Peristaltic motion is also widely used in engineering devices and systems where the role of muscles plays the mechanisms compressed the walls of flexible tubes. Some of the most important application include: (i) an artificial heart-lung device that maintains the circulation of the blood and the oxygen content of the body repeatedly drawing off the blood from the veins, re-oxygenates it, and pumps it into the arterial system, (ii) the artificial kidney machine (hemodialysis), which provides the diffusion of waste products, such as urea and creatinine, through the semipermeable membrane into the special dialysis fluid discarding them, (iii) devices for pumping biomedical fluids, like blood, clean and sterile stuff, pharmaceutical production and food to prevent the transported substance from contact with the parts of the mechanical pump as well as to isolate environment from conveying materials, such as corrosive fluid, slurries, and sewage, (iv) microdevices for improving the mixing process of chemical reagents and facilitate preparing biological and other mixtures, and (v) devices for enhancing a mass transfer rate though porous walls.

Comment 5.1 (i) Dialysis is a process of separation of smaller molecules from larger ones resulting in dissolved substance from colloidal particles (a mixture of dispersed minute particles) in a solution by selective diffusion through a semipermeable membrane, (ii) homodialysis is a mechanized system used to perform the dialysis, (iii) urea or carbamide is an organic nitrogen-containing substance, (iv) creatinine is a breakdown product of natural substance named creatine, that the body used to store energy for muscles.

To consider the peristaltic flow as a conjugate problem, we use the problem domain similar to that in the heat transfer conjugate problem described in details above in Part I. Such domain involves two subdomains and conjugate conditions on the interface consisting in the case of peristaltic flow: (i) the fluid domain governed in general case by Navier-Stokes equation or by simplified versions of this equation, that is, creeping (S. 7.4.1) or boundary layer (S. 7.4.4) equations for low and high Reynolds numbers, respectively, (ii) the wall domain governed in general case by dynamic equation for flexible wall, or by simplified equation for thin flexible wall, or by some approximation approach, such as by given propagation wave imposed on rigid wall, and (iii) the conjugate relations consisting of no-slip condition and the balance of forces on the interface instead of equalities of temperatures and heat fluxes in the case of heat transfer

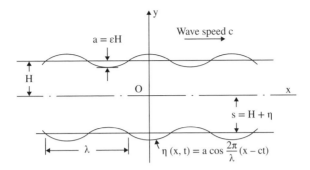

Figure 5.1   Scheme of two-dimensional channel of semi-conjugate model

conjugate problem. Although the formulations of both conjugate problems heat transfer and peristaltic flow are similar, these problems differ significantly because the latter, in contrast to former is nonlinear and due to that is much complicated to be solved. In particular, the efficient superposition method widely used in the case of conjugate heat transfer is not applicable to the nonlinear peristaltic flow problems.

Because of complexity, the known results are or approximate analytical or numerical solutions (Exer. 5.3). Moreover, the majority solutions of peristaltic problems are not full conjugate, rather those are a semi-conjugate solutions. We use this term to specify the approach when for simplicity instead of flexible wall one considers a rigid boundary with assigned wave propagating along a wall or transversely flexible wall of prescribed form. In such a semi-conjugate problem, the only effect of the propagation wave on fluid flow is studied, whereas the backward effect of fluid flow on the wall motion is completely ignored (Exer. 5.4).

A typical semi-conjugate model presents a peristaltic flow as a stream in a plane two-dimensional channel with flexible in transverse direction walls on which the longitudinal progressive sinusoidal waves are imposed (Fig. 5.1). Such a model specifies a particular problem by four dimensionless parameters: (i) the amplitude ratio $\varepsilon = a/H$, where a is amplitude of the wave and H is a half of channel cross-section, (ii) the wave number $\alpha = 2\pi H/\lambda$, where λ is the wavelength, (iii) the Reynolds number $R = cH/v$, where c is a wave velocity, and (iv) the dimensionless pressure gradient or the dimensionless time-mean flow rate. The flow is unsteady in a fixed coordinate system (the laboratory frame) because of moving boundary. However, in coordinate system moving with the wave (the wave frame), the flow is steady one. The longitudinal velocities and coordinates in both systems are connected as $u = \hat{u} + c$ and $x = \hat{x} + ct$, $y = \hat{y}$, where overscore denotes the values in the wave frame (Exer. 5.5).

There is no analytical solution of a problem with arbitrary values of all four parameters, rather the known results are found on the assumption that at least one of parameters is zero or small. Most studies, especially in early years, are performed for small or zero Reynolds number and long wavelength. The typical values of parameters are [178]: $R \approx 1, \alpha = 0.02, \varepsilon < 1$ and $R \approx 10, \alpha = 2, \varepsilon \approx 0.3$ for ureter and gastrointestinal tract, respectively, $R \approx 1, \alpha = 0.5, \varepsilon \approx 1$ for the pumps.

The available literature of peristaltic flows is expansive, however, due to nonlinearity of the problems, the methods of solution do not vary much. Here, we consider examples presenting the basic methods that are different in principle of each other and are frequently used.

5.1.2   The First Investigations

The first studies of peristaltic flows published up to the 1970s of the last century were reviewed by Jaffrin and Shapiro [178]. These works model the peristaltic flow in ureter using simple linear models. In such a model, the peristaltic fluid motion in an infinite two-dimensional channel is induced by propagating sinusoidal wave. The main objective of the early studies was the understanding of the peristalsis mechanism in order to get some insight into physical processes in the ureter. A special interest had the flow reflux, because this occasion might be a reason why bacteria sometimes travel from the bladder to the kidneys agents the mean urine flow. Early researches contributed also to engineering applications considering the flow in the roller pumps, which are used, in particular, in medicine procedures, and performing the first specific experiments confirming the premature theories of peristaltic flows [178].

■**Example 5.1: Peristaltic Motion in the Two-Dimensional Channel at low Reynolds Number and a Long Wavelength (Linear Model) [178, 345]**

In the frame moving with wave, the peristaltic flow is steady, and in the case of low Reynolds number, the solution of the problem in question is similar to well known simple steady flow in the two-dimensional channel or in a tube (S. 7.3.2). Nevertheless, there is a difference that becomes clear if one relates the flow direction to pressure gradient trend. In both flows in the fixed frame, the pressure decreases in flow direction due to energy losses. At the same time, in the wave frame, the peristaltic flow is directed opposite to that in the fixed frame, whereas the pressure gradient remains unchanged. It follows from these notes that in a moving frame, where the peristaltic is a steady flow, the Navier-Stokes equation simplifies, taking the same form (7.22) as the exact solution of Navier-Stokes equation for the creeping flow in channel, but with different sign at the pressure gradient (Exer. 5.6, Fig. 5.1)

$$\frac{\partial p}{\partial \hat{x}} + \mu \frac{\partial^2 \hat{u}}{\partial \hat{y}} = 0, \quad u = \frac{1}{2\mu}\left(\frac{dp}{dx}\right)(y^2 - s^2), \quad s = H + \eta, \quad \eta = a\cos\frac{2\pi}{\lambda}(x - ct) \quad (5.1)$$

Double integration of differential equation (5.1) subjected to boundary conditions $\hat{u} = -c$ (wave velocity) at $\hat{y} = s$ and $(\partial \hat{u}/\partial \hat{y})_{\hat{y}=0} = 0$ (symmetry condition) results, after return to stationary frame, in the second relation (5.1) defining the distribution of flow velocity. It is seen that this distribution is of Poiseuille type, such as, for example, velocity (7.23) or (7.24). However, profile (5.1), unlike the Poiseuille one, depends not only on transverse coordinate but is also a function on the longitudinal coordinate $\xi = 2\pi x/\lambda$ and time $\tau = 2\pi ct/\lambda$ through variable s (and, hence, through η, Fig. 5.1) (Exer. 5.7).

Comment 5.2   The differential equations (5.1) and (7.23), except signs, look identical. Despite that, these equations differ in essence because the latter is an exact Navier-Stokes equation for the case of creeping flow (when the inertia terms vanish) inside the two-dimensional channel, whereas the former is an approximate Navier-Stokes equation obtained under assumptions of small Reynolds number and long wavelength. This difference is mathematically expressed using different type of derivatives: (i) equation (7.22) is written in ordinary derivatives because in this case, the pressure depends only on x as well as the velocity depends only on y, whereas (ii) equation (5.1) at first is presented in partial derivatives and then, comparing the order of equation terms (such procedure is described in S. 7.4.4.1), it is shown [345] that for ureter the Reynolds number is of order ~ 1, and due to that, the equation (5.1) and it approximate solution may by written in the form similar to equation (7.22), using ordinary derivative for the pressure (Exer. 5.8).

The instantaneous dimensionless flow rate $Q(\xi, \tau)$ in the fixed frame is obtained by integration of profile (5.1). Then, the corresponding flow rate $\widehat{Q}(\xi, \tau)$ in the wave frame is found using relation $u = \widehat{u} + c$ and expression for dimensionless pressure $\overline{p}$

$$Q(\xi, \tau) = \frac{1}{Hc} \int_0^s u\, dy = -\frac{d\overline{p}}{3\, d\xi} s^3, \quad \overline{p} = \frac{2\pi H^2}{\mu \lambda c} p, \quad \widehat{Q}(\tau) = Q(\xi, \tau) - s, \quad \frac{d\overline{p}}{d\xi} = -\frac{3(\widehat{Q} + s)}{s^3}$$

$$(5.2)$$

It may be shown that the flow rate in the moving frame (the diversity) depends on time, but not on distance. Physically, that is because according to continuity equation, the flow rate of incompressible fluid is constant lengthwise the tube or channel. At the same time, the flow rate in the fixed frame depends on both time and distance because in the fixed coordinates, the peristaltic flow is unsteady. From the last equation (5.2), it follows that the pressure gradient strongly depends on the wave characteristics, so that high gradients occurs at $s \ll 1$ when the wave is intensive and contractions are severe (Exer. 5.9).

Analysis of the Lagrangian trajectories of flow particles shows that under certain conditions reflux occurs near the walls. However, the observed negative Eulerian time-mean velocity could not be surely interpreted as a reflux because calculation and experimental data show that sometimes despite the negative Eulerian average velocity, the particle Lagrangian trajectories undergo positive displacement. Comparison with experimental data indicates that the validity of this simple model is surprisingly wide.

Comment 5.3 The Lagrangian specification of flow field is the way when an observer moves with flow looking at a particular particle (moving frame). The trace of such particle gives a trajectory. The Eulerian specification of flow field is the way when an unmoving observer is focused on a specific location in a space recording the parameters of passing flow at this location (laboratory frame).

OTHER EARLY WORKS: The majority of the first peristaltic flow investigations was motivated by insight into uretral function. In particular, the visco-uretral reflux studied in one of the earliest article [169] was basically of researches interest. The studies [250, 432, 446, 228], and [346] are several examples of early works considered the reflux and other urethral problems. The peristalsis phenomena in others physiology organs are also simulated by some early authors. The vasomotion of small blood vessels was studied in article [139], in paper [32], the chime flow in small intestine was theoretically analyzed, and an influence of stomach peristalsis on blood flow in the gastrosplenic vein was examined in work [342]. Early researches contributed also to more general results regardless of particular physiology organ. Thus, the two key methods of peristaltic flow simulation, the progressive wave imposed on the channel rigid walls and the transversal bending of the flexible walls are analyzed in the articles [54] and [158], respectively. In a thesis [257], the inertia-free peristaltic pumping theory was applied to study a blood flow in roller pump. The results of early experimental investigations were presented in theses [214, 419], and [124]. As one of the earliest numerical studies of peristaltic flow we mention the work [391].

5.1.3   Semi-Conjugate Solutions

■ **Example 5.2: Peristaltic Motion in Two-Dimensional Channel at Finite Reynolds Number and Moderate Amplitude (Nonlinear Model) [139]**

The nonlinear approach of perturbation series developed in this early work is widely used up to now for applications, including studying of peristaltic motion of Newtonian and non-Newtonian fluids (Exam. 5.4-5.7). In the reviewed article, the mathematical model consists of Navier-Stokes equation in the form of stream function (7.16), and the perturbation series in power of $\varepsilon = a/H$ is used for it solution

$$\frac{\partial}{\partial t}\nabla^2\psi + \psi_y\nabla^2\psi_x - \psi_x\nabla^2\psi_y = (1/R)\nabla^2\nabla^2\psi, \quad \psi = \psi_0 + \varepsilon\psi_1 + \varepsilon^2\psi_2 + \varepsilon^3\psi_3 \ldots \quad (5.3)$$

Here, all linear and velocity parameters are scaled by half wide of channel H and wave speed c, respectively, and for other quantities, the scales in terms of these two are used: cH for ψ, ρc^2 for p, H/c for t, and Reynolds number is defined as $R = cH/v$. The sinusoidal wave (5.1) in the same scales has the form $\eta = \varepsilon\cos\alpha(x-t)$. Solution of equation (5.3) should satisfy boundary conditions on the upper and lower waves (the upper signs pertain to upper wave)

$$\psi_y = 0, \quad \psi_x = \mp\alpha\varepsilon\sin\alpha(x-t) \text{ at } y = \pm(1+\eta), \quad (5.4)$$

Substituting series (5.3) into equation (5.3) and collecting terms with the same powers of ε yield equations for coefficients ψ_n of series (5.3) (Exer. 5.10)

$$(1/R)\nabla^2\nabla^2\psi_0 = \frac{\partial}{\partial t}\nabla^2\psi_0 + \psi_{0y}\nabla^2\psi_{0x} - \psi_{0x}\nabla^2\psi_{0y} \quad (5.5)$$

$$(1/R)\nabla^2\nabla^2\psi_1 = \frac{\partial}{\partial t}\nabla^2\psi_1 + \psi_{0y}\nabla^2\psi_{1x} + \psi_{1y}\nabla^2\psi_{0x} - \psi_{0x}\nabla^2\psi_{1y} - \psi_{1x}\nabla^2\psi_{0y}, \text{ etc.} \quad (5.6)$$

To get the boundary conditions for these equations, relations (5.4) are presented in Taylor series in power of η. For the first relation one gets $\psi_y + \eta\psi_{yy} + (\eta^2/2)\psi_{yyy} + \ldots = 0$. Substituting expressions for η and for ψ (i. e. series (5.3)) into this and similar relations in Taylor series yields boundary conditions for equations (5.5) and (5.6) (Exer. 5.11)

$$\psi_{0y} = 0, \quad \psi_{1y} + \psi_{0yy}\cos\alpha(x-t) = 0 \quad (5.7)$$

$$\psi_{0x} = 0, \quad \psi_{1x} + \psi_{0xy}\cos\alpha(x-t) = -\sin\alpha(x-t), \text{ etc.} \quad (5.8)$$

For the case of constant pressure gradient $\partial p/\partial x$, equation (5.5) with boundary conditions (5.7) and (5.8) and symmetry condition give the Poiseuille profile (5.9) for ψ_0, whereas equations (5.6) under the same conditions leads to the Orr-Sommerfeld equation

$$\psi_0 = -R(\partial p/\partial x)[y - (y^3/3)] \quad (5.9)$$

$$\left\{\frac{d^2\Phi}{dy^2} - \alpha^2 + i\alpha R\left[1 + R\frac{dp}{dx}\left(1 - y^2\right)\right]\right\}\left(\frac{d^2\Phi}{dy^2} - \alpha^2\right)\Phi + 2i\frac{dp}{dx}R^2\Phi = 0 \quad (5.10)$$

The second term of series (5.3) is defined as $2\psi_1 = \Phi \exp[i\alpha(x-t)] + \tilde{\Phi} \exp[-i\alpha(x-t)]$, where $\tilde{\Phi}(y)$ is the conjugate function to function $\Phi(y)$.

Comment 5.4 Another common term conjugate means that two functions of complex variables are conjugate if they differ by sign, for example, function $z = x - iy$ is conjugate to function $z = x + iy$ (Exer. 5.12).

Comment 5.5 The Orr-Sommerfeld differential equation named after two scholars plays a fundamental role in the stability theory of laminar flow considering the transition to turbulent flow as a passage from stable to unstable regime.

In addition to the first two terms of series (5.3) defined by equation (5.9) and (5.10), the third term ψ_2 is found in [139]. Calculation of this term as well as defining the second one ψ_1 requires a solution of Orr-Sommerfeld differential equation (Exer. 5.13). Instead of numerical integration of this complex nonlinear fourth order differential equation, authors considered the pumping regime at zero initial pressure gradient $(dp/dx = 0)$ and obtained an analytical solution of Orr-Sommerfeld equation (5.10) for this case. Then, an expression for the averaged over time axial velocity is computed and is shown that it distribution significantly depends on average pressure gradient $(dp/dx)_2$ of order ε^2 induced by peristaltic flow. Analysis indicates that this velocity distribution: (i) at negative pressure gradient $(dp/dx)_2 < 0$ is close to the Poiseuille profile, (ii) as the pressure gradient grows, it deforms becoming concave at $(dp/dx)_2 \approx 0$, (iii) as the pressure gradient continue increasing and reaches some critical positive value, the velocity of a caved-in profile becomes zero on the central line of a channel and then turns to be negative when the pressure gradient $(dp/dx)_2$ exceeds critical value.

It follows from such velocity profile that there is a reversal moving flow in the neighborhood of the channel central line, which authors interpreted as a reflux showing a possible bacteria pass from the bladder to the kidney. However later on, it was shown in [345] that such backward flow could not be seen as a reason of bacteria moving past, rather considering the Lagrangian trajectories of particles is more reliable way to define the bacteria transmission (see the previous example and comment 5.3).

■ **Example 5.3: Peristaltic Motion in Two-Dimensional Channel at Moderate Reynolds Number (Numerical Solution) [382].**

The peristaltic flow problems of moderate Reynolds number are usually solved numerically, because in this case both inertia and viscose terms are of the same order, and the full Navier-Stokes equation should be considered. In this model, the steady Navier-Stokes equation is used in forticity-stream function $(\omega - \psi)$ variables (S. 7.1.2.3)

$$\frac{\partial \psi}{\partial y}\frac{\partial \omega}{\partial x} - \frac{\partial \psi}{\partial x}\frac{\partial \omega}{\partial y} = \frac{1}{Re}\left(\alpha^2 \frac{\partial^2 \omega}{\partial x^2} + \frac{\partial^2 \omega}{\partial y^2}\right), \quad \alpha^2 \frac{\partial^2 \psi}{\partial x^2} + \frac{\partial^2 \psi}{\partial y^2} = -\omega \quad (5.11)$$

The boundary conditions are specified in wave frame: at central axis $\psi = 0, \partial\psi/\partial y = -1$, $\partial^2\psi/\partial y^2 = 0$, at wave $\psi = q, \partial\psi/\partial y = -1$, for input-output $\partial\psi/\partial x = 0$. These values defined variables on all boundaries of considering domain (upper half of channel) as it is required for Navier-Stokes equation (S.1.1): (i) no-slip condition on the lower (centre axis $\psi = 0$) and on the upper (wave $\psi = q$) boundaries, (ii) symmetry condition on the lower boundary ($\psi = 0$), (iii) assuming that there is no-flow at the leading and trailing edges of domain as if it were infinitely long (Exer. 5.14). Here, dimensionless variables are scaled using: λ for x, H for y, c for velocities, cH for $\psi, c/H$ for ω, cH for flow rate q in wave frame and Reynolds number is defined as $\mathrm{Re} = (cH/v)\alpha$, where $\alpha = H/\lambda$.

Comment 5.6 To carry out practically an input-output condition $\partial \psi / \partial x = 0$, one obtains numerical results for increasing number of longitudinal points before and after channel or tube. Comparing the desired accuracy with the difference of results at adjacent points, one estimates the location where both values are of the same order showing that this condition is approximately realized. Actually, such a procedure is a way to extend the computing domain at the entrance and at the exit.

The problem was solved numerically applying the finite-difference approach (S. 9.6). The details of numerical scheme may be found in [382]. The longitudinal velocity profile in the ending section was compared with experimental data from [23] showing qualitatively agreement of calculation and measured values. The basic results are:

- In the major central part with forward flow, the longitudinal velocity profiles are almost parabolical and close to those obtained in a simple linear model in Example 5.1; at the leading and trailing edges relatively small regions with retrograde flow exit.
- At $\mathrm{Re} < 1$, the velocity profiles are independent of Re, whereas as the Reynolds number grows, they change and reach asymptotically the parabolic profiles.
- The Lagrangian particles trajectories indicate that the reflux occurs near the wall at $\mathrm{Re} < 1$, but for $\mathrm{Re} > 1$ the reflux exists near the axis.
- The pressure rise per wavelength is independent of Reynolds at $\mathrm{Re} < 1$, and monotonically decreases as Reynolds grows.
- The relation between the dimensionless pressure rise per wave length and the time-mean flow rate is close to linear, which is important to know for applications.
- The shearing stresses vary slightly along the wall for small values of $\alpha = H/\lambda$ and are steeply distributed, having a remarkably large maximum for large values of α, which also should be taken into account in applications.
- The results for the pressure rise and shearing stresses plotted via $\alpha\varepsilon == a/\lambda$ show that curves for different ε are close, forming one general relation indicating that this parameter controls the peristaltic flow behavior.

■ Example 5.4: Peristaltic Motion in Rectangular Container (Perturbation Solution) [341]

Practical interest in efficient microelectromechanical systems (MEMS) with small scales results in a variety of design suggestions for such devices. This study is also aimed to understand the possibility of using the peristaltic flow for mixing processes in small channels in order to design MEMS without mechanical elements. In conformity with the prototype of MEMS, the model consists of closed long rectangular container with typical for MEMS high oscillating peristaltic flow. To simulate the peristaltic flow, it is assumed that one of the walls undergoes sinusoidal oscillation in the form of traveling wave $y = 2\varepsilon H \cos(kx - \omega t)$, where k is the wave number and ω is the frequency.

The mathematical model contains the continuity and Navier-Stokes equations in vector form (7.8) and boundary conditions on the wave and on unmoved wall in special form:

$\mathbf{V}(x, y = 2\varepsilon H \cos(kx - \omega t), t) = 0, 2\varepsilon H\omega \sin(kx - \omega t)$ and $\mathbf{V}(x, y = 2H, t) = 0$, respectively (Com. 5.7). In dimensionless variables this system becomes (Exer. 5.15)

$$\nabla \cdot \mathbf{V} = 0, \quad \alpha^2 \frac{\partial \mathbf{V}}{\partial t} + \varepsilon \alpha^2 \mathbf{V} \cdot \nabla \mathbf{V} = -\nabla p + \nabla^2 \mathbf{V} \tag{5.12}$$

$$\mathbf{V}(x, \ \varepsilon \cos(\kappa x - t), \ t) = 0, \ \sin(\kappa x - t), \quad \mathbf{V}(x, 1, t) = 0$$

Here, lengths are scaled by $2H$, velocities by $2H\varepsilon\omega$, time by $1/\omega$, pressure by $\varepsilon\mu\omega$, whereas $\kappa = 2Hk$, and $\alpha = 4H^2\rho\omega/\mu$ are the dimensionless wave number and frequency.

Comment 5.7 Boundary conditions (5.12) are written in a vector form as $\mathbf{V}(x, y, t) = u, v$ which means: vector $\mathbf{V}$ depends on variables x, y, t and has components u, v. Thus, the first condition (5.12) tells us that: vector $\mathbf{V}$ at coordinates and time $x, y = \varepsilon \cos(\kappa x - t), t$, has components $u = 0, v = \sin(\kappa x - t)$.

The perturbation series are employed for solution taken into account that $\varepsilon << 1$

$$\mathbf{V} = \mathbf{V}_0 + \varepsilon \mathbf{V}_1 + \varepsilon^2 \mathbf{V}_2 + \ldots, \quad \int_{\varepsilon \cos(\kappa x - t)}^{1} u\, dy = \frac{\cos t - \cos(\kappa x - t)}{\kappa} \tag{5.13}$$

Because the model simulates the prototype of MEMS, the high frequency required considering $\alpha >> 1$. The last equation (5.13), defining the flow rate, should be added to usual governing system for peristaltic flow, because of a closed domain (container). This relation is obtained knowing that velocities on the walls are zero including side walls (assuming that in a long container the effect of flow at side walls is negligible). It is shown that the dominant part of solution $\mathbf{V}_0$ satisfies the unsteady Stokes equation (S. 7.3.1). Although this problem is similar to Stokes' second problem (Exer. 5.16), analysis reveals that in this case an asymptotic solution should be constructed of inner and outer perturbation series. That is because the problem is specified by two small parameters ε and $1/\alpha$. Such technique was developed by Van Dyke [404]. Using Van Dyke's approach requires long mathematical manipulations (given in [341]), but authors obtain the inner and outer expansions for close to and away from walls areas of flow and after matching those get the leading term $\mathbf{V}_0$ of series for velocity.

Analysis of final expressions results in the following specific features of peristaltic flows with high frequency oscillations:

- The time-averaged velocity consist of two components: one independent of position x and another that is periodic in x; the x-independent component is dominant for a wave with a short dimensionless wavelength ($\kappa >> 1$), and the x-periodic component is dominant for a wave with a long wavelength ($\kappa << 1$); for moderate wavelengths ($\kappa \approx O(1)$) both components are comparable (Com. 3.8).
- The time-average velocity along the channel near the oscillating boundary of order $O(2H\alpha\varepsilon^2\omega)$ is in direction of the traveling wave; in this low the thickness of boundary layer is of order $O(2H/\alpha)$.
- The opposite to this wave direction flow in the considering closed channel for $\alpha >> 1$ is independent of α and is of order $O(2H\varepsilon^2\omega)$.
- For high frequencies, the Eulerian and Lagrangian descriptions are in consistent, whereas at low frequencies, the results of this study confirm the conclusions (Exam. 5.1) obtained by Jaffrin and Shapiro (Exer. 5.17).

■ **Example 5.5: Peristaltic Flow in Closed Cavity (Series and Numerical Solutions) [431]**

This investigation was as well motivated by application of peristaltic flows for enhance mixing process in microdevices. The model is similar to that in a study reviewed in previous example, but: (i) the peristaltic motion is produced by both upper and lower vibrating walls moving with the same amplitude and frequency in opposite phases, and (ii) the cavity is shorter, which requires accounting for side walls effects. The problem is solved analytically for small amplitudes using perturbation approach and numerically for arbitrary amplitude employing the finite-element approximation (S. 9.6).

The problem is governed by the Navier-Stokes equation in the same form as in the previous example using the same dimensionless variables for length, time, velocities, and different variable for pressure scaled by $\varepsilon R^2 \mu \omega$, where $R = 2H \sqrt{\omega/\nu}$ is the Strouhal number applied in this study instead of Reynolds number. Thus, governing equations and the boundary conditions are similar to system (5.12), but with regard to literal walls using two third equations and to both moving walls by the last two expressions (Exer. 5.18)

$$\nabla \cdot \mathbf{V} = 0, \quad \frac{\partial \mathbf{V}}{\partial t} + \varepsilon \mathbf{V} \cdot \nabla \mathbf{V} = -\nabla p + (1/R^2)\nabla^2 \mathbf{V}, \quad \mathbf{V}(0, y, t) = \mathbf{V}(L, y, t) = 0 \quad (5.14)$$

$$\mathbf{V}(x, \ \varepsilon \cos(\kappa x - t), \ t) = 0, \ \sin(\kappa x - t), \quad \mathbf{V}(x, 1 - \varepsilon \cos(\kappa x - t), t) = 0, - \sin(\kappa x - t)$$

Using perturbation series and integral condition (5.13) with upper limit $1 - \varepsilon \cos(\kappa x - t)$ instead of 1 and following the tedious procedure from [139], authors obtained solution of order $O(\varepsilon^2)$ containing two first terms. Detailed description of the procedure of this solution as well as of the scheme and technique of numerical solution are given in [431].

Some conclusions specific for closed short cavities are derived:

- The first order solution consists of two terms: one the same as in open channel (Exam. 5.2) and another induced by the pressure oscillating occurring due to sidewalls; this is a contrary conclusion to previous results based on an assumption that the side walls effects are of the second order.
- The displacement of passive traces over one period of oscillation predicted by the second order approximation is of the same order of magnitude as that given by the first order approximation, so that both data is necessary to taken into account; however, only the time-independent term of the second approximation is of order ε^2, whereas the time-dependent term of this solution is of order ε^3, and hence, it may be neglected in this order of approximation.
- In contrast to the case of infinity long channel, the peristaltic flow in closed conduit is unsteady even in the moving frame.
- Comparison with numerical results shows that the analytical solution is valid in the middle cavity zone located one wavelength distance away from both sidewalls.
- The Eulerian flow next to the vibration walls is in the wave direction, whereas the flow in the central area is in opposite direction.
- For finite amplitudes, the particles trajectories indicate the existence of global circulation in the entire cavity where the fluid moves in the direction of traveling waves next to vibrating walls, turning around next to the sidewalls, and moves in the opposite direction in the central region.

- The flow forms also stationary circulation bubbles, which next to vibration walls rotate in a clockwise direction, whereas far away from the vibrating walls, they rotate in the counter-clockwise direction (Exer. 5.19).

OTHER WORKS: Numerous semi-conjugate solutions of different peristaltic flows have been published over the years starting from the 1970s of the last century. These studies consider the peristaltic flows of various non-Newtonian and other types of fluids, different forms of channels and tubes, effects of heat and mass transfer, electrical and magnetic field, impact of particles on the flow and some other factors influence. However, the majority of these solutions are obtained under the same assumptions of a long wavelength and small Reynolds number using the same perturbation approach which were first developed by early researchers. As well, the numerical solutions of these studies differ from early investigations basically by details of integration. We review some examples mainly selected from the latest results to confirm just-stated characteristics of semi-conjugate solutions. In particular, many recently published articles consider the peristaltic flows of different non-Newtonian fluids: the Carreau-Yasuda fluid with nanoparticles in asymmetric channel [1], the electrically conducting Jeffrey fluid in inclined channel [400] and in asymmetric rotating channel under magnetic field influence [3], the Oldroyd fluid in planar channel and tube at presence of magnetic field and heat transfer [203], and the Burgers fluid in channel with compliant walls and heat transfer [180].. The last article models the earth's mantle motion, the third one studies the MHD (magnetohydrodynamic) flows useful for applications, and the three others simulate flows in human organs. The first solution is obtained numerically, and the analytical solution in closed form is attained in the third study, whereas the others used the perturbation approach. Similar problems are examined in the papers published over the two last decades: the flow of Carreau fluid in divergent tube with combined effects of heat and chemical reactions [10], the Newtonian fluid flow in the two-dimensional coiled channel studying the curvature effect [13], the Phan-Thien-Tanner fluid creeping flow (S. 7.4.1) in a tube using linear and exponential rheology models [157], the nonlinear magnetohydrodynamic flow of electrically conducting Oldroyd fluid affected by transversely located magnetic field [159], the flow of rarefied (low density) gas in microchannel with slip and no-slip conditions [75], the unsteady Newtonian fluid flow in finite-length tube analyzing effect of wave shape [229].

Comment 5.8 A non-Newtonian fluid is a fluid whose properties differ in any way from those of Newtonian fluid. Most commonly, the viscosity of non-Newtonian fluid depends on spatial or/and on time deformation, whereas the viscosity of Newtonian fluid depends only on the temperature. The viscous stress of Newtonian fluid is linearly proportional to strain rate of deformation. In contrast to that, the similar relations for non-Newtonian fluids are much complicated and usually are named after researchers who first employed corresponding rheological law. With relatively simple the power law non-Newtonian fluid we encountered in Section 1.9 considering the universal function for a fluid with another than a Newtonian rheology behavior. Many non-Newtonian models that we just mentioned, for example, and many others exist, because each substance or a group of it shows individual rheological behavior that describes specific phenomenon and corresponding mathematical model.

Exercises

5.1 Explain how the tubular human organs work transporting the blood, urine, and other physiological fluids.

5.2 Learn more about artificial human organs using, for example, Wikipeda

5.3 Compare the formulation of the heat transfer and peristaltic flow conjugate problems. Explain why the latter is much complicated than the former. Recall or study what is a nonlinear problem.

5.4 Think about the difference between conjugate and semi-conjugate peristaltic flow problems. Why the second is much simpler? What is basically missing in a semi-conjugate solution?

5.5 Explain why the peristaltic flow is an unsteady problem in fixed frame, but in the frame attached to the wave, the same problem is a steady one. Hint: imagine that you are sitting on the wave and are looking at the flow passes inside of a channel.

5.6 Show that the difference between similar equations (7.22) and (5.1) is caused by the opposite peristaltic flow directions in fixed and moving frames.

5.7 Derive the velocity distribution (5.1) using integration of differential equation (5.1). Compare the result with a profile of Poiseuille type. Hint: use Figure 5.1 and profile (7.23) or (7.24).

5.8 Study carefully Comment 5.2 to understand in essence the difference between equations (7.22) and (5.1), distinction between their solutions, as well as different style of mathematical presentation of the corresponding expressions. Explain why one solution is an exact result, whereas another is an approximate solution.

5.9 Analyze equations (5.2) and following discourse to understand the conclusions formulated in the text. Hint: also use the dependence between parameters in the moving and laboratory frames and explanation from Comment 5.2.

5.10 Derive equations (5.5) and (5.6) using equations (5.3) and collecting terms with the same powers of ε.

5.11 Obtain the Taylor series in power of η for first relations (5.4). Compare the result with expression from text. Hint: Recall that coefficients of Taylor series are found by differentiation of the function that is expanded in the series.

5.12 Recall or study the basic features of functions of complex variable using *Advanced Engineering Mathematics*.

5.13 As it indicated above, the model 5.2 is nonlinear in contrast to model 5.1. Show that this is true considering the initial equation (5.3) or the Orr-Sommerfeld equation (5.10). Specify what terms are responsible for the nonlinearity; see Exercise 2.43.

5.14 Explain physical reasons of boundary conditions for equations (5.11). Recall what was said about such boundary conditions for Navier-Stokes equation in previous chapters and analyze the last condition taking into account the comment 5.6.

5.15 Analyze the boundary conditions (5.12) on the wave and on the unmoving wall of container written in the vector form to understand this type of presentation.

5.16 Learn about the second Stokes' problem using, for example "Fluid Mechanics, SG2214,HT2010" on the internet.

5.17 Compare the basic features of two peristaltic flows: at low frequency in channels observed in the first examples and at high frequency in closed container described in Example 5.4.

5.18 Derive the system (5.14) from the system (5.12) taken into account the changes indicated in the text.

5.19 Compare results obtained for close containers in two studies reviewed in Examples 5.4 and 5.5 to understand physically the observed difference caused by sidewalls.

5.1.4   Conjugate Solutions

Very few solutions are full conjugate investigating both effects (wall on fluid and backward one) of interaction between flexible walls and fluid inside the channel. We analyze two results of that type, one of which is an early publication, and another presents an example of more contemporary studies. Whereas the former research is performed assuming that the peristaltic flow is produced by a sinusoidal wave traveling along the walls, the latter is based on a more realistic model considering the interaction between oscillating walls and fluid flow.

■ **Example 5.6: Two-Dimensional Peristaltic Flow Induced by Sinusoidal Wave [270]**

This early research is an extension of Fung and Yih study (Exam. 5.2) by including the effect of walls in purpose to understand the inherent dynamic solid/fluid interaction. This is achieved by considering equations of motion of both of the fluid and of flexible walls. It is assumed that a peristaltic flow in an infinite long channel is produced by traveling sinusoidal waves of moderate amplitude $\eta = a\cos(2\pi/\lambda)(x - ct)$ imposed on the walls, which are considered as thin elastic plates or membranes. This relatively simple approach is applied by other authors (see other works below).

The same mathematical model as in Example 5.2 is used. Thus, the problem is governed by Navier-Stokes equation (5.3) in the form of stream function with boundary conditions (5.4) written in the same dimensionless variables. The additional conjugate condition is presented as an equality of derivatives of forces acting on the wave, which is considered as an interface: the pressure gradient $\partial p/\partial x$ acting from the flow and the derivative of wall stresses defining by a special operator $\tilde{L}(\eta)$

$$\frac{\partial}{\partial x}\tilde{L}(\eta) = \frac{\partial p}{\partial x} = \mu\nabla^2\psi_y - \rho(\psi_{yt} + \psi_y\psi_{yx} - \psi_x\psi_{yy}) \quad \text{at} \quad y = \pm H \pm \eta \qquad (5.15)$$

$$\tilde{L}(\eta) = -T\frac{\partial^2}{\partial x^2} + m\frac{\partial^2}{\partial t^2} + C\frac{\partial}{\partial t} \qquad \tilde{L}(\eta) = D\frac{\partial^4}{\partial x^4} + m\frac{\partial^2}{\partial t^2} + C\frac{\partial}{\partial t}$$

The second expression (5.15) for the pressure gradient is obtained from Navier-Stokes equation (5.3) with additional term $\partial p/\partial x$ modified using velocity components $u = \psi_y$ and

$v = -\psi_x$ (S. 7.1.2.3) (Exer. 5.20). Two types of wall stresses acting as a membrane or a flexural elastic plate are specified applying the relations (5.15) for operator $\tilde{L}(\eta)$, where T is the tension of the membrane, D is the flexural rigidity of the plate, m is the mass per unit area, and C is the coefficient of viscous damping.

Comment 5.9 The operator is the means to indicate the procedures that should be performed (Exer. 5.21).

Employing these equations and dimensionless variables from Example 5.2 yields the following condition on the interface at $y = \pm(1 + \eta)$ instead of (5.15) (Exer. 5.22)

$$D\frac{\partial^5 \eta}{\partial x^5} - T\frac{\partial^3 \eta}{\partial x^3} + m\,\frac{\partial^3 \eta}{\partial x \partial t^2} + C\frac{\partial^2 \eta}{\partial x \partial t} = \frac{1}{R}\nabla^2 \psi_y - (\psi_{yt} + \psi_y \psi_{yx} - \psi_x \psi_{yy}) \tag{5.16}$$

Here, coefficients are scaled: D by $\rho c^2 H^3$, T by $\rho c^2 H$, m by ρH, and C by $\rho H c^2 / \nu$. Comparing relations (5.16) and (5.15) makes clear that equation (5.16) is valid for a membrane and for a plate. To see this, set in the first case $D = 0$ and in the second $T = 0$.

In the case of pure peristaltic flow that occurs at the zero initial pressure gradient dp/dx, the solution of this problem is presented by the Orr-Sommerfild equation (5.10) (Exam. 5.2). However, the part of solution depending on the wall effects leads to some differences. Using the solution for the time-average axial velocity, the authors investigate the effect of a thin elastic plate in cases of negligible and significant dissipation occurring due to viscous damping. The interaction between the Poiseuille flow, which presents as a part of whole flow when the initial pressure gradient is not zero, and peristaltic motion, is also studied. The following conclusions are stated:

- The mean velocity is maximum at the centre and remains constant over some range, which increases as the Reynolds number grows.
- At the boundaries the mean velocity decreases with increasing the damping, but increases with increasing the wall tension and elastance.
- The mean velocity perturbation increases with increasing the wall damping, wall tension and wall elastance.
- At higher wavelength a reversal in the flow direction may occur for very high rigidity of the walls; the damping may also cause the mean flow reversal at the walls, which is not possible when the reversal at the centre occurs in the case of pure peristalsis.
- Higher level of tension enhances the efficiency of peristaltic pumping and at lower tension, the possibility of flow reversal increases.

■ **Example 5.7: Steaming Flows in a Channel with Elastic Vibrating Walls [350]**

As mentioned above, this study is based on a more realistic conjugate model analyzing the interaction between vibrating walls and flowing fluid without assumption of the waves imposed on the walls. The problem is governed by two-dimensional Navier-Stokes equations (1.5)–(1.6) in Cartesian coordinates and conjugate dynamic equations for a thin elastic plate. Because the elongate deformations of thin elastic plate are usually small in comparison with those of bending, the thin plate is modeled employing surface $y = \eta(t, x)$, which is supposed to vibrate in the transverse direction. The conjugate dynamic equations are formulated using theory from the book [390]

$$\sigma_+ - \sigma_- = 2\rho_w \Delta \frac{\partial^2 \eta}{\partial t^2} + \frac{2E\Delta^3}{3(1 - v_w^2)} \frac{\partial^4 \eta}{\partial x^4}, \quad \frac{\partial \eta}{\partial t} = v, \quad u = 0, \quad -\tau_{xy} \frac{\partial \eta}{\partial x} + \tau_{yy} - p = \sigma_-$$

(5.17)

In the first relation, Δ is the wall thickness, E is the Young modulus, v_w is the Poisson coefficient, σ_+ and σ_- are normal stresses at the external and internal wall surfaces, so that difference $(\sigma_+ - \sigma_-)$ specifies the surface loading. The second and third relations (5.17) are the kinematic (equality of velocities) and the dynamic (equality of forces) conjugate conditions on the solid/fluid interface $y = \eta(t, x)$, where τ_{ij} are the viscous stresses in the fluid (Exer. 5.23). Three additional relations

$$\partial u / \partial y = v = 0 \text{ at } y = 0, \quad \eta(0, t) = H + a \cos \omega t, \quad p(x = L) - p(x = 0) = \Delta p \quad (5.18)$$

are needed to complete the problem statement. The first condition follows from the symmetry of a channel, the second determines the harmonic oscillations imposed on the left wall edges, and the third specifies the time-average pressure drop in the channel.

The following dimensionless variables are introduced: $x/L, y/H, u/\omega L, v/\omega H, \eta/H$, $p/\rho \omega^2 L^2 \text{Re}$ and time ωt, where $\text{Re} = H^2 \omega / v$ is the vibration Reynolds number, which is a main dimensionless parameter in this study. Using the same notations for dimensionless variables, yields the system of two-dimensional equations (1.4)–(1.6) for fluid and conjugate conditions (5.17) on the interface written in the dimensionless variables (except dimensional scales H, L, ω and ρ) (Exer. 5.24 and 5.25)

$$\frac{\partial u}{\partial x} + \frac{\partial v}{\partial y} = 0, \quad \text{Re} \left(\frac{\partial u}{\partial t} + u \frac{\partial u}{\partial x} + v \frac{\partial u}{\partial y} \right) = -\frac{\partial p}{\partial x} + \frac{\partial^2 u}{\partial y^2} + \left(\frac{H}{L} \right)^2 \frac{\partial^2 u}{\partial x^2}$$

$$\text{Re} \left(\frac{H}{L} \right)^2 \left(\frac{\partial v}{\partial t} + u \frac{\partial v}{\partial x} + v \frac{\partial v}{\partial y} \right) = -\frac{\partial p}{\partial y} + \left(\frac{H}{L} \right)^2 \frac{\partial^2 v}{\partial y^2} + \left(\frac{H}{L} \right)^4 \frac{\partial^2 v}{\partial x^2} \quad (5.19)$$

$$-\tilde{\gamma} \frac{\partial^2 \eta}{\partial t^2} = -p + \left(\frac{H}{L} \right)^2 \frac{\partial v}{\partial y} - \left(\frac{H}{L} \right)^3 \left(\frac{\partial u}{\partial y} + \frac{H}{L} \frac{\partial v}{\partial x} \right) \frac{\partial \eta}{\partial x} + \tilde{\beta} \frac{\partial^4 \eta}{\partial x^4}, \quad \tilde{\beta} = \frac{2E\Delta^3 H^3}{3(1 - v_w^2)\omega \mu L^6}$$

where $\tilde{\gamma} = 2\text{Re}\rho_w H\Delta / \rho L^2, \eta(0, t) = 1 + \varepsilon \cos t, \varepsilon = a/H$. If the channel is relatively long such that $H/L << 1$, the system (5.19) may be significantly simplified by neglecting the terms with ratio H/L in power two and higher to obtain

$$\text{Re} \left(\frac{\partial u}{\partial t} + u \frac{\partial u}{\partial x} + v \frac{\partial u}{\partial y} \right) = -\frac{\partial p}{\partial x} + \frac{\partial^2 u}{\partial y^2}, \quad \frac{\partial p}{\partial y} = 0, \quad -\tilde{\gamma} \frac{\partial^2 \eta}{\partial t^2} = -p + \tilde{\beta} \frac{\partial^4 \eta}{\partial x^4} \quad (5.20)$$

Comment 5.10 The approach of simplifying the equations (5.19) is similar to that in boundary layer theory (S. 7.4.4.1). Both approaches are based on using the ratio of scales proportional to main sizes of domain: a small transverse scale (boundary thickness or channel height) and a large longitudinal gage so that the boundary layer procedure is applicable to flows over bodies, and another similar way is suitable to the channel flows.

To obtain an analytical solution, the simplified problem (5.20) containing the nonlinear first equation and complicated boundary condition is linearized. This is achieved by neglecting in the first equation the nonlinear terms, which are small in comparison with pressure in the long channel with $H/L \ll 1$. The solution of such linear problem is constructed using the perturbation series (5.3) for velocities (Exerc. 5.26)

$$u = \varepsilon u_1(x, y, t) + \varepsilon^2 u_2(x, y, t) + \ldots, \quad v = \varepsilon v_1(x, y, t) + \varepsilon^2 v_2(x, y, t) + \ldots . \tag{5.21}$$

and similar series for oscillations $\eta = 1 + \varepsilon \eta_1(x, t) + \varepsilon^2 \eta_2(x, t) + \ldots$ Differentiating the linearized (without the second and third nonlinear terms) first equation (5.20) with respect to y, leads to the partial differential equation (Exer. 5.27)

$$\mathrm{Re}\frac{\partial^2 u_1}{\partial t \partial y} = \frac{\partial^3 u_1}{\partial y^3}, \quad u_1 = e^{it} F(\tilde{x})(e^{\varphi y} + e^{-\varphi y} - e^{\varphi} - e^{-\varphi}), \quad F^{(2)} = -\gamma \frac{\varphi}{\tanh \varphi - \varphi} F + \beta F^{(6)}$$

$$\tag{5.22}$$

$$\varphi = \sqrt{\frac{\mathrm{Re}}{2}}(1 + i), \quad \gamma = \frac{\rho H}{2\rho_w \Delta}, \quad \beta = \frac{\Delta^2 E}{3H^4 \rho_w (1 - v_w^2)\omega^2}, \quad \beta k^6 - k^2 - \gamma \frac{\varphi}{\tanh \varphi - \varphi} = 0$$

It is shown [350] that the first order term of the solution of differential equation (5.22) for u_1 satisfying the second (5.17) and the first (5.18) conditions, has the form of second relation (5.22). In this solution, the function $F(\tilde{x})$ depends on variable $\tilde{x} = xL/H$ (here x is dimensionless) and parameters γ and β, which characterize the ratio of the fluid and the walls masses and the walls rigidity, respectively (Exer. 5.28).

Ordinary linear differential equation (5.22) (the upper signs in parentheses denote the derivative number) with constant coefficients for function $F(\tilde{x})$ is analyzed using it characteristic equation given by the last equation (5.22) (Exer. 5.29). Authors show that the roots of this characteristic equation determine a variety of oscillation regimes depending on three parameters: vibrational Reynolds number $\mathrm{Re} = H^2 \omega / v$ and two ratios γ and β determined by relations (5.22). Some of these regimes principally differ from oscillations usually observed in the problems of this type. Analysis of the first order approximation of order $\mathrm{Re}^{-1/2}$ reveals that the three roots of a cubic (in new variable k^2) characteristic equation (5.20) correspond to three feasible regimes. One root pertains to pair of traveling waves and two others describe the standing waves (Exer. 5.30). Similar analysis for low Reynolds numbers ($\mathrm{Re} \ll 1$) indicates also the three possible regimes. One describes a relatively slow damping behavior of the system; others correspond to the standing waves. All three modes weak depend on parameters, such as initial frequency, the walls rigidity, fluid and walls densities, and so on. The drift velocities and mass transfer are investigated as well. Detailed data are given in [350].

The following basic conclusions are formulated:

- The two-phase model of peristaltic motion (say conjugate) considered here is basically different from the usual models assuming unchangeable channel walls (semi-conjugate).
- The walls-fluid interaction leads to waves damping and to essential decreasing of the bending walls oscillations.
- The fluid motion is not periodic in space as it is assumed in usual models, and the cross-section velocity profiles are not constant changing from boundary layer type at the entrance to Poiseuille profiles at the end of the channel.

- Three different regimes are possible: one is the traveling wave and two others are standing waves.
- In general, the mass flux intensity is much higher at high Reynolds numbers; at high Reynolds numbers, the standing waves damp rapidly along the channel and provides the mass flux surpassing the mass flux induced by traveling waves; the drift velocity profile exposes in this case a counter-flow stream near the walls at the entrance and the Poiseuille flow at the outlet; at low Reynolds number, the traveling waves with rather low damping give the maximal fluid mass flux, and the drift velocity profile exposes an intense stream along the walls and counter flow at the axis.
- The average pressure gradient consists of two terms: one is the constant value of an external pressure drop providing the Poiseuille flow, and another is a variable pressure gradient existing due to vibrating walls with zero pressure difference at the channel ends.
- The study leads to the conclusion of efficiency of the mass flux at a standing wave regime rapidly damping along the channel, which is contrary to existing opinion that the traveling wave regime is the basic mechanism for fluid pumping (Exer. 5.31).

OTHER WORKS: The first simple way to take into account both effects of wall/fluid interaction early developed in [270] was used by many authors for different fluids, channel configurations and for other specific conditions. Twenty-four articles that used this approach are listed on the Web of Science, but actually this study has much more followers. We mention some examples starting from recently published results. In article [126], the three-dimensional flow of non-Newtonian Williamson fluid in rectangular channel is investigated simulating the chyme moving in intestine. The influence of a slip boundary condition, magnetic field, heat and mass transfer on the flow of conducting Carreau fluid in a nonuniform channel is studied in [399] in a view of MHD application. The combined effect of Brownian motion and thermophoretic diffusion (the Soret effect observed in a mixture of particles) in the nanofluid flow is analyzed in [278]. The flow of the couple-stress non-Newtonian fluid that models the behavior of complex liquids such as blood, infected urine or liquid crystals is studied in [300]. In study [279], the flow of micropolar fluid motion in a flexible tube is considered studying effects of fluid parameters on elastic wall properties. Works [374] and [373] are examples of researches which among others early extended the Mittra and Prasad approach [270] to more complicated flows such as flow of particle-fluid mixture and flow with heat transfer in nonuniform channel in the first and the second cases, respectively.

Besides considering followers, we also refer to similar methods proposed in papers [273] and [266]. Both proposed models are different from the initial approach [270] by establishing the progressive sinusoidal wave. In the first paper, the sinusoidal wave is imposed on walls taken as compliant instead of rigid walls considered by Mittra and Prasad. In the second version, the sinusoidal pressure wave is used instead of usually applied velocity wave. Similar approach is employed also in the study [384] where the sinusoidal wave of tension is imposed on the moving boundaries. The basic shortage of approaches of this type is that the walls/flow interaction process defining the form and motion of the solid is substituted by a progressive wave a priori giving walls attributes.

Much less researchers followed the more realistic but rather complex model [350] based on the dynamic conjugate conditions. On the other hand, several attempts to take into account the wall effects were initiated starting from early time. Despite the muscular structure and its working mechanism were of interest of early investigators (see, e.g., [396]), Fung [140] was apparently the first who studied the walls/fluid interaction influence on the peristaltic flow. He considered a solitary bolus moving along an axsymmetric slender channel simulating the ureter. A two-dimensional model based on creeping flow and equations of normal stresses equilibrium in the wall were used but no predetermined wall configuration was assumed.

Over the next two decades up to the 1990s, the wall effects are studied by many researchers but, as we just discussed, assuming (like Mittra and Prasad) that due to one or another enforcement, the channel walls are sinusoidal. The importance of the wall properties effects and the insufficiency of these studies for precise understanding the wall/flow interaction are analyzed in papers [55, 56], and [397]. The studies considering the fluid/solid interaction naturally using equilibrium equations instead of giving walls are reviewed in papers [56] and [398]. The relatively early results are analyzed in the first review, and more contemporary publications are discussed in the second one. It is seen that not much studies of this type, which we view as the conjugate simulations, are performed, especially in comparison with the amount of semi-conjugate solutions and researches with given walls. We will analyze below some typical conjugate solutions. The attempt to study peristaltic flow mechanism by detailed copying the small part of inherent intestine was made in [265]. Authors formulated a system of 15 complex equations and five relations for initial and boundary conditions simulating not only walls and fluid kinematics and dynamics but also muscle composed of two electromechanical coupling layers, and relationships for passive and active components of muscle. A model consisting such complicated system of nonlinear differential equations with numerous constants and functions, part of which is difficult to estimate, is unreliable due to doubt of the approximation of natural organs operation. Similar but much simpler model of ureter is proposed in [56]. However, this model contains also too many assumptions and experimental functions including activation sinusoidal wave.

The physically grounded mathematical models based on equations of the fundamental laws (S.7.1) and equilibrium equations on the interface are simpler and more reliable. We encountered two such models considering an early relatively simple study by Fung [140] and published thirty years later advanced result [350] (Exam. 5.7). In a recent article [398], uretral flow model of this type is developed. The model is constructed using geometrical data of real uretral lumen, and an advanced approach known as arbitrary Lagrangian-Eulerian (Com. 5.3) formulation (ALE) is employed for fluid domain. The Navier-Stokes and continuity equations are used in the form containing additional terms caused by moving channel walls, as it follows from the advanced ALE analysis presented in [34]. The Newton laws of mechanics are applied as the governing equations for solid domain. On the fluid/ walls interface, the no-slip condition at the walls and the equalities of fluid and solid stresses and displacements are specified. Similar models are utilized to study other biomechanical processes such as propulsion or motility of solids in flow. A review of early works studied solid motility one may find in article [41] in which the peristaltic propulsion of spherical bolus is investigated. The model in this study consists of a segment of a circular cylinder with a membrane simulating the muscle. The creeping approximation equations are used for flow domain. Papers [264] and [383] are examples of modern analyses of peristalsis with solids. The first work simulates the segment of the gut modeling it as cylindrical shell with muscular flexible walls. An imposed propagation wave activates the construction/relaxation of shell surface and the pellet propulsion. The problem is solved numerically. The second article presents a model for peristaltic waves propelling the infinite long rod or lozenge of finite length in the fluid-filled axisymmetric tube. The fluid flow is described employing creeping flow equations, and the walls deformation is simulated taken the linearized equations of elasticity theory.

Exercises

5.20 Derive equation (5.15) as described in the text. Hint: use formulae for velocity components to modify the equation (5.3) with additional term $\partial p/\partial x$ to get as, for example, for the first term $\partial(\nabla^2\psi)/\partial t = \partial[(\partial\psi/\partial y)\nabla\psi]/\partial t = \partial(\psi_y\nabla\psi) = \psi_{yt}$.

5.21 Learn more about operator method using, for instance, the article on the internet: http://qedinsight.wordpress.com/2011/02/25/an-operator-method-for-solving-second-order-differential-equations/.

5.22 Show that equation (5.16) is valid for both for a membrane and for an elastic plate and determine the scales for coefficients at derivatives knowing that all variables in equation (5.16) are dimensionless. Hint: take into account that in relations (5.15) written in dimension variables all terms are in the same units.

5.23 Study or recall basic terms of strength materials from Wikipedia or from more fundamental book, for example, [390].

5.24 Convert the two-dimension Navier-Stokes equations (1.5) and (1.6) in the dimensionless variables given in the text to obtain the second and third equations (5.19).

5.25 Simplify system (5.19) by comparing order of terms to get the set of equations (5.20). Compare two methods of simplifying equations of flows over bodies and in the long channels.

5.26 Explain way series (5.21) for velocity components begin from the terms with ε unlike to serial (5.3) for stream function that starts from the term without ε. Hint: use formulae (7.12) for velocity components.

5.27 Obtain the linearized first equation (5.20) (without nonlinear terms) as described in the text. Prove that the derivative $\partial^2 p/\partial x\partial y$ that appears in the linearized equation (5.20) equals zero. Hint: take into account the second equation (5.20) and the fact that the derivative is of a mixed kind.

5.28 Show that the solution (5.22) u_1 satisfied the second (5.17) and the first (5.18) boundary conditions for velocity component u. Hint: note that all variables in expression (5.22) are dimensionless.

5.29 Derive the characteristic equation (5.22) for ordinary differential equation defining the function $F(\tilde{x})$. Hint: recall or study in *Advanced Engineering Mathematics* the integration of linear ordinary differential equations with constant coefficients applying characteristic equations.

5.30 Transform the characteristic equation for function $F(\tilde{x})$ (the last equation (5.22)) in the cubic equation by using a new variable.

5.31 Compare conjugate and semi-conjugate formulation of peristaltic flow models to understand what procedure basically makes the conjugate problem much more complicated but also more reliable than the semi-conjugate one.

5.2   Methods of Turbulence Simulation

5.2.1   Introduction

The classical numerical methods (Chap. 9) are applicable to both laminar and turbulent flows. However, the level of accuracy of governing equations for both flows is different. Although the unsteady three-dimensional Navier-Stokes equations describes in principle the turbulent flow as well as laminar flow, the range of eddies sizes is so wide that before the 1960s of the last century, only Reynolds-average Navier-Stokes equations and models such as $k - \varepsilon$ or $k - \omega$ (S. 8.4.3) might be used for studying and application. The computer advent changed the situation, and in the last fifty years, the numerical methods of solving the exact Navier-Stokes equations without averaging are developed and used. The basic values of such methods is that they provide insight into physics of turbulence increasing our understanding of its nature and opened a new possibilities in modeling at close to real Reynolds numbers in applications.

In this chapter, we consider three known methods of this type: Direct Numerical Simulation (DNS), Large Eddy Simulation (LES), and Detached Eddy Simulation (DES). Because the range of eddies sizes is enormous, each method uses a special restriction in order to be in line with possibilities of nowadays computer resources. In DNS the exact Navier-Stokes equations are solved for all range of scales of eddies, but the possible solutions are confined to moderate Reynolds numbers. Two other methods achieved the restriction by separating large and small eddies, computing exactly the first part and modeling the others employing Reynolds-averaging. The difference between LES and DES lies in variation of the small size eddies' treatment.

Three characteristic scales are usually used in analyses of turbulence: integral length l, which is appropriate to the energy-bearing eddies, the Kolmogorov scale η, characterizing the smallest eddies, and the Taylor microscale λ, which is relevant to median size eddies. These scales are related to each other as follows: [421, 422]

$$\eta/l \sim \mathrm{Re}_{tb}^{-3/4}, \quad \lambda/l \sim \mathrm{Re}_{tb}^{-1/2}, \quad \eta = (v^3/\varepsilon)^{1/4}, \quad \lambda \sim (l\eta)^{1/3}, \quad \mathrm{Re}_{tb} = k^{1/2}l/v \qquad (5.23)$$

where ε is dissipation (S. 8.4.1). A comparison shows that $\eta << \lambda << l$, and, since $l \sim 0.1\,\delta$, one estimates that the Kolmogorov length scale outside the viscous wall region (S. 8.3.2) is one ten-thousandth times less than the thickness of the boundary layer.

In this chapter, we present basic features of the three mentioned above methods of direct turbulence simulation. Examples of application are reviewed in the next chapter.

5.2.2   Direct Numerical Simulation

A direct numerical simulation is a method to solve the Navier-Stokes equations in order to obtain the complete space and time-dependent field of turbulent flow. The practical and fundamental importance of such simulation is eminent because these solutions give numerically exact results, which may be taken as experimental data. These accurate results on the microscale level may be used for different studies: for analyzing the turbulence structure and instantaneous characteristics including fluctuations correlations, to test the other approximate approaches, to create new turbulence models for applications, to study specific phenomena of turbulent flows that usually are difficult or impossible obtain in laboratory (Exer. 5.32).

However, performing DNS requires fine grid sizes. Estimation of the number of grid points and time steps needed for getting accurate results shows that such calculation is a complex computation problem. If the increment along the mesh direction is h, the number N of points should satisfy the inequality $Nh > l$, providing that the integral scale l is contained within the computational domain. On the other hand, to resolve the Kolmogorov scale, it is necessary to have $h \leq \eta$. Using these two inequalities, relation (5.23) for η and knowing that $\varepsilon \approx (u')^3/l$ (u' -fluctuation velocity, S. 8.2.2) leads to expression determining the number of points $N \geq (u'l/v)^{9/4} = \mathrm{Re}_{tb}^{9/4}$ that is required to perform the three-dimensional DNS. In addition, to obtain accurate results applying the explicit method of integration, the time step Δt should satisfy inequality for the Courant number $C = u'\Delta t/h < 1$, which ensures that a fluid particle path in each step is less than a mesh spacing h. Because the turbulence time scale is of order l/u' and $h \sim \eta$, a simple calculation yields the order of the number of time steps as $N_t \sim l/u'\Delta t = l/hC = l/\eta C$, which according to (5.23) gives the following estimation $N_t \sim \mathrm{Re}_{tb}^{3/4}/C$ (Exer. 5.33).

More specific examples of estimation are presented in [422] using formulae $N \approx (3\mathrm{Re}_{tb})^{9/4}, N_t \approx 0.006H/u_\tau \sqrt{\mathrm{Re}_{tb}}$, where $\mathrm{Re}_{tb} = u_\tau H/v$ and $2H$ is a channel height. According to these formulae, the numbers of grid points and time steps for a case of one of Laufer's experiments [215] are: $N = 6.7 \cdot 10^6 - 2.1 \cdot 10^9$ and $N_t = 32 \cdot 10^3 - 114 \cdot 10^3$ for $\mathrm{Re}_{tb} = 360\,(\mathrm{Re}_H = 0.61 \cdot 10^4) - 4650\,(1.15 \cdot 10^5)$, respectively. These estimations show that in DNS the required number of grid points and time steps grows fast with Reynolds number, resulting in large computer memory and time essential for calculation and high cost.

Because the Reynolds numbers of the most engineering applications are higher than moderate ones, some years ago, the only relatively simple problems might be investigated applying DNS. However, starting with simple geometric flows in channels and in boundary layers, the DNS availability grows so that the more complicated problems of more complex geometric flows, high-speed flows, two-phase and reacting flows become possible to consider. Analysis of the latest publications shows that during the last several years significant progress is achieved particularly in studying the reacting flows and different form of combustion. Examples are presented in the next chapter.

The importance of direct Navier-Stokes equation solutions in turbulence theory is difficult to overestimate: significantly improving the exactness of results, this approach has opened a new chapter in turbulence understanding, leading to fresh ideas and thoughts. As the more powerful computers appear, and higher Reynolds numbers become possible, the DNS would be widely used, resulting in fundamental studies in different science and technology areas [272] (Exer. 5.34).

5.2.3   Large Eddy Simulation

A large eddy simulation is a method of reduction the number of grid points and time steps in order to solve directly Navier-Stokes equation for higher Reynolds numbers. Estimations show that usually the number of grid points required in LES is abut ten times less than that for DNS [422]. The procedure of LES first was proposed in 1963 by Smagorinsky for the atmospheric motion study [358]. The main idea of such procedure is to separate the treatment of large and small eddies. Because the large eddies carry the majority of the energy and are mostly affected by boundary condition, they should be computed directly using DNS approach. At the same time, the small turbulence eddies are weaker, nearly isotropic, having almost universal characteristics and therefore are more tractable for modeling, applying Reynolds-average models.

To perform the scale separation, a special method known as filtering was developed. A simple example [422] helps to understand the notion of filtering and, in particular, to see that the integration may be used to perform filtering. Consider the finite-difference formula for the first derivative of some function $f(x)$. Using the Leibniz rule of interchanging the integration and differentiation, this relation may be presented in the following integral form (Exer. 5.35)

$$\frac{df}{dx} = \frac{f(x + \Delta x) - f(x - \Delta x)}{2\Delta x} = \frac{1}{2\Delta x} \int_{x-\Delta x}^{x+\Delta x} \frac{df(\xi)}{d\xi} d\xi = \frac{d}{dx} \left[\frac{1}{2\Delta x} \int_{x-\Delta x}^{x+\Delta x} f(\xi) d\xi \right] \qquad (5.24)$$

This expression gives an average value of the derivative of function $f(x)$. That derivative is constant within one grid step Δx, but its value is variable changing from one to another step. Due to that, the relation of type (5.24) may be considered as a filter of the derivative sizes. One of the first filters employed in 1970 for three-dimensional turbulence model, known as volume-average box filter was of this type (Exer. 5.36)

$$\bar{u}_i(\mathbf{x}, t) = \frac{1}{\Delta^3} \int_{x-(1/2)\Delta x}^{x+(1/2)\Delta x} \int_{y-(1/2)\Delta y}^{y+(1/2)\Delta y} \int_{z-(1/2)\Delta z}^{z+(1/2)\Delta z} u_i(\xi, t) d\xi d\eta d\zeta \qquad (5.25)$$

Since that time other types of filters are suggested and used. In relation (5.25), $\mathbf{x}$ and ξ are vectors defining the coordinates, Δ denotes the smallest of turbulence scales rated by filter as the large scales, which are computed, whereas the others, under scale Δ, termed as subgrid scales (SGS), are eliminated and modeled. Because relation (5.25) determines the filter velocity $\bar{u}_i$, the subgrid velocity is found as a difference $\tilde{u}_i = u_i - \bar{u}_i$ (Exer. 5.37).

Filtering the Navier-Stokes equation in Einstein notation (S. 7.1.2.2) gives

$$\frac{\partial \bar{u}_i}{\partial t} + \bar{u}_j \frac{\partial \bar{u}_i}{\partial x_j} = -\frac{1}{\rho} \frac{\partial \bar{p}}{\partial x_i} + 2v \frac{\partial}{\partial x_j} \bar{S}_{ij} - \frac{\partial \tau_{ij}}{\partial x_i} \qquad \bar{S}_{ij} = \frac{1}{2} \left(\frac{\partial \bar{u}_i}{\partial x_j} + \frac{\partial \bar{u}_j}{\partial x_i} \right) \qquad (5.26)$$

This equation presents the field of filtered large scales modified by the subgrid scales stresses through tensors S_{ij} and τ_{ij} (Com. 1.9). The rate-of-strain tensor S_{ij} describes the rate of change of local deformation (Exer. 5.38). The tensor τ_{ij} represents the process of interaction among and between large and small eddies and plays a fundamental role in LES. Determining this tensor requires modeling the subgrid scales stresses, which is a challenge, and during the past half-century, many models from the first simple to the more complicate nonlinear models were suggested [143, 223, 422].

The simplest Smagorinsky model is created using analogy with mixing-length formula (S. 8.3.1). In this case, the tensor τ_{ij} is proportional to the strain-rate tensor S_{ij} and is defined as [421, 422] $\tau_{ij} = 2v_{tb}\bar{S}_{ij}$, where $v_{tb} = (C_s \Delta)^2 \sqrt{2\bar{S}_{ij}\bar{S}_{ij}}$ is Smagorinsky eddy viscosity, Δ is the grade size, and C_s is a constant coefficient. This model as well as mixing-length approach is not universal, rather it is calibrated by adjusting the coefficient that varies: $C_s \approx 0.1 - 0.2$. Some other models used Smagorinsky formula with different eddy viscosity. For example, the Lilly model [231] with turbulence viscosity depending on SGS kinetic energy is similar to Prandtl's one-equation model (S. 8.4.2). The dynamic SGS models apply variable coefficients C_s, defining its spatial [146] or temporal [262] average value using additional filtering (Exer. 5.39).

5.2.4   Detached Eddy Simulation

Large eddy simulation significantly widened the application of the direct solutions of Navier-Stokes equation due to increasing the possible Reynolds numbers. This was achieved by using the Reynolds-average modeling for small eddies instead of computing those in DNS. However, the important engineering applications such as airfoil ground or marine vehicle require much higher Reynolds number and so demand great numbers of grid points and time steps that lie far beyond the resources of current computers. The basic reason for the growth of such grid number is related to near-wall region with the smallest eddies, whose role increases as the Reynolds number grows. Estimation shows that Reynolds number increasing by factor 10 leads to increase in the computer work by factor about 30 [422] (Exer. 5.40).

These problems motivated the attempts to develop improved LES methods by changing the treatment of small eddies. Different approaches were proposed (see, e.g., review [17], p. 715). The detached eddy simulation (DES) proposed by Spalart et al. in 1997 [365] is the most promising method of improved LES. This approach gained an acceptance of research community due to showing successful applications of computing separation flows past bluff bodies and vehicles with blunt forebody.

Detached eddy simulation is a hybrid approach combining the RANS (Reynolds-average Navier-Stokes equation) and LES methods in united technique by using the former procedure for near-wall region and the latter one for the rest part of domain with large eddies. Because DES is a single solution for entire computation domain, in this method as well as in LES, the essential issue is the means of distinguishing between RANS and LES for treatment in corresponding regions. This is achieved using a special operation, called blending function $f = \min(d, C_{DES}\Delta)$, where d is distance to the closest surface, Δ is the largest grid cell and C_{DES} is constant of proportionality. The blending function impels the model behave as RANS in regions close to walls, where $d << \Delta$, and perform as a subgrid model, such a Smagorinsky model, away from the walls at $\Delta << d$. That feature of changeable procedure in DES instead of fixed one in LES basically furnishes the better simulation and reduced computing cost (Exer. 5.41).

One of the most important accomplishments of DES is the successful simulation of massive flow separation at high Reynolds numbers. In this technique, the RANS model is applied to study the attached boundary layer, whereas the LES approach is used in the separation region. Some early results in the form of patterns showing the contours and isosurfaces of instantaneous vorticity and graphs of pressure coefficient distribution may be seen in the fundamental articles [365, 368], and survey [372]. These first results for sub- and super-critical flows over a sphere are typical examples of flows past bluff bodies which resistance significantly depends on separation. The early flow separation before sphere equator (at 82° for laminar boundary layer) and below it (at 120° for turbulent flow) obtained by computation in both cases in agreement with experimental data and corresponding drag coefficients shown the reliability of DES for application at actual Reynolds numbers (Exer. 5.42). These and other results of simulation of this type listed in [368] demonstrated the wide applicability of DES in different areas: cavities with various Mach numbers, wing high-lift systems, ground vehicles, space launchers, air inlets, buildings, flow control, combustors, cavitations in jets, aerodynamic noise, and more. The patterns of flow past aircraft presented in [372] around forebody at 90° angle of attack, and over fighters: F-15E at 65° of attack, F-18E in the case of the abrupt wing stall, and F-18C at 30° with the vortex burst reveal that DES provides the successful modeling of the aerodynamics of the real industrial objects.

Nevertheless, scholarly continue to improve DES. Spalart et al [367]. and Deck [90] proposed modifications of original DES improving the treatment of the area where the model switches from RANS to LES, and where the rapid decrease of the RANS eddy viscosity might result in the strong instabilities. In order to get rid of this weakness, the authors of the first study developed a new version Delayed Detached Eddy Simulation (DDES) where the switch into LES is delayed preventing the undesired depletion of the model strength. In a different way, Deck resolved this problem by proposing the Zonal Detached Eddy Simulation (ZDES) in which the RANS and LES domains are separated so that regime in each zone is selected individually according to the requiring conditions.

Later on, both new approaches were developed farther. To improve the treatment of the flow at the near-wall area, in study [352], the wall-modeling in LES (WMLES) is adopted in which RANS is used only at much closer distance from a wall than in DDES, whereas LES is employed for the rest part of attached boundary layer and separation. In this version known as Improved Delayed Detached Eddy Simulation (IDDES), two cases depending of inflow conditions are considered. If the inflow has turbulent content, the model reduces to WMLES so that most of the turbulence is resolved except the small region near the wall where the RANS is used. In the other case when the inflow does not have turbulent content, the model performs as DDES using RANS for whole attached boundary layer and LES for separation area. It is obvious that in the case of applying WMLES, this model essentially improves the accuracy of modeling providing LES treatment for much greater part of flow than that in DDES. In the next study [353], the same authors introduce a procedure that stipulates the treatment by RANS only in thinnest near-wall area in both cases of inflow with and without turbulent content. This is achieved by injecting the resolved turbulent content in an overlap of RANS and LES domain where the RANS treatment is switched to LES one.

The alternative aforementioned improved DES, the Zonal Detached Eddy Simulation (ZDES), is also farther enhanced in [91] using combination of zonal approach [90] and the best features of DDES [367]. In this case instead of two different inflow conditions considered in IDDES, three separate zones are specified in which the flow regime may be selected according to practical needs. In conformity with that, three possible situations, depending on the cause of separation are considered. The first and the second zones concern flows where the point of separation is fixed by the geometry (like terrace or stepped surface) or is induced by a pressure gradient on a smooth surface, respectively. These are similar to flows of relatively simple first case considered in IDDES study [352] and are treated as well using wall-modeling in LES (WMLES) as described in [91]. The flows pertained to the third zone are more involved by influence of incoming boundary layer and, like the flows of the second case in [352], require special treatment to set up the desired inflow conditions. This is done in study [213] applying the Synthetic Eddy Method, which similar to injection of turbulent content, used in [353] for IDDES, generates the switch from RANS to LES quite close to the wall (Exer. 5.43).

Both IDDES and ZDES are the most promising advanced approaches providing accurate simulation of complex turbulent flows past real objects including industrial prototypes at natural Reynolds numbers under current computational resources. In the next chapter, we review examples of simulating flows past various bodies at different conditions performed by DNS, LES, DES, IDDES, and ZDES including the latest.

For further reading about turbulence simulation methods a reader is referred to the just mentioned latest original articles. Earlier original publications may be found in [422]. Relatively simple presentation of modern methods in turbulence is given in articles [386] and [368].

5.2.5   Chaos Theory

Chaos theory is a field in mathematics that studies behavior of dynamical systems highly sensitive to initial conditions. This feature known as the butterfly effect means that small differences in initial conditions yield widely different outcomes: a butterfly wings flapping today may result in storm tomorrow far away from here. Chaotic behavior can be observed in some natural and engineering systems, such as weather. One remarkable example of chaos theory success is a qualitatively simulation of Rayleigh-Benard flow by simple system of three ordinary differential equations. This phenomenon occurs in a fluid present between two horizontal plates in gravitational field if the lower plate is heated (Exam. 7.16). Currently, the chaos theory is not a tool for turbulence modeling; however, some turbulence characteristics are of the conditions that specify the chaotic regime. This indicates that there are hopes of using the chaos theory for attacking the turbulence problems. At the same time, some researches think that wavelengths spectrum in turbulence, ranging from Kolmogorov length scale to the dimension size of flow, is so broad that describing turbulence by chaotic methods would require a system of several hundred differential equations and hence is unrealistic [422] (Exer. 5.44).

Exercises

5.32 What is the principle difference between classical Reynolds-average models and modern computation methods of simulation turbulence?

5.33 Obtain estimations of numbers of grid point and time steps required for three-dimensional DNS as it described in text to understand the difficulties arising in performing this type of simulation.

5.34 Explain why DNS is critically important even if it may be used basically for relatively small Reynolds numbers.

5.35 Show that expression (5.24) is correct and explain why such relation may be used as a filter. Hint: recall the Leibniz rule and perform the integration and differentiation as formulae (5.24) indicate.

5.36 What is the concept of filtering? Why the filtering process is used in LES? Read about filter from Wikipedia

5.37 What is Δ in relations (5.25) and for what stands the abbreviation SGS? What quantities divided the filtering procedure?

5.38 Read about rate-of-strain tensor on Wikipedia or on Google the article "Velocity vector and strain rate tensor". You will see that this tensor is of second order (Com. 1.9) defined by nine components. Why nine? Hint: think about components of deformation.

5.39 Why modeling SGS is very important in LES. Hint: consider the relation between eddies (S. 8.4.1). Describe the Smagarinsky model. Read "Large eddy simulation" on Wikipedia or simpler article [386] (at least two first paragraphs) to understand the SGS model properties.

5.40 What limited the increase of Reynolds number in LES and stimulated attempts to improve this technique?

5.41 Explain the basic features of DES. What are the properties of DES that differ in essence this approach from the pure LES? What is blending function, how does it work?

5.42 Think: is it always true that turbulent flow separates later than the laminar? Recall how Prandtl use the later separation of turbulent flow around bluff bodies to reduce the resistance (Com. 8.1). Read detailed explanation of separation role in the bluff body resistance at laminar and turbulent boundary layer [338, 1979, p. 41].

5.43 Compare IDDES and ZDES and specify the similarity and dissimilarity of both approaches. To better understanding, consider examples of simulation presented in Chapter 6.

5.44 Read the article on Wikipedia to learn more about chaos theory.

6

Applications of Fluid Flow Modern Models

In this chapter, the examples of modern fluid flow applications in two areas are considered. In biology and medicine, the applications of peristaltic flow and of $k - \varepsilon$ and $k - \omega$ turbulence models are presented. In engineering, the applications of peristaltic flow models and of direct simulation of turbulence methods are analyzed. Examples are discussed briefly in the same way as it done for heat transfer models in Chapter 4, giving only the mathematical model of problem formulation, conception of solution approach and basic results. For convenience, examples are marked by letters **a** (analytical) or **n** (numerical) showing the type of solution employed. Additional information from Part III or at least from some *Advanced Engineering Mathematics* course should be sufficient for comprehension as well as the previous text.

6.1 Applications of Fluid Flow Models in Biology and Medicine

6.1.1 Blood Flow in Normal and Pathologic Vessels

■ **Example 6.1n: Arterial Stenosis Modeling [144]**

The stenosis is a constriction of the blood vessel cross-section that occurs due to formation of plaques in an arterial wall, leading to limitation of blood flow and to separation of flow once the stenosis becomes large. The disturbance of the blood flow by stenosis produces abnormal circulation, resulting in vascular disorders such as post-stenoic dilatation, losses in pressure, abnormally high shear stresses that may provoke blood problems (in particular, associated with red cells and platelets).

The model consists of a tube with constriction (stenosis) of the length $2D$ in the form $r(z)/D = 0.5 - A(1 + \cos \pi z/D)$ where A is a constant, r and z are coordinates with origin $z = 0$ at the throat of the stenosis, and D is the unobstructed tube diameter. The stenosis is located at $15D$ from the inflow and $20D$ prior the outflow sections. It is assumed that:

Applications of Mathematical Heat Transfer and Fluid Flow Models in Engineering and Medicine,
First Edition. Abram S. Dorfman.
© 2017 John Wiley & Sons Ltd. Published 2017 by John Wiley & Sons Ltd.

(i) the blood is a homogeneous, incompressible Newtonian fluid with constant kinematic viscosity $v = 0.035 \text{ cm}^2/\text{sec}$ and density $\rho = 1.06 \text{ g/cm}^3$, (ii) the flow is steady laminar current with parabolic profile at the inflow that remains laminar proximal to the stenosis, (iii) the disturbed blood flow after stenosis is turbulent and may be simulated using a turbulence model for moderate Reynolds number ranging from 400 (human carotid artery) to 1,500 (human ascending aorta).

The mathematical model described the disturbed flow includes the steady state Navier-Stokes equation, and two transport equations (8.26) and (8.27) determined k and ω in $k - \omega$ turbulence model (S. 8.4.3.1) specified for low Reynolds numbers [422]. The boundary conditions are prescribed around an entire computational domain for Dirichlet problem (S. 1.1): (i) at the wall, the usual non-slip conditions for the velocities, $k = 0$ and $\omega = 6\mu/\beta_\omega\Delta^2$ for $k - \omega$ model, where $\beta_\omega = 0.8333$ is a constant in equation for ω and Δ is the height of the first node above the wall, (ii) the symmetry conditions at the central axis, (iii) at the channel inlet for laminar flow very small values are taken $k = 0.0001$, $\omega = 0.45$, which corresponds to $\varepsilon = 0.000045$ (see (8.27)), (iv) at the outlet, the normal and tangential stresses are prescribed as zero, the flow is fully developed and properties are no longer varied with distance so that $\partial k/\partial z = \partial\omega/\partial z = 0$.

The above-mentioned system of three equations including boundary conditions is solved by a finite element method using the Fluid Dynamic Analysis Package (FIDAP) software. Three stenosis models with lumen area reduction: 50% ($A = 0.073$), 75% (0.125), and 86% (0.1565), defined as $1 - (r/R)^2$, show the following results:

- The stenosis leads to separation of flow and to formation of a zone with vortex which length increases almost linearly with Reynolds number until it reaches maximum at critical Reynolds number indicating that flow distal to the stenosis is laminar.
- When Reynolds exceeds critical values of 1,100, 400, and 230 for the first, second, and third models, respectively, the flow becomes transitional or turbulent; comparison with data of laminar flow modeling shows that in laminar flow range, both results are in agreement, but for Reynolds numbers larger than the critical value, the laminar flow model overestimates the vortex length.
- The wall static pressure distribution obtained for second model at Re $= 2,000$ well agrees with experimental data and corresponds to streamlines pattern showing sharp pressure dropping in the stenosis throat; comparison with data given by $k - \varepsilon$ model at Re $= 15,000$ indicates that $k - \varepsilon$ model predicts much higher pressure post-stenosis.
- The wall shear stress distribution reveals that the highest value takes place at the throat, and the lowest negative pressure appears at the reattachment point of the vortex; in the vortex region, the wall share stresses are negative, indicating that they are acting in opposite directions upstream and downstream of the reattachment point.
- In the case of laminar flow modeling, the wall share stresses downstream of the reattachment point approach, but never exceed the corresponding value of fully developed Poiseuille flow; in contrast, when the flow is transient or turbulent in this area, the shear stresses are higher than the fully developed Poiseuille flow values further decreasing to the laminar flow rate, which shows that relaminarization takes place.
- The turbulence intensity (a rate of turbulent fluctuations as a percentage of average velocity) along the central line are computed for second model at Re $= 2,000$, and the results are compared with experimental measurements and data obtained by $k - \varepsilon$ model at Re $= 15,000$;

the computed results as well as experimental data reveal that the turbulent intensity increases after the throat and reaches the peak at the vortex center in the separation zone, but the computed prediction underestimates the turbulence intensity, whereas the $k - \varepsilon$ model forecast differs much more from the experimental data.

■ Example 6.2n: Blood Flow Through Stenosis Series [221]

Multiple stenoses occur because the primary stenosis resulting in disturbed downstream circulation in time forms a secondary stenosis. This may lead to a third one and so on, yielding series of stenoses. The paper presents a detailed analysis on the flow dynamics with double bell-shaped stenoses at the relatively low, realistic Reynolds numbers, ranging from 100 to 4000. The dynamic characteristics including separation, reattachment, the formation of recirculation eddy, and the kinetic energy distribution are investigated for the cases of one and two stenoses at different distances between those.

The model consists of a tube in cylindrical coordinates z, r with two constrictions modeling the stenoses. The walls along the constriction have the bell-shaped profile

$$f(z) = 1 - c_i \exp[-c_s(z - s)^2]$$ (6.1)

which in curvilinear coordinates, used in this study, transforms in the rectangular domain. In this expression, c_s is the shape constant, $c_i = (D - d_i)/D$ is the constriction ratio, and s_i is the distance of stenosis from the inlet section. The problem is governed by two-dimensional Reynolds averaged Navier-Stokes equation (RANS) and by $k - \omega$ model (S. 8.4.3.1). Both systems of equations for RANS and for $k - \omega$ model are transformed in the curvilinear coordinates applying formulae given in the paper.

Using curvilinear coordinates simplifies the integration of governing equations. The boundary conditions for Dirichlet problem specified parameters at inflow as follows: $u = 1 - r$, $k = 1.5I_{tb}^2 u^2$, $\omega = \sqrt{k}/C_\mu^{1/4}l$, $l = \min(\kappa y_w, 0.1\text{Re})$. At the outflow, the velocities are extrapolated from interior, and a constant static pressure is imposed. The gradients of k and ω at the exit are assumed to be zero $\partial k/\partial z = \partial \omega/\partial z = 0$. The zero gradients along the axis of symmetry for k and ω with respect to r also are applied. At the walls, the no-slip condition for velocities, zero pressure gradient, $k = 0$ and $\omega = 6\nu/\beta_\omega l$ are used. In these conditions, turbulence intensity of inflow is $I_{tb} \approx 1\%$, y_w is a normal distance from the wall, $C_\mu = 0.09$, $\beta_\omega = 0.075$, $\kappa = 0.41$. Other dimensionless variables are scaled by r_0, u_0 for lengths and velocities, ρu_0^2 for p, u_0^2 for k, and u_0/r_0 for ω.

The solution procedure is based on modifying the governing elliptic equation in hyperbolic type. Two tests are performed to verify the accuracy of produced software: (i) the results of fully developed steady channel flow calculation are compared with data obtained by direct numerical simulation, and (ii) the computed steady turbulent flow in a circular tube with a constriction is compared with known experimental data.

Comment 6.1 The elliptic equation is transformed into hyperbolic type by adding an artificial unsteady term. Such modified unsteady equation, unlike the elliptic equation, may be solved straightforward starting from initial condition by using some implicit numerical method to reduce the solution to a system of linear algebraic equations. The desired steady solution of elliptic equation is obtained as a limit at $t \to \infty$.

The following results are obtained:

- The flow-through stenosis is complicated because the laminar, transitional, and turbulent regimes coexist there; since the type of the flow is unknown in advance, it is important to have a model capable to simulate at least laminar and turbulent flows. The developed approach for this purpose applies the $k - \omega$ turbulence model.
- The streamlines pattern for turbulent flow through stenosis shows two zones: the circulation zone behind the stenosis and the main flow zone near the centre of the tube with relatively straight and parallel streamlines.
- The length of the vortex in laminar flow increases as Reynolds number grows until it reaches a critical Reynolds number, whose value is about 300, and then when the flow becomes transitional or turbulent it decreases as the Reynolds number grows.
- The dimensionless vorticity at the wall, which relates to wall shear stress, increases rapidly as flow approaches stenosis and reaches a peak value slightly upstream of the stenosis area; then downstream, it value decreases and becomes negative where the separation at the wall occurs; the value of the vorticity peak increases and tends to shift upstream as the Reynolds number increases; the negative magnitude of the vorticity at wall in recirculation zone also increases as the Reynolds number grows.
- The wall pressure and centerline velocity distributions for Re = 100 and 300 in comparison with data obtained for laminar flow in the same problem [220] indicate that for Re = 100, $k - \omega$ turbulence model gives the same results as those in the case of laminar flow modeling; for Re = 300, the centerline velocity obtained by laminar flow simulation is much slower because the flow distal to the stenosis becomes transitional in this case, and the laminar model cannot take this into account.
- In the case of two stenoses, the recirculation eddies are formed downstream of each stenosis; the separation streamline divides the flow in recirculation region distal to each stenosis and a region with main flow near the centre line, as in the case of the one stenosis, when the relative distance between stenoses S/D is less than 3, a recirculation zone fills the valley region between the two stenoses, and the reattachment occurs at the front of the second stenosis; the recirculation zone distal to the second stenosis reduces as S/D grows so that for $S/D = 4$, it becomes much smaller than that for a first stenosis;
- The two peaks exist in the wall vorticity distribution such that the peak induced by second stenosis is smaller than that generated by the first stenosis; as the distance between stenoses increases the second vorticity peak grows, and at $S/D = 4$, they become almost equal; the negative wall vorticity peak occurs proximal to the second stenosis when S/D is les than 3, whereas it does not appear when S/D is 3 or 4; the maximum centerline velocity develops slightly downstream of the stenosis because the formation of a recirculation zone behind each stenosis reduces the cross-sectional area;
- The maximum of centerline disturbance intensity, which characterized the turbulence intensity, near the second stenosis is higher than that near the first stenosis for all the spacing ratios; at the same time, the downstream peak value of wall vorticity increases with spacing ratio growing until the spacing ratio S/D reaches 4; thus, it may be deduced that the double stenosis have the strongest effect on the distribution of the turbulence intensity in downstream region when the spacing ratio is $S/D = 4$.

- As the Reynolds number increases, the value of two wall vorticity peaks grow as well as their difference, the centerline velocity decreases, and the peak of centerline disturbance of turbulence intensity goes up, whereas the distance between the peaks and stenoses decreases.
- The data obtained for different values of ratio c_i of the stenosis profile (6.1) shows that $c_i = 0.5$ is a critical value, which blocks the vessel leading to abrupt changes in the flow properties; this result is in the line with the clinic practice usually based the treatment of the artery disease depending on whether the artery is more than 75% stenotic, which corresponds to inequality $c_i > 0.5$.

■ Example 6.3a: Blood Flow Through Multi-Stenoses Under Magnetic Field [258]

This paper studied the effects of several parameters on the blood flow in the elastic artery with multi-stenosis, basically focusing on the influence of an external magnetic field. Such analysis simulates the blood flow during the magnetic resonance angiography (MRA) or magnetic resonance imaging (MRI). The model is constructed as anisotropically elastic cylindrical tube with multi-stenosis filled with incompressible viscous electrically conducting fluid modeling the blood. The model with elastic wall is more realistic than the most other much simpler models of this type considering the vessel wall as a rigid tube. The form of the stenoses is adopted from [61]

$$R(z) = \begin{cases} a\left\{1 - a_1 \left[s_l^{m-1}\left(z - d_l\right) - (z - d_l)^m\right]\right\} & d_l \le z \le d_l + s_l \quad a_1 = \dfrac{\delta_l m^{m/(m-1)}}{a s_l^m (m-1)} \\ a & \text{otherwise} \end{cases}$$

$$\tag{6.2}$$

where $m \ge 2$ is the shape parameter, a is the constant radius of the normal artery, s_l is the length of the stenosis and d_l measures the location of stenosis, where $l = 1, 2, 3 \ldots, \delta_l$ is the maximum height of the stenosis located at $z = a + s_l/m^{1/(m-1)}$, and the ratio δ/a of the height of the stenosis δ to the radius a of artery is much less than unity.

The governing equations for fluid are two-dimensional unsteady Navier-Stokes equations in cylindrical coordinates with additional terms taken into account the effects of magnetic field $(\sigma B^2 + \mu/k)u$ and permeability $(\sigma B^2 + \mu/k)w$, where u and w are the radial and axial velocity components, σ, k and B are electrical conductivity, coefficient of permeability, and magnetic flux density.

The equilibrium relations for elastic tube, adopted from [21], are formulated on the base of the membrane theory of shells. In this case, a mathematical model represents balances of longitudinal and radial components of three forces: inertial, surface forces, and forces of constraint associated with surrounding connective tissues

$$(T_t - T_\theta)\frac{dR}{dz} + R\frac{\partial T_t}{\partial z} - R\left[M_0\frac{\partial^2 \xi}{\partial t^2} + C_1\frac{\partial \xi}{\partial t} + K_1\xi + \left(M_0\frac{\partial^2 \eta}{\partial t^2} + C_r\frac{\partial \eta}{\partial t} + K_r\eta\right)\frac{dR}{dz}\right] +$$

$$+ \frac{R}{\sqrt{1 + (dR/dz)^2}}\left\{\frac{dR}{dz}\left(T_{zz} - T_{rr}\right) + \left[\left(\frac{dR}{dz}\right)^2 - 1\right]T_{rz}\right\}_{R-\Delta/2} = 0 \tag{6.3}$$

$$
\frac{T_\theta}{R\sqrt{1+(dR/dz)^2}} - \frac{d^2R}{dz^2} \frac{T_t}{[1+(dR/dz)^2]^{3/2}} - \frac{1}{\sqrt{1+(dR/dz)^2}}
$$

$$
\left[\frac{dR}{dz} \left(M_0 \frac{\partial^2 \xi}{\partial t^2} + C_1 \frac{\partial \xi}{\partial t} + K_1 \xi \right) - \left(M_0 \frac{\partial^2 \eta}{\partial t^2} + C_r \frac{\partial \eta}{\partial t} + K_r \eta \right) \right]
$$

$$
- [1+(dR/dz)^2] \left[2\frac{dR}{dz}T_{rz} - T_{rr} - \left(\frac{dR}{dz} \right)^2 T_{zz} \right]_{R-\Delta/2} = 0 \qquad (6.4)
$$

Here, $M_0 = \rho_0 \Delta + M_a$, where ρ_0 and Δ are the density and thickness of the arterial wall, ξ and η are the displacement components, T_t and T_θ are the components of viscoelastic stress acting along the longitudinal and the circumferential directions, K_1, C_1 and M_a represent (per unit area) the spring coefficient, the frictional coefficient of the dashpot, and the additional mass of the mechanical model in longitudinal tethering and K_r and C_r are those in radial direction (Com. 6.2). The boundary conditions are given on the arterial wall as the velocities defined through displacement derivatives (first two relations) and on the symmetry axis as zero radial and gradient of axial velocities

$$
u(r,z,t) = \frac{\partial \eta}{\partial t}, \quad w(r,z,t) = \frac{\partial \xi}{\partial t} \quad \text{on} \ r = R(z), \quad (r,z,t) = 0, \quad \frac{\partial w(r,z,t)}{\partial r} = 0 \ \text{on} \ r = 0 \ (6.5)
$$

Comment 6.2 The simulation of flow in a flexible tube is a complicated problem. Atabek in the early work [21] proposed simplified model of blood vessel as elastic tube with fluid. This model was used by many researchers (Web of Science indicates 100 followers). Three forces mentioned above are taken into account in this model. The two first, the inertia and surface forces, are considered in his earlier study [20]. The inertia force is determined using the Newton second law, and the surface forces are calculated as the reaction of the fluid to its container. The basic difficulty in model formulation of this type is the definition of the third force arising due to vessel surrounding, which in biology is known as the connective tissue. In study [21], the effect of tethering force (as such force is called) is described by a simple mechanical model constructed of an additional mass, a spring for motion, and a dashpot for deceleration.

The equations governing the motion of both the fluid and the arterial wall are nonlinear. Approximate solution of this complicated system is found considering for all variables only the first term of perturbation series in power of small parameter $\varepsilon = \delta/a$. In order to eliminate the nonlinear terms and to get the linear differential equations, this term is further approximated presenting it as the first term of Taylor series for middle wall section at $r = R(z) - \Delta/2$ in the form

$$
f(r,z,t)|_{R(z)-\Delta/2} = f_0(a,z,t) + \varepsilon \left[f_1(a,z,t) + R_1(z)\frac{\partial f_0(a,z,t)}{\partial r} \right] + O(\varepsilon^2) \qquad (6.6)
$$

Here, $f_0(a,z,t)$ is a known solutions (for example, for artery without stenosis), and the last term indicates the order of the next terms of Taylor series (Com. 3.8). Substituting relation (6.6) for each variable in governing equations for fluid, arterial wall (6.3) and (6.4) and boundary conditions (6.5) after gathering terms with ε and ignoring the others yields the system of linear differential equation governing in this approximation the studied problem. A corresponding tedious procedure is given in the reviewed paper [258].

Calculations for the artery with three stenosis and relevant experimental data are presented in paper. The following basic results are observed:

- The resistance to flow (resistive impedance) at the wall surface is influenced by the unsteady behavior of the flowing blood as well as by the vessel wall distensibility, by the Hartman number (Ha $= BL\sqrt{\sigma/\mu}$), the permeability coefficient k, the maximum height of stenosis δ_1, and shape parameter m in (6.2) defined stenosis form.
- In the first stenosis region ($0.375 \leq z \leq 0.875$), the resistance impedance increases at the onset of the stenosis until its maximum constriction at the throat, then it decreases steeply to reach the end point of the constriction. This observation holds true for the second stenosis region ($1.25 \leq z \leq 1.75$) and for the third one ($2.125 \leq z \leq 2.625$); a resistance increases also as the Hartman number and maximum height of stenosis grow, whereas it decreases with shape parameter and permeability increasing, attaining its maximum in the symmetrical stenosis with $m = 2$.
- The wall shear stress τ_{rz}, unlike the characteristic of resistance impedance, in the first stenosis area decreases at the onset until the throat, then it decreases steeply to reach the end point of the constriction; this observation is similar to that in the second and the third stenoses; stress τ_{rz} increases as the longitudinal viscoelastic stress component and anisotropy parameter grow, whereas it decreases with the increasing the viscoelastic stress in the circumferential direction, the total mass of the vessel, the surrounding tissues, and the contributions of the viscous and elastic constraints to the total tethering.
- The studying the effect of stenosis shape shows that the magnitude of the wall shear stress decreases in the converging zones as shape parameter increases, whereas it increases in the diverging zones in similar situation; for any given stenosis shape, the wall shear stress steeply decreases in the upstream from an approached magnitude to the peak value at the throat, and then increases in the downstream of the throat to the its value at the end point of the constriction profiles.
- The transmission of wall shear stress distribution through a tethered tube is substantially lower than that through the free tube, whereas the shearing stress distribution at the stenosis throat have the inverse character through totally tethered and free tubes.
- The stream function at the surface wall increases along the vessel and steeply decreases in the stenotic region. The stream function decreases as the number of stenosis increases. It also decreases in the three stenotic region with increasing shape parameter m and height of stenosis δ.
- The radial velocity at the surface wall decreases with increasing the anisotropy and the contribution elastic constraints of the total tethering, whereas it increases as both longitudinal and circumferential initial stresses components grow; at the same time, the effects of the contribution viscous constraints of the total tethering, total mass of the vessel, and surrounding tissues of the radial velocity are negligible small. Similarly, the axial velocity at the surface wall decreases with increasing the contributions of the viscous and elastic constraints to the total tethering, the total mass of the vessel, and the surrounding tissues, whereas it increases with the increasing the anisotropy and the longitudinal and circumferential initial stresses components.
- The formed bolus (called trapped) defined as a volume of fluid bounded by closet streamlines in the wave frame is transported at the wave; the trapped bolus at the central line

increases in the size as the permeability increases, whereas it decreases in size by increasing the Hartman number; the trapped bolus appear in the non-symmetric stenosis, but they disappear in the case of symmetric stenosis; they also are smaller in the free tube than in tethered tube.

■ Example 6.4a: Simulation of Blood Flow in Small Vessels [269]

In his article, the small blood vessels are modeled by considering the flow of the Herschel-Bulkley non-Newtonian fluid (Com. 5.8) in channels with varying cross-sections. It is assumed that the progressive sinusoidal wave of the form

$$H = H_0 + \Lambda x + a \sin[(2\pi/\lambda)(x - ct)] \qquad \mu = \frac{\tau_0 + \alpha[\dot{\gamma}^n - (\tau_0/\mu_0)^n]}{\dot{\gamma}} \qquad (6.7)$$

propagates with constant speed c along such channel filled with Herschel-Bulkley fluid, which takes into account the combined effect of Bingham plastic and power-law fluids behavior according to second expression (6.7). In equations (6.7), H_0 is the half-width at the inlet (for small vessel the values of H_0 are $10 - 60\,\mu m$), a is an amplitude, and Λ is a constant defining a channel form so that for converging tube (simulating arterioles) the width of the outlet for one wave length is 25% less and for divergent tube (venules) it is for one wave length 25% more than width of inlet. At low strain rate $\dot{\gamma} < (\tau_0/\mu_0)$, and the Herschel-Bulkley fluid behaves as viscose fluid with constant viscosity μ_0, whereas when the strain rate grows and passes the threshold, τ_0, it behaves as the power law fluid with viscosity determined by relation (6.7), where α is the consistency factor and n is the power law index defining the thinning ($n < 1$) or thickening ($n > 1$) fluid (S. 1.9). The other typical for small blood vessels parameters used here are: $n = 1/3 - 2$, $H_0/\lambda = 0.01 - 0.02$, $\Delta p = -300 - 50$, $\tau = 0 - 0.2$, and $\phi = a/H_0 = 0.1 - 0.9$, is the amplitude ratio.

It is shown that the governing system of Navier-Stokes equations and relevant boundary conditions under usual assumptions of small Reynolds number and long wavelength simplifies and in the fixed frame becomes

$$\frac{\partial p}{\partial x} = \frac{\partial \tau_{yx}}{\partial y}, \qquad \frac{\partial p}{\partial y} = 0, \qquad \tau_{yx} = \left(\tau_0 + \left|\frac{\partial u}{\partial y}\right|^n\right) \text{sgn}\left(\frac{\partial u}{\partial y}\right)$$

$$y = 0, \qquad \frac{\partial u}{\partial y} = 0, \qquad \tau_{yx} = 0, \qquad y = H, \quad u = 0 \qquad (6.8)$$

The dimensionless variables are scaled by $\lambda, H_0, c, \mu c^n \lambda/H_0^{n+1}$, and $\mu(c/H_0)^n$ for x and y, H, u, p and τ_{yx}, respectively. The expression $\text{sgn}(\partial u/\partial y) = \partial u/\partial y/|\partial u/\partial y|$ is the signum function. The boundary conditions are the symmetry conditions on the central axis of a channel and no-slip one on the walls. The solution of the first equation (6.8), which is independent of y, under conditions (6.8) is presented in the form

$$\frac{\left(H\frac{\partial p}{\partial x} - \tau_0\right)^{\frac{1}{n}-1} - \left(y\frac{\partial p}{\partial x} - \tau_0\right)^{\frac{1}{n}-1}}{\frac{\partial p}{\partial x}\left(\frac{1}{n} - 1\right)} \quad \text{and} \quad \frac{\left(H\frac{\partial p}{\partial x} - \tau_0\right)^{\frac{1}{n}-1} - \left(-y\frac{\partial p}{\partial x} - \tau_0\right)^{\frac{1}{n}-1}}{\frac{\partial p}{\partial x}\left(\frac{1}{n} - 1\right)} \qquad (6.9)$$

for $y \geq 0$ and for $y < 0$. On the basis of these solutions, the flow characteristics are computed for data typical for small blood vessels, and following results are formulated:

- The data for velocity distribution in the case of Newtonian fluid ($n = 1$) agree with results obtained before by Takabatake and Ayukawa [382]: the effect of non-Newtonian rheology leads to disturbed parabolic profile, which disturbance increases as the index n grows.
- For a converging channel, the magnitude of velocity is greater than that for a uniform channel, and for a diverging channel the result is altogether different; the data of influence of the pressure on the velocity distribution show that in the case of $\Delta p = -1$, for a shear thinning fluids ($n < 1$), flow reversal is totally absent in the uniform/divergent channel; for a converging channel, although there is a reduction in the region of flow reversal, it does not vanish altogether, irrespective of whether the fluid is of shear thinning or of shear thickening fluids ($n > 1$) type.
- The relationship between the pressure difference and the mean flow rate averaged over one wave is nonlinear for converging and diverging channels for both values of power index $n \neq 1$, whereas it becomes linear for Newtonian fluid ($n = 1$), which is in line with before observed results [345]; it is found that the mean flow rate increases as the pressure difference decreases.
- The pumping region ($\Delta p > 0$) increases with the growing of amplitude ratio φ for both shear thinning and shear thickening fluids as well as with increasing the power index n; in the co-pumping region ($\Delta p < 0$), the pressure rise decreases when the flow rate exceeds a certain value; pumping region increases also with τ increasing, and this effect is greater in the case of shear thickening fluids.
- The data for wall shear stress distribution at four instants of wave period show that at each of these instants, there exists two peaks: a negative τ_{min} and a maximum τ_{max}; transition from τ_{min} to τ_{max} occurs between the maximal and minimal channel heights.

■ **Example 6.5a/n: Simulation of Blood Flow During Electromagnetic Hyperthermia [268]**

The electromagnetic hyperthermia is a procedure for cancer treatment that is based on the experimental fact that at temperature higher than $41°C$, the transformed malignant cells are more sensitive than the normal cells. The procedure consists of injecting the magnetic fluid into an artery supplying the malignant tissues or directly into tumor and then subjecting the system to an alternating current magnetic field. The elevating of $45 - 47°C$ temperature generated in the injected magnetic fluid owing to the imposed magnetic field results in destroying the cancer sells. This method is applicable for treatment some tumors sites, including brain, soft tissues, liver, abdominal, pancreatic cancer, and head/neck tumors.

To simulate the blood flow under the action of the magnetic field, the model is constructed as two-dimensional channel with a flow of a biomagnetic non-Newtonian viscoelastic fluid whose rheology behavior is described by second-grade equation (Com. 5.8). The external magnetic field is generated by dipole located above a channel wall at the distance b. The channel walls are assumed to be porous and the flow is assumed to be driven by the stretching walls such that the velocity of each wall is proportional to the axial coordinate. These assumptions correspond to the physical conditions on the vessels.

The problem is governed by two-dimensional Navier-Stokes and energy equations with the following three expressions in the right-hand parts accounting for additional effects

$$-k_0 \left[u \left(\frac{\partial^3 u}{\partial x^3} + \frac{\partial^3 u}{\partial x \partial y^2} \right) + v \left(\frac{\partial^3 u}{\partial x^2 \partial y} + \frac{\partial^3 u}{\partial y^3} \right) - \frac{\partial u}{\partial x} \frac{\partial^2 u}{\partial y^2} - \frac{\partial u}{\partial y} \frac{\partial^2 v}{\partial y^2} \right.$$

$$- 2 \left(\frac{\partial u}{\partial x} \frac{\partial^2 u}{\partial x^2} + \frac{\partial v}{\partial y} \frac{\partial^2 u}{\partial y^2} \right) \Bigg] + \mu_0 M \frac{\partial \Omega}{\partial x} - \frac{\mu u}{k}, \quad -k_0 \Bigg[u \left(\frac{\partial^3 v}{\partial x^3} + \frac{\partial^3 v}{\partial x \partial y^2} \right)$$

$$+ v \left(\frac{\partial^3 v}{\partial x^2 \partial y} + \frac{\partial^3 v}{\partial y^3} \right) - \frac{\partial v}{\partial x} \frac{\partial^2 u}{\partial y^2} - \frac{\partial v}{\partial y} \frac{\partial^2 v}{\partial y^2} - 2 \left(\frac{\partial v}{\partial y} \frac{\partial^2 v}{\partial y^2} + \frac{\partial u}{\partial x} \frac{\partial^2 v}{\partial x^2} \right) \Bigg]$$

$$+ \mu_0 M \frac{\partial \Omega}{\partial y} - \frac{\mu v}{k}, \quad -k_0 \frac{\partial u}{\partial y} \frac{\partial}{\partial y} \left(u \frac{\partial u}{\partial x} + v \frac{\partial u}{\partial y} \right) + \mu_0 T \frac{\partial M}{\partial T} \left(u \frac{\partial \Omega}{\partial x} + v \frac{\partial \Omega}{\partial y} \right) \quad (6.10)$$

The two first expressions are terms of Navier-Stokes equations, and the third one presents terms of energy equations. The first term of each expression, which starts with coefficient k_0 takes into account the viscoelastic effect associated with stretched walls behavior. The terms $\mu_0 M (\partial \Omega / \partial x)$ and $\mu_0 M (\partial \Omega / \partial y)$ in two first expressions define components of magnetic force per unit volume, where μ_0, M and Ω are the magnetic permeability parameter, magnetic moment, and magnetic field strength. The second term in the third expression associated with magnetic field represents the thermal power per unit volume due to magnetocaloric effect, and the last two terms $\mu u / k$ and $\mu v / k$ in the first two equations (6.10) determine according to the Darcy law (Com. 4.25) the additional pressure gradients arising due to the permeability of channel walls.

Comment 6.3 The magnetocaloric effect is the magneto-thermodynamic phenomenon in which a temperature change of a suitable material (calling magnetocaloric) is caused by exposing the material to a changing magnetic field. In such a process, a decrease in the strength of external magnetic field owning to increased phonons thermal energy leads to reorientation of the material magnetic field resulting in changes of material temperature. This effect differs from well-known thermomagnetic (former term pyromagnetic) effect observed when a sample allowing electrical conduction (for instance, semiconductor) with temperature gradient is subjected to magnetic field, and as a result, the electrical field establishes in a sample (Nernst effect).

The boundary conditions consist of symmetry conditions on the central axis x, the proportional to x longitudinal stretching velocity component $u = \tilde{c}x$ and zero the transversal component $v = 0$ on the wall, the total pressure $p_0 = p + (\rho/2)(u^2 + v^2)$, and the temperature T_w. The governing system of equations is reduced to ordinary differential equations using five dimensionless functions: $f(\eta) = \psi / \tilde{c} H^2 \xi$, $\theta_1(\eta) + \xi^2 \theta_2(\eta) = T/T_w$, and $P_1(\eta) + \xi P_2(\eta) = -p/\rho \tilde{c}^2 H^2$, depending on two variables $\xi = x/H$ and $\eta = y/H$, and seven dimensionless quantities: distance to dipole $\tilde{b} = 1 + b/H$, Prandtl $\mathrm{Pr} = \mu c_p / \lambda$ and Reynolds $\mathrm{Re} = \rho \tilde{c} H^2 / \mu$ numbers, and coefficients of viscoelastic $K_0 = k_0 \tilde{c} / \mu$, viscous dissipation $K = \tilde{c} \mu^2 / \rho k_{pm} T_w$, permeability $\tilde{K} = k/H^2$, and ferromagnetic interaction $B = \gamma \mu_0 k_{pm} T_w \rho / 2\pi \mu^2$, where γ and k_{pm} are coefficients of magnetic strength and pyromagnetic (Com. 6.3). The system of five ordinary differential equations determining functions P_1, P_2, θ_1, θ_2 and f, specifies velocity components, pressure, and temperature. This system was solved applying perturbation series in small viscoelastic parameter K_0 and numerically by the finite difference method.

The effect of various parameters on the velocity components, pressure, skin friction coefficient and temperature of the arteries blood flow during electromagnetic hyperthermia was investigated for the following ranges of parameters: $\mathrm{Re} = 1 - 2, \mathrm{Pr} = 5 - 8, \tilde{b} = 2.5 - 4$, $K = 0.05 - 0.1, \tilde{K} = 0.08 - 2, B = 0 - 10$. Analysis of graphs presenting results gives the following basic conclusions:

- The axial velocity decreases up to central line reaching negative minimum close to it and then increases; such distribution is obtained for all studied values of parameters showing that the back flow occurs near the central line; it develops due to the stretching walls and may be reduced by applying strong external magnetic field.
- Close to the lower wall, the separation is observed for large Reynolds and Prandtl numbers and small porosity permeability $\widetilde{K}$, whereas it decreases as distance $\widetilde{b}$ to the magnetic dipole increases.
- For all studied cases, the transverse velocity increases with distance from lower wall reaching maximum at $\eta \approx 0.7$ and then monotonically decreases; it decreases as the Reynolds and Prandtl numbers increases; for large Prandtl numbers the reversal flow takes place in the vicinity of the central line; the same tendency is observed for small distances to magnetic dipole and porosity permeability.
- The data obtained for all values of parameters for function $\theta_1(\eta)$ (θ_2 was not studied) indicates that temperature monotonically decreases across the channel section and grows as the Reynolds, or Prandtl number, or the ferromagnetic interaction parameter increases, but the temperature decreases as permeability parameter increases.
- In all cases studied, the distribution of function $P_2(\eta)$ (the behavior of function $P_1(\eta)$ was not studied) shows that pressure decreases across the channel section, reaches minimum close to the upper wall ($\eta \approx 0.8$) and then remains almost constant. The pressure increases as the Reynolds or Prandtl number increases.
- Skin friction decreases for any values of Pr, $K, \widetilde{K}$ and $\widetilde{b}$ as the ferromagnetic interaction parameter B grows; the skin friction increases as Reynolds number increases, and at high Prandl numbers it grows for the values of B between one and two and then decreases as B becomes larger.

OTHER WORKS: The review of the early works studied blood flow was published in [186]. The analysis of results obtained up to the first years of 21 century is given in review [154]. The three-dimensional turbulence model of stenosed arterial bifurcation (division in branches) is developed in study [29]. Analysis of the blood flow considered as two-phase substance in a narrow catheterized artery is presented in article [375]. In paper [160], the accuracy of the velocities in carotid artery obtained during the diagnosis tests by ultrasound Doppler is estimated via comparison with numerical simulation data. Articles [236] and [137] are examples of recently published works. In the first paper, the pulsatile spiral blood flow through arterial stenosis is investigated. The other research work presents a review of studies of blood clot formation.

6.1.2 Abnormal Flows in Disordered Human Organs

■**Example 6.6a: Particle Motion in Peristaltic Flow with Application to Ureter [184]**

Simulation of interaction process of small particle with peristaltic flow is important for applications, such as moving cells in blood, or stones, or bacterium in ureteral flow, or microorganisms in solutions. In engineering the results of these researches are applicable, for example, in hydraulic particles transport systems design.

In the article reviewed, the interaction between small particle and incompressible Newtonian two-dimensional peristaltic flow is investigated. The flow in channel is described by mathematical model consisting two-dimensional steady-state Navier-Stokes equations in the moving frame (S. 5.1.1). The sinusoidal waves are imposed on the walls that are assumed to be flexible in transverse direction. The Navier-Stokes equation for stream function is used in the form (5.5) often used in studying. Solution of this equation is obtained by perturbation series up to ε^2, similar to Fung and Yih approach (Exam. 5.2).

The momentum equation for suspended in a moving fluid particle is formulated by Basset-Boussinesq-Oseen (BBO) mathematical model in the dimensionless form. This equation is applicable to small spherical particle in fluid flow at low Reynolds number and takes into account specific forces significant in this case

$$\frac{d\bar{u}_p}{dt} = \frac{2\tilde{\rho}(\bar{u} - \bar{u}_p)}{\text{Stk}(2\tilde{\rho} + 1)} + \frac{3}{2\tilde{\rho} + 1}\frac{d\bar{u}}{dt} + \frac{r^2}{40(2\tilde{\rho} + 1)}\frac{d}{dt}\nabla^2\bar{u} + \frac{\tilde{\rho}r^2}{12\,\text{Stk}\,(2\tilde{\rho} + 1)}\nabla^2\bar{u} \times$$

$$\sqrt{\frac{9}{2\pi\tilde{\rho}\,\text{Stk}}}\frac{2\tilde{\rho}}{2\tilde{\rho} + 1}\left[\int_0^t \frac{d\left(\bar{u} - \bar{u}_p\right)}{dt}\frac{dt'}{\sqrt{t - t'}} + \frac{\bar{u}_0 - \bar{u}_{p0}}{\sqrt{t}}\right] + \left[\frac{2\left(\tilde{\rho}\text{-}1\right)}{2\tilde{\rho} + 1}\right]\frac{g\tau_c}{c}\bar{g}$$

$$(6.11)$$

Here, $\bar{u}, \bar{u}_p$ and $\bar{g}$ are velocity and gravity vectors, $\tilde{\rho} = \rho_p/\rho$, r is the particle radius, $\text{Stk} = \tau_p/\tau_c$ is the Stokes number, $\tau_p = 2\rho_p r^2/9\mu$ is the particle relaxation time, c is the wave speed, $\tau_c = \lambda/\pi c$, $\nabla^2 = \varepsilon^2(\partial^2/\partial x^2) + \partial^2/\partial y^2$, $\varepsilon = \pi R/\lambda$, R is the half of channel height, and subscript p refers to particle. Variables are scaled as: the axial coordinate by λ/π, the normal coordinate and r by R, velocities by the wave speed c, and time by τ_c. The terms of sum in the right-hand side of BBO equation, which defines the particle acceleration $d\bar{u}_p/dt$, represent the steady-state Stokes drags force, the virtual mass forces (second and third terms), the Faxen, Basset, and gravity forces.

Comment 6.4 (i) The Basset force arises due to temporary delay of the boundary layer development when a body accelerates in Stokes flow. The corresponding term in the dynamic equation is known as Boussinesq-Basset (or history) term, who proposed take to account this force independently at the end of the nineteenth century; (ii) The Faxen force, named after a scholar who suggested it in 1922, is a correction to the Stokes law for a sphere (S. 7.4.2) that is valid when the body is moving close to the wall; (iii) The correction to the creeping flow developed by Oseen in 1910 (S. 7.4.3) takes into account neglected at the low Reynolds numbers inertia effect; (iv) The virtual mass is some mass of fluid added to a moving body to account an inertia effect arising at changing the body velocity.

Analysis of the analytical solution in series up to ε^2 for flow and numerical solution of BBO equation (6.11) for particle leads to the following basic results:

- The streamlines and velocity profiles indicate that flow is nonuniform near the wave trough in which the flow enters; below the wave crest, the downward-upward region grows with the flow rate increasing; the pressure distribution shows adverse pressure gradient in opposite to peristaltic flow direction in the upper half plane and favorable pressure gradient in the flow direction in the lower half plane.
- The satisfactory agreement of these results obtained from solution up to order ε^2 with experimental data [420] is observed for $\varepsilon\text{Re} = \omega R^2/\nu \leq 10$ where $\omega = \pi c/\lambda$ is the frequency; this is better than it is expected since the Reynolds number is assumed to be of order unity; the explanation is the sufficient convergence of perturbation series.

- Two real situations are analyzed by modeling the flow of mixture along the ureter with (i) stones or small their pieces after acoustic break up the stones and (ii) urine with bacteria that usually is a result of the stones or of a treatment; the results obtained for velocities components from a numerical solution of the BBO equation (6.11) give trajectories of simulated stones and bacteria using particles of corresponding sizes and characteristics. Analyzing of trajectories provides the insight into individual particles behavior showing the domains with normal particles motion and reflux.
- The trajectories of spherical and nonspherical particles in the case of retrograde motion near the longitudinal axis are similar; for zero pumping regime, the particles near the longitudinal axis have positive displacement, whereas near the wall the net displacement is negative. This corresponds to the reflux situation reported in the early studies (Exam. 5.1) showing that there is a possibility of transport of bacteria to the upper urinary tract.
- The behavior of a group of particles is investigated using a simplified equation (6.11) and the same assumptions as in the whole steady that: (i) each particle acts independently of the others and (ii) only flow affects a particle behavior but the back effect may be neglected.
- The groups of particles initially distributed uniformly near the center move forward, whereas those near the wall are delayed; particles simulating groups of stones and bacteria behave similar: some particles get closer to the wall, and others participate in formation of a bolus creating satellite structures from which after a while the bolus are detached; these observed particles tending to move to the wall may be considered as a possible explanation for the failure of calculi after successful ESWL (exstracorporeal shock wave lithotripsy) for acoustic break up of the stones.
- It is shown by comparative calculations that the most significant contribution in BBO equation are made by the Stokes drag, gravity, and Basset forces.

■ Example 6.7a: Simulation of Chyme Flow During Gastrointestinal Endoscopy [328]

The endoscope is a powerful means to diagnosis and management of intestinal illnesses. The interaction between intestinal, chyme flow, and endoscope during thegastrointestinal endoscope process is studied in this article. The pressure, pressure drop, velocity, and forces acting by endoscope and the intestine on chyme flow are calculated.

The model consists of cylindrical annulus bounded at the outer boundary by small intestine and at the inner boundary by the inserted endoscope. The sinusoidal wave is imposed on the outer tube-small intestine. The two-dimensional Navier-Stokes equations in cylindrical coordinates governed the problem. In dimensionless variables in the frame moving with wave at speed c, these equations and boundary conditions are presented in the form containing ε as a small parameter

$$\frac{\partial ru}{\partial r} + \frac{\partial rw}{\partial z} = 0 \quad \varepsilon^3 \mathrm{Re}\left(u\frac{\partial u}{\partial r} + w\frac{\partial u}{\partial z}\right) = -\frac{\partial p}{\partial r} + \varepsilon^2 \left[\frac{1}{r}\frac{\partial}{\partial r}\left(r\frac{\partial u}{\partial r}\right) - \frac{u}{r^2} + \varepsilon^2\frac{\partial^2 u}{\partial z^2}\right]$$

$$\varepsilon \mathrm{Re}\left(u\frac{\partial w}{\partial r} + w\frac{\partial w}{\partial z}\right) = -\frac{\partial p}{\partial z} + \varepsilon^2\left[\frac{1}{r}\frac{\partial}{\partial r}\left(r\frac{\partial w}{\partial r}\right) - \frac{u}{r^2} + \varepsilon^2\frac{\partial^2 w}{\partial z^2}\right]$$

$$u = 0, \ w = -1 \quad \text{at} \quad r = \Delta \quad w = -1, \quad \text{at} \quad \eta = 1 + \varphi \sin 2\pi z \qquad (6.12)$$

The variables are scaled by: r_0, λ, $\lambda/r_o c$, c, λ/c, $\lambda c\mu/r_0$ for radial and axial distances, radial and axial velocities, time, and pressure, respectively, $\mathrm{Re} = \rho cr_0/\mu$, $\varepsilon = r_0/\lambda$, $\varphi = a/r_0$,

$\Delta = r_i/r_0$, $\eta = \tilde{r}/r_0$, where r_0 and r_i are radii of outer and inner (endoscope) tubes, a and $\tilde{r}$ are an amplitude and radial coordinate of sinusoidal wave (6.12).

Under assumptions of small Reynolds number and long wavelength that are relevant in this case because the intestine is small as compared to wavelength, the ratio $\varepsilon < 1$ is small. Therefore, from simplified equations (6.12) containing only the term of order ε follows that: (i) according to first equation the pressure is independent on the radial variable, and (ii) the second equation may be integrated twice with respect to r giving expressions for the dimensionless radial velocity and flow rate in the wave frame

$$w = -1 - 0.25\frac{dp}{dz}\left[\Delta^2 - r^2 + (\Delta^2 - \eta^2)\ln\frac{r}{\Delta}/\ln\frac{\Delta}{\eta}\right]$$

$$q = \pi(\Delta^2 - \eta^2)\left[1 + 0.125\frac{dp}{dz}\left(\Delta^2 + \eta^2 - \Delta^2\eta^2/\ln\frac{\Delta}{\eta}\right)\right]$$

(6.13)

Solving the first equation for dp/dz and knowing that $q = Q - \pi[1 + (\phi^2/2) - \Delta^2]$, where Q is the flow rate in fixed frame, one finds basic characteristics: pressure drop Δp across one wavelength and the frictional forces F_e and F_0 over one wavelength acting on endoscope and intestine

$$\Delta p = \int_0^1 G(z)dz, \qquad G(z) = -\frac{8}{\Delta^4 - \eta^4 - (\Delta^2 - \eta^2)^2/\ln(\Delta/\eta)}\frac{[(Q/\pi) + \eta^2 - 1 - \varphi^2/2]}{}$$

$$F_e = \pi\int_0^1 G(z)\frac{\Delta^2 - (\Delta^2 - \eta^2)}{2\ln(\Delta/\eta)}dz, \qquad F_0 = \pi\int_0^1 G(z)\frac{\eta^2 - (\Delta^2 - \eta^2)}{2\ln(\Delta/\eta)}dz$$

(6.14)

These expressions present an exact solution of the problem in question in the case of small Reynolds number and long wavelength and lead to the following conclusions:

- The pressure drop is generally positive (which means that the pressure in the chyme flow decreases with increasing the axial coordinate) when the flow rate is not sufficiently small, and it is negative otherwise; thus, there is the pressure gradient acting on the flow, if the flow rate is not too small, whereas this pressure gradient acts in opposite direction in the case of too small flow rates; the magnitude of the pressure drop increases as the wave amplitude or endoscope diameter increases.
- The inner friction force is generally negative (the force acts by the chyme flow on the endoscope), and provided the flow rate is not sufficiently small being otherwise the positive one; in any regime, the magnitude of inner force increases with either the wave amplitude or endoscope cross-section size growing; thus, for small flow rates, the force acts by the endoscope on chyme flow and may become strong in the case of large amplitude or endoscope size, whereas at large flow rate this force acts in opposite direction so that the chyme flow affects the endoscope.
- The outer friction force is generally positive (the force acts by the chime flow on the intestine), and provided flow rate is not too small, whereas it is negative otherwise; at sufficiently large flow rate, the outer force is greater for larger amplitude or aspect ratio of an annulus; thus, for large flow rate, the chyme exerts on the intestine, but at the small flow rate, the intestine acts on the chyme flow; this effect increases as amplitude or endoscope grows.

- The leading order term of velocity is higher for higher flow rate, has higher amplitude and decreases with increasing the radial variable in the absence on endoscope, whereas it increases as the radial variable grows in the presence of endoscope.

■ **Example 6.8a: Simulation of Bile Flow in a Duct with Stones [251]**

The article presents a study of the motion of a mixture of the fluid (simulate the bile) and the solid particles (simulate the stones) that forms a dense porous mass. The model consists of a channel with wave $y = (2\pi ac/\lambda)\sin(2\pi/\lambda)(x - ct)$ propagating along walls. The problem is governed by Brinkman equations which are Navier-Stokes equations with viscous terms $\mu\nabla^2 u/e$ and $\mu\nabla^2 v/e$ instead of usual ones $\mu\nabla^2 u$ and $\mu\nabla^2 v$ and additional pressure gradients $\mu u/k$ and $\mu v/k$ taken into account the porosity and permeability of bile with stones. In these expressions e and k are proper coefficients and $\nabla^2 = \partial^2/\partial x^2 + \partial^2/\partial y^2$ is the Laplace operator (Com. 1.1). Additional pressure gradients are defined according to Darcy law (Com. 4.25).

In this study, the Brinkman equation is considered in streamline form, which is obtained after pressure eliminating (like Navier-Stokes equation derived in S. 7.1.2.4). The Saffman slip boundary conditions is applied on the walls

$$\text{Re}\,(\psi_{tyy} + \psi_{txx} + \psi_y\psi_{yyx} - \psi_x\psi_{yyy} + \psi_y\psi_{xxx} - \psi_x\psi_{xxy}) = (1/e)(\psi_{yyyy} + \psi_{xxyy} + \psi_{xxxx}) -$$

$$(1/k)(\psi_{yy} + \psi_{xx}), \quad \psi_y = \mp s\psi_{yy}, \quad s = b/\sqrt{k} \quad \text{at} \quad y = \pm H \pm \eta \tag{6.15}$$

Here, the variables are scaled as follows: x and y, u and v, ψ, η, p, t, k by $H, c, cH, H, \rho c^2$, $H/c, H^2$, respectively, and s is slip parameter in Saffman slip condition.

Comment 6.5 Saffman slip condition is used for smooth or permeable boundaries and has the form $\partial u/\partial y = (b/\sqrt{k})u$ where b is a constant depending only on the properties of a porous material and k is permeability coefficient [43].

The problem (6.15) is solved using perturbation series up to ε^2 in standard form

$$\psi = \psi_0 + \varepsilon\psi_1 + \varepsilon^2\psi_2 + \ldots, \quad \frac{\partial p}{\partial x} = \left(\frac{\partial p}{\partial x}\right)_0 + \varepsilon\left(\frac{\partial p}{\partial x}\right)_1 + \varepsilon^2\left(\frac{\partial p}{\partial x}\right)_2 + \ldots \tag{6.16}$$

where $\varepsilon = a/H$ is the dimensionless amplitude considered as a small parameter, and the first term in the last equation is the imposed pressure gradient, whereas the others are arising due to peristaltic motion. In general case, the solution is reduced to the fourth-order differential equation, which may be solved numerically. In a special case, for initially stagnation fluid, a close solution is obtained. It is observed that in such a case, the maximum pressure gradient that can be created by one wave of small amplitude is of the order ε^2. Using this solution, the analytical relations are derived for velocity, time-averaged velocity, and for critical pressure of reflux.

Analyzing of computation results yields the following results and conclusions:

- The time average bile velocity in porous medium is determined by three terms: the parabolic mean velocity distribution term that arises out of time, constant term, and the term of perturbation of the velocity, which controls the peristaltic mean flow.
- To validate the results, the values of perturbation velocity term are compared with data for the case without porosity and permeability effects from an early Fung and Yih article [139];

it is found that this term as well as in the previous study decreases as the Reynolds number grows; it is also estimated that the perturbation velocity term decreases as coefficients of porosity and permeability decrease, but the wave number $2\pi H/\lambda$ and slip parameter s only slightly affect the value of this term.

- The contribution of the constant term in magnitude of bile velocity increases as Reynolds number and porosity coefficient increase and decreases with slip parameter and permeability coefficient growing.
- The averaged mean bile velocity increases as the Reynolds number grows, showing that the reflux occurs in the central region when the pressure gradient attains a critical value 0.220966; the bile velocity strongly depends on the wave amplitude ratio increasing as the amplitude ratio increases and at high values of it becomes reversal near the boundaries; these results are in conformity with known experimental data [380].
- The velocity in central region reduces as the Darcy number (which is the same as dimensionless permeability coefficient, Com. 4.25) decreases indicating that the velocity profiles in this case are quite different from the familiar Poiseuille profile, which means that the bile velocity decreases as the number of stones increases.
- The velocity increases as the porosity parameter grows under other fixed parameters in the case when the Darcy number exceeds the value 0.05, although its parabolic nature changes for small Darcy numbers showing that velocity increases as the number of stones decreases reaching maximum in the case of the absence of stones.
- The velocity value strongly depends on the slip parameter in the case of porous medium, whereas the wave number affects the velocity only slightly, more prominently in the vicinity of the boundaries.
- The critical pressure of reflux decreases as the Reynolds and wave numbers increase, whereas it first increases with Darcy number growing and then maintains nearly constant; the critical pressure significantly decreases as the porosity parameter increases showing that at high amount of stones, the reflux occurs at low pressure; it also decreases considerably when the slip parameter grows.

Comment 6.6 One may think that some conclusions here and in other similar studies are trivial (like "the bile velocity decreases as the number of stones increases"). Such results obtained by mathematical simulation differ in principle from usual qualitative statements due to quantitative data giving the dependence between characteristics ($u = f(e)$ in this example), which could not be gained before in the descriptive disciplines.

OTHER WORKS: The reviews of studies of peristaltic flows in normal and abnormal human organs from early works to the results published up to first decade of the twenty-first century one may find reading introduction of articles [184] and [251]. The pyeloureteral function, which is responsible for transport urine from kidney to bladder is investigated in [294]. The kidney stones drift and some other biological processes are simulated in [74] by particles moving in a viscoelastic peristaltic flow. Fluid structures in a tissue during Hyperthermia (disordered thermal regulation) are analyzed in [11]. The effect of inserted endoscope evaluated in reviewed above article [328] is also examined in [2] modeling a real situation by peristaltic flow in cylindrical tube with endoscope and taking into account rotation and magnetic field influence. The results of experimental and CFD (calculation fluid dynamic) investigation of pathological bile flow in the binary system are presented in paper [207]. The stents (small expendable tubes used for inserting in blocked vessels) properties for bile channels and for gastric acid environment are evaluated in vitro (not in living organ) in studies [28] and [303], respectively, using experimental units.

6.1.3   Simulation of Biological Transport Processes

■ **Example 6.9a: Modeling Transport Processes in Cerebral Perivascular Space [414]**

Perivascular space (PVS) within the brain is an important pathway for interstitial fluid (ISF) and solute transport in the cerebral cortex that has significant impacts on physiology. In this paper, a model of fluid flow in the cerebral perivascular space induced by peristaltic motion of the blood vessel is developed. The model is used to study effects of various physiological parameters on the perivasclar fluid transport and particularly to investigate the interaction of special method (convection-enhanced delivery, Com. 6.7) of infusion compounds into the brain with peristaltic motion of close located blood vessel.

The model is a thin annual fluid-filled porous medium surrounding a blood vessel. The outer wall is fixed at the distance $r = R_2$ from the blood vessel central line, which is taken as axis z of cylindrical coordinates. It is assumed that vessel oscillates sinusoidal as $\eta = R_1 + a\sin(2\pi/\lambda)(z - ct)$ where R_1 is a mean radius. The mathematical model is based on the same as in the last example Brinkman equation, but with modified terms defined additional pressure gradients in the form $\mu eu/k$ and $\mu ev/k$. This form follows from Darcy law if one takes into account that due to e, the velocity components are enhanced by q/e, where q is the flow rate per unit area.

It is shown that in moving frame under usual assumptions of small Reynolds number and long wavelength, the Brinkman equations reduce to simple system for pressure gradients leading to the following solutions for axial velocity and the flow rate

$$\frac{dp}{dr} = 0, \quad \frac{dp}{dz} = -\frac{e\mu}{k}u - \frac{e\mu c}{k}, \quad u = -\frac{k}{e\mu}\frac{dp}{dz} - c, \quad Q = \pi\left[R_2^2 - \eta^2(z)\right]u \quad (6.17)$$

It is clear from (6.17) that pressure depends only on z, and due to that eliminating u from last two relations gives equations for dp/dz and for pressure drop on the one wavelength

$$\frac{dp}{dz} = -\frac{e\mu}{k}\left\{\frac{Q}{\pi\left[R_2^2 - \eta^2(z)\right]} + c\right\}, \quad \Delta p_\lambda = \frac{ec\mu\lambda}{k}\left[-\frac{Q}{4\pi cR_2^2}\left(\frac{1}{\sqrt{R_+}} + \frac{1}{\sqrt{R_-}}\right) - 1\right] \quad (6.18)$$

where $R_\pm = [1 \pm (R_1/R_2)]^2 + (a/R_2)^2$.

Analysis yields the following conclusions:

- The total volumetric flow rate is a sum of contributions from the pressure gradient and from the peristaltic movement of the boundary (vessel walls), which are coupled in solution; the first part is the flow rate of pressure-driven flow through an annulus filled with a porous medium, while the second part comes from peristaltic flow.
- The time-averaged displacement of a tracer particle is always positive regardless of the initial position of the particle, which means that there is no reverse transport in the perivascular space; the reverse perivascular transport observed by Schley *et al.* [337] is found using

other model for simulating flow in non-porous space based on Navier-Stokes equation so that there are no contradictions between two different results.

- As mentioned above, the one of the goals of this study is to investigate the effect of blood vessel peristaltic motion on the convection-enhanced delivery (CED). It is found that interaction between the peristaltic motion of a close blood vessel and CED infusion pressure gradient depends on the orientation of the vessel; the maximum effect is achieved when the blood vessel is oriented in radial direction resulting in the peristaltic wave traveling in the outward radial direction as well.

Comment 6.7 Convection-enhanced delivery is the method in which drugs are infused directly into the brain tissue through a needle or catheter.

- The comparative calculations of the effects of CED therapy in the presence and in the absence of peristaltic blood vessel show that the fluid transport in the perivascular space is predominantly depends on the distance from the needle and on the values of the permeability coefficient k and the wave amplitude a; at sufficiently large distances, the main contribution to fluid transport comes from the peristaltic wave, whereas near the infusion source, the importance of peristaltic wave depends on the values of k and a and in general is greater for the lower values of permeability and higher wave amplitudes.

■ **Example 6.10a/n: Simulation of Macromolecules Transport in Tumors [36]**

The therapeutic efficiency of various genetically engineered macromolecules, which also are known as monoclonal antibodies, depends on their delivery and distribution in tumors during the treatment. The reviewer study analysis effects of various physiological parameters on the transport of interstitial fluid and macromolecules in tumors and uses the results for testing the basic assumptions in current models.

Comment 6.8 The term "genetically engineered" stands for subjects changed by genetic engineering methods. Monoclonal antibodies are products developed following idea of a "magic bullet" coined by Paul Ehrlich at the beginning of the twentieth century: if a compound selectively target against a disease-causing organism, then a toxin for that organism could be delivered along with this compound. In the 1970s, this idea was realized by production of monoclonal antibodies that now widely used in biochemistry, molecular biology and medicine. One possible treatment for cancer involves monoclonal antibodies that bind only cancer cell-specific antigens and induce an immunological response against the target cancer cells.

Comment 6.9 The interstitial fluid (ISF) and blood plasma are two major parts of the all human body fluid outside of the cells called extracelluar fluid. The interstitial space between cells in a tissue is called the interstitium

The model consists of a cylindrical region surrounding an individual blood vessel of radius r_b and of intercapillary half-distance L streamlined by fluid with velocity u_∞. The pressure and velocity profiles around a cylinder simulating an isolated capillary are estimated. In contrast to previous studied, the nonuniform filtration and convection resulting from a heterogeneous pressure distribution are taken into account. The flow of interstitial fluid is modeled using Darcy's law for flow through a porous medium.

The problem is governed by a creeping flow equation for small Reynolds number (S. 7.4. 1) in cylindrical coordinates subjected to boundary conditions defining normal pressure gradient on the vessel wall and velocity components far away from the vessel

$$\mathbf{u} = -K\nabla p = -K\left(\frac{\partial p}{\partial r}\mathbf{i}_r + \frac{1}{r}\frac{\partial p}{\partial \theta}\mathbf{i}_\theta\right), \quad u_r|_{r=r_b} = -K\frac{\partial p}{\partial r}\bigg|_{r=r_b} = L_p(p_e - p) \tag{6.19}$$

$$u_r|_{r\to\infty} = -u_\infty \cos\theta \quad u_\theta|_{r\to\infty} = -u_\infty \sin\theta \ \ (-\pi/2 \le \theta \le \pi/2)$$

Here, K and L_p are hydraulic conductivity of the interstitium (Com. 6.9) and of the vessel wall, respectively, $p_e = p_v - \sigma(\pi_v - \pi_i)$ is the effective pressure, p_v is the vascular pressure of the vessel, $(\pi_v - \pi_i)$ is the osmotic pressure difference of capillary and interstitial fluid, and σ is the osmotic coefficient.

Comment 6.10 The effective pressure is defined using the law formulated by Starling in 1896, which states that net filtration (or net fluid movement) is proportional to driving force determined as $p_v - p_i - \sigma(\pi_v - \pi_i)$. The osmotic pressure is a force of flow exerted through a semipermeable membrane separating two solutions with diverse concentrations The solution of the problem (6.19) for the pressure and velocity has the form [35]

$$p = p_i^0 - \frac{L_p r_b}{K}(p_e - p_i^0)\ln\frac{r}{r_b} + \frac{u_\infty}{K}Nr\cos\theta \tag{6.20}$$

$$u_r = -u_\infty N\cos\theta + L_p(p_e - p_i^0)\frac{r_b}{r}, \quad u_\theta = u_\infty N\sin\theta, \quad N = 1 + \frac{r_b^2}{r^2}\left(\frac{1 - L_p r_b/K}{1 + L_p r_b/K}\right)$$

where p_i^0 is interstitial pressure at $r = r_b$ and $\theta = \pi/2$. The profiles of macromolecular concentration are found via diffusion equation with convection and binding (Com. 6.11)

$$\frac{\partial C_i}{\partial t} = \nabla\cdot(D\nabla C_i) - \mathbf{u}\cdot\nabla C_i - k_f C_i(B_{max} - B_i) + k_r B_i \tag{6.21}$$

satisfying two boundary conditions: no- flux far away from the blood vessel at one-half of the intercapillary distance L and the fluxes equilibrium at the vessel wall

$$[-D(\partial C_i/\partial r) + u_r C_i]_{r=L} = 0, \quad \left(-D\frac{\partial C_i}{\partial r} + u_r C_i\right)_{r=r_b} = u_r|_{r=r_b}(1 - \sigma)\left(\frac{C_p\exp\text{Pe} - C_i}{\exp\text{Pe} - 1}\right) \tag{6.22}$$

Here, C_i and B_i are the interstitial free solute and the bound solute concentrations determined as $\partial B_i/\partial t = k_f C_i(B_{max} - B_i) - (k_r + k_e)B_i$, where B_{max} is the concentration of binding sites available, D is the interstitial diffusion coefficient, k_f is the forward binding rate constant, k_r is the reverse (dissociation) rate constant, k_e is the elimination (or metabolism) rate constant, Pe is the transcapillary Peclet number defined as a ratio of convective to diffusive fluxes in transverse direction

Comment 6.11 The diffusion equation with convection and binding (6.21) is a sum of a passive diffusion equation and irreversible binding of the antibodies (Com. 6.8) with binding sites on the cell surface.

The equilibrium condition (6.22) physically means that the sum of diffusion and convective fluxes at the wall from capillary inside defined by the left-hand part of second equation (6.22) is equal to the flux at the capillary wall from surrounding medium side given by the right hand part of this equation. The first boundary condition (6.22) states that the flux defined by either part (here by the left one) vanishes at one-half of the intercapillary distance L, which is considered as far away from capillary distance.

The problem (6.21)–(6.22) is solved numerically using finite element approach (S. 9.6). Calculations are performed employing typical for tumors values of parameters. The following results and conclusions are derived:

- The interstitial pressure profile for baseline parameters corresponding to those in a lymphatic tumor surrounded by normal tissue is not symmetric due to the filtration from blood vessel; this leads to a stagnation point upstream to the vessel, but none downstream; when the far away velocity increases by factor 100, the profile resembles a classical solution of inviscid flow around cylinder; the interstitial velocity streamlines are perpendicular to the pressure isobars, that is parallel to the pressure gradients.
- These results confirm two assumption used in macroscopic models: the first, although the pressure is not symmetric around the vessel, the extravasation (leakage from capillary to surrounding) is a weak function of θ, and the second, the contribution of convection to interstitial transport and the effect of nonuniform velocities are limited on a macroscopic scale; binding reduces the diffusive and convective transport rates by the same amount, keeping their relative contributions equal;
- Unlike the distorted pressure profiles, the concentration profiles for nonbinding macromolecules at the vessel wall corresponding to the same baseline parameters of a lymphatic tumor are relatively unaffected. This is because the transcapillary Peclet number is small. Since the influence of convection (small Pe) is small, the angular dependence on profiles is weak, and as a result, the one-dimension model may be used for sensitivity analysis, which, in particular, has shown that for early times, here considered, there is insufficient material to saturate the binding sites; therefore, increasing the plasma concentration moderately has no effect on the dimensionless concentration profile; only significant increasing in plasma concentration approaching a level close to saturation results in profile change; conversely, if the forward rate constant k_f and the binding affinity ratio k_f/k_r are increased, then penetration into regions far away from the blood vessel is diminished, and larger concentrations are found outside of the blood vessel. This reduction in the capability to penetrate is known as the "binding site barrier".
- The major limitation in the presented microscopic analysis is that interaction between blood vessels are neglected so that pressure and velocity fields are computed for a single vessel in an infinite medium; however, for the physiological parameters chosen, the perturbation of the uniform velocity field is limited essentially to a couple of vessel radii and is nearly uniform at a distance $r = L$; neglecting the axial variations in pressure, permeability, or vessel diameter is the second limitation because there is known to be gradients of these characteristics; however, for the physiological parameters chosen, the perturbation of the velocity field is limited essentially as well to a couple of vessel radii, and is nearly uniform at a distance $r = L$ since even the smallest capillaries are $100 \, \mu m$ long, which is an order of magnitude greater than the radius of the vessel.

■ Example 6.11n: Simulation of Embryo Transport [430]

Within three days after fertilization, the embryo is transported along the uterine in the upper part of the uterus. Unlike sperm, the embryo does not have a self-propelling mechanism, and therefore, it is transported by intrauterine fluid flow patterns to it final site of implantation during the early process of human reproduction. In the outlined paper, the transport characteristics of this peristaltic motion, which serves as a vehicle for embryo, are investigated. The model of the uterine cavity is simulated by uniform two-dimensional channel, which is closed at the rigid end and open towards the cervix. Such a model is more realistic than existing models with open end due to more adequate imitation of uterine cavity geometry. The fluid motion is induced by two trains of sinusoidal waves propagating along the flexible in y-direction channel walls. The sinusoidal wall motility decreases towards the rigid ends of both upper ($+$) and lower ($-$) walls according to the expression

$$\eta = \pm H + a \cos\{2\pi[(x/\lambda) - (t/T)] \pm (\varphi/2)\} \tanh[\phi(L - x)] \qquad (6.23)$$

Here, $2H$ is the unperturbed channel width, T and $0 \le \varphi \le \pi$ are the period and the phase difference between upper and lower walls. The hyperbolic tangent is introduced to enforce the conditions of anchored boundary conditions near the rigid end that is assumed to be a simple rigid circular curve. The angle ϕ controls the slope of the hyperbolic function and ensures smooth transition between a sinusoidal wave and rigid circular end.

The problem is governed by mathematical model consisting two-dimensional Navier-Stokes equations, no-slip and no-penetration boundary conditions on the walls. It is assumed that at the open inlet fluid may flow into and out of the channel. Computing was performed using the finite volume package FLUENT for moving boundaries according to equations (6.23). The mesh was composed of 20,000 triangle cells. Analysis of velocity components depending of time and positions lead to the following results:

- The present study reproduces the previous data obtained for uterus modeled as an open channel [130]; as before, it is found that the velocity profiles are depended on wall motility, level of asymmetry and frequency of peristalsis.
- At the same time, the present study shows that flow characteristics in a closed model are affected by the closed end indicating, in particular, that the magnitude of the axial velocity is increasing towards the open end of the channel.
- The trajectories of particles revealed the periodic motions in small moving loops; the particles initially separated by wavelength are transported in almost identical pattern; this outcome is profound for the simulation with a small wavelength; the particles initially separated at full wavelength ($x = n\lambda$) experience small velocities so that their displacement is negligible.
- Particles within the channel recirculate around their initial location along and across the channel; trajectories pattern illustrate the overall transport of embryo after it entering the uterine cavity, where in the idealized conditions, the embryo should recirculte around its initial location until it will be ready for implantation; this result shows that the real embryo may never reach the end being implanted in the anterior or posterior walls at some distance from end. These finding support the observations that implantation of the embryo occurs in the area where it was placed naturally or artificially.

- Peristaltic motions due to uterine contractions towards the fundus practically "lock the embryos" within a small area around the location where they were deposited during the transport process; this result is in conformity with observation that standing after transport procedure does not affect the final position of the embryo.

■ **Example 6.12a: Modeling the Bioheat Transfer in Human tissues [162]**

The bioheat processes are important since heat transfer in human body determines the performance of thermoregulation system and the efficiency of various procedures in thermotherapy. Heat transfer in human body is complicated involving different processes, like conduction in human tissues, perfusion of the arterial-venous blood through the tissue pores, metabolic heat generation and interaction of blood flow with external magnetohydrodynamic (MHD) and electromagnetic fields.

To study these effects, the relevant model is used that consists of electrically conducting fluid filling the porous space in an asymmetrical channel in the presence of transversely directed magnetic field B. The induced magnetic field is neglected. Two asymmetric sinusoidal waves are imposed on the upper (η_1) and lower (η_2) walls

$$\eta_1 = d_1 + a_1 \cos[(2\pi/\lambda)(x - ct)], \qquad \eta_2 = -d_2 - a_2 \cos[(2\pi/\lambda)(x - ct) + \phi], \qquad (6.24)$$

where d_1 and d_2 are distances from x axis to upper and lower walls, and $0 \le \phi \le \pi$ is the phase difference satisfying the inequality $a_1^2 + a_2^2 + 2a_1 a_2 \le (d_1 + d_2)^2$.

The governing model consists of Navier-Stokes and energy equations with additional terms $\mu u/k$, $\mu v/k$ and $\sigma u B^2$ accounting for permeability and magnetic effects in the same way as in example (6.3) using the same notations k, σ and B for the permeability coefficient, electrical conductivity and magnetic flux density, respectively. Under assumptions of small Reynolds number and long wavelength, the governing system in wave frame reduces to three simple equations

$$\frac{\partial^4 \psi}{\partial y^4} - N\frac{\partial^2 \psi}{\partial y^2} = 0, \quad \frac{dp}{dx} = \frac{\partial^3 \psi}{\partial y^3} - N\left(\frac{\partial \psi}{\partial y} - 1\right), \quad \frac{\partial^2 \theta}{\partial y^2} + \Pr \mathrm{Ec}\left(\frac{\partial^2 \psi}{\partial y^2}\right)^2 = 0, \quad (6.25)$$

where $N = (1/\mathrm{Da} + \mathrm{Ha}^2)$, and two boundary conditions: $\psi = q/2, \partial\psi/\partial y = -1, \theta = 0$ and $\psi = -q/2, \partial\psi/\partial y = -1, \theta = 1$ on the upper and lower waves at $\eta_1 = 1 + a_2 \cos 2\pi x$ and $\eta_2 = -d_2 - a_2 \cos(2\pi x + \phi)$, respectively. The variables are scaled by: cd_1 for stream function and flow flux q, λ for x, d_1 for y, η, d_2 and a, c for u, and $\lambda\mu c/d_1^2$ for pressure, $\theta = (T - T_1)/(T_2 - T_1)$, $\mathrm{Ha} = Bd_1\sqrt{\sigma/\mu}$, $\mathrm{Ec} = c^2/c_v(T_2 - T_1)$, and $\mathrm{Da} = k/d_1^2$ are Hartmann, Eckert, and Darcy numbers. Solutions of equations (6.25) has the form

$$\psi = C_1 y + C_2 + C_3 \cosh\sqrt{N}y + C_4 \sinh\sqrt{N}y, \quad dp/dx = -N(C_1 + 1) \qquad (6.26)$$

Constants are found satisfying indicated above boundary conditions. Expressions for constants and temperature are awkward [162]. The following conclusions are stated:

- The variation of the pressure rise per wavelength against flow $\Delta p_\lambda(Q)$ shows three regions: peristaltic pumping ($\Delta p_\lambda > 0$ and $Q > 0$), free pumping ($\Delta p_\lambda = 0$) and augmented pumping ($\Delta p_\lambda < 0$ and $Q > 0$). In the first region, the pumping rate increases as Hartmann number increases, reaches maximum critical value at $Q = 0.6$ and decreases as Ha grows farther; similar behavior is observed in the both other regions.

- The effect of permeability in all regions is opposite to that of Ha with the same critical value for Q; an increase in distance between waves leads to decreasing in Δp_λ in peristaltic and free pumping regions; in augmented pumping region a reverse is observed.
- The velocity profile is parabolic at the inlet; for large permeability, increasing in Ha results in increased axial velocities in the neighborhood of the walls and their decreasing close to the channel centre.
- Significant variations in temperature profiles occur near the lower wall and in the channel centre, where a reduction in θ is noticed with increasing in Ha; an increase in permeability affects the temperature profile in opposite way to that of Ha: the porous medium resists the heat flow and this resistance increases as permeability decreases; an increase in the Brinkman number Br = EcPr leads to increase in heat transfer rate.
- In the wave frame, the stream lines split to trap a bolus; in a symmetrical channel, the bolus is symmetrical about centerline, and the bolus size is reduced as Ha increases; in asymmetric case, the boluses tend to shift to the left side of the channel; size of bolus increases if permeability grows in both symmetric and asymmetric cases.
- The heat transfer coefficient for MHD flow is greater when compared with that for the hydrodynamic flow; heat transfer coefficient increases as Hartmann number increases and when permeability (Darcy number) or Brinkman number decreases.

OTHER WORKS: The physiological analysis showing the role of peristaltic fluid flow in modeling the intrauterine fluid motion of the uterus is given in [430]. The embryo transport simulated in this article is also considered in other works, in particular, in [131] by studying the cyclic uterine peristalsis using a model with two-dimensional tapered channel, in [132] by employing the laboratory model for in vitro simulation, and in [299] by applying the non-Newtonian Maxwell fluid flow in channel with varying cross section. A review [68] presents the biomechanical and molecular aspects of intrauterine embryo distribution (spacing and orientation) studied by genetically engineered mouse models (Com. 6.8). Articles [359] and [326] as well as study [414], which we reviewed in Example 6.9, contributed in convection-enhanced delivery (Com. 6.7). In the first article, a model of convection-enhanced delivery in brain is developed, and in second one, an in-dwelling cannula for this procedure targeted for neuro-oncological trials is proposed. The rabbit model with human hormone is used in [293] to investigate the ovum transport in the oviduct. The spermatozoa swimming stability near a surface is studied in paper [174] by direct numerical computation (S. 5.2.2) via the boundary element approach (S. 9.6). Bioheat transfer considered in Example 6.12 is investigated also in recent researches [4] and [187] more specifically simulating numerically cooling heart by pumping cold liquid through blood vessels and studying thermal effects in eye during treatment by lasers, respectively.

6.2   Application of Fluid Flow Models in Engineering

6.2.1   Application of Peristaltic Flow Models

In Chapter 5, in analyzing methods of problem solution, we considered two examples of peristaltic flow application to microelectromechanical systems (MEMS) (Exam. 5.4 and 5.5) to show analysis of flow in short closed channels typical to MEMS. Here, we review some specific engineering peristaltic flow applications and listen, as at the end of each section, the other works in this area. We consider in detail four recent results obtained during the last years, including a complex robot design published in 2015. These examples show the role and efficiency of mathematical models in engineering applications of peristaltic motion changing in principle the study methods in this area.

■ Example 6.13a: Effects of Chemical Reaction, Heat and Mass Transfer in a Tube Flow [128]

In this research, the effects of mass diffusion of species, of chemical reaction, and heat transfer on peristaltic motion of a non-Newtonian Jeffrey fluid flow are investigated. It is assumed that the fluid flows through porous medium between vertical concentric tubes. The inner tube is uniform with radius $y = r_i$, and the outer tube is conical with radius $r_0 = R - x \tan \varphi$, where R and φ are the maximum radius and conicity angel. Along the outer tube travels a sinusoidal wave $y = r = r_0 + a \sin[(2\pi/\lambda)(x - ct)]$ determining the peristaltic flow. Such a model simulates different engineering and geophysical processes: geothermal reservoirs, drying of porous solids, thermal insulation, enhanced oil recovery, packed-bed reactors, cooling of nuclear reactors, and underground energy transport.

The problem is governed by mathematical model of a complex system consisting of two-dimensional continuity, Navier-Stokes, and temperature and concentration equations for Jeffrey fluid flow with chemical reaction. The complex system after employing the usual assumption of long-wavelength and low Reynolds number is presented in the form

$$\frac{\partial p}{\partial x} = \frac{1}{1 + \lambda_1} \frac{\partial^2 u}{\partial y^2} - \varepsilon^2 u + G\Theta + Gr_C\Phi, \quad \frac{\partial^2 \Theta}{\partial y^2} = -\varepsilon^2 Ec\, u^2, \quad \frac{\partial^2 \Phi}{\partial y^2} = Sc\,\gamma\,\Phi \quad (6.27)$$

The forth equation $\partial p/\partial y = 0$ that follows from the second Navier-Stokes equation states that the pressure is independent on transverse coordinate. The dimensionless variables in equations (6.27) are scaled by: $\lambda, R, c, \lambda\mu c/R^2$ for x, y, u and p, respectively, a porosity parameter is defined as $\varepsilon^2 = R^2/k_2(1 + \lambda_1)$, where λ_1 is a relaxation time, k, k_1 and k_2 are thermal conductivity, constant of chemical reaction, and permeability coefficient of porous media, $\Theta = (T - \overline{T})/(T_i - \overline{T})$ and $\Phi = (C - \overline{C})/(C_i - \overline{C})$ are dimensionless temperature and concentration determined via mean values marked by overbar. Grashof, Eckert, and Schmidt numbers, and chemical reaction parameter γ are defined as follows

$$Gr = \frac{g\beta(T_i - \overline{T})R^2}{vc}, \quad Gr_C = \frac{g\beta_C(C_i - \overline{C})R^2}{vc}, \quad Ec = \frac{\mu c^2}{k(T_i - \overline{T})}, \quad Sc = \frac{v}{D}, \quad \gamma = \frac{k_1 R^2}{v}$$
$$(6.28)$$

The problem is solved following Fang and Yih (Exam. 5.2) by perturbation series in small parameter ε^2 under boundary conditions given at the inner surface $(y = r_i)$ and at the wave $y = r = r_0 + a \sin[(2\pi/\lambda)(x - t)]$, respectively as

$$u = v = 0, \quad \Theta = \Theta_0, \quad \Phi = \Phi_0 \qquad u = 0, \quad v = \partial y/\partial t, \quad \Theta = \Phi = 1 \quad (6.29)$$

Two first series terms for velocity and pressure are obtained. The pressure rise and pressure forces for the inner and outer tubes per wavelength, the skin friction, the heat and mass transfer coefficients are computed as functions of time-averaged over a period flow rate $q = Q/cR$ using the following expressions

$$\Delta P_\lambda = \int_0^1 \frac{\partial p}{\partial x} dx, \quad F_\lambda = \int_0^1 y^2 \frac{\partial p}{\partial x} dx, \quad c_f = \frac{1}{1+\lambda_1} \left(\frac{\partial u}{\partial y} \right)_w, \quad \text{Nu} = \left(\frac{\partial \Theta}{\partial y} \right)_w, \quad \text{Sh} = \left(\frac{\partial \Phi}{\partial y} \right)_w$$

$$(6.30)$$

The following conclusions are formulated:

- The variation of the pressure rise per wavelength Δp_λ as a function of the time averaged over a period flow q for various values of parameters indicates that in the region $\Delta p_\lambda > 0$, an increase of Grashof number Gr increases the pumping rate Δp_λ, and in the region $\Delta p_\lambda < 0$, the pumping rate decreases with increasing Gr; an inversely linear relation between Δp_λ and average flow rate q is observed.
- The effects of the parameters φ, a, r_i, Ec, and ε on Δp_λ are similar to the effect of Grashof on Δp_λ; in the region $-1 \le q < 0$, the rate Δp_λ decreases as λ_1 grows, while in the region $0 \le q \le 1$, Δp_λ increases with increasing λ_1; the effects of Grashof Gr_C (defined via concentration) and Shmidt Sc numbers on Δp_λ are similar to effects of Gr and of λ_1, respectively, whereas the parameter γ affects pumping rate Δp_λ as well as Sc.
- The variations of pressure forces per wavelength of the outer and inner surfaces as functions of flow rate q indicates that the absolute value of the outer surface forces are greater than that of the inner surface; the effects of Gr, $\varphi, a, \lambda_1, \varepsilon$, Sc, γ, Ec, Gr_C and r_i on both forces are similar to the according effects of these parameters on pressure rise.
- The variation of the skin friction coefficient at the inner and outer tubes with the flow rate indicates that skin friction decreases or increases with increasing Gr_C depending on $c_f < 0$ or $c_f > 0$, respectively, whereas for fixed Gr_C, skin friction increases for inner and decreases for outer tubes as the flow rate grows; the effects of Sc, γ and λ_1 on skin friction are similar to just described effect of Gr_C, whereas the effects of other parameters Ec, Gr, ε, φ, a and r_i on c_f are similar to foregoing effects on the forces.
- The variation of the Nusselt number at both tubes as functions of the flow rate q shows that the effects of different parameters on heat transfer are similar to the effects on skin friction, but with different points of corresponding curves interaction with q axis; for both Grashof numbers Gr and Gr_C this results in an increasing of negative values of Nusselt number with their increasing; the effect of λ_1 is similar to the effect of Gr, and the effects of Sc and γ are found to be similar to that of Gr_C on c_f, whereas the effects of $\text{Gr}_C, \varepsilon, r_i, a, \varphi$ are similar to the effect of r_i on skin friction coefficient c_f.
- The variation of the Nusselt number at the tubes with time indicates that Nu increases for inner and decreases for outer tubes as Sc increases; the effect of γ on Nu is similar to that effect, while the effect of λ_1 on Nu is opposite to the effect of Sc on Nu; the effects of r_i, Gr, φ, and a on Nu are similar to the effects of λ_1.
- The variation of the Sherwood number at the inner and outer tubes as functions of time for fixed parameters indicates that Sh increases for inner and decreases for outer tubes as r_i increases; the effects of Sc, φ and a on Sh are similar to the effect of r_i; the effect of γ on Sh is similar to the effect of Sc on Sh. Sherwood number decreases at both tubes as the parameter Φ increases.

■ **Example 6.14* n: Optimization of Micropumping Systems [193]**

This study presents optimization design of micropump with improved discharge efficiency and reduced reverse flow of its fluid chamber. Devices with such high characteristics are important for diagnostics, particularly, for blood cell sorting assays (counting cells) where a full delivery of fluid containment of chamber is critical. The optimized design version was achieved by changing usual circular camera in two steps. The aim of the first step was to find a chamber profile, which creates the maximum forward discharge and the minimum retainability and reverse flow. The flow simulation inside a chamber was performed using solution of two-dimensional Navier- Stokes equations with uniform inlet velocity, constant pressure outlet, and no-slip boundary conditions. The FLUENT software was applied to perform this step. The reverse flow of species at different inlet velocities towards the exit of a circular chamber was monitored, and the design of the containment was changed iteratively until no reverse flow took place. Two pattern images are presented showing velocity distribution inside both initial circular camera and camera of optimized profile with inlet velocity 0.002 m/s providing desired forward discharge and back flow rate.

In the second step, the tree-dimensional distributions of fluid particles in the form of velocity vectors that occur due to membrane impact in both initial and optimized chambers are simulated. This was performed using the COMSOL multiphysics program for solving 3D Navier-Stokes equations under zero velocity and pressure values at the inlet and outlet as the boundary conditions and freely moving mesh constituting the computation domain. These results are presented in an article showing much higher deliverability per unit pumping volume up to 85% for the constructed chamber against 25–30% for the conventional cylindrical chamber.

Both considered chambers were extensively tested to investigate the differences between initial and final versions in the basic characteristics and in the operation performance to show high efficiency of suggested design. To perform these tests, various fluids like fluorescent bead solutions, DI (deionized) water, blood, and PBS (phosphate buffer solution) are employed. A special set of trials was performed for retention analysis of blood samples and velocimetry image was used to estimate the real time flow behavior for the fluid containment. Finally, the flow rate against the atmospheric pressure head was evaluated to check the operating parameters such as membrane deflection pressure and actuation frequency. The article consists of detailed descriptions about using apertures as well as the methodology of investigation.

■ **Example 6.15n: A Valve-Less Microfluidic Peristaltic Pumping Method [442]**

The article presents two constructions of valve-less micropumps: the primal linear shape micropump with continuous outflow but with intrinsic back-flow and strokes and improved a round shape one with circular micro-channel squeezed by bearings. The final improved version equips continuous, steady, and precise fluidic perfusion with optimized channel layout. Both pumps may be used for simultaneous control on multiple flows by squeezing parallel channels of different width.

Experiments have shown the following pumps characteristics:

• The cam-driven linear micropump is easy to fabricate and operate; however, the backflow pattern makes the pumping inefficient.

- The round-shape layout of fluidic channel greatly improves the continuity and stability of the flow, and makes possible to use miniaturizes motors in portable devices.
- Using a direct pressing of round-shape channel rather than previously reported magnetic driven rolling ball ensures robust driving; however, the liquid flow still shows small fluctuations resulting in backflow when the bearings approaches the channel outlet.
- The buffer chamber placed outside and upstream of the outlet compensates these adverse effects and effectively reduces the backflows.
- The velocity of pump is controlled by rotation speed of the motor providing delivery error not larger than ± 3 nl (nanoliter is 10^{-9} liter).
- The pump shows that at backpressure of 300 kPa, the flow rate drops by 14.7% is much better than similar characteristics of most commercial peristaltic devices.
- The pump is low-cost, easy-to-fabricate microdevice with precise flow control.

■ **Example 6.16n: The Biomimetic Peristaltic Swallowing Robot [94]**

The proposed biomimetic robot is designed to investigate the transport of boluses due to peristaltic motion such as in esophagus and in similar intestine organs. The robot mimics the biological process of swallowing smoothly and continuously to achieve the peristaltic contraction trajectories propagating like those in human body. Synonymous with architecture of the esophagus, the robot has no skeletal structures at the artificial conduit and exhibits distributed actuation in a similar arrangement to muscles around the biological conduit. Peristaltic waves are specified by inflammation of a series of twelve adjacent stacked vertically whorls of pneumatic chambers. Such robot manifests as a benchtop engineering rheometric (to measure rheological parameters, S. 1.9) instrument for investigation relationship among bolus formulation, interluminal pressure signature (ILPS) and peristaltic transport effects. The article presents detailed instructions for bolus preparing and experimental determining swallowing trajectories, intrabolus pressure signatures (IBPS), and other manometric (esophageal motility study) features.

The following results are obtained using the robot to discern between IBPS in response to manipulating transport parameters such as peristaltic wave velocity (20, 30, and 40 mm/s), wave-front length (40, 50, and 60 mm), and starch-based bolus (from Nutulis, particularly) formulation concentration (25, 50, 75, 100, and 150 g/L).

- The wave velocity and starch thickener concentration exhibits the most profound changes in the intrabolus pressure signatures, and the last parameter is the most sensitive perturbation to the swallowing signature across clinically significant ranges of variables.
- The highest bolus tail pressure gradient of 0.33 kPa/mm is achieved at 150 g/L bolus formulation being transported at 40 mm/s with a wave-front length of 60 mm.
- The relationships between the parameters and features of the manometric pressure signature are nonlinear due to the complex shear field and non-Newtonian nature of the model bolus materials
- The swallowing robot, augmented with monometric investigation capability, has demonstrated good sensitivity to bolus transport parameters and displayed the relationship between mechanical features of process and the resulting ILPS.
- The manometric signature for the robotic swallowing model exhibits the same salient features of those captured in the clinical setting; using such a robot overcomes the limitations of current swallow investigations suffering from intrasubject variability.

OTHER WORKS: A faithful review of MEMS biologically oriented for drug delivery is presented in [287]. Shorter reviews (up to the first years of the twenty-first century) of applied peristaltic flow investigations and of different versions of microfluidic devices may be found in introductions of reviewed above papers [128] and [193], respectively. Many types of peristaltic pumps and micropumps were proposed: early models operating by electric motor [292] or rotary solenoid [57] and more contemporary constructions using various driving approaches: thermo-pneumatic (pipette-like) [364], pneumatic [181], piezoelectric [37], electro-pneumatic [64], electrostatic [42], and electromagnetic [38]. Some recent publications considered the complex engineering problems. For example, the peristaltic magnetohydrodynamic flow with Hall, ion slip, and Ohmic heating effects is investigated in [18], and the influence of the variable viscosity and thermal conductivity on the peristaltic flow characteristics is studied in [259].

6.2.2   Applications of Direct Simulation of Turbulence

In Chapter 5, we considered three methods of direct numerical simulation of turbulence: DNS, LES, and DES including the most improved IDDES and ZDES versions of DES. We explained the importance of such simulation, the basic distinction of new direct methods from existing means, and discussed the difference between these three approaches of simulation. It was underlined that the direct numerical simulation provides the information, which is viewed as experimental data obtained computationally. The turbulence characteristics such as the instantaneous velocity components, stresses, fluctuation correlations, and so on, may be calculated instead of or in addition to experimental results. Moreover, each direct numerical solution of Navier-Stokes equations is a challenge similar in complications to physical experiment, when computing accuracy of results may slightly differ from each other owning to specific numerical schemes and prescribed boundary conditions as well as data of different experimental investigations exhibiting some discrepancies due to variations in the setup and experimental procedures.

Here, we present some examples of direct simulation applications including the latest showing significant improvement of our understanding of turbulence nature due to new methods and fast-growing applicability of these methods up to industrial prototypes.

6.2.2.1   Direct Numerical Simulation

■ **Example 6.17: Direct Numerical Simulation of a Turbulent Boundary Layer [218]**

The article presents the characteristics of turbulent boundary layer obtained by direct numerical simulation at $\mathrm{Re}_\theta = 570 - 2560$, where θ is the momentum thickness (S. 7.5.1.1). The model consist of three-dimensional zero pressure gradient flow along a flat plat described by Navier-Stokes equations in Einstein notations (S. 7.1.2.2) and variables scaled by the free-stream velocity and momentum thickness θ at the inlet

$$\frac{\partial u_i}{\partial t} + \frac{\partial u_i u_j}{\partial x_j} = -\frac{\partial p}{\partial x_i} + \frac{1}{\mathrm{Re}}\frac{\partial^2 u_i}{\partial x_j x_j}, \qquad \frac{\partial u_i}{\partial x_i} = 0 \qquad\qquad (6.31)$$

A basic goal of this study is to increase the value of Reynolds number of turbulent zero pressure gradient flow over a flat plate achieved in previous simulations as, for example, the range $80 \leq \mathrm{Re}_\theta \leq 940$ attained by Wu and Moin [423]. The main reason of the Reynolds number restriction in that study and others of this type is associated with starting laminar inflow that requires simulation of the transition process to get the developed turbulent boundary layer. The problem arising in such approach is a costly simulation of transient process becoming unacceptable with growing Reynolds number.

During the last two decades, several methods for overcoming this problem are suggested. The analysis presented in the reviewed article shows that a method proposed by Lund *et al.* [248] for generating a turbulent inflow boundary conditions is a realistic, reliable means. The idea consists of using an auxiliary turbulent inflow to displace the region of interest with simulating main turbulent boundary layer for a relatively short distance upstream from it. To avoid the artificial numerical periodicity induced in the Lund *et al.* approach, authors used a long streamwise domain of length $x/\theta_{in,i} = 1000$, where $\theta_{in,i}$ is the inlet momentum thickness for the development boundary layer adjustment. The other domain parameters and mesh resolution are given in the reviewed article. The ratio of the main and inflow thicknesses is $\theta_{in}/\theta_{in,i} \cong 2.5$. The height and width of domain are about three times larger than the maximum momentum boundary thickness at the maximal $\mathrm{Re}_\theta = 2560$. The total number of grid points is 315 million.

The governing equations are integrated in time using the fractional step method with implicit velocity decoupling procedure proposed by Kim *et al.* [204]. The results obtained at $\mathrm{Re}_\theta = 2500, 2000, 1410$ are compared with data of five other simulations and of five experiments for Reynolds numbers in the range at $\mathrm{Re}_\theta = 1410 - 2900$. The study results are summarized in the following basic conclusions:

- The calculating velocity profiles and two point correlations of velocity fluctuations are in good agreement with previous simulations and experimental data. In particular, well-known universal wall and defect velocity laws for inner and outer regions (S. 8.3.2) found previously from experimental data are confirmed computationally. Some deviations between compared data are associated with experimental uncertainties, numerical inaccuracy, and insufficiently long domains. The famous logarithmic formula $U^+ = (1/k)\ln y^+ + B$ for inner velocities follows from calculating profiles as well.
- The present simulation showed the consistent behavior with the recent DNS data of Schlatter *et al.* [335, 336], despite of the different numerical schemes and inflow conditions. These observations indicate that: (i) the sufficiently long streamwise domain and high grid resolution in simulations are essential to obtain the reliable turbulent flow, and (ii) the influences of the numerical scheme and inflow generation method on the properties of the turbulent boundary layer quantities are relatively weak.
- The peak values and locations of all the velocity and pressure fluctuations showed Reynolds-number dependent behaviors: as the Reynolds number increases, the peaks of the velocity and pressure fluctuations increase; the Reynolds number similarity is achieved near the wall when the fluctuations are scaled by the friction velocity u_τ; the similarity of the r. m. s. (root mean square value is the square root of arithmetic mean of the squares) of vorticity fluctuations is observed in the range $5 < y^+ < 70$ (S. 8.3.2), and the lack of similarity is discovered very near the wall and in the wake region.
- Inspection of the instantaneous field and two-points correlation revealed the existence of the very large-scale motions with the characteristic widths of $(0.1–0.2)\,\delta$ and the flow structures

for a length of approximately 6δ fully occupied a streamwise domain statistically. These motions are the coherent structures residing in the turbulent boundary layer and are consistent with the 'superstructure' (extended or developed from basic structure) observed before.

- The results from the two-point correlations of the streamwise velocity fluctuations displayed that the outer scaling variable (δ) is an appropriate length scale for normalizing the turbulent structure in the logarithmic layer (S. 8.3.2) with respect to the Reynolds number.

■ **Example 6.18: Heat Transfer in Turbulent Boundary Layer Flow in a Channel [199]**

The effects of Reynolds and Prandtl numbers on the heat transfer characteristics of turbulent boundary layer are investigated. The model consist of fully developed flow in rectangular channel with dimensions $6.4\delta \times 3.2\delta \times 2\delta$, where δ is the half channel width, heated by the uniform heat flux $q_w = const$. The governing system involves the continuity and Navier-Stokes equations in the form (6.31), as in the last example, the energy equation for statistically averaged temperature (i.e., with regard to random errors)

$$\frac{\partial \theta}{\partial t} + u_j \frac{\partial \theta}{\partial x_j} = \frac{1}{\mathrm{Re}_\tau \mathrm{Pr}} \frac{\partial \theta}{\partial x_j \partial x_j} + \frac{u_1}{\langle \overline{u} \rangle}, \quad T(x,y,z) = \left[d \left\langle \overline{T}_m \right\rangle / dx \right] x - \theta(x,y,z), \quad (6.32)$$

and boundary conditions $u_i = \theta = 0$ at $y = 0$ and $y = 2$. In the case considered, the statistically averaged temperature increases linearly with respect to x. Therefore, the instantaneous temperature is defined by second equation (6.32), where the mixed mean temperature is estimated as $\left\langle \overline{T}_m \right\rangle = \left\langle \overline{uT} \right\rangle / \langle \overline{u} \rangle$, and the derivative of it for the case of considered configuration is inversely proportional to velocity $d \left\langle \overline{T}_m \right\rangle / dx = 1 / \langle \overline{u} \rangle$. In this study, equations (6.31) and (6.32) are written in dimensionless coordinates, velocities, and temperatures scaled by δ, $u_\tau = \sqrt{\tau_w / \rho}$, and $T_\tau = q_w / \rho c_p u_\tau$ (friction velocity and temperature, S. 8.3.2). The ovescore $(-)$ and brackets $\langle \rangle$ defined the variables averaged statistically and over the channel section, respectively, the Reynolds number is specified as $\mathrm{Re}_\tau = u_\tau \delta / v$, and the term $u_1 / \langle \overline{u} \rangle$ in equation (6.32) represents the last term of boundary layer energy equation in dimensionless form.

The simulation was performed by the finite difference approach for 180 and 395 Reynolds numbers and 0.025, 0.2, and 0.71 Prandtl numbers using grids $128 \times 66 \times 128$ and $256 \times 128 \times 256$ for first and second Reynolds numbers. The presented results including analysis of visualization pattern are summarized as follows:

- For both Reynolds and three Prandtl considered numbers, the region with logarithmic profile $\overline{\theta}^+ = (1/k_\theta) \ln y^+ + B_\theta$ exists with value of k_θ close to Karman constant 0.4; the peak in the r.m.s temperature distributions weakly depends on Re for $\mathrm{Pr} = 0.71$, but for smaller Prandtl numbers, the peak increases with Re growing.
- The peak of total and the wall-normal turbulent heat fluxes arise at around $y^+ = 30 - 60$ for $\mathrm{Pr} \geq 0.2$ and at $y^+ > 50$ for $\mathrm{Pr} = 0.025$ and increase with Re and Pr increasing; the peak of the streamwise turbulent heat flux increases as the Prandtl number grows, its value dependence on Reynolds number is negligible for $\mathrm{Pr} = 0.71$, but is appreciable for $\mathrm{Pr} = 0.2$ and $\mathrm{Pr} = 0.025$; in the central channel region for $\mathrm{Pr} \geq 0.2$, the streamwise heat flux does not depend on Prandtl number, and its value is much larger than the wall-normal heat flux, which is a result of much larger streamwise velocity fluctuations u'^+ than corresponding wall-normal velocity fluctuations v'^+.

- The turbulent Prandtl number is calculated as $\text{Pr}_{tb} = \nu_{tb}/\alpha_{tb}$, where turbulent viscosity ν_{tb} and diffusivity α_{tb} are computed from relations for statistically averaged turbulent stress $\overline{u'^+ v'^+} = \nu_{tb}(d\overline{u}_1^+/dy^+)$ and heat flux $\overline{v'^+\theta'^+} = \alpha_{tb}(d\overline{\theta}^+/dy^+)$; It is known that the contradictory experimental data leads to a long discussion about dependence of the turbulent Prandtl number on transverse coordinate, in particular, at the wall vicinity (Com. 2.5). The calculation shows that close to the surface for $\text{Pr} \geq 0.2$, the turbulent Prandl number is close to unity and is independent on y^+ and Prandtl and Reynolds numbers; although this conclusion is in line with results obtained before in [16], for small Prandtl number $\text{Pr} = 0.025$ calculation indicates that Pr_{tb} is higher for small Reynolds and becomes the same as for larger Prandtl numbers when Re increases.
- Both the production and dissipation terms (S. 8.4.1) of the budget of the transport equation for temperature, scaled by $u_\tau^3 T_\tau/\nu$, increase as the Reynolds number increases; the dominating terms of the budget for wall-normal turbulent heat flux increase with the Reynolds number growing as well and this effect is more pronounced for $\text{Pr} = 0.025$; the production and the temperature pressure gradient (TPG) correlation terms are dominant at $\text{Pr} = 0.71$ and the corresponding dissipation term is considerable small, whereas at $\text{Pr} = 0.025$, the production and dissipation terms are prominent leading to the negligible TRG term; at the same time at $\text{Pr} = 0.2$, both the TRG and dissipation terms are comparable, and as the Reynolds number increases, the dissipation term stays unchanged and the TRG term increases contributing more dominantly to budget.
- Instantaneous flow and temperature fields are visualized to investigate the streaky and vertical structures and to compare the patterns for both Reynolds numbers; in the case of low Reynolds number, only limited types of vertical structures are observed, but with the increase of the Reynolds number, various shapes of the vortices appear; in the case of $\text{Pr} = 0.71$, the velocity and thermal streaky structures show a strong similarity, while at $\text{Pr} = 0.025$, the thermal streaks are not so elongated in streamwise direction and their spanwise spacing are larger than for $\text{Pr} = 0.71$.

■ **Example 6.19*: Exothermic Gas-Phase Reaction in Packed Bed with Particles [277]**

The paper presents a simulation of an exothermic gas-phase reaction between ethylene and oxygen in a two-dimensional packed bed reactor with 600 cylindrical particles with diameter $d_p = 2.9\,\text{mm}$. The model is constructed by arranging 10 particles randomly perturbed on a square lattice spanning the width of reactor of 38.3 mm and by putting 60 of these lattices next to each other defining the axial reactor size of 240.6 mm. 70 grid nodes on the surface of each particle and 15 cells around it together with 3000 grid points along the reactor wall and 500 points along the inlet and outlet are provided resulting in 1.8×10^6 grid cells in the whole domain. The flow is considered as laminar with Reynolds number 3.5 based on particle diameter and interstitial velocity $v_{\text{inlet}}/\varepsilon$, where ε is porosity. Mathematical model of four equations in the vector form (S. 7.1.2.1) governed the problem

$$\frac{\partial \rho}{\partial t} + \nabla \cdot (\rho \mathbf{v}) = 0, \quad \frac{\partial (\rho \mathbf{v})}{\partial t} + \nabla \cdot (\rho \mathbf{v}\mathbf{v}) = -\nabla p + \nabla \cdot \boldsymbol{\tau} \cdot \rho \mathbf{g}$$

$$\frac{\partial (\rho h)}{\partial t} + \nabla \cdot (\rho \mathbf{v} h) = -\nabla \cdot \left(\lambda \nabla T - \sum_{i-0}^{N} h_i \mathbf{J}_i \right) + S_h, \quad \frac{\partial (\rho Y_i)}{\partial t} + \nabla \cdot (\rho \mathbf{v} Y_i)$$
$$= -\nabla \cdot \mathbf{J}_i + R_i + S_i \tag{6.33}$$

In these mass, momentum, energy and species conservation equations, the variables are: τ is stress tensor (Com. 1.9), $\mathbf{g}$ is gravity vector, h_i, Y_i, S_i, $\mathbf{J}_i$, and R_i are enthalpy, mass fraction, rate of source creation, diffusion flux, and net rate of production of chemical species i. The diffusion flux of species i that arises due to concentration and temperature gradients is calculated using Ficks' law (S. 7.1.1) as $\mathbf{J}_i = -\rho D_{i,mix} \nabla Y_i - D_{T,i}(\nabla T/T)$, where $D_{T,i}$ and $D_{i,mix}$ are the thermal and mass diffusion coefficients estimated by kinetic theory as it suggested in [256]. The source of chemical species i due to ethylene-oxide reaction $C_2H_4 + (1/2)O_2 = C_2H_4O$ is computed as the Arrhenius type reaction (Com. 4.15) using relation $R_i = M_{w,i}\widehat{R}_i$, where $M_{w,i}$ is a molecular weight of species i, and a molar rate of ethylene reaction is given as $\widehat{R}_i = -2.66 \cdot 10^{13} \exp(-15107/T)c_{Et}$, where c_{Et} is the concentration of ethylene.

The system of equations (6.33) was solved under the following boundary conditions: (i) the constant velocity profile at inlet and the zero gauge pressure at the outlet, (ii) no slip and zero heat transfer on the reactor walls and on particles surfaces, (iii) the 0.01 m/s velocity and 450 K temperature of the mixture of ethylene and oxygen in mass fraction 0.7×0.3 at the inlet. Simulation was performed using Computational Fluid Dynamic (CFD) code FLUENT and the SIMPLE software for pressure/velocity uncoupled procedure (S. 9.7). The elliptical equations (6.33) are transformed into hyperbolic type by adding artificial unsteady terms and solved straightforward with time step 0.01 s until the steady state is achieved (Com. 6.1).

The comparison of simulation results with data obtained by one-dimensional with plug velocity (S. 7.7) solution shows agreement. The following conclusions are stated:

- The local axial velocities between the particles are approximately 2-8 times higher than the superficial or inlet velocity and are the highest in the regions with high porosity; the average axial velocities are low near the walls becoming higher as the distance from the wall increases; the mean velocity is small $v_{inlet}/\varepsilon = 0.0175\,m/s$. The reason of that is the depletion of the reactant at about 40 mm from the inlet resulting in "dead zones" of about half of the space available for gas.
- Both simulation and one-dimensional solution results show that the ethylene and oxygen concentrations decrease along reactor due to reaction and reach zero at around 68 mm (65 mm according to one-dimensional data) from inlet where the reaction is finished resulting in temperature 580 K and constant concentration of ethylene oxide.
- The temperature distributions along the reactor at $y = 10, 20$ and 30 mm from the right hand wall side show the existence of high radial and axial temperature gradients at the distance of 55-60 mm from the inlet; the comparison of those three distributions reveals the large difference between temperatures of these three curves at the same axial location; the reason of that is the mentioned above slow fluid flowing at the wall, which leads to a longer residence time for reactants moving near the wall than the corresponding time for mixture flowing far from the wall; as a result, the more intensive reaction and consequent heat production occur closer to the inlet in the wall region than in central part of the reactor; the same observation follows from the temperature contours pattern.
- The existence of large gradients indicates that the convection in this model is small leading to slight dispersion and mixing processes; this conclusion is in line with small heat and mass Peclet numbers, which values are in the ranges: $Pe = vd_p/\alpha = 1.96 - 16$, and $Pe_m = vd_p/D = 1.73 - 22.8$; this shortage of the model arises due to arrangement of particles distributed less randomly than in real packed bed reactor.

OTHER WORKS: In the most recently published in 2016 three articles, DNS is used: to study the dynamics of new hydroelastic solitary waves in deep water [141], to analyze the transport of solid particles immersed in a viscous gas [142], and to evaluate the effect of the size of spanwise domain on the transitional flow past airfoil NACA 0012 at 5 degree attack [441]. In several other recent DNS articles published during the last three years are investigated: laminar-turbulent transition in hypersonic boundary layer on a sharp cone at Mach 6 [357], drag and forces acting on a bubble near a plane wall [379], flow in pipes with an arbitrary roughness topography [406], ignition of pulverized coal particle-laden mixture [51], droplet-laden heated and humid flow with phase transition [52], and ignition of lean biodisel fuel/air mixture [249]. For comparison, we mention some of earlier publications considered the following topics: heat and mass transfer in particulate flow [84], transport of scalars in turbulent channel flow at high Schmidt numbers [340], stable and unstable flows affected by buoyancy [161], transitional flow obstructed by rectangular prisms [198], and flow in a rod-roughened channel [19]. Review of very early results may be found in [195].

6.2.2.2   Large Eddy Simulation

■ **Example 6.20:   Vortex and Pressure Fluctuation in Aerostatic Bearings [443]**

Aerostatic bearings are widely used in precision stationary and moving equipment due to their merit near-zero friction and low heat generation. However, the inherent small vibration damages stability and precision of bearing. Recently many studies were fulfilled to understand the mechanism and suppress those vibrations considering the flow inside the bearing as a steady and applying the RANS equations or CFD simulation. In this paper to further insight, the transient compressible turbulent flow is investigated using the LES simulation. The model is constructed as 1/12 sector of cylindrical bearing of $d = 20$ mm that is composed of two components modeling the carrier (upper) and the base (lower) of a real bearing prototype. The air flow domain is divided in three parts: orifice ($d_0 = 0.15$ mm), cylindrical recess on the carrier bottom ($d_1 = 3$ mm and depth of 0.1 mm), and the gap between the carrier and base for air film of thickness $h = 10$ μm.

To reduce the computational cost, different approaches are used for these parts: the $k - \varepsilon$ model (RANS) for orifice and the LES for the recess are adopted, whereas the air film flow in a gap is assumed to be laminar, and the time-dependent compressible Favre filtered continuity and Navier–Stokes equations are employed to model this film flow

$$\frac{\partial \widetilde{\rho}}{\partial t} + \frac{\partial(\widetilde{\rho}\,\overline{u}_i)}{\partial x_i} = 0, \qquad \frac{\partial(\widetilde{\rho}\,\overline{u}_i)}{\partial t} + \frac{\partial(\widetilde{\rho}\,\overline{u}_i\overline{u}_j)}{\partial x_i} = -\frac{\partial \widetilde{p}}{\partial x_i} + \frac{\partial \overline{\sigma}_{ij}}{\partial x_j} - \frac{\partial \overline{\tau}_{ij}}{\partial x_j}$$

$$\overline{\sigma}_{ij} = \frac{\partial \overline{u}_i}{\partial x_j} + \frac{\partial \overline{u}_j}{\partial x_i} - \frac{2}{3}\delta_{ij}\frac{\partial \overline{\sigma}_{ij}}{\partial x_j}, \quad \overline{\tau}_{ij} = \widetilde{\rho}(\widetilde{u_i u_j} - \overline{u}_i\overline{u}_j) \qquad (6.34)$$

Equations (6.34) are written in Einstein notations, in which $\overline{\tau}_{ij}$ is the subgrid scale (SGS) stress, σ_{ij} is the viscous stress tensor, $\widetilde{\rho}$ and $\widetilde{p}$ are spatial filtered density and pressure (S. 5.7), overscore $(-)$ denotes the Favre density-weighted (S. 9.6) filtering, and δ_{ij} is the Kronecker delta (S. 7.1.2.2).

Comment 6.12 In the case of compressible fluid, the Reynolds averaging of Navier-Stokes equations leads to extra unknown terms that require additional hypotheses for closure (S. 8.2.3), further involving the problem. Density-weighted averaging suggested by Favre [133], which in fact changes the dependent variable from velocity u_i to ρu_i, leads to equations in the form close to the Reynolds incompressible equations [422].

The following boundary conditions are used: flow with initial turbulent intensity of 1.5 or 10% at the orifice inlet, atmospheric pressure at the outlet, symmetric conditions on both gap surfaces in the circumferential direction, no-sleep and no heat transfer on the walls, which are assumed to be perfectly smooth. The air is considered as ideal gas with viscosity $1.7894 \cdot 10^{-5}\,\mathrm{kg/m\,s}$, molecular weight $28.966 \cdot 10^{-3}\,\mathrm{kg/mol}$, specific heat $1006 \cdot 43\,\mathrm{J/kg\,K}$, and thermal conductivity $0.0242\,\mathrm{W/m\,K}$.

The CFD software ANSYS FLUENT with finite volume method (S. 9.6) is used for performing LES. The total numbers of mesh volumes are 268075 and 519644 for coarse and fine refinement studies, respectively. For the pressure-velocity decomposition the PISO algorithm is applied (S. 9.7). The second order upwind interpolation is adopted for density, turbulent kinetic energy and dissipation, and the central differencing is used for interpolation in LES. The implicit second order scheme is employed with time step $10^{-8}s$, satisfying the condition $u\Delta t/\Delta x < 1$, where Δx is a control volume size.

The following results and conclusions are obtained:

- The validation of model is justified by comparison of the numerical pressure distribution with experimental data [435] showing agreement of both results almost everywhere, except a small region near the orifice outlet.
- The instantaneous flow field in recess reveals series of vortices of varying sizes and shapes resulting in vortex shedding phenomenon when the vortices downstream break into small eddies and finally dissipated due to viscosity;
- The pressure distributions at three locations in the recess show the pressure fluctuations, which weaken in radial direction.
- The comparison of the results for three intensities of initial air turbulence indicates that influence of initial turbulence on the level of fluctuations is very small.
- The results obtained at 2, 3 and 4 atm. of air supply show that the air flow in recess is steady and remains laminar at 2 atm., becoming unsteady with vortex shedding at two other values of air pressure.
- The repeated pressure depression in space and fluctuations in time are observed in the bearing clearance when vortex shedding occurs; this pattern is not resolvable for RANS because of its statistical averaging.
- As the pressure of air supply and corresponding mass flow rate $\dot{m}$ increase, the Reynolds number grows from $\mathrm{Re} = \dot{m}/\pi r\mu = 631$ at $p = 2$ atm. to $\mathrm{Re} = 1516$ and 2515 at $p = 3$ and 4 atm., resulting at some critical Reynolds number between 1000 and 1500 in the transition of laminar flow in turbulent, and an increasing of vortex shedding.
- Comparison of flow fields in two models with and without recess reveals that in the second case the flow is laminar, steady and practically without vortices, which is a result of small Reynolds numbers.
- The vibration of both recessed and non-recessed models associated with pressure fluctuations are measured by accelerometer showing that amplitude of vibration in recessed

bearing increases as the supply pressure grows, and in non-recessed bearing, the amplitude of vibration is much weaker; this confirms the conclusions obtained from computational results.

■ **Example 6.21: Dynamic Flame Response in Gas Turbine Combustion Chamber [166]**

The lean premixed combustion systems providing low pollutant emissions are susceptible to thermo-acoustic instabilities. The article review shows that many investigations performed to understand a mechanism and to prevent the appearance of these instabilities are basically studied simple laboratory premixed flame. In contrast, the describing here results are obtained by LES of a real gas turbine burner at an actual Reynolds number considering the effect of equivalence (air/fuel) ratio fluctuations on the dynamic flame response.

The burner is of hybrid type where air is injected through diagonal and axial coaxial swirlers containing 24 and 8 vanes, respectively. Methane is injected through small holes in the vanes and mixes with air before the combustion chamber. The pilot methane injection is used for the flame stabilization and the cooling air inlets are applied to shield a cylindrical burner outlet. The burner is mounted on a 15-degree section of an annular combustion chamber constituting the computation domain. The mesh is buildup of 1.921.370 nodes and 10.472.070 tetrahedral elements, the time step is $9 \cdot 10^{-8}$s.

Two simulations are performed on the same geometry. In the first one, called TECH (technical), the burner operates as a premixed mode where fuel is injected through small holes in the vanes of the diagonal swirler and mixes downstream with air prior combustion. In the second simulation (FULL), the fully premixed flow enters through the diagonal passage. The LES were performed using the fully explicit code for the compressible reactive multi-species Navier-Stokes equations adopted from [339]. Other details, including the boundary conditions, are given in reviewed paper.

The obtained results yield the following basic conclusions:

- The equivalence ratio distribution and temperature isolines show that both flames are very similar, although the equivalence ratio of mixture equals unity in both cases, and the overall heat release is practically the same, the TECH flame produces slightly less heat release than FULL one.
- Both cases exhibit similar mean pulsated and non-pulsated flame shapes but with some differences in the combustion regime. Phase average solutions show that for technically premixed case, the flame transfer function (FTF) delay is 1.5 times larger than that for the FULL case, showing that fluctuations in the diagonal swirler modifies the FTF.

Comment 6.13 Flame transfer function is one of the key parameters in the flame dynamic analysis defined as a ratio of heat release to the velocity of thermo-acoustic fluctuations.

- Because fuel and air jets oscillate with different phases at the injection point, the velocity and trajectory of the jets also oscillate resulting in the mixing oscillations
- The local FTF fields indicate that the mixture oscillations propagating in the diagonal swirler lead to locally different responses along the flame showing that mixing process in the diagonal passage is not sufficient to damp perturbation induced by unsteady fuel flow. Two mechanisms are responsible for that: the pulsating injected fuel flow rate and the fluctuating

trajectory of the fuel jets. Both effects occur due to pressure fluctuations at the fuel injection holes caused by forcing.

- These mixing fluctuations are not damped at the combustion chamber inlet; they are phased with velocity oscillations, combined with them and lead to different FTF modifying the flame response to forcing; local fields of delay and the interaction indices reveal that the flame is not compact and is affected by fluctuations.
- The obtained data indicate that normal mixing is achieved only for steady flames; as soon as the flame is pulsated, the fuel injection system produces a response different from that in a fully premixed system.

■ Example 6.22: Pebble Bed in the High Temperature Nuclear Reactor [344]

An inherent safety advantage of high temperature nuclear reactor (HTR) is related to the very high temperature that the fuel can sustain preventing the fuel from melting even in the loss of cooling. Generally, the core is designed using graphite pebble bed, which usually provides efficient operation. However, heat transfer around curved surfaces, turbulent flow through the gaps between pebbles, turbulent flow pressure gradients may result in local hot spots affecting the pebble integrity.

Because the real pebbles bed has complex geometry, the existing investigations were performed on simplified models considering basically single pebble of different configuration and using RANS. In the reviewed work, the study of flow and heat transfer in the single cubic pebble of spherical type is performed. Such configuration is called Face cubic centered (FCC) arrangement consisting of a cube with half-spherical pebbles on each face and 1/8 of such sphere on each corner. For model validation, the results of quasi-direct numerical simulation (q-DNS) are used. In this quasi-DNS simulation, the model is considered as a porous medium with corresponding porosity that for model in question is estimated as 0.42. Helium is applied as working fluid with the following parameters: $\rho = 5.36 \, \text{kg/m}^3$, $\mu = 3.69 \cdot 10^{-5} \, \text{kg/ms}$, $\lambda = 0.3047 \, \text{W/mK}$, $c_p = 5441.6 \, J/kgK$, $\dot{m} = 0.01606 \, \text{kg/s}$, $T_{\text{inlet}} = 737 \, \text{K}$. A fully polyhedral mesh is employed, which consists of about 6 million dimensionless grid cells of size smaller than unity.

As usually in LES, the large eddy motion is simulated, whereas the sub-grid scale motion is modeled using the wall-adapting local eddy-viscosity model [283] as follows:

$$\tau_{ij} - \frac{1}{3}\tau_{kk}\delta_{ij} = -2\mu_t S_{ij}, \quad \mu_t = \frac{\rho L_s^2 (S_{ij}S_{ij})^{3/2}}{(S_{ij}S_{ij})^{5/2} + (S_{ij}S_{ij})^{5/4}}, \quad S_{ij} = \frac{1}{2}(g_{ij}^{-2} - g_{ji}^{-2} - \delta_{ij}g_{kk}^{-2})$$

(6.35)

where $g_{ij} = \partial U_i / \partial U_j$, δ_{ij} is Kronecker delta (S. 7.1.2.2), $L_s = 0.544 \, V^{1/3}$, and V is the cell volume. The simulation was performed using commercial software STAR-CCM+ with time step $5 \cdot 10^{-5} \, s$ and 8 interactions per step, which requires for complete simulation 1.47 million interactions and the computational time around 1272 hours.

The following results and conclusion are summarized:

- Both predictions by LES and q-DNS are in agreement, except some details for heat fluxes distribution indicated below.
- The flow field is asymmetric despite the symmetric pebble configuration, which is a result of formation vortices of different scales on both sides of the pebble; occurrence of such

asymmetry may be associated with processes similar to the Coanda effect; the iso-contours indicate the stagnant regions at the front and rear of the pebbles and the shear layers near gaps between them; the high velocity gradients near the sphere surfaces and broad regions of low velocities away from spheres are observed.

Comment 6.14 The Coanda effect is a property of fluid jet to be attached to nearby surface.

- The mean temperature distribution shows the high and low temperature zones along the pebble surfaces that correspond to the maximum and minimum regions of wall shear stresses; these hot and cold spots display a symmetric pattern indicating an opposite behavior of flow near pebble top and bottom; quantitative results reveal that r.m.s (root mean square) temperatures obtained by LES are over-predicted near the wall region and under-predicted in the center area with difference of 2% and 6%, respectively.
- The uneven behavior with various peaks is observed from the heat flux profiles indicating a strongly thermal activity fluctuating of three heat flux components; for the principle component, both LES and q-DNS results well agree with difference of 4%, whereas for two other, less-intensive components, the discrepancy is higher consisting of 33-50 %; due to the small components intensity, the overall heat fluxes are almost equal.
- The LES was six time faster than q-DNS with respect to the computation power.

OTHER WORKS: Some examples of the latest studies using LES during two last years present: structure of the transient cavitation (formation of vapor cavities in a liquid) of vortical flow around a NACA 66 hydrofoil (similar to fairwater (Com. 6.15)) used for reducing the drag coefficient [182], Reynolds number effect in the wake flow behind a circular cylinder [205], analysis of extinction in non-premixed flame [97], application of the stretched-vortex model to the atmospheric boundary layer [255], and flow in-cylinder in a DISI (direct injection spark ignition) gasoline engine [315]. Over the last decade, LES was employed to investigate: heat transfer at supercritical pressures [282], plasma-based boundary layer separation control [324], flow and heat transfer in rotating ribbed channel [9], shock-wave induced turbulent mixing [389], and laser induced surface-tension driven flow [66]. In the just published in 2016 articles, the two phase transient airflow in the indoor environment is simulated [194], and a specific approach for simulation of dispersion around cubical building is proposed [177].

6.2.2.3 Detached Eddy Simulation

■ **Example 6.23: Sub-Critical and Super-Critical Flows Over Sphere [81]**

The main goal of this study is the prediction of drag crisis in flow around sphere leading to well-known significant difference in resistance in the cases of laminar (sub-critical regime) and turbulent (super-critical regime) boundary layer separations, which first was demonstrated by Prandtl (Com. 8.1). Prediction of such flows with massive separation is a challenge requiring accurately capture of complicated flow structure possessing transition from laminar to turbulent flow, large-scale vortex shedding, and turbulent wake with random and periodic Reynolds stresses.

The three-dimensional and time-dependent DES formulation [366] is based on a modification to the Spalart-Allmaras one-equation model (S. 8.4.2) such that it reduces to RANS close to the solid surfaces and to the LES away from the walls as it requires for DES approach (S. 5.8). In this model, the transport equation is used to compute the working variable $\tilde{v}$, which is applied to determine the turbulent eddy viscosity

$$\frac{D\tilde{v}}{Dt} = c_{b1}\tilde{S}\tilde{v} - c_{w1}f_w\left(\frac{\tilde{v}}{d}\right)^2 + \frac{1}{\sigma}[\nabla \cdot ((v+\tilde{v})\nabla\tilde{v}) + c_{b2}(\nabla\tilde{v})^2], \quad v_t = \tilde{v}f_{v1}$$

$$f_{v1} = \frac{\chi^3}{\chi^3 + c_{v1}^3}, \quad \tilde{S} = f_{v3}S + \frac{\tilde{v}}{\kappa^2 d^2}f_{v2}, \quad f_{v2} = \left(1 + \frac{\chi}{c_{v2}}\right)^{-3}, \quad f_{v3} = \frac{(1+\chi f_{v1})(1-f_{v2})}{\chi}$$

$$f_w = g\left(\frac{1+c_{w3}^6}{g^6 + c_{w3}^6}\right)^{1/6}, \quad g = r + c_{w2}(r^6 - r), \quad r = \frac{\tilde{v}}{\tilde{S}\kappa^2 d^2}, \quad \chi = \frac{\tilde{v}}{v} \qquad (6.36)$$

Here, S is the magnitude of the vorticity, and the constants are: $c_{b1} = 0.1355$, $\sigma = 2/3$, $c_{b2} = 0.622$, $\kappa = 0.41$, $c_{w1} = c_{b1}/\kappa^2 + (1+c_{b2})/\sigma$, $c_{w2} = 0.3$, $c_{w3} = 2$, $c_{v1} = 7.1, c_{v2} = 5$, and d is the distance to the nearest wall.

Solution are performed in polar coordinates for extended domain of 10 diameters from the sphere using uniform velocity as the boundary condition upstream and velocities and turbulence values obtained by extrapolation from interior domain as boundary condition downstream. The inflow eddy viscosity is set to zero for laminar separation case and $3v$ for the flow with turbulent separation. No-slip conditions on the sphere and periodic boundary condition in the azimuthal direction are employed. The calculation mesh was $141 \times 41 \times 101$ points in the three directions with refinement of $141 \times 82 \times 101$ in the azimuth. The time step was $0.02\, D/U$ with 60 iterations per each step. The residual values of velocity and pressure for convergence was set as 10^{-4}.

The following conclusions are summarized:

- Contours of the instantaneous vorticity magnitude in the sphere wake for the flow with laminar boundary layer separation at $Re = 10^5$ and for turbulent separation at $Re = 1.1 \cdot 10^6$ show the marked difference in wake structure for the sub- and super-critical solutions indicating that in the fist case the flow detachment occurs at a polar angle $80 - 82°$ and in the second one at $110 - 114°$.
- In sub-critical regime, the predicted shedding frequencies are in reasonable agreement with measured values; contributions from the rollup of the detached shear layers to the shedding were resolved, along with the low frequency f shedding mode at Strouhal number $St = fL/U \approx 2$; the drag coefficient ($0.4 - 0.51$ for $Re = 10^5$), pressure (the value and minimum position) and skin friction coefficients are also in reasonable agreement with the measurement data from [7].
- In the case of super-critical regime, the predicted skin friction does not agree well with measured values; this occurs because the turbulent separation was established by igniting the turbulence model over the entire surface of the sphere leading to markedly increasing the skin friction due to arising plentiful regions of laminar flow; nevertheless, the drag coefficients in that case, $0.096 - 0.106$ for $Re = 4.2 \cdot 10^6$ and $0.07 - 0.1$ for $Re = 1.1 \cdot 10^6$, is adequate, which is a result of small contribution of the skin friction to the total drag at high Reynolds number. The pressure prediction is also accurate.

- The super-critical solutions are chaotic and unsteady, although vortex shedding is quite different compared to that observed at lower Reynolds numbers in flows with laminar boundary layer separation; in particular, the wake is dominated by a pair of streamwise vortices that are 'locked' in the sense of producing lateral forces that are of the same sign over substantial sampling intervals; longer sampling periods than the 60 time units over which the super-critical solution at $Re = 4.2 \cdot 10^5$ was averaged is required to determine the period over which the lateral forces will average to zero.
- As first observed in DES of the flow over a circular cylinder, fully turbulent simulations such as those used to predict the super-critical flows are not the definitive approach, even for rather large Reynolds numbers.

■ Example 6.24: Reentry-F Vehicle Flight Experiment [27]

Reentry-F was a flight test conducted by NASA in 1968 to investigate vehicle heating in a turbulent environment. For the experiment, the spacecraft was launched outside the atmosphere and then accelerated back at reentry velocity on a ballistic trajectory. The body consisted of 3.96-m-long 5 deg half-angle cone constructed from beryllium with a 0.254-cm-radius graphite nose tip [332].

The aim of this paper is to predict the unsteady wake dynamic and its impact of the surface heating rates of Reentry-F. For this study, two trajectory points were selected for which data indicated the transitional or turbulent flow on the base, an altitude of 70 and 80 kft and the following flow parameters: Mach number 19.93 and 20.01, Reynolds number $30.1 \cdot 10^6$ and $18.5 \cdot 10^6$ per meter, $\rho_\infty = 0.07092$, $T_\infty = 218$, $U_\infty = 5.9$, $T_{base} = 354$ and $\rho_\infty = 0.043523 \, kg/m^3$, $T_\infty = 221$ K, $U_\infty = 5.965 \, km/s$, $T_{base} = 354$ K, respectively. The employed mathematical model is similar to described in previous example consisting the transport equation for the same working variable $\tilde{\nu}$ determining the eddy viscosity ν_t. Five-species finite rate chemical kinetics model is used to take into account the dissociation of nitrogen and oxygen occurring due to high temperatures in the stagnation region of the nose and shear layer downstream of the base. Three grid meshes: coarse, medium, and fine consisting of 1.7, 4.2, and 9.7 million cells are applied to adequately resolve the shear layers and recirculation zones. The computational time step was $10^{-7} s$.

The simulation was performed using US3D finite volume code for hypersonic reacting flows developed at the University of Minnesota [284].

The following result and conclusion are formulated:

- The DES shows the superior ability over the conventional Reynolds-averaged Navier-Stokes (RANS) models in capture unsteady flow behavior.
- The presented structure of turbulent base flow colored by temperature clearly shows the shear layer, recompressing point, and recirculation zone; the shear layer separates from a forebody and expands to enclose the recirculationg region, recompresses heating significantly gas downstream, which results in air dissociation.
- The laminar flow is dominated by a singe large toroidal vortex; the transonic flow at the vortex center impinges on the base, compressing the flow and creating the peaks in pressure and heating profiles near the base center; the vortex begins to break down through transition until the flow becomes turbulent showing very unsteady, chaotic structure with eddies breaking down into large range of length scales; more uniform pressure and heating profiles are formed in the wake due to enhanced turbulent mixing.

- The performed three-dimensional simulation indicates that the small angles of attack are responsible for the asymmetry experienced during the Reentry-F flight; comparing the computational and measured data reveals superior agreement confirming that the slight pitch and yaw angles play critical role in explaining an apparent asymmetry of the flight data.
- The predicted heating rates display a strong sensitivity to the vehicle orientation; it is shown that small errors in angle position might lead to significantly erroneous results; by accounting for this sensitivity, the base heat transfer to the vehicle was predicted within the experiment uncertainty.

■ **Example 6.25: Free-Surface Flow Around Submerged Submarine Fairwater [170]**

The paper presents results of investigation of a free-surface flow around the submarine fairwater and the effects of reducing its depth on the flow characteristics.

Comment 6.15 A fairwater is a device that improves the ship streamlining through water.

The prototype of the submarine fairwater in the form of cylindrical cross-section airfoil is simulated; the computational domain stretched around the model is of size $9l_c \times 4l_c \times dl_c$ in xyz directions with $3 \cdot 10^6$ grid notes. Three versions of domain with heights $d/l_c = 1.02113,\ 0.96237$, and 0.90474 are considered that correspond to different submergence ratios $(d - h)/h = 0.44373,\ 0.38610$, and 0.32847, where l_c is the profile chord length in y-direction and $(d - h)$ is the submergence depth in z-direction. The conforming Froude numbers defined ahead of a model are $\mathrm{Fr} = U/\sqrt{gL} = 0.4,\ 0.42$, and 0.44. Typical length, speed and submergence conditions are used giving $\mathrm{Re} = 11 \cdot 10^6$.

Comment 6.16 Froude number is defined as a ratio of characteristic velocity U to the velocity of gravitational wave free-surface $\sqrt{gL}$ where L is the characteristic length.

The model is located at $3l_c$ from the inflow side, at $5l_c$ before the outflow section and is centered in spanwise direction. The both inflow and outflow are prescribed in streamwise direction; the first one is specified through constant flux conditions, whereas the second is set using the continuative conditions. Periodical boundary conditions are employed in lateral direction. The submerged body is modeled using the ghost-cell immersed boundary method proposed in [394]. In this method, the ghost cells are obtained extrapolating pressure and velocity to the nearby cells that fall just within the prescribed boundary, whereas the cells located within studied body are neglected. This is achieved in two steps. First, the value is interpolated at a mirror point located within the fluid part of the domain and second, the interpolated value is reflected beck to the ghost cell. For such interpolation, a 10-point stencil is required to get the second order accuracy. As a result, the mesh generation around complex geometric body greatly simplifies and solution time decreases compared to alternative methods.

The free-surface flow associated with wave is modeled using moving mesh and governing equation adopted from [40]

$$\int_v \frac{\partial \bar{u}_i}{\partial t} dv + \int_s \bar{u}_i (\bar{u}_j - v) \cdot nds = \int_s \left(v \frac{\partial \bar{u}_j}{\partial x_i} - \frac{\bar{p}}{\rho} \right) \cdot nds + \int_v gdv - \int_v \frac{\partial \tau_{ij}}{\partial x_j} dv \qquad (6.37)$$

A parallelized collocated code developed for modeling an incompressible, free-surface flow [388] is used. The governing equations are spatially discretized applying a central differencing finite volume method (S. 9.6). A special scheme is used to insure that the blending (S. 5.8) between the first order upwinding and central differencing is minimal.

The paper presents detailed results of the flow behavior around the fairwater including: profiles of time averaged velocity components and pressure, turbulent intensity and Reynolds stresses, streamline instantaneous traces along the body, instantaneous vertical structures for different depth, vortex alignment and persistence, effects of submergence depth on forces, shedding frequencies, and turbulent energy budget. On the base of these data, the following conclusions are formulated:

- Time-averaged Reynolds stresses and turbulent kinetic energy distributions have shown that the major part of the turbulent energy and flow variation is confined to the near wake region of the fairwater; the wake is found to grow behind the fairwater whereas the level of turbulence decreased and continues to do so further downstream.
- Within the measured range (aft of the trailing edge) the flow is found not to have fully recovered even though, on average, the wake/separation region is found to be small; for all cases and for all positions the turbulent kinetic energy spectrums have shown that the wake is fully developed and that with reducing height along the fairwater the level of turbulent kinetic energy in the wake is increased
- Similarly, the effects of reducing the submergence depth have been shown to increase the turbulent kinetic energy for all wave numbers, whereas these effects were less pronounced on the time averaged velocity components; the pressure, streamwise intensity, and Reynolds stresses are influenced by the submergence depth as well; similarly, the turbulent kinetic energy budget terms are found to show the greatest variation in the near tip and near wake region, while further downstream the contribution of the turbulent kinetic energy budget terms substantially decreased.
- Vortical structures are found to show no significant rise or interaction with the free surface, whereas in the wake region, the results show that vorticity is present for over 50% of the monitored time across the fairwater height and monitored positions; reducing the submergence depth resulted in the tip vortex shedding being influenced by the wake.
- Time-averaged forces and the variation in forces showed that the reduction in the submergence depth resulted in an increase in the coefficients of both the pressure and total drag.
- For all of the considered cases, the Strouhal number (Exam. 6.23) ranged between 0.31–0.35; the enclosing free-surface wave angle was measured to be between 38-40 deg, showing that the surface waves are that of a Kelvin kind and are dominated by gravity and local inertial effects. To know more about Kelvin waves, see [440].

OTHER WORKS: During the last three years, the most improved versions of DES were used to investigate by IDDES: the supersonic combustion [413], cavity-induced transition in hypersonic boundary layer [424], vortex breakdown past double-delta wing [238], and flows around rudimentary landing gear (simplified four-wheel lending apparatus) [425], and by ZDES: the inlet condition effects on the tip clearance flow of an axial compressor [323], flat plate turbulent boundary layer over the Reynolds number up to $Re_\theta = 14000$ [92], and flow of airfoil in poststall condition [226]. During the same time, the original DES was used as well, in particular, to study: the sub- and supersonic flows past bluff body with comparison to wind-tunnel data [348], flow control by synthetic jet around airfoil at high angles attack [173], flow past Delft-372 catamaran including vortical and wake break structures [98]. Similar problems were considered using DES in earlier publications such as: the vertical field about the VFF-2 delta wing [83], effects of different wind gust on aerodynamic of road vehicles [134], unsteady flow of abrupt wing stall [138], and ability of stall flow of iced airfoils [298].

Part III

Foundations of Fluid Flow and Heat Transfer

This part is the third portion of the book containing information intended to help a reader to understand the applications and to posses the methods outlined in two basic parts. In conformity with this purpose, the majority of equations and some other details presented in the main text are not repeated here, whereas the physical analysis of these equations, additional explanation of application results, and farther specification of some terms are given in three chapters of his part: laminar fluid flow and heat transfer (Chapter 7), turbulent fluid flow and heat transfer (Chapter 8), and analytic and numerical methods in fluid flow and heat transfer (Chapter 9).

Applications of Mathematical Heat Transfer and Fluid Flow Models in Engineering and Medicine,
First Edition. Abram S. Dorfman.
© 2017 John Wiley & Sons Ltd. Published 2017 by John Wiley & Sons Ltd.

Part III

Fundamentals of Fluid Flow and Heat Transfer

7

Laminar Fluid Flow and Heat Transfer

7.1 Navier-Stokes, Energy, and Mass Transfer Equations

The mathematical models describing the transfer processes are based on the system of the Navier-Stokes equations for momentum transfer and similar equations for energy and mass transfer. These equations are conservation laws expressed in terms of the velocity components, temperature, and concentration. For incompressible fluid flows with constant properties, these expressions are given by equations 1.4–1.8 (without dissipation function S) from Chapter 1 supplemented by equation for mass transfer similar to energy equation, where C is dimensionless concentration and ρD_m is the coefficient similar to μ.

$$\rho \left(\frac{\partial C}{\partial t} + u \frac{\partial C}{\partial x} + \mathrm{v} \frac{\partial C}{\partial y} + w \frac{\partial C}{\partial z} \right) = \rho D_m \left(\frac{\partial^2 C}{\partial x^2} + \frac{\partial^2 C}{\partial y^2} + \frac{\partial^2 C}{\partial z^2} \right) \tag{7.1}$$

Comment 7.1 The term "Navier-Stokes equations" strictly implies the entire system of equations (1.4)–(1.7) including the continuity equation (1.4). However, for simplicity, the same term is used when only the momentum equations (1.5)–(1.7) are considered.

7.1.1 Two Types of Transport Mechanism: Analogy Between Transfer Processes

Consider a system of equations (1.4)–(1.8) for a steady-state regime together with equation (7.1). This system of six equations determines six flow characteristics: three components u, v, w of velocity, pressure p, temperature T, and concentration C. The continuity and the three Navier-Stokes equations describe the momentum transfer, whereas the heat and mass transfer are specified by energy and diffusion equations (1.8) and (7.1), respectively. Analysis of these fundamental equations shows that these three transfer processes are similar, being based on the same physical principles. The major part (without pressure gradients and dissipation function) of each equation has identical structure being composed of two

Applications of Mathematical Heat Transfer and Fluid Flow Models in Engineering and Medicine,
First Edition. Abram S. Dorfman.
© 2017 John Wiley & Sons Ltd. Published 2017 by John Wiley & Sons Ltd.

analogous groups of terms. Such groups are built of derivatives of a quantity that corresponds to the driving force of relevant transfer process. In Navier-Stokes equations, these groups are constructed of velocity derivatives because the velocity gradient is the driving force in Newton's viscosity law. In the energy and mass transfer equations, the analogous groups are made up of the derivatives of temperature or concentration according to the pertinent gradients forming driving forces in Fourier's and in Fick's laws of heat and diffusion, respectively.

Two groups of terms in these transfer equations correspond to two basic transport mechanisms. The group of terms at the right-hand side of each equation corresponds to the molecular transport, whereas the other group at the left-hand side of each equation represents the convective transport. According to that, the structure of each group is in line with transport nature. In each Navier-Stokes equation, the molecular transport group has the form of viscous force in conformity with molecular mechanism. Such group consists of a sum of the second derivatives of velocity components with respect to relevant coordinate multiplied by viscosity coefficient μ. The molecular transport groups in energy and mass transfer equations have analogous structure. Each group is composed of similar product of sum of the second derivatives of temperature or of concentration and corresponding molecular transfer coefficient, thermal conductivity λ or similar to viscosity diffusion coefficient ρD_m.

Another structure is essential to the convective transport groups. Since in this case, the momentum, heat, and species are transported by the fluid flow, the mechanism of this process is defined by hydrodynamic laws. Therefore, in each Navier-Stokes equation, the convective group has a structure of inertia force, which in conformity with the second Newton's law is a driving force of the fluid flow. According to that, each of these groups is composed as a product of a mass and acceleration. Because the changes may occur in time and in space, the result is presented as a product of the substantial derivative of velocity that takes into account both changes (Exam. 7.2) multiplied by unit mass defined by density. Similarly, the convective groups in energy and in mass transfer equations are composed. They consist of the substantial derivative of temperature or of concentration instead of velocity derivative and the specific thermal capacity ρc_p or the mixture density instead of fluid density in the first and the second cases, respectively.

To express mathematically the similarity of transfer processes, we transform the considering system of equations to dimensionless form. Let ϕ is any dependent (general) variable: velocity components, temperature, or concentration transformed by pertinent scale to dimensionless form $(u, v, w)/U$, $(T - T_\infty)/(T_w - T_\infty)$ or $(C - C_\infty)/(C_w - C_\infty)$, respectively. Introducing as well dimensionless time $\bar{t}$ and coordinates $\bar{x}, \bar{y}, \bar{z}$ scaled by L/U and by characteristic length L, respectively, we obtain an equation in dimensionless general form valid for any of three considering transfer processes

$$\frac{\partial \phi}{\partial \bar{t}} + \bar{u}\frac{\partial \phi}{\partial \bar{x}} + \bar{v}\frac{\partial \phi}{\partial \bar{y}} + \bar{w}\frac{\partial \phi}{\partial \bar{z}} = \frac{1}{N}\left(\frac{\partial^2 \phi}{\partial \bar{x}^2} + \frac{\partial^2 \phi}{\partial \bar{y}^2} + \frac{\partial^2 \phi}{\partial \bar{z}^2}\right), \quad N = \frac{UL}{\sigma} \qquad (7.2)$$

where velocity components $\bar{u}, \bar{v}, \bar{w}$ are scaled by U, σ is the kinematic viscosity v, thermal diffusivity α, or diffusion coefficient D_m, and the dimensionless number is $N =$ Re, Pe or $Re_m = ReSc$, whereas $Sc = v/D_m$ is the Schmidt number. Equation (7.2) shows that three transfer processes are greatly similar and in dimensionless variables are described by the same equation that in the case of zero pressure gradient and neglecting thermal dissipation differs only by Reynolds, Peclet, or mass Reynolds numbers.

Equation (7.2) shows that transfer processes of momentum, heat, and mass are described in dimensionless variables by one equation containing only one dimensionless number Re, Pe or $Re_m = ReSc$ specifies the type of process. This mathematical result confirms the physical analysis considered above justifying that three basic transport processes are greatly similar.

Considering analogy of processes is useful in understanding the mechanism of transfer processes by comparing similar effects. Another advantage of such an idea is the possibility of using results obtained in studying one phenomenon for investigating similar others. In particular, due to this principle the basic laws and methods of problem solutions developed in fluid mechanics are used in heat transfer and diffusion theory.

Equation (7.2) also indicates that all geometrically similar objects having the same characteristic number Re, Pe, or Re_m behave similarly if the boundary conditions are identical. This similarity principle established by Osborne Reynolds is important in modeling, especially in experimental investigations, because it makes possible to extrapolate dimensionless characteristics obtained on the models to natural objects and systems. For example, the drag coefficient measured on car or plane model in a wind tunnel may be used to estimate the resistance force for real prototype.

7.1.2   Different Forms of Navier-Stokes, Energy, and Diffusion Equations

7.1.2.1   Vector Form

To present Navier-Stokes equation in vector form, two operators are needed: the Laplace operator

$$\nabla^2 = \frac{\partial^2}{\partial x^2} + \frac{\partial^2}{\partial y^2} + \frac{\partial^2}{\partial z^2} \tag{7.3}$$

and the Hamilton operator ∇ called del. or nabla. Using nabla as a vector makes it possible to express the three basic field characteristics: the gradient, the divergence (as a dot product) and the curl (as a cross product)

$$\nabla = \frac{\partial}{\partial x}\mathbf{i} + \frac{\partial}{\partial y}\mathbf{j} + \frac{\partial}{\partial z}\mathbf{k}, \quad \text{grad } V = \nabla V = \frac{\partial V}{\partial x}\mathbf{i} + \frac{\partial V}{\partial y}\mathbf{j} + \frac{\partial V}{\partial z}\mathbf{k}$$

$$\text{div}\,\mathbf{V} = \nabla \cdot \mathbf{V} = \frac{\partial u}{\partial x} + \frac{\partial v}{\partial y} + \frac{\partial w}{\partial z} \tag{7.4}$$

$$\text{curl}\,\mathbf{V} = \nabla \times \mathbf{V} = \left(\frac{\partial w}{\partial y} - \frac{\partial v}{\partial z}\right)\mathbf{i} + \left(\frac{\partial u}{\partial z} - \frac{\partial w}{\partial x}\right)\mathbf{j} + \left(\frac{\partial v}{\partial x} - \frac{\partial u}{\partial y}\right)\mathbf{k}$$

where $\mathbf{i}$, $\mathbf{j}$, and $\mathbf{k}$ are outward unit normal vectors in x, y, and z directions. To use these equations, recall that the scalar products of the same unit vectors equal unity, whereas the other scalar products of unit vectors are zero and vice versa: the vector products equal zero of the same units but equal to one for product of two others, for example, $\mathbf{i} \times \mathbf{j} = \mathbf{k}$, $\mathbf{j} \times \mathbf{i} = -\mathbf{k}$ and similar others.

Comment 7.2 Nabla is not usual vector, rather it is a symbolic vector that simplifies some mathematical operations.

■ **Example 7.1: Deriving Continuum Equation (1.4) Via the Expressions For the Divergence**

According to (7.4), the divergence is determined as a scalar product resulting in the expression for continuity equation (1.4) as follows

$$\nabla \cdot \mathbf{V} = \left(\frac{\partial}{\partial x}\mathbf{i} + \frac{\partial}{\partial y}\mathbf{j} + \frac{\partial}{\partial z}\mathbf{k}\right) \cdot (u\mathbf{i} + v\mathbf{j} + w\mathbf{k}) = \frac{\partial u}{\partial x} + \frac{\partial v}{\partial y} + \frac{\partial w}{\partial z}, \quad \nabla \cdot \mathbf{V} = 0 \qquad (7.5)$$

■ **Example 7.2: Deriving the Expression For the Substantial Derivative**

The sum of time and space derivatives like that in the parentheses at the left-hand side of each Navier-Stokes equation determines the time derivative for observer moving with the flow. Such a derivative known as a substantial derivative is used as an operator

$$\frac{D}{Dt} = \frac{\partial}{\partial t} + u\frac{\partial}{\partial x} + v\frac{\partial}{\partial y} + w\frac{\partial}{\partial z}, \quad \frac{D\mathbf{V}}{Dt} = \frac{\partial \mathbf{V}}{\partial t} + \mathbf{V} \cdot \nabla\mathbf{V} \qquad (7.6)$$

The spatial part $\mathbf{V} \cdot \nabla\mathbf{V}$ called convective derivative has x, y and z components. Since they are similar, we show the deriving only for the x-component

$$[\mathbf{V} \cdot \nabla\mathbf{V}]_x = (u\mathbf{i} + v\mathbf{j} + w\mathbf{k}) \cdot \left(\frac{\partial u}{\partial x}\mathbf{i} + \frac{\partial u}{\partial y}\mathbf{j} + \frac{\partial u}{\partial z}\mathbf{k}\right) = u\frac{\partial u}{\partial x} + v\frac{\partial u}{\partial y} + w\frac{\partial u}{\partial z} \qquad (7.7)$$

In terms of field vectors just considered, the continuity equation (1.4) and Navier-Stokes equations (1.5)–(1.7) became the compact form

$$\nabla \cdot \mathbf{V} = 0, \quad \rho\frac{D\mathbf{V}}{Dt} = -\nabla p + \mu\nabla^2\mathbf{V}, \quad \frac{D\mathbf{V}}{Dt} = \frac{\partial \mathbf{V}}{\partial t} + \mathbf{V} \cdot \nabla\mathbf{V} \qquad (7.8)$$

Vector heat and mass transfer equations are similar to vector Navier-Stokes equation

$$\rho c\frac{DT}{Dt} = \rho c\left(\frac{\partial T}{\partial t} + \mathbf{V} \cdot \nabla T\right) = \lambda\nabla^2 T + \mu S, \quad \rho\frac{DC}{Dt} = \rho\left(\frac{\partial C}{\partial t} + \mathbf{V} \cdot \nabla C\right) = \rho D_m\nabla^2 C \quad (7.9)$$

7.1.2.2 Einstein and Other Index Notations

In Einstein notations equations (7.8) simplifies farther and takes the form

$$\frac{\partial V_i}{\partial x_i} = 0 \quad \rho\left(\frac{\partial V_i}{\partial t} + V_j\frac{\partial V_i}{\partial x_j}\right) = -\frac{\partial p}{\partial x_i} + \mu\frac{\partial^2 V_i}{\partial x_j\partial x_j} \qquad (7.10)$$

The idea of index notations known as Einstein's convention came from a sum. Since sum implies a repeated index, one may omit the sign of summation if the number of terms is known. According to Einstein's convention when an index variable appears twice in a single term, it implies that we are summing over all of indicated values. In particular, for coordinate components, each index should be repeated three times. Thus, for continuity equation (7.10), we take $i = 1, 2, 3$, and $V_1 = u, V_2 = v, V_3 = w, x_1 = x, x_2 = y, x_3 = z$, whereas for Navier-Stokes equation we should put $i = 1, j = 1, 2, 3$, which results in the first equation (1.5). Similar, putting $i = 2, i = 3$ and $j = 1, 2, 3$, we obtain the second and third Navier-Stokes equations (1.6) and (1.7).

Other index notations often used are Kronecker delta δ_{ij} and Levi-Civita symbol

$$\delta_{ij} = \begin{cases} 1 & if \ \ i = j \\ 0 & if \ \ i \neq j \end{cases} \qquad \varepsilon_{ijk} = \begin{cases} +1 & if \ \ i,j,k = 1,2,3,3,1,2 \ \ or \ \ 2,3,1 \\ -1 & if \ \ i,j,k = 2,1,3,3,2,1 \ \ or \ \ 1,3,2 \\ 0 & if \ \ any \ two \ equal \end{cases} \qquad (7.11)$$

The first symbol means that a combination of two variables is 1 if the indices equal and is zero if they are different. The second symbol ε_{ijk} indicates that the value of a combination of three variables is $+1$ or -1 depending on indices order, or is zero if two of indices are equal. To understand the indices order, start with 123 and than move 3 in the front of 1 to get 312 and again move 2 in the front of 3 to get 231. Do the same to obtain the second row of indices. Start once more with 123 and put 2 in the front of 1 to get 213 and finally use displacement similar to that done in first row to gain 321 and 132.

7.1.2.3   Vorticity Form of Navier-Stokes Equation

This form is usually used for two-dimensional flows for which the stream function ψ may be introduced, and the expression (7.4) for curl simplifies to one term

$$u = \frac{\partial \psi}{\partial y}, \quad v = -\frac{\partial \psi}{\partial x}, \quad curl = \omega = \left(\frac{\partial v}{\partial x} - \frac{\partial u}{\partial y} \right) = -\left(\frac{\partial^2 \psi}{\partial x^2} + \frac{\partial^2 \psi}{\partial y^2} \right) = -\nabla^2 \psi \quad (7.12)$$

where the last expression is the Laplacian of ψ (Com. 1.1).

Stream function introduced in such a way satisfies the continuity equation (1.4), and the system of Navier-Stokes equations reduces to one equation after differentiating equations (1.5) and (1.6) with respect to y and to x, respectively, to obtain

$$\frac{\partial^2 u}{\partial t \partial y} + u\frac{\partial^2 u}{\partial x \partial y} + \frac{\partial u}{\partial y}\frac{\partial u}{\partial x} + v\frac{\partial^2 u}{\partial y^2} + \frac{\partial u}{\partial y}\frac{\partial v}{\partial y} = -\frac{1}{\rho}\frac{\partial^2 p}{\partial x \partial y} + v\left(\frac{\partial^3 u}{\partial x^2 \partial y} + \frac{\partial^3 u}{\partial y^3} \right) \quad (7.13)$$

$$\frac{\partial^2 v}{\partial t \partial y} + u\frac{\partial^2 v}{\partial x^2} + \frac{\partial u}{\partial x}\frac{\partial v}{\partial x} + v\frac{\partial^2 v}{\partial x \partial y} + \frac{\partial v}{\partial x}\frac{\partial v}{\partial y} = -\frac{1}{\rho}\frac{\partial^2 p}{\partial y \partial x} + v\left(\frac{\partial^3 v}{\partial x^3} + \frac{\partial^3 v}{\partial x \partial y^2} \right) \quad (7.14)$$

Because mixed derivatives in both equations are equal, subtracting the first equation from the second eliminates the pressure leading to the Navier-Stokes equation in vorticity form

$$\frac{\partial \omega}{\partial t} + u\frac{\partial \omega}{\partial x} + v\frac{\partial \omega}{\partial y} = v\left(\frac{\partial^2 \omega}{\partial x^2} + \frac{\partial^2 \omega}{\partial y^2} \right) \quad or \quad \frac{D\omega}{Dt} = v\nabla^2 \omega \quad (7.15)$$

The first term of the first equation is obtained as a difference between the first terms of equations (7.14) and (7.13), the second, third, and the last terms are differences of the second, fourth, and the last terms of these equations. Other terms vanish because they can be arranged in the form $\omega(\partial u/\partial x + \partial v/\partial y)$ where the sum in parentheses is zero, which follows from the continuity equation (1.4). Equation (7.15) describes the vorticity transport and is useful in studying some general properties of lows. For example, it follows from (7.15) that the vorticity of a fluid flow does not change with time if $D\omega/Dt = 0$. Thus, the initially irrotational inviscid flow remains irrotational in space and time (S. 7.1.2.5).

7.1.2.4   Stream Function Form of Navier-Stokes Equation

Substituting the last relations (7.12) into equation (7.15) leads to another form of Navier-Stokes equation containing only stream function as unknown variable

$$\frac{\partial \nabla^2 \psi}{\partial t} + \frac{\partial \psi}{\partial y}\frac{\partial \nabla^2 \psi}{\partial x} - \frac{\partial \psi}{\partial x}\frac{\partial \nabla^2 \psi}{\partial y} = v\nabla^4\psi = 0, \tag{7.16}$$

where the last expression is a Laplacian of Laplacian (Com. 1.1) defined as follows

$$\nabla^4\psi = \nabla^2(\nabla^2\psi) = \frac{\partial^4\psi}{\partial x^4} + 2\frac{\partial^4\psi}{\partial x^2\partial y^2} + \frac{\partial^4\psi}{\partial y^4} \tag{7.17}$$

■ **Example 7.3: Determining Streamlines in a Two-Dimensional Flow Field**

Streamlines are curves plotted in flow field such that they are tangent to the direction of the flow at each point of the field. If γ is an angle of a slope of a streamline, one finds from vector triangle of velocity components that $\tan\gamma = v/u = dy/dx$. Thus, the streamline equation is $vdx = udy$ or

$$udy - vdx = 0, \qquad \frac{\partial\psi}{\partial x}dx + \frac{\partial\psi}{\partial y}dy = 0, \qquad d\psi = 0, \qquad \psi = const. \tag{7.18}$$

Here, the second relation is obtained from the first one after substitution equations (7.12) for velocity components. Two other results follow from the fact that the left hand part of the second relation determines the exact differential of stream function $d\psi$, which equals zero indicating that a stream function is constant along the streamline. Then, since ψ is constant along the streamlines, the volume of flow rate is defined as difference $\psi_2 - \psi_1$. This may be shown by integration of an elementary volume rate $d(\dot m/\rho) = udy = d\psi$, whereas physically this is obvious because streamlines indicate the flow direction, and hence, there is no flow across the streamlines.

7.1.2.5   Irrotational Inviscid Two-Dimensional Flows

It follows from equation (7.12) that in the case of irrotational flow ($\omega = 0$), the stream function satisfies the Laplace equation $\nabla^2\psi = 0$. In this case, the another useful function satisfying the Laplace equation exists. This function called potential of flow field is defined by similar Laplace equation $\nabla^2\varphi = 0$. Such a function exists only in the case of irrotational flow (without any rotation) when $\omega = 0$ which is called a potential flow as well. Because of that one gets according to the last equation (7.12) two first relations (7.19) and then comparing this result with velocity components (7.12) obtains

$$u = \frac{\partial\varphi}{\partial x}, \quad v = \frac{\partial\varphi}{\partial y}, \qquad \frac{\partial\varphi}{\partial x} = \frac{\partial\psi}{\partial y}, \qquad \frac{\partial\varphi}{\partial y} = -\frac{\partial\psi}{\partial x} \tag{7.19}$$

Although both functions ψ and φ are similar and both are field characteristics, they are different. In particular, the potential function in contrast to stream function does not satisfy the continuity equation (1.4). The other distinction between these functions is that the potential function exists in two- and three-dimensional flows, whereas the stream function exists only in two-dimensional flow field. On the other hand, the stream functions are applicable in viscous real fluid flows as well as in potential ideal inviscid fluids flows, whereas potential functions are useful only in fields of potential inviscid flows.

Comment 7.3 Functions satisfying the Laplace equation describe many physical processes and are known as harmonic functions.

The last two relations (7.19) are known as Cauchy-Riemann conditions of function differentiability. Any complex function $w(z) = \varphi(x, y) + i\psi(x, y)$ where $z = x + iy$ is differentiable (in other words: is analytic) if it satisfied Cauchy-Riemann conditions.

Comment 7.4 Strongly speaking, the satisfaction of conditions (7.19) may be not sufficient, but except some special examples, the functions that we frequently use in applications are analytic functions.

Any analytic function $w(z)$ gives two harmonic functions $\psi(x, y)$ and $\varphi(x, y)$ representing stream functions and velocity potentials of some irrotational inviscid fluid flow. Two families of curves $\psi(x, y) = c_1$ and $\varphi(x, y) = c_2$ constitute the streamlines and equipotential lines, which are orthogonal to each other. The pattern of these curves is used for graphical presentation and analyzing potential flows. One of the streamlines may be considered as a body surface because the normal component of the velocity, which coincides with tangent to equipotential line at the surface is zero. The last conclusion follows from the fact that streamlines and equipotential lines are perpendicular to each other.

The characteristics of such potential flow may be calculated if the corresponding complex function $w(z)$ (complex potential) is known. Dividing the complex potential in the form $w(z) = w(x + iy) = \varphi(x, y) + i\psi(x, y)$ in real and imaginary parts yields functions $\varphi(x, y)$ and $\psi(x, y)$. Differentiating these functions and using equations (7.19) give the velocity components. The pressure may be determined from the Bernoulli equation using the knowing velocity. Different forms of Bernoulli equation follow from Navier-Stokes equations at zero viscosity. For the case of a steady potential flow the Navier-Stokes equation (1.5) simplifies after omitting time-dependent and viscosity terms. Considering such potential flow as one-dimensional flow along the streamline leads to the second equation 1.12, which after integration gives the Bernoulli equation

$$\rho U dU + dp = 0 \quad \frac{1}{2}\rho U^2 + p = const \quad \frac{1}{2}\rho(u^2 + v^2) + p = const. \qquad (7.20)$$

Despite all real fluids possess viscosity, models based on idealized inviscid fluid give useful and meaningful results for several kinds of problems. At the same time, the assumption of negligible viscosity significantly simplifies the investigation. That is why the potential theory has been extensively developed. Potential theory methods are applicable in studying low-viscosity flows. In particular, such methods are widely used in boundary layer theory because potential flows adequately describe the flow outside the boundary layer. In many cases of flow with favorable (negative) pressure gradient, the potential theory description is almost entirely close to reality. For example, the flows through cylindrical and contractive short channels or flows past relatively short plates are cases of this type. The reason of such close results is that in these flows the viscose forces are small and, what is most important, the flow structures in real and potential flows are practically the same. Thus, the potential pattern gives understanding of flow forms, but the energy losses could not be estimated by potential approaches. The situation is completely different in the opposite cases when the pressure gradient is unfavorable (positive) such that the pressure increases in flows direction as, for example, in diffuser. Flows of this type only at small pressure gradients have continue structure (like in the case of favorable pressure gradient), which becomes destroyed by flow separation as pressure increases. This flow pattern could not be modeled by potential theory. However, until the flow is unseparated, the inviscid potential flow theory may be used, and even in separated flow there are parts to which potential methods may be applied [33].

7.2　Initial and Boundary Counditions

The initial and boundary conditions are considered in conjugate problem formulation (S. 1.1). Here, are added some important details:

- The major problem in formulation of boundary conditions for Navier-Stokes equation consists of no-slip condition on the surface. This condition was suggested by establishers of this equation. Today, after more than 150 years, we could not exactly prove that the Navier-Stokes equation with such boundary condition describes correctly the fluid motion. The reason of this is that due to huge mathematical difficulties, there is no even single exact analytical solution of the full Navier-Stokes equations, which is required for comparing it with corresponding experimental data to get the desired proof. Nevertheless, the available information, in particular, analytical solution of simple problem like flow in a channel or in a tube, analytical solution of other reduced Navier-Stokes equations, numbers of numerical solution of full Navier-Stokes equations, including direct numerical simulations, results in the boundary layer theory so well agree with experimental data that there are no reasons to be in doubt about validity of the Navier-Stokes equation and no-slip boundary condition [338].
- As mentioned in Section 1.1, the basic problems with formulation of the boundary conditions for energy equation are the same as the difficulties arising in solving the Navier-Stokes equation (Com. 1.2). The reason for this lies in the disturbed temperature profiles at the leading and trailing edges of an embedded object that are unknown in advance as well as the similar velocity profiles. Therefore, additional experimental information is needed to create the boundary conditions in this case also.

We present two Dirichlet problems (S. 11) as examples of such extra information.

■ **Example 7.4: Horizontal Channel Heated From Below Considered in Example 3.21 [71]**

Two-dimensional Navier-Stokes and energy equations with no-slip and conjugate boundary conditions on the interface are solved. The velocity and ambient temperature parabolic profiles at the entrance are assumed and experimentally checked. To take into account the recirculation effects at the exit section, some experimental known data were adopted using: $\partial u/\partial x = v = 0$ for velocity and $\theta = T - T_\infty = 0$ if $u < 0$ (inflow in the channel), $\partial \theta/\partial x = 0$ if $u > 0$ (outflow) for temperature.

■ **Example 7.5: Rectangular Slab Heated From One Surface and Isolated Others (Exam. 3.9)**

Two-dimensional Navier-Stokes equations in variables $\psi - \omega$ and energy equation in variable $\theta = (T - T_\infty)/(T_{bt} - T_\infty)$ are solved (T_{bt} is the temperature of a bottom). No-slip and conjugate conditions are used. Uniform velocity and temperature profiles before and behind far away from the slab are assumed. Special investigations were performed to estimate perturbed profiles of ψ, ω and θ close to the slab (Exam. 3.9) [412].

- In contrast to just considered similarity of boundary conditions for Navier-Stokes and energy equations for both fluid domains, the boundary conditions on the channel walls or on the body surface for the energy equation are inherently different from those for the Navier-Stokes equation. To understand the physical reasons of this contrast, note that by crossing the interface, the dynamical characteristics experience infinite changes: the velocity becomes zero and the viscosity turns into infinity, whereas the thermal properties after crossing interface remain finite.

Taking this into account, the no-slip boundary condition is used for Navier-Stokes equation on the solid. Much more complicated is the situation in the case of energy equation. In this case, the conductivity and other physical properties experienced discontinuity across the interface according to their different values of fluid and of a body. As a result, the two unknown in advance temperatures and heat fluxes arise, which in determining as well as the interface temperature, requires conjugate solution. That is the reason why the simple approach based on heat transfer coefficient was common before computers came to use.

Comment 7.5 It was just said that because the solid viscosity is infinite, the no-slip boundary condition is used. This is not always the case. The low-density gas slips on the body surface, so that no-slip boundary condition does not hold. Such an effect occurs when the Knudsen number $\text{Kn} = l/L$ is of order unity or greater. Knudsen number is defined as a ratio of the molecular mean free path l (average distance that molecule travels between collision with other molecules) to characteristic scale length L.

7.3   Exact Solutions of Navier-Stokes and Energy Equations

There are problems when the Navier-Stokes equation simplifies, and it becomes possible to obtain the exact solution. The importance of such solutions we discuss above showing the fundamental role that the exact results play. Here, we consider examples of such solutions of two type: (i) unsteady and steady problems where the nonlinear inertia terms identically vanish resulting in linear equations and (ii) steady problems governed by full nonlinear Navier-Stokes equations that may be reduced to ordinary differential equations. Other exact solutions of both types may be found in [338].

7.3.1   Two Stokes Problems

Stokes was the first given the two exact solutions of the first type. The first solution presents a flow near a plate suddenly accelerated from rest in its own plane. The second Stokes problem describes the flow near an infinite plate, which harmonic oscillates parallel to itself. Both problems are governed by the same simplified Navier-Stokes equation containing only unsteady and viscous terms and relevant boundary conditions

$$\frac{\partial u}{\partial t} = v \frac{\partial^2 u}{\partial y^2}, \quad t \le 0, \quad u = 0, \quad t > 0, \quad u(0) = U_0, \quad u(\infty) = 0, \quad u(0, t) = U_0 \cos nt$$

$$(7.21)$$

The first two conditions (7.21), the starting plate velocity U_0 and zero velocity far from the plate, pertain to the first problem, and the last equation (7.21) specifies the boundary condition for the second problem. Differential equation (7.21) is the same as one-dimensional conduction equation (1.1) without source. Thus, the solution of the first problem is given by error function (S. 9. 1) as $u/U_0 = erfc(y/2\sqrt{vt})$. The solution of the second problem should satisfy the same equation (7.21), the last condition (7.21), and an extra asymptotical zero condition at $y \to \infty$, resulting in $u(y, t) = U_0 \exp(-\xi) \cos(nt - \xi)$, where $\xi = \sqrt{n/2v}$.

7.3.2   Steady Flow in Channels and in a Circular Tube

Parallel flows containing only longitudinal velocity component in plane channel and steady Couette flow (flow between two parallel plates, one of which is moving relative to the other) are described by simple differential equation like equation (7.21) but with pressure gradient instead of unsteady term. Such equation, boundary conditions and solutions are as follows

$$\frac{dp}{dx} = \mu \frac{d^2u}{dy^2}, \quad y = \pm H, \ u = 0, \quad y = 0, \ u = 0, \quad y = 2H, \ u = U \tag{7.22}$$

$$u = -\frac{1}{2\mu}\frac{dp}{dx}(H^2 - y^2) \quad u = \frac{y}{2H}U - \frac{H^2}{\mu}\frac{dp}{dx}\frac{y}{H}\left(1 - \frac{y}{2H}\right) \tag{7.23}$$

Here, $2H$ is the distance between channel walls. The first solution (7.23) determines the parabolic velocity profile in the plane channel, whereas the second one shows how the pressure gradient in Couette flow deforms the linear velocity distribution $u = (y/2H)U$, which exists at zero pressure gradient. According to this expression, the decreasing in the flow direction pressure leads to positive velocity profile over the whole channel cross-section, whereas the increasing pressure results in the profile containing near the unmoving wall the negative velocities, which represent the back flow domain.

The meaningful exact solution of Navier-Stokes equation is the flow in circular tube, which is known as Hagen-Poiseuille flow. At first in the 1930s of the nineteenth century, it was found experimentally that in this practically important case of circular tube, coefficient of resistance ζ depends inversely proportional on Reynolds num $Re = 2\bar{u}R/v$. Later on, it was shown that such dependency obtained from exact solution of the reduced Navier-Stokes equation in cylindrical coordinates with no-slip boundary conditions on walls $u = 0$ at $r = R$ as follows

$$\mu\left(\frac{d^2u}{dr^2} + \frac{1}{r}\frac{du}{dr}\right) = \frac{dp}{dx}, \quad u(r) = -\frac{1}{4\mu}\frac{dp}{dx}(R^2 - r^2), \quad \zeta = -\frac{dp}{dx}\frac{4R}{\rho\bar{u}^2} = \frac{64}{Re}, \tag{7.24}$$

well agrees with experimental data.

7.3.3   Stagnation Point Flow (Hiemenz Flow)

This is an example of steady nonlinear problem of type (ii), described by the full Navier-Stokes equation, which is reduced to the ordinary differential equations in the form applicable for numerical solution and tabulation. To obtain such form of equations, the governing system is transformed using similarity variables. One example of such form of equation we just considered as solution of the first Stokes problem $u/U_0 = erfcz$, where $z = x/2\sqrt{\alpha t}$ is similarity variable. Physically, the existing of similarity variable in this case means that velocity distributions for different times and locations are similar so that they construct one curve in similarity variables, which here are u/U_0 and z. The more general self-similar solutions for the case of velocity distribution $U = cx^m$ that we encountered in Section 1.6.1 are analyzed below in Section 7.5.2.

Example of another type of similarity equation gives the dependence (7.24) ζ (Re) indicating that using a Reynolds number as a similarity variable leads to universal curve for coefficient ζ instead of different curves for each combination of $\bar{u}, R$ and v.

The stagnation flow occurs at blunt nose of any cylindrical body at the stagnation point. In the considering prototype, the two-dimension flow arrives in a perpendicular direction to a plate and impinges on some point, which is taken as origin $x = y = 0$. Since the potential flow (S. 7.1.2. 5) slips at the wall, it leaves in both directions along the plate so that the velocity components of this potential flow become to be proportional to corresponding coordinates. Then, the Bernoulli's equation (7.20) gives the pressure

$$U = cx, \quad V = -cy, \quad p_o - p = (1/2)\rho(U^2 + V^2) = (1/2)\rho c^2(x^2 + y^2) \tag{7.25}$$

It is assumed that solution of the Navier-Stokes equation has the form $u = xf'(y)$ and $v = -f(y)$. It is easy to check that such form of solution satisfied the continuity equation (1.4) and give after substituting into the first Navier-Stokes equation (1.5) an ordinary differential equation $f'^2 - ff'' = c^2 + vf'''$.

This nonlinear equation could not be solved analytically. To solve this equation numerically, it is reasonable to eliminate two parameters c and v to make possible to tabulate the calculation results. Otherwise, the calculation would be necessary to be perform for each pair of these parameters. One way to find the proper similarity variables is to introduce new constants putting $\eta = c_1 y$ and $\varphi = c_2 f$. Substituting the new variables into considering ordinary differential equation and equating coefficients to eliminate parameters c and v give two algebraic equations $(c_1/c_2)^2 = c^2 = (c_1^3 v/c_2)$ determining c_1 and c_2. Then, the similarity variables $\eta = y\sqrt{c/v}$, $\varphi = f\sqrt{cv}$ and corresponding differential equation with boundary conditions in these variables are obtained

$$\phi''' + \phi\phi'' - \phi'^2 - 1 = 0, \quad \eta = 0, \quad \phi = \phi' = 0, \quad \eta \to \infty, \quad \phi' \to 1 \tag{7.26}$$

To find the last condition, one starts with $\phi' = d\phi/d\eta = (c_2/c_1)df/dy = (1/c)f'(y)$ and after using the first assuming relations $u = xf'(y)$ and $U = cx$ obtains $\phi' = u/U$, which for $\eta \to \infty$ gives the last condition (7.26).

Solution of this problem was obtained in 1911 by Hiemenz in his thesis and is known as the Hiemenz flow. Equation (7.26) was tabulated by Howarth as the velocity profile in boundary layer $u/U = \varphi'(\eta)$ (Table 5.1 in [338, 1079]). At $\eta = 2.4$ the ratio $u/U = 0.99$, which approximately determines the boundary layer thickness (Exam. 7.7).

The other exact solutions of this type are [338]: three-dimensional stagnation flow, the flow induced by a disk rotating about axis perpendicular to its plane, and flow in convergent and divergent channels. For each case, the similarity variables are found, and Navier-Stokes equations are reduced to ordinary differential equations, which are solved numerically and the results are tabulated.

7.3.4   Couette Flow in a Channel with Heated Walls

Let the temperatures of resting and moving walls of a plane channel are T_0 and $T_1 > T_0$, respectively. In the absence of pressure gradient when the velocity distribution contains only the first linear term $u = yU/2H$ of the second equation (7.23), the dissipation function S in relation (1.8) simplifies to one term $(du/dy)^2$. In this case, the energy equation simplifies as well because only one velocity component u depends on y, while the other component v is zero. As the result, the energy equation reduces to the form containing only one viscous term and one term defining the dissipation function. For the considering Couette flow with linear velocity profile this equation and its solution satisfying the prescribed wall temperatures mentioned above are $(\eta = y/2H)$

$$\lambda \frac{d^2T}{dy^2} + \mu \left(\frac{du}{dy}\right)^2 = \lambda \frac{d^2T}{dy^2} + \mu \frac{U^2}{4H^2} = 0, \quad \frac{T - T_0}{T_1 - T_0} = \eta + \eta(1 - \eta)\frac{\mu U^2}{2\lambda(T_1 - T_0)} \quad (7.27)$$

It follows from the last expression that fluid cools the heated moving wall only until $(\mu U^2/2\lambda) < (T_1 - T_0)$, whereas with further velocity increasing, the fluid starts to heat the moving wall despite $T_1 > T_0$. The reason of this is that the heat generated due to the friction exceeds the effect produced by cooling fluid owing to a temperature difference.

7.3.5   Adiabatic Wall Temperature

The effect of the adiabatic wall temperature may be analyzed using the solution of equation (7.27) subjected to boundary condition for thermally isolated wall. Considering an isolated unmoving wall $(dT/dy = 0)$ and another at constant temperature T_0, one gets the following boundary conditions, corresponding solution of equation (7.27), and the value of adiabatic temperature T_{ad}

$$y = 0, \quad \frac{dT}{dy} = 0, \quad y = 2H, \quad T = T_0, \quad T(y) - T_0 = \mu \frac{U^2}{2\lambda}\left(1 - \frac{y^2}{4H^2}\right), \quad T_{ad} = T_0 + \mu \frac{U^2}{2\lambda}$$
$$(7.28)$$

The last result is obtained from the temperature distribution at $y = 0$, that pertains to isolated wall. This expression shows that adiabatic is the temperature that thermally isolated surface reaches when the whole released friction heat is adopted because the fluid cannot cool the isolated wall. Comparing the last formula (7.28) with just gained condition $(\mu U^2/2\lambda) < (T_1 - T_0)$, of cooling-heating process for hot wall, one sees that fluid cools the wall until it reaches the adiabatic temperature, so that an inequality $\Delta T_w > \Delta T_{ad}$ (or $\Delta T_w < \Delta T_{ad}$) determines what process, cooling or heating, takes place. In these inequalities, Δ is a difference between wall and reference $(T_0$ in this example) temperatures, and according to (7.28) $\Delta T_{ad} = \mu U^2/2\lambda$.

We considered the estimation of adiabatic temperature in the case of flow impingent on an isolated wall in Section 2.1.4.3.

7.3.6   Temperature Distributions in Channels and in a Tube

In the case of equal walls temperatures $(T_0 = T_1)$, the last equation (7.27) gives the symmetrical temperature distribution in Couette flow. Analogous symmetrical temperature distribution is obtained from solution of similar to (7.27) differential equation for Poiseuille flow, which is a flow through a channel with plane walls. The temperature profiles for both Couette and Poiseuille flows are represented by parabolas of the second and of the four degrees, respectively

$$T(y) - T_0 = \frac{\mu U^2}{4\lambda} \frac{y}{H} \left(1 - \frac{y}{2H}\right), \quad T(y) - T_0 = \frac{\mu u_m^2}{3\lambda} \left[1 - \left(\frac{y}{H}\right)^4\right] \quad (7.29)$$

As we just mentioned, the last expression is a result of solution of the equation similar to equation (7.27). Such differential equation differs from (7.27) by the second term that is $4\mu(u_m y)^2/H^4$ instead of $\mu U^2/4H^2$. This term accounts for dissipation heat produced by Poiseuille flow with maximal velocity on a symmetry axis $u_m = -(H^2/2\mu)(dp/dx)$.

Comment 7.6 One may be confused by a seeming contradiction: first it was said that there is no exact solutions of Navier-Stokes equation, and then some exact solutions are discussed. The answer is that the first statement pertains to solutions of Dirichlet problem of the full Navier-Stokes equations, whereas the considered exact results are examples of simple solutions of different parts of Navier-Stokes equation. Indeed, we study simple solutions because we could not obtain the exact ones. Nevertheless, there is no doubt of usefulness of the known exact solutions, and we explained the importance of such results in Chapters 5 and 6 considering direct turbulence simulation. We return to this question again in Chapter 9 in discussing the relation between numerical and analytical methods.

7.4   Cases of Small and Large Reynolds and Peclet Numbers

The two limiting cases of small and large Reynolds or Peclet numbers correspond to situation when one of two basic groups of terms in Navier-Stokes and energy equations are negligible small relatively to other (S. 7.1.1). In particular, the inertia terms defining the rate of convective momentum transfer are proportional to the square of the velocity components, whereas the viscous terms that specify another, molecular part of transfer process are proportional only to first power of velocity. Proceeding from this fact, it is easy to understand that in the case of small Reynolds numbers, the inertia terms may be neglected, whereas in the opposite case of large Reynolds numbers, the viscous terms are negligible small relatively to inertia ones. Analogously considering the energy equation, one sees that the convective group of terms contains the velocity components as factors, but the terms presenting heat conduction are independent on velocity. Thus again, the conductive terms are dominated at small Peclet numbers, whereas the convective group may be omitted in this case, and vice versa, under large Peclet numbers.

The more simple relations obtained in this way unlike the reduced exact equations discussed before are approximate since in this case, the terms are omitted not due to physical sense, but are neglected causing some inaccuracy.

7.4.1   Creeping Approximation (Small Reynolds and Peclet Numbers)

Creeping flow is a very slow motion at small Reynolds number (Re $\ll$ 1), that is, at small velocity, or small object size, or for very viscous fluid. As mentioned above, in this case, the inertia terms are small and they may be neglected significantly simplifying the Navier-Stokes equations to the form $\nabla p = \mu \nabla^2 \mathbf{V}$ for a steady flow. This type of flow is also known as Stokes flow since he first considered such a flow. The energy and diffusion equations may be simplified to similar Laplace equations as well neglecting small convective terms.

The two-dimensional creeping flow equations may be transformed to a single equation for the stream function, such as general two-dimensional Navier-Stokes equation (7.16). Neglecting the inertia terms (the left-hand side of this equation), one obtains the Stokes flow equation in stream function form. Because in this case, the pressure is defined by Laplace equation, the creeping flow problems are governed by the Laplace and simplified (7.16) equations, whereas the similar thermal and diffusion problems are governed as mentioned above only by the Laplace equation. Thus, we have

$$\nabla^4\psi = \nabla^2(\nabla^2\psi) = \frac{\partial^4\psi}{\partial x^4} + 2\frac{\partial^4\psi}{\partial x^2\partial y^2} + \frac{\partial^4\psi}{\partial y^4} = 0 \quad \nabla^2 p = 0 \quad \nabla^2 T = 0 \quad \nabla^2 C = 0 \quad (7.30)$$

Note that creeping flow is irrotational because it follows from equations (7.12) and (7.30) that $\omega = \nabla^2\psi = 0$. The creeping approximation approach has wide applications for small objects in different areas from a charging electron to pollution of environment. In nature, this type of flow occurs in the moving bacteria and other microorganisms, in the motile sperm, and in a lava flow. In engineering, the application includes polymer and other high viscosity substances flows, film production, studying of small size, micro- and nano-technology systems.

7.4.2   Stokes Flow Past Sphere

This problem of parallel flow past sphere was as well first considered by Stokes. Considering creeping flow equation in spherical coordinates r, γ, ϕ leads to following expressions for stream function, pressure, and drag coefficient (e.g., [147, 44])

$$\psi = Ur^2\sin^2\gamma\left[\frac{1}{2} - \frac{3}{4}\frac{R}{r} + \frac{1}{4}\left(\frac{R}{r}\right)^3\right], \quad p - p_\infty = -\frac{3}{2}\frac{\mu U}{R}\cos\gamma, \quad F = 6\pi\mu UR, \quad C_f = \frac{24}{\mathrm{Re}}$$
$$(7.31)$$

These results show that: (i) the pressure reaches the extreme values $\mp(3/2)(\mu U/R)$ at $\gamma = \pi/2$ (minimum) and $\gamma = (3/2)\pi$, (maximum), (ii) the drag is proportional to the first power of velocity and consists of two parts $F = 2\pi\mu UR + 4\pi\mu UR$, where first one comes from pressure and the other results from share stress, (iii) the drag coefficient is inversely proportional to Reynolds Number $\mathrm{Re} = 2\rho UR/\mu$.

Comment 7.7 The solution of similar problem for liquid drop in immiscible liquid medium indicates that the drag coefficient (7.31) should be multiplied by factor $(2\mu + 3\mu_w)/(\mu + \mu_w)$, where μ and μ_w are viscosity of medium and drop [227].

7.4.3   Oseen's Approximation

As the distance from surface increases, the accuracy of creeping approximation decreases due to growing flow velocity. Oseen took into account the effect of the inertia using the perturbation approach. Presenting the velocity field as $U + u'$, v', w', where U is the constant velocity far from the sphere and u', v', w' are a small field, its perturbations leads to the Navier-Stokes and continuity equations with inertia terms of the first and second order. Neglecting the relatively small second order terms, containing products of perturbation values, gives the Oseen linearized equations. For u' component such Navier-Stokes and Oseen linearized equations are as follows

$$U\frac{\partial u'}{\partial x} + v'\frac{\partial u'}{\partial y} + w'\frac{\partial u'}{\partial z} + \frac{1}{\rho}\frac{\partial p}{\partial x} = v\nabla^2 u', \quad U\frac{\partial u'}{\partial x} + \frac{1}{\rho}\frac{\partial p}{\partial x} = v\nabla^2 u' \qquad (7.32)$$

Two other Navier-Stokes and corresponding linearized equations for components v' and w' are similar. The system of Oseen equations that takes into account the inertia effects in the first approximation consists of the second equation (7.32), two similar others with v' and $\partial p/\partial y$ or w' and $\partial p/\partial z$ instead of u' and $\partial p/\partial x$, and continuity equation (1.4) with components u', v', w' instead of $u, v,$ and w.

Solution of the system of Oseen equations under the no-slip boundary conditions $U + u' = v' = w' = 0$ gives for a sphere the velocity field that in the front is close to Stokes pattern, whereas the velocities behind the sphere are larger than in Stokes case. The Oseen drag coefficient may be obtained by multiplying the Stokes drag coefficient (7.31) by $[1 + (3/16)\text{Re}]$. Experimental data show that formula (7.31) is applicable for Re < 1, whereas Oseen approximation is accurate up to Re ≈ 5 [338].

7.4.4   Boundary Layer Approximation (Large Reynolds and Peclet Numbers)

Two fundamental results published after Euler's work in the next about hundred years: Navier-Stokes equations and Reynolds procedure of averaging Navier-Stokes equations for turbulent flow practically completed classical hydrodynamics—our basic mathematical means of understanding and modeling flow processes. However, nonlinear, sophisticated Navier-Stokes and Reynolds equations could not be solved before computers came into use. Thus, despite the known equations of the real viscous fluid motion, only the Euler's equations for perfect fluid could actually be used at the end of the nineteenth century. Because it was clear that the friction forces in the air or water around a moving body are small in comparison with pressure and gravity forces, it was expected that solutions of Euler's equations should be close to the authentic pattern. In fact, the theory of perfect fluid motion leads to satisfactory results in many problems such as formatting waves, jets, or in determining the pressure distribution around a moving body, but this theory fails to predict the pressure drags. That contradiction between theory and reality is known as a D'Alembert's paradox: the pressure losses in a flow moving around a body are zero.

Ludwig Prandtl was the first who understand why the solution of Euler equations for a perfect fluid do not show any pressure losses and resistance forces of a moving body, even in the case of small viscosity fluid. In his article published in 1904, he explained that the perfect flow model describes a major part of the real flow field except the small layer adjusting to surface called the boundary layer. Therefore, the model that describes the real fluid motion should consist of two parts: a thin boundary layer where the friction forces are significant, and another flow part where the friction effects are negligible, and hence, the perfect fluid model is applicable. The practical significance of the Prandtl approach follows from the fact that the majority of technical fluids including water, air, and oil are low viscosity and conductivity liquids.

7.4.4.1 Derivation of Boundary Layer Equations

The procedure of simplifying the Navier-Stokes and energy equations in boundary layer equations is based on comparing the order of magnitude of equations terms. To make the terms comparable, the dimensionless variables, that scaled by maximum value of each variable, are used. Dividing the longitudinal velocities by U_∞, longitudinal coordinates by length L, transverse coordinates by thicknesses δ and δ_t for Navier-Stokes and energy equations, respectively, pressure by ρU_∞^2, time by L/U_∞, and temperatures by temperature head $\theta = T_w - T_\infty$, gives evidence that the value of each dimensionless variable and derivatives do not exceed a unity. The scale for transverse velocity is obtained from continuity equation. Taking into account that both terms in this equation are of the same order and knowing that the scales of the first and second terms are U_∞/L and v/δ, respectively, we get $U_\infty/L \sim v/\delta$, and then define the scale for transverse velocity $v \sim U_\infty \delta/L$.

We begin from estimating the order of magnitude of boundary layer thicknesses. Proceeding from the fact that both group of terms in the Navier- Stokes and energy equations presenting convective and molecular transport (S. 7.1.1) are of the same order, we compare the order of magnitude of these groups using the variable scales. For inertia, terms of two-dimensional Navier-Stokes equation (1.5) we have $\rho\left(\frac{\partial u}{\partial t} + u\frac{\partial u}{\partial x} + v\frac{\partial u}{\partial y}\right) \sim \rho\left(\frac{U_\infty}{L/U_\infty}\right) \sim$ $U_\infty\frac{U_\infty}{L} \sim \delta\frac{U_\infty}{L}\cdot\frac{U_\infty}{\delta}$. Thus, all inertia terms are of the same order $\rho U_\infty^2/L$. Analogously, one obtains an order $\rho c_p U_\infty \theta/L$ for convective terms of energy equation.

Quite a different result is attained by comparing terms responsible for the molecular transport. In this case, the orders of magnitude of two viscous terms of Navier-Stokes equation differ markedly as well as two conductive terms of energy equation, which is clear from the following two expressions

$$\mu\left(\frac{\partial^2 u}{\partial x^2} + \frac{\partial^2 u}{\partial y^2}\right) \sim \mu\left(\frac{U_\infty}{L^2} + \frac{U_\infty}{\delta^2}\right) = \mu\frac{U_\infty}{L^2}\left(1 + \frac{L^2}{\delta^2}\right) \tag{7.33}$$

$$\lambda\left(\frac{\partial^2 T}{\partial x^2} + \frac{\partial^2 T}{\partial y^2}\right) \sim \lambda\left(\frac{\theta}{L^2} + \frac{\theta}{\delta_t^2}\right) = \lambda\frac{\theta}{L^2}\left(1 + \frac{L^2}{\delta_t^2}\right) \tag{7.34}$$

Since $(L/\delta)^2 \gg 1$, we get from these equations the first statement of Prandtl model: the one viscous term in the Navier-Stokes equation and the one conduction term in the energy equation are small compared with another and can be omitted.

Comparing just obtained orders of inertia and viscous or convective and conductive terms, we accomplish our intent to estimate the order of dynamic and thermal boundary layer thicknesses magnitude, and show that those are inversely proportional to square root of Reynolds or Peclet number

$$\rho\frac{U_\infty^2}{L} \sim \mu\frac{U_\infty}{L^2}\frac{L^2}{\delta^2}, \quad \frac{\delta}{L} \sim \frac{1}{\sqrt{\text{Re}}}, \quad \rho c_p\frac{U_\infty\theta}{L} \sim \lambda\frac{\theta}{L^2}\frac{L^2}{\delta_t^2}, \quad \frac{\delta_t}{L} \sim \frac{1}{\sqrt{\text{Pe}}} \tag{7.35}$$

The second statement of Prandtl model is obtained under comparison of Navier-Stokes equations (1.5) and (1.6). Comparing magnitude of similar terms of both equations, one sees that the ratio of magnitude of each pair of terms is the same. For example, this ratio for first inertia terms is: $v\partial v/\partial x/u\partial u/\partial x \sim v/u \sim \delta/L \sim 1/\sqrt{\text{Re}}$, and similarly, it may be shown that the ratio of magnitude of the others pair of terms is the same. Thus, each term in the second Navier-Stokes equation is $1/\sqrt{Re}$ times smaller than the analogous term in the first equation. Because the magnitude of all terms in each equation is of the same order, it follows from the last result that the magnitudes of the pressure gradients $\partial p/\partial x$ and $\partial p/\partial y$ are of the same order as those of the other terms in the first and second Navier-Stokes equations, respectively. Thus, the ratio of orders of the pressure gradients is also the same $(\partial p/\partial y)/(\partial p/\partial x) \sim 1/\sqrt{Re}$. At the same time, inside the boundary layer, the transverse coordinate y is of the order of the thickness, which is $\delta \sim 1/\sqrt{Re}$, and hence, within the boundary layer the order of the ratio dy/dx is as well $1/\sqrt{Re}$. Therefore, the order of the ratio of the pressure drops inside the boundary layer in y and x directions is $(\partial p/\partial y)dy/(\partial p/\partial x)dx \sim 1/\sqrt{Re} \cdot 1/\sqrt{Re} = 1/Re$.

Two conclusions of Prandtl model follow from the last analysis: in the case of high Reynolds number, (i) the second Navier-Stokes equation may be neglected because it order is $1/\sqrt{Re}$, and (ii) the pressure may be considered as unchanged across the boundary layer being equal to the pressure in external flow at the outer edge because the pressure variation in normal direction is of order $1/Re$. One more conclusion gives the comparison of the terms of dissipation function (1.8), which indicates that the order of only one term $(\partial u/\partial y)^2$ is $1/Re$, whereas others are of smaller orders and may be omitted.

Thus, at high Reynolds and Peclet numbers, Prandtl model reduces the system of continuity, two Navier-Stokes equations (1.4)–(1.6) and energy equations (1.8) for two dimensional flow to the system of equations (1.9)–(1.11) governing the velocity and temperature fields inside the boundary layer and Bernoulli equation (1.12) determining the unchanged across the boundary layer pressure in the external potential flow at the outer edge of boundary layer.

Comment 7.8 In the case of flow past a flat plat, the external potential flow is a parallel stream with given velocity U_∞. Therefore, at the outer edge of the boundary layer, the velocity U in Bernoulli equation (1.12) is equal to U_∞, whereas v -component is zero. In the case of flow past another body shape $U \neq U_\infty$, and to get the same result, one uses the natural coordinates measuring x and y along a body surface and in the normal direction to it, respectively.

The system of boundary layer equations is much simpler than the initial system of Navier-Stokes and energy equations because this system of partial differential equations contains three equations with one second derivative instead of four equations with two second derivatives. The reducing of the number of the second derivatives from two to one transforms the elliptic Navier-Stokes and energy equations into the parabolic boundary layer equations that require significantly simpler boundary conditions than complex Dirichlet or Neumann problems for elliptic equations (S. 1.1).

Comment 7.9 For the first 25 years, the boundary layer theory was developed only by Prandtl and his students, so that about one or two articles were published per year. The situation changed after Prandtl's lecture at the meeting of the Royal Aeronautical Society in London in 1927, and in the following years, the amount of publication in boundary layer theory was grown steadily and reached about 100 papers per year in the middle of the last century increasing to almost 300 articles yearly twenty years later [338]. This historical fact shows how long it takes a new idea or new result (even as practical important as the boundary layer theory) to become widely known and be used.

7.4.4.2 Prandtl-Mises and Görtler Transformations

Prandtl and Mises independently show that boundary layer equations in the case of steady flow can be transformed to the form closed to one-dimensional heat transfer equation (1.2). This form is obtained using stream function and longitudinal coordinate x as independent variables and $Z = U^2 - u^2$ and T as unknown. Applying chain rule for differentiation yields the equations and boundary conditions for velocity in new variables, but for temperature the same boundary conditions as for usual form (S. 1.1) remain valid

$$\frac{\partial Z}{\partial x} = vu\frac{\partial^2 Z}{\partial \psi^2}, \quad \frac{\partial T}{\partial x} = \alpha\frac{\partial}{\partial \psi}\left(u\frac{\partial T}{\partial \psi}\right), \quad \psi = 0 \quad Z = U^2(x), \quad \psi \to \infty \quad Z \to 0 \quad (7.36)$$

In contrast to the one-dimensional conduction, the first equation (7.36) for velocity is nonlinear because the factor u at the second derivative depends on unknown function Z.

Boundary layer equations (7.36) have a singularity at the surface because the derivative with respect to ψ become infinite at $\psi = 0$.

Comment 7.10 A function is singular at some point if this function is not analytical at this point, that is, if it is not differentiable (S. 7.1.2.5). Since such function does not have some derivatives at singular point, it could not be expended in Taylor series at this point.

To see that equations (7.36) are singular at point $\psi = 0$, note that the left part of both equations is finite on the surface. At the same time, velocity u is zero on the surface, and consequently, the second derivative in the right part of each equation should be infinite. The type of singularity may be estimated by considering the analytic solution of the boundary layer equation for velocity. Presenting such solution near the surface by the Taylor series in power of y, one gets corresponding expansion for stream function and then, the series in power of stream function ψ for velocity

$$u = c_1 y + c_2 y^2 + \ldots, \quad \psi = \int_0^y u d\xi = \frac{c_1}{2}y^2 + \frac{c_2}{3}y^3 \ldots,$$

$$u = b_1 \psi^{1/2} + b_2 \psi + b_3 \psi^{3/2} + \ldots \quad (7.37)$$

The last series is obtained analyzing the second expansion, from which follows that near the surface at small values of stream function, we have $\psi \approx y^2$, and hence $y \approx \psi^{1/2}$. Substituting $\psi^{1/2}$ for y in the first series gives the third expansion.

This expansion is not Taylor series since it contains fractional exponents and is singular at $\psi = 0$ resulting in the infinite first derivative $\partial u/\partial \psi \approx \psi^{-1/2}$ and in the infinite higher derivatives as well. The situation may be changed by introducing a new variable $\zeta = \psi^{1/2}$, which transforms the singular expression (7.37) for velocity in a Taylor series and finite derivatives with respect to a new variable $u = b_1\zeta + b_2\zeta^2 + b_3\zeta^3 + \ldots$

Another form of the boundary layer equations is obtained using Görtler variables

$$\Phi = \frac{1}{v}\int_0^x U(\xi)d\xi, \quad \eta = \frac{yU}{v\sqrt{2\Phi}}, \quad \varphi = \frac{\psi}{v\sqrt{2\Phi}}, \quad \theta = \frac{T - T_w}{T_\infty - T_w} \quad (7.38)$$

that after using chain rule transform boundary layer equations into following [338, 1968]

$$\frac{\partial^3 \varphi}{\partial \eta^3} + \varphi\frac{\partial^2 \varphi}{\partial \eta^2} + \beta(\Phi)\left[1 - \left(\frac{\partial \varphi}{\partial \eta}\right)^2\right] = 2\Phi\left[\frac{\partial^2 \varphi}{\partial \Phi \partial \eta}\frac{\partial \varphi}{\partial \eta} - \frac{\partial \varphi}{\partial \Phi}\frac{\partial^2 \varphi}{\partial \eta^2}\right]$$

$$\frac{1}{Pr}\frac{\partial^2\theta}{\partial\eta^2} + \varphi\frac{\partial\theta}{\partial\eta} + \Phi\beta_t(\Phi)\frac{\partial\varphi}{\partial\eta}(1-\theta) = 2\Phi\left[\frac{\partial\varphi}{\partial\eta}\frac{\partial\theta}{\partial\Phi} - \frac{\partial\varphi}{\partial\Phi}\frac{\partial\theta}{\partial\eta^2}\right] \tag{7.39}$$

$$\beta(\Phi) = \frac{2}{U^2}\frac{dU}{dx}\int_0^x U(\xi)d\xi \qquad \beta_t(\Phi) = \frac{4}{U\theta_w}\frac{d\theta_w}{dx}\int_0^x U(\xi)d\xi$$

Equations (7.39) are obtained taken Φ and η as independent variables and ϕ and θ as unknown. Görtler variables may be applied to transform Prandtl-Mises equation (7.36). In this case, variables Φ and ϕ (similar to x and ψ in Prandtl-Mises form) are considered as independent variables and the unknown are the same, $Z = U^2 - u^2$ and the temperature or temperature excess $\theta = T - T_\infty$. Using the chain rule leads to boundary layer equations and boundary conditions in the Prandtl-Mises-Görtler form

$$2\Phi\frac{\partial Z}{\partial\Phi} - \varphi\frac{\partial Z}{\partial\varphi} - \frac{u}{U}\frac{\partial^2 Z}{\partial\varphi^2} = 0, \quad 2\Phi\frac{\partial\theta}{\partial\Phi} - \varphi\frac{\partial\theta}{\partial\varphi} - \frac{1}{Pr}\frac{\partial}{\partial\varphi}\left(\frac{u}{U}\frac{\partial\theta}{\partial\varphi}\right) = 0 \tag{7.40}$$

$$\varphi = 0, \qquad \theta = \theta_w(\Phi) \qquad \varphi\to\infty, \qquad \theta = 0.$$

Properties of equations (7.40) are the same as these of Prandtl-Mises equations (7.36), in particular, the first equation (7.40) is nonlinear and on the surface at $\varphi = 0$ both relations have a singularity that may be removed employing the variable $\varphi^{1/2}$.

7.4.4.3 Theory of Similarity and Dimensionless Numbers

Similarity theory is intended for reducing the number of variables describing the physical processes by combining parameters in the similarity variables or dimensionless numbers. There are areas where the results of similarity theory have crucial importance. In particular, the similarity principle enables investigate natural processes using models of small size at different values of other parameters on condition of equal dimensionless numbers. For example, to keep the same Reynolds number on a small model that the nature object has, one uses in the experiment a larger velocity or a fluid with lesser viscosity. This results in the same dimensionless number because $Re = UL/\nu$, providing in experiment the same conditions as exist in nature

In theoretical studies, similarity ideas permit to use the data gained for the dimensionless number or for similarity variables for each of the parameters built into this dimensionless combination. The similarity theory methods are also useful in presentations showing the research results, because applying dimensionless numbers enlarged the number of variables to which the tabulated data from the tablets or graphs are valid.

The dimensionless numbers are produced employing the technique similar to that we use in deriving the boundary layer equations. We show this procedure considering the heat transfer between body and flow past it. The two-dimensional problem for this case is governed by the system of two-dimensional Laplace equation (1.1) for a body and of the boundary layer equations (1.9)–(1.12) for fluid. To attain the dimensionless numbers that follows from this system, we transform it, converting the variables to dimensionless form by applying their greatest values as the scales. Using scales similar to those applying in Section 7.4.4.1 such as: L for coordinates, U_∞ for velocities, L/U_∞ for time in fluid and L^2/α for a body, and $(T_w - T_\infty)$ for temperatures, we obtain, for example, for velocity $u = \bar{u}U_\infty$, where $\bar{u}$ is a dimensionless variable. Substituting this and analogous relations for other variables with relevant scales in governing equations gives the transformed system. Then, in order to get the dimensionless numbers in traditional form, we divide the first transformed equations by U_∞^2/L, the second one by $U_\infty(T_w - T_\infty)/L$, and the third transformed equations by $\alpha(T_w - T_\infty)/L^2$, and finally obtain the following results

$$\frac{\alpha}{U_\infty L}\frac{\partial u}{\partial t} + u\frac{\partial u}{\partial x} + v\frac{\partial u}{\partial y} - \frac{\partial U}{\partial t} - U\frac{\partial U}{\partial x} + \frac{g\beta L(T_w - T_\infty)}{U_\infty^2}\frac{T - T_\infty}{T_w - T_\infty} - \frac{v}{U_\infty L}\frac{\partial^2 u}{\partial y^2} = 0 \quad (7.41)$$

$$\frac{\alpha}{U_\infty L}\frac{\partial T}{\partial t} + u\frac{\partial T}{\partial x} + v\frac{\partial T}{\partial y} - \frac{\alpha}{U_\infty L}\frac{\partial^2 T}{\partial y^2} - \frac{vU_\infty}{c_p L(T_w - T_\infty)}\left(\frac{\partial u}{\partial y}\right)^2 = 0 \quad (7.42)$$

$$\frac{\partial T_s}{\partial t} - \frac{\partial^2 T_s}{\partial x^2} - \frac{\partial^2 T_s}{\partial y^2} - \frac{q_v L^2}{\lambda_s(T_w - T_\infty)} = 0 \qquad \frac{q_w L}{\lambda(T_w - T_\infty)} = \frac{\lambda_s}{\lambda}\left(\frac{\partial T}{\partial y}\right)_{y=0} \quad (7.43)$$

This system of equations differs from initial one as follows: (i) although variables in both equations look identical (we use the same notations) here, unlike the variables in initial system, they are dimensionless, (ii) there is no continuity equation (1.9) in this system because transformed dimensionless continuity equation remains unchanged, (iii) there also is no pressure because it is eliminated by Bernoulli equation (1.12) resulting in two terms with derivatives of external velocity U instead of pressure, (iv) system (7.41)–(7.43) consist of additional dimensionless groups as factors at the derivatives, and (v) two extra terms: the sixth term in the first equation that takes into account natural heat transfer effects (S. 7.8), and the last equation (7.43), which is obtained from conjugate condition (1.18) and determines the dimensionless heat flux on the body/fluid interface.

The six dimensionless groups of parameters from system (7.41)–(7.43) determine eight dimensionless numbers that are named (except the group with heat source) after pioneers such as Peclet and Fourier (first term in (7.41)), Grashof (sixth term in (7.41)), Reynolds (last term in (7.41)), Eckert (last term in (7.42)), Source number (last term in first (7.43)), Nusselt (the second equation (7.43)), and Prandtl (last and first terms in (7.41))

$$Pe = \frac{U_\infty L}{\alpha}, \quad Fo = \frac{\alpha}{UL} = \frac{\alpha t}{L^2}, \quad Gr = \frac{g\beta L^3(T_w - T_\infty)}{v^2}, \quad Re = \frac{U_\infty L}{v},$$

$$Ec = \frac{U_\infty^2}{c_p(T_w - T_\infty)}, \quad \bar{q}_v\frac{q_v L^2}{\lambda_s(T_w - T_\infty)}, \quad Nu = \frac{q_w L}{\lambda(T_w - T_\infty)}, \quad Pr = \frac{Pe}{Re} = \frac{v}{\alpha} \quad (7.44)$$

In a physical sense, these dimensionless numbers define the ratio of orders of: convection to conduction heats (Peclet), conduction to energy storage heats (Fourier), buoyancy to viscous forces (Grashof), inertia to viscous forces (Reynolds), kinetic energy to enthalpy (Eckert), source to conduction heats (source dimensionless number), convective to conduction heat transfer across the interface (Nusselt), and kinematic viscosity to thermal diffusivity (Prandtl). Seven numbers (7.44) are independent variables, whereas the Nusselt number is an unknown dimensionless heat flux through the interface that should be estimated. Thus, in a problem of convective heat transfer between body and flowing past it fluid, the seven dimensionless numbers (Gr, Re, Pe, Ec, Fo, $\bar{q}_v$, and Nu) substitute thirteen dimensional parameters ($L, U_\infty, T_\infty, T_w, v, \alpha, \lambda, \alpha_s, \lambda_s, g, \beta, c_p, q_v$), which cut almost in half reduces numbers of independent parameters. As it noticed above, this lowers the number of variations that are necessary to study the effects of different parameters on the final results, and simplifies presentation of obtained data.

In practice, usual effects of two parameters are studied. If, for example, the effects of Reynolds and Eckert numbers are investigated, and three different values of each parameter are considered, it is necessary to perform $3 \cdot 3 = 9$ studies (calculations or experiments). On the other hand, if we try to get the same information using dimensional parameters, it would be necessary to perform $3 \cdot 3 \cdot 3 \cdot 3 \cdot 3 = 3^5 = 243$ studies because Re and Ec substitute five dimensional parameters $U_\infty, L, v, c_p,$ and ΔT.

Numbers (7.44) are used basically in heat transfer researches. Many other dimensionless numbers are used in studying different phenomena and processes. Some of them are listed in nomenclature, and a more complete table may be seen on Wikipedia.

7.5   Exact Solutions of Boundary Layer Equations

Although both boundary layer equations are simpler than the Navier-Stokes and full energy equations, their solutions as well encounter considerable difficulties. The dynamic boundary layer equation is difficult to solve due to its nonlinearity, whereas the number of exact solutions of even linear energy equation is also restricted by the same problem since these solutions depend on the velocity components. Therefore, only a few exact solutions of both velocity and temperature boundary layer problems are found. We consider well-known solutions of velocity and thermal boundary layer for the plate and self-similar boundary layer solution. Other exact solutions of boundary layer equations may be found in [338]. Different exact solutions of thermal boundary layer equation in the forms of universal functions we presented in Chapter 1.

7.5.1   Flow and Heat Transfer on Isothermal Semi-infinite Flat Plate

Both indicated problems are governed by steady-state boundary layer system (1.9)–(1.12) at zero pressure gradient and simple boundary conditions

$$y = 0, \quad u = v = 0, \quad T = T_w, \quad y \to \infty, \quad u \to U_\infty, \quad T \to T_\infty \qquad (7.45)$$

Because the plate is semi-infinite, the solution cannot depend on specific length so that the similar variables should be used when the resulting dependences do not depend on location (S. 7.3.3). Analysis leads to following similar variables: $\eta = y/\delta$, stream function $f(\eta) = \psi/\sqrt{vxU_\infty}$, and temperature $\theta(\eta) = (T - T_\infty)/(T_w - T_\infty)$ two of which $f(\eta)$ and $\theta(\eta)$ should be determined. Substituting velocity components obtained via stream function by equations (7.12)

$$u = \frac{\partial \psi}{\partial y} = \frac{\partial \psi}{\partial \eta}\frac{\partial \eta}{\partial y} = U_\infty f', \qquad v = -\frac{\partial \psi}{\partial x} = \sqrt{vxU_\infty} f' \frac{\partial \eta}{\partial x} + \frac{1}{2}\sqrt{\frac{vU_\infty}{x}} f$$

$$= \frac{1}{2}\sqrt{\frac{vU_\infty}{x}}(\eta f' - f) \qquad (7.46)$$

into equations (1.9)–(1.11) results in two ordinary differential equations

$$ff'' + 2f''' = 0, \quad \theta'' + \frac{\Pr}{2}f\theta' = -\Pr \mathrm{Ec} f''^2 \tag{7.47}$$

known as Blasius and Pohlhausen equations who, respectively, solved these problems.

7.5.1.1   Solution of Blasius Equation

The first nonlinear equation (7.47) was solved first by Prandtl's Ph. D. student Blasius in 1908. Since there were no computers at that time, he constructed the solution using two expansions: an inner series at the surface and the outer expansion for external flow matching them at some point inside the boundary layer.[338, 1951] In the following years, several other solutions of this basic equation were published, and finally Howarth gave in 1938, the numerical solution of Blasius equation ([338, 1979, Table 7.1). The gained velocity distribution $u/U_\infty = f'(\eta)$ excellent agrees with experimental data [338].

■ **Example 7.6: Determining the Skin Friction**

The local skin friction is found by Newton law applying the value of the second derivative of the Blasius function on the surface $f''(0) = 0.332$ ([338, 1979] Table 7.1). Integrating the local data gives the total skin friction for the plate of unit width

$$\frac{c_f}{2} = \frac{\tau}{\rho U_\infty^2} = \frac{\nu}{U_\infty}\sqrt{\frac{U_\infty}{\nu x}}f''(0) = \frac{0.332}{\sqrt{\mathrm{Re}_x}}, \quad C_f = \frac{2}{\rho U_\infty^2 L}\int_0^L \tau\, dx = \frac{1.328}{\sqrt{\mathrm{Re}_L}} \tag{7.48}$$

■ **Example 7.7: The Boundary Layer Thickness Estimation**

Asymptotically, the boundary layer thickness is infinite, however, in many cases, it is more reasonable to consider the boundary layer of finite thickness. In such a model as it mentioned above (S 7.3.3), the boundary layer thickness δ is approximately determined as a distance from surface to the point where $u = 0.99\,U_\infty$. The two other thicknesses of boundary layer using in application are displacement δ_1 and momentum δ_2 thicknesses. In contrast to the approximately defined finite thickness δ, these two, known as integral thicknesses, are determined strongly. The first characteristic δ_1 shows how much of the potential flow displaced the established boundary layer, which took it place, whereas the second one δ_2 defines how much momentum this displacement thickness takes away from the flow.

The volume of potential flow displaced at some point of boundary layer is proportional to difference between potential and boundary layer velocities as $(U_\infty - u)dy$. On the other hand, if δ_1 is displacement thickness, this amount should be equal to the product $U_\infty d\delta_1$ that also defines the displaced volume of potential flow. This results in $U_\infty d\delta_1 = (U_\infty - u)dy$ and $d\delta_1/dy = (U_\infty - u)/U_\infty$. Similarly, the local amount of momentum that is taken away from boundary layer $u(U_\infty - u)dy$ is equal to the same amount defined through momentum thickness δ_2, which gives $U_\infty d\delta_2 = u(U_\infty - u)dy$ and $d\delta_2/dy = u(U_\infty - u)/U_\infty$. Integration and data from the Howarth numerical solution of Blasius equation [338] lead to formulae determining the boundary layer thicknesses

$$\delta_1 = \int_0^\infty \left(1 - \frac{u}{U_\infty}\right) dy = 1.72\sqrt{\frac{vx}{U_\infty}}, \quad \delta_2 = \int_0^\infty \frac{u}{U_\infty}\left(1 - \frac{u}{U_\infty}\right) dy = 0.664\sqrt{\frac{vx}{U_\infty}},$$

$$\delta = 5\sqrt{\frac{vx}{U_\infty}} \tag{7.49}$$

The factor $\sqrt{vx/U_\infty}$ comes from similarity variables $u/U_\infty = f'(\eta)$ during differentiation. The last formula for finite boundary layer thickness obtained using Howarth data as well according which $u/U_\infty = 0.99$ at $\eta = y/\delta = 5$.

7.5.1.2 Solution of Pohlhausen Equation

The second equation (7.47) is a linear inhomogeneous equation of second order governing the problem of heat transfer on flat plate. Solution of this problem obtained by Pohlhausen in 1921 is presented as a sum $\theta = C\theta_1 + (Ec/2)\,\theta_2$ of a general solution θ_1 and particular solution θ_2 of the inhomogeneous equation (7.47) (S. 9.2). The general solution under boundary conditions (7.45) $\eta = 0$, $\theta_1 = 1$, $\eta \to \infty$, $\theta_1 = 0$ covers the cooling process, whereas a particular solution subjected to boundary conditions $\eta = 0$, $\theta_2' = 0$, $\eta \to \infty, \theta_2 = 0$ determines the adiabatic temperature (S. 7.3.5).

Because Pohlhausen equation (7.47), unlike the Blasius one, is linear, both θ_1 and θ_2, solutions may be found using standard technique.

■ Example 7.8: The Thermal Boundary Layer Thickness Estimation

The homogeneous Pohlhausen equation $\theta'' + (Pr/2)f\theta' = 0$ is integrated by separation of variables (S. 9.2). After satisfying the boundary conditions in form (7.45) just indicated, the solution θ_1 defining cooling process is obtained

$$\theta_1 = \int_\eta^\infty \exp\left(- Pr \int \frac{f}{2}d\xi\right) d\xi = \int_\eta^\infty [f''(\xi)]^{Pr}d\xi / \int_0^\infty [f''(\xi)]^{Pr}d\xi \qquad \int(-f/2)d\xi = \ln f'' \tag{7.50}$$

Here, the final expression is attained after using the last equality, which follows from Blasius equation (7.47) and which validity is easy to check by differentiation.

It follows from the first equation (7.50) that the temperature distribution and, hence, the thermal boundary layer thickness depend on Prandtl number. In the case of $Pr = 1$, evaluation of integral (7.50) gives $\theta_1 = 1 - f' = 1 - u/U_\infty$ or $(T - T_w)/(T_\infty - T_w) = u/U_\infty$.

This means that temperature head and velocity distributions are identical, and therefore, both dynamic and thermal boundary layer thicknesses are equal $\delta_t/\delta = 1$ (Reynolds analogy, S. 2.1.2.3). Physical reason of such result is that in the case of $Pr = 1$, the flow kinematic viscosity and thermal diffusivity are equal. In the case of $Pr > 1$, the fluid viscosity is larger, and due to that wall braking effect extends farther into flow. As a result, the dynamic boundary layer thickness becomes larger than the thermal one, $\delta > \delta_t$. Consequently, in the opposite case of $Pr < 1$ when the fluid viscosity is smaller, the dynamis boundary layer is thinner, and a contrary inequality $\delta < \delta_t$ is valid.

■ Example 7.9: The Adiabatic Wall Temperature Estimation

As it mentioned above, an adiabatic wall temperature θ_2 is defined by a particular solution of inhomogeneous full Pohlhausen equation (7.47). Such a linear equation of the second order, which does not contain an independent variable (in this case η) is reduced by substitution of a new variable $z = \theta'_2$ to the first order equation for the new variable

$$z' + (\text{Pr}/2)fz = -\text{Pr}\,\text{Ec}f''^2, \quad \theta_2(\text{Pr}) = 2\,\text{Pr}\int_\eta^\infty [f''(\xi)]^{\text{Pr}}\int_0^\xi [f''(\zeta)]^{2-\text{Pr}}d\zeta\,d\xi,$$

$$\theta_{ad} = \frac{\text{Ec}}{2}\theta_2(\text{Pr}) \tag{7.51}$$

Employing a standard method for solving the first order linear differential equation, one finds expression for θ'_2, which integration under the second boundary condition (7.45) for adiabatic temperature results in solution for θ_2. Dimensionless temperature in the form $\theta_{ad} = (T_{ad} - T_\infty)/(T_w - T_\infty)$ is defined then using an Eckert number (7.51). For $\text{Pr} = 1$, after evaluation of second integral, the integrant may be presented as a full differential $2f'(\xi)f''(\xi)d\xi = d(f'^2)$, and hence, the second integration leads to relation $\theta_2(1) = f'^2|_0^\infty = 1$ (recall that $f' = u/U_\infty$). This result together with Eckert number (7.44) indicates that for $\text{Pr} = 1$, the adiabatic wall temperature is defined as $T_{ad} = T_\infty + U_\infty^2/2c_p$ (compare to (7.28)). It follows from this formula that in the case of $\text{Pr} = 1$, the adiabatic temperature is equal to temperature rise produced by velocity change from U_∞ to zero. The values of $\theta_2(\text{Pr})$ are greater for $\text{Pr} > 1$ and lesser for $\text{Pr} < 1$ than for $\text{Pr} = 1$. Approximate formulae presented these values for moderate and large Prandtl number as follows: $\theta_2(\text{Pr}) = \text{Pr}^{1.2}$ and $1.9\text{Pr}^{1/3}$ for $0.6 < \text{Pr} < 10$ and $\text{Pr} \to \infty$, respectively [338]. Then, the adiabatic wall temperature is estimated as $T_{ad} = T_\infty + \theta_2(\text{Pr})U_\infty^2/2c_p$.

■ Example 7.10: Estimation of the Heat Flux at the Surface

According to similarity analysis (S. 7.4.4.3) the heat flux on a surface is defined by Nusselt number (7.43) via temperature derivative on surface $(\partial T/\partial y)_{y=0}$. To estimate this derivative from full solution of Pohlhausen equation $\theta = C\theta_1 + (\text{Ec}/2)\theta_2$, given as a sum of general θ_1 and particular θ_2 solutions, the constant C should be specified. From boundary conditions for general solution it is known that on the surface $\theta_1 = 1$ (see also Example 7.8)). Using this result, and solving equation $1 = C + (\text{Ec}/2)\theta_2(\text{Pr})$ for C, one substitutes the obtained relation for C in full solution and finds the expressions first for θ, and then for $(\partial T/\partial y)_{y=0}$, and finally for Nusselt number

$$\theta = \left[1 - \frac{\text{Ec}}{2}\theta_2(\text{Pr})\right]\theta_1 + \frac{\text{Ec}}{2}\theta_2(\text{Pr}), \quad \text{Nu}_x = -\left(\frac{\partial\theta_1}{\partial\eta}\right)_{\eta=0}\sqrt{\text{Re}_x}\left[1 - \frac{\text{Ec}}{2}\theta_2(\text{Pr})\right] \tag{7.52}$$

Here, $\sqrt{\text{Re}_x}$ comes from differentiation with respect to $\eta = y/\delta$ and a link $\delta \sim 1/\sqrt{\text{Re}_x}$. A minus at the derivative in (7.52) defines according to Fourier law the positive heat flux. Analysis of equation (7.52) shows that $\text{Nu}_x > 0$ if $\text{Ec}\theta_2(\text{Pr}) < 2$ and $\text{Nu}_x < 0$ when $\text{Ec}\theta_2(\text{Pr}) > 2$. At the same time, it follows from the last equation (7.51) and expression for θ_{ad} from Example 7.8 that $[1 - (\text{Ec}/2)\theta_2(\text{Pr})] = 1 - \theta_{ad} = (T_w - T_{ad})/(T_w - T_\infty)$. Comparing this result with equation (7.52) shows that the Nusselt number is proportional to difference $T_w - T_{ad}$ so that positive or negative Nu_x corresponds to $T_w - T_{ad} > 0$ or $T_w - T_{ad} < 0$, respectively. This means that fluid cools a plate only until the plate temperature is greater than the adiabatic plate temperature $T_w > T_{ad}$, and then at $T_w < T_{ad}$ the plate heats the fluid.

Thus, in the case of isolated plate, the cooling process is reduced comparing to that in a usual case when the plate at given surface temperature T_w is cooled as long as $T_w > T_\infty$. Similar results we obtained in Section 7.3.5 considering the Couette flow. As there we mentioned, the physical reason of the cooling process restriction is that in the case of adiabatic wall, the whole released friction heat is adopted because the fluid cannot cool the isolated wall.

For practical applying relation (7.52), the values of derivative $(\partial\theta_1/\partial\eta)_{\eta=0}$ are needed. As well as for $\theta_2(\mathrm{Pr})$, we present approximate relations for this derivative from [338]: $\theta_1(\mathrm{Pr}) = 0.564\,\mathrm{Pr}^{1/2}$, $0.332\,\mathrm{Pr}^{1/3}$ and $0.339\mathrm{Pr}^{1/3}$ for $\mathrm{Pr} \to 0$, $0.6 < \mathrm{Pr} < 10$ and $\mathrm{Pr} \to \infty$, relatively.

Another useful result becomes clear if one writes equation (7.52) for heat flux and substitutes the expression in brackets using relation for adiabatic wall temperature to get $q_w = -\lambda(\partial\theta_1/\partial\eta)_{\eta=0}\sqrt{U/vx}(T_w - T_{ad})$. This expression differs from common relation for heat flux only by temperature difference $(T_w - T_{ad})$, which substitutes the usual difference $(T_w - T_\infty)$. By employing such substitution of adiabatic temperature head $\theta_{ad} = T_w - T_{ad}$ for usual temperature head $\theta_w = T_w - T_\infty$ in known relations yields the expressions valid for the case with heat dissipation. We discussed this question in more detail, especially considering university functions for the recovery factor (S.1.14, Com. 1.11).

7.5.2 Self-Similar Flows of Dynamic and Thermal Boundary Layers

As indicated in Section 1.6.1, the velocity distribution of a potential flow over the wedge with opening angle $\pi\beta$ has the form of power function $U = Cx^m$ with exponent $m = \beta/(2 - \beta)$. If the temperature head distribution along the wedge surface is also the power function $T_w - T_\infty = C_1 x^{m_1}$, then both boundary layer partial differential equations for velocity and for temperature head can be reduced to ordinary differential equations, which have similarity, called self-similar, solutions. As we discussed in Section 7.3.3, such solutions describe flows in which the velocity or the temperature head profiles at different locations are similar forming one profile in similarity variables. Some problems of that type we consider above as exact solutions of Navier-Stokes equation. The just discussed Blasius and Pohlhausen solutions of dynamic and thermal boundary layers on the plate are particular cases of self-similar problems with $m = m_1 = 0$. Applying similar to employing in the case of plate variables gives equations analogous to equations (7.47)

$$\eta = y\sqrt{(m+1)Cx^{m-1}/2v}, \qquad f = \psi\sqrt{\frac{m+1}{2vCx^{m+1}}}, \qquad \theta = \frac{T - T_w}{T_\infty - T_w} \qquad (7.53)$$

$$f''' + ff'' + \beta(1 - f'^2) = 0, \qquad \theta'' + \mathrm{Pr}f\theta' + \frac{2m_1}{m+1}\mathrm{Pr}f'(1 - \theta) = 0 \qquad (7.54)$$

Comment 7.11 Variable (7.53) are known as variables of Falkner and Skan who first studied self-similar boundary layers. These variables for the case of plate differ from Blasius variable by factor $\sqrt{2}$.

The numerical solution for different β obtained by Hartree (Fig. 9.1) [338] shows that the behavior of accelerated and decelerated flows is essentially different: in flows with exponents from $m = 0$, $\beta = 0$ (plate) to $m = 1$, $\beta = 1$ (stagnation point), which correspond to constant and decreased pressure, the velocity profiles in boundary layer are regular without specific features, whereas in flows with negative exponent $m = -0.091$, $\beta = -0.199$, the pressure increases, and the velocity profiles first exhibits a point of inflexion and then separation occurs (compare to S. 2.3, Exam. 2.14).

Numerical solution of the second equation (7.54) for thermal boundary layer was performed by several authors. Employing Hartree and other authors' numerical results, the friction and heat transfer characteristics are obtained. Here, we present some approximate data for stagnation point flows in the form similar to that given for the flows around the plate in Section 7.5.1.2: for friction coefficient $c_f/2 = 1.233/\sqrt{Re_x}$, and for heat transfer coefficient $Nu_x/\sqrt{Re_x} = 0.791 \, Pr^{1/2}, \, 0.570 \, Pr^{1/3}, \, 0.661 \, Pr^{1/3}$ valid for $Pr \to 0$, $Pr = 1$ and $Pr \to \infty$, respectively. These relations are useful, in particular, for calculating the dynamic and thermal characteristics of flows around the blunt noses of cylindrical bodies.

7.6 Approximate Karman-Pohlhausen Integral Method

The idea of the integral method is based on transforming the two-dimensional boundary layer problem to one-dimensional task by integrating the boundary layer equations in the y-direction. Since the integration in fact is an averaging process, this procedure results in losing the information across the boundary layer. Therefore, the solution of such one-dimensional integral equations requires choosing some functions satisfying the given boundary conditions for describing the profiles in boundary layer. Substituting these functions into integral equation yields the one-dimensional differential equation defining the boundary layer characteristics.

To obtain the integral equations for steady state, the system (1.9)–(1.11) modified by Bernoulli equation (1.12) is presented in the form

$$\frac{\partial(Uv)}{\partial x} + \frac{\partial(Uv)}{\partial y} - u\frac{dU}{dx} = 0, \quad \frac{\partial(T_\infty u)}{\partial x} + \frac{\partial(T_\infty v)}{\partial y} = 0, \quad \frac{\partial u^2}{\partial x} + \frac{\partial(uv)}{\partial y} - U\frac{dU}{dx} - v\frac{\partial^2 u}{\partial y^2} = 0$$

$$(7.55)$$

The two first equations are obtained by multiplying the continuity equation (1.9) by U or T_∞ in the first and second cases, respectively. The third equation is a sum of the continuity equation multiplied by u and the modified (by Bernoulli equation (1.12)) the momentum equation (1.10). Integrating a difference between the first and third relations (7.55) and using obvious conditions $\partial u/\partial y = 0$ at $y \to \infty$ and $v(U-u) = 0$ at $y = 0$ and $y \to \infty$ leads to final form of integral momentum equation in terms of thicknesses (7.49)

$$\frac{\partial}{\partial x}(U^2 \delta_2) + U\frac{\partial U}{\partial x}\delta_1 = \frac{\tau_w}{\rho}, \quad \frac{\partial}{\partial x}(U\theta_w \delta_{2t}) = \frac{q_w}{\rho c_p} \quad \delta_{2t} = \int_0^\infty \frac{u}{U}\frac{T - T_\infty}{T_w - T_\infty}dy \quad (7.56)$$

Analogously, integrating the difference between the second equation (7.55) and energy boundary layer equation (1.11) (without dissipative term) is obtained the thermal integral equation using specific thermal integral thickness δ_{2t} defined by the last equation (7.56).

7.6.1 Approximate Friction and Heat Transfer on a Flat Plate

■ **Example 7.11: Friction on a Flat Plate**

We consider three functions for velocity profiles $u/U = f(\eta)$, where $\eta = y/\delta$,

$$f_1(\eta) = \eta, \quad f_2(\eta) = a_0 + a_1\eta + a_2\eta^2 + a_3\eta^3, \quad f_3(\eta) = a_0 + a_1\eta + a_2\eta^2 + a_3\eta^3 + a_4\eta^4$$

$$(7.57)$$

in a purpose to illustrate the effect of a chosen function on the accuracy of an integral method. Functions (7.57) should satisfy some of or all following boundary conditions

$$\eta = 0, \quad f = 0, \quad \nu \frac{\partial^2 u}{\partial y^2} = \frac{dp}{dx}, \quad \frac{\partial^2 f}{\partial y^2} = 0 \quad \eta = 1 \quad f = 1, \quad \frac{\partial f}{\partial y} = \frac{\partial^2 f}{\partial y^2} = 0 \qquad (7.58)$$

Here, the second condition follows from steady boundary layer equation (1.10) since on the surface $u = v = 0$ and shows that for zero pressure gradient the third condition is true. Other conditions are evident. The first function (7.57) satisfies two first conditions (7.58) at $\eta = 0$ and $\eta = 1$. The coefficients of two other profiles are found satisfying the same two conditions, the third condition at $\eta = 0$ and the condition for first derivative at $\eta = 1$. The condition for the second derivative at $\eta = 1$ is used in addition to find coefficients of the third profile. Solutions of corresponding algebraic equations give the values of the coefficients in the polynomials (7.57): $a_0 = a_2 = 0, a_1 = 3/2, a_3 = -1/2$ for the second profile and $a_0 = a_2 = 0, a_1 = -a_3 = 2, a_4 = 1$ for the third one. Using these data, one finds profiles (7.57) and then obtains the thicknesses δ_1 and δ_2 by formulae (7.49) and shear stress by Newton law. Substitution of relations (7.59) into integral momentum equation (7.56) gives differential equation for δ and finally leads to formulae for $\delta(x)$ and $c_f(x)$

$$\frac{\delta_1}{\delta} = \int_0^1 (1 - f)d\eta, \quad \frac{\delta_2}{\delta} = \int_0^1 f(1 - f)d\eta, \quad \frac{\tau}{\rho} = \mu \left(\frac{\partial u}{\partial y} \right)_{y=0} = va_1 \frac{U_\infty}{\delta} \qquad (7.59)$$

$$\delta \frac{d\delta}{dx} = \frac{\delta a_1 v}{\delta_2 U_\infty}, \quad \delta(x) = \sqrt{\frac{2\delta vx}{\delta_2 U_\infty}}, \quad c_f(x) = \sqrt{\frac{2va_1\delta_2}{\delta x U_\infty}} \qquad (7.60)$$

Here, final results are expressed in terms of δ_2/δ determined by second formula (7.59).

The calculation yields: $\delta\sqrt{U_\infty/vx} = 3.46, 4.64, 5.84,$ and $c_f\sqrt{U_\infty x/v} = 0.577, 0.646,$ 0.577 for three profiles, respectively, whereas the exact solution indicates that yields $\delta\sqrt{U_\infty/vx} = 5$ and $c_f\sqrt{U_\infty x/v} = 0.664.$ Thus, the results for friction with accuracy $\mp 3\%$ are achieved with the second and third profiles (7.57). Less exactness with difference of 7% and 17% is achieved in the thickness estimation. That is because of conventional definition of the boundary layer thickness. In contrast, the results for strongly defined integral thicknesses are almost the same $\delta_1\sqrt{U_\infty/vx} = 1.73, 1.74, 1.75,$ as exact data 1.72.

■ Example 7.12: Heat Transfer From Nonisothermal Plate [241]

The same third polynomial (7.57) $f_3(\eta)$ is used for the temperature profile in the form $(T - T_w)/(T_\infty - T_w) = f_t(\eta_t)$ and $\eta_t = y/\delta_t$. Then, the heat flux on the surface is defined by Fourier formula, and an integral energy equation (7.56) (the second one) gives the similar to (7.59) differential equation for thermal boundary layer thickness $\delta_t(\eta_t)$

$$\frac{q_w}{\rho c_p} = \frac{\alpha a_1 \theta_w}{\delta_t} \quad \delta_t \frac{d}{dx}(\theta_w \delta_{2t}) = \frac{\alpha a_1 \theta_w}{U_\infty} \qquad (7.61)$$

Two cases depending on whether the thermal boundary layer is thinner ($Pr \geq 1$) or thicker ($Pr \rightarrow 0$) than the velocity boundary layer are considered, and the following results for the ratio δ_t/δ of both thicknesses are obtained for $\text{Pr} \geq 1$ and $(\text{Pr} \to 0)$ in [241]

$$\frac{\delta_t}{\delta} = \frac{0.871}{\text{Pr}^{1/3}\theta_w^{1/2}(x)x^{1/4}}\left[\int_0^x \theta_w^{3/2}(x)x^{-1/4}dx\right]^{1/3}, \quad \frac{\delta_t}{\delta} = \frac{0.626}{\text{Pr}^{1/2}\theta_w(x)x^{1/2}}\left[\int_0^x \theta_w^2(x)dx\right]^{1/2}$$
(7.62)

Using these relations and the data for δ obtained with third polynomial $(7.57)\,f_3(\eta)$ in the last example, one calculates the heat flux via first equation (7.61) and then gets other heat transfer characteristics. The accuracy of this approach is estimated comparing the Nusselt numbers for power law temperature obtained by relations (7.62) using formulae

$$\text{Nu}_x = 0.358(2m+1)^{1/3}\,\text{Pr}^{1/3}\text{Re}_x^{1/2}, \quad \text{Nu}_x = 0.548(2m+1)^{1/2}\,\text{Pr}^{1/2}\text{Re}_x^{1/2}$$
(7.63)

with corresponding data for self-similar solutions. For surfaces with $\theta_w = \text{const.}$ $(m = m_1 = 0)$ and $q_w = \text{const.}$ $(m = 0, m_1 = 1/2)$ equations (7.62) give: $\text{Nu}_x = 0.358, 0.548$ and 0.452, 0.775 for $\text{Pr} \geq 1$ and for $\text{Pr} \to 0$, respectively. The largest difference of these values from exact data from [338] is about 12% for $q_w = \text{const.}$ and $\text{Pr} \to 0$.

7.6.2 Flows with Pressure Gradients

The effect of pressure gradient is taken into account applying the parameter Λ based on the second boundary condition (7.58) at $\eta = 0$

$$\Lambda = \frac{\partial^2 f}{\partial \eta^2} = \frac{\delta^2}{v}\frac{dU}{dx}, \quad f = 2\eta - 2\eta^3 + \eta^4 + \frac{\Lambda}{6}(\eta - 3\eta^2 + 3\eta^3 - \eta^4)$$
(7.64)

This parameter, the no-slip condition at $\eta = 0$, and the first conditions (7.58) $f = 1$ at $\eta = 1$ define the four coefficients of polynomial (7.57) giving the velocity profile (7.64). Physically parameter Λ may be interpreted as a ratio of pressure to viscous forces. This becomes clear if the first formula (7.64) is modified to form $\Lambda = (-dp/dx)\delta/v(U/\delta)$ by multiplying and dividing it by U. This expression shows that flows with increasing and decreasing pressure gradient are described by negative and positive Λ, respectively. The zero pressure gradient flow corresponds to $\Lambda = 0$. Separation occurs at $\tau_w \sim (\partial f/\partial \eta)_{\eta=0} = 0$, which according to (7.64) is achieved at $2 + (\Lambda/6) = 0$, i.e. at $\Lambda = -12$.

The friction coefficient is obtained by determining boundary layer thickness from the momentum integral equation (7.56) in the same way that was used for the flat plate (Exam. 7.11). Comparison between approximate and exact results shows that the results are accurate enough for the accelerated flows with negative pressure gradients. Accuracy achieved for the flows with increasing pressure, especially close to separation point, is significantly less. For example, in the case of transverse flow past circular cylinder compared with the numerical data shows that both results are practically the same in the range of the angle $0 - 90°$ where the pressure decreases, whereas after a pressure minimum, the approximate data differ more significantly so that exact angle of separation is $104.5°$ against approximate result $109.5°$. Similar results give the other approximate approaches. For example, an integral method based on self-similar profiles described in the last edition of Schlichting's book [338, 2000] or approach grounded on linearized boundary layer equation that we used considering the gradient analogy in Section 2.3.

7.7 Limiting Cases of Prandtl Number

In both limiting cases, for fluids with small or large Prandtl numbers, the thermal boundary layer equation may be simplified. That is because Prandtl number determining the ratio of boundary layer thicknesses shows that in the first case the thermal thickness is large and in the second one it is small compared to the velocity boundary layer thickness. Due to that in both cases, the velocity distribution across the thermal boundary layer may be considered as known. When the thermal boundary layer is large, a velocity distribution across it may be considered as the same as in external flow since the relatively small velocity layer covers only a miner part of the thermal layer at vicinity of a surface. Thus, in this case, the major part of thermal boundary layer appears to be in external flow. In the other case, the situation is opposite because a small thermal boundary layer lies at the surface covering only minor part of the large velocity boundary layer. Therefore, in that case, the velocity distribution across the thermal boundary layer may be considered as a linear using tangent to velocity profile at the surface. The physical reasons of Prandtl number effects we analyzed determined the thermal boundary layer thickness (Exam. 7.8).

There are at least three situations when the real velocity distribution across thermal boundary layer is close to a linear. The first one is the case of fluids with large Prandtl numbers just discussed. For example, the high Prandtl numbers are typical for non-Newtonian fluids (S. 1.14). The other is the case when an unheated part of a surface precedes the heated zone so that the thermal boundary layer starts to develop inside the velocity boundary layer. This results in a thin thermal boundary layer inside the thicker velocity layer. Such a situation takes place, in particular, in a special problem when the step change in temperature follows the unheated zone. As indicated in Section 1.3.1, the solution of this problem is usually used as a standard influence function in the Duhamel integral. The third case with practically linear velocity distribution is the entrance of a tube or a channel when the thermal boundary layer grows inside the fully developed flow resulting as well in a thin thermal boundary layer inside the thicker velocity layer.

■ Example 7.13: Deriving an Influence Function

The deriving of a general expression for influence function is a complicated problem (S. 1.3.2). For a simplest case of a plate with zero pressure gradient and $Pr \geq 1$, the influence function was found by integral method [123, 201]. Third power polynomial (7.57) $f_2(\eta)$ is applied for describing the velocity and temperature profiles. This leads to first equation (7.60) for δ and to second equation (7.61) modified for the ratio $\varepsilon = \delta_t / \delta$

$$\delta \frac{d\delta}{dx} = \frac{140}{13} \frac{v}{U_\infty}, \quad \delta\varepsilon \frac{d(\delta\varepsilon^2)}{dx} = 10 \frac{\alpha}{U_\infty}, \quad \varepsilon^3 + 4\varepsilon^2 x \frac{d\varepsilon}{dx} = \frac{13}{14\,Pr} \tag{7.65}$$

Defining δ from the first equation (7.64) and substituting the result into the second one yields the third differential equation for the ratio of thicknesses. Assuming $13/14 \approx 1$ and integrating this differential equation gives an expression for ε containing integration constant. Determining this constant knowing that at $x = \xi$ the ratio of thicknesses is zero (S.1.3.1), which finally results in influence function (1.24).

7.8   Natural Convection

Natural convection occurs whenever there are the density differences in gravitational field. The considered above forced convection exists due to external forces such as pressure differences, which drives the flow, or other external action, for example, in a moving body. In contrast to that in natural convection, the buoyancy driving force is produced naturally by the density difference, and that is why this type of heat transfer is called natural or free convection. The density difference arises usually due to temperature gradients and decreases with temperature increasing. The velocities in free convention are small, which result in the much smaller heat transfer rates than that in forced convection. Since in this case, there is no establishing velocity, the Grashof number (7.44) or related to it Rayleigh number Ra = Gr Pr is used instead of Reynolds number.

As well as in general in forced heat transfer, the majority problems of natural convection require solution of Navier-Stokes and full energy equations. A small group of problems associated with natural heat transfer on vertical surfaces such as a plate or a vertical cylinder may be considered in terms of boundary layer theory. Some of those have analytical solutions (Exam. 7.14), whereas the others are usually treated numerically (Exam. 7.15). To understand physical reasons of this difference of two types of problems, note that in free convection the driving force is of gravitational nature so that flow is directed vertically. Such a flow along the vertical surface forms a typical boundary layer structure similar to that in the case of longitudinal forced flow along the flat plate. A completely different structure has a natural convective flow on horizontal plate. In that case, the whole heated surface is covered by small rising flows in the form of plumes without any dominated direction typical to a boundary layer structure.

In solutions of Navier-Stokes and energy equations for natural convection flows, some specific issues should be taken into account in addition to general requirements to similar problems considered above. In particular, since the rates of natural convection are small, the radiation heat transfer is often on the same order so that it should be considered along with convection in such a case (Exam. 7.15). The other phenomenon significant for some free convection flows is the stability conditions (Exam. 7.16).

■ Example 7.14: Free Convection on a Vertical Plate

Solutions of this problem include early attempts of solution and comparison with experimental data are reviewed in [125]. Here, we consider the solution given by Pohlhausen in 1921, which is now regarded as classical. The problem is governed by the steady-state boundary layer system (1.9)–(1.11) without pressure and dissipation terms in the second and the third equations, respectively. A term $g\beta(T - T_\infty)$ is added in the second equation to take into account the buoyancy effects. In this problem as well as in the case of the forced heat transfer on a flat plate, employing similarity variables reduces the partial differential equations to system of two ordinary differential equations [125]

$$\eta = c\frac{y}{x^{1/4}}, \quad \zeta = \left[\frac{\psi}{4vcx^{3/4}}\right]^{1/4}, \quad \zeta''' + 3\zeta\zeta'' - 2\zeta'^2 + \theta = 0, \quad \theta'' + 3\Pr\zeta\theta' = 0, \quad (7.66)$$

where $c = [g(T - T_\infty)/4v^2 T_\infty]^{1/4}$ and $\theta = (T - T_\infty)/(T_w - T_\infty)$. The boundary conditions cover: $\theta = 1$ at $\eta = 0$, no-slip $\zeta = \zeta' = 0$, and asymptotic $\zeta' = \theta = 0$ at $\eta \to \infty$ conditions.

It follows from relations (7.66) that: (i) according to similarity variable η both boundary layer thicknesses are proportional to $x^{1/4}$ (unlike to $x^{1/2}$ in forced convection), (ii) both velocities on the surface and far away from the plate are zero (unlike that in forced convection) and (iii) dynamical and thermal differential equations (7.66) (unlike Blasius and Pohlhausen equations (7.47)) are coupled, so that the temperature and velocity depend on each other, and hence, both equations should be solved concurrently. Another difference between both types of convection follows from similarity analysis, which shows that the heat transfer rate depends on Grashof and Prandtl (or Rayleigh) numbers in the case on free convection instead of Reynolds and Prandtl (or Peclet) numbers for forced convection. Relevant formulae for natural convection are given [125]:

$$\mathrm{Nu}_x = \frac{3A}{4}(\mathrm{Gr}_x\,\mathrm{Pr})^{1/4}, \qquad A = \left[\frac{2\,\mathrm{Pr}}{5\left(1 + 2\mathrm{Pr}^{1/2} + 2\,\mathrm{Pr}\right)}\right]^{1/4}, \qquad \mathrm{Gr}_x = \frac{g\beta x^3(T_w - T_\infty)}{v^2}$$

$$(7.67)$$

where $A = 0.8\mathrm{Pr}^{1/4}$ and $A = 0.670$ for $\mathrm{Pr} \to 0$ and $\mathrm{Pr} \to \infty$, relatively. Expression (7.67) for A is an approximation of tabulated numerical solution of equations (7.66), whereas the limiting data are asymptotic values for $\mathrm{Pr} \to 0$ and $\mathrm{Pr} \to \infty$.

■ Example 7.15: Free Convection with Radiation From Horizontal Fin Array [319]

This complex problem is discussed here without detailed solution presenting only a formulation of such type of problems, the way of solution and some of the basic results. More complete information one may find in the original paper or in review in book [119]. The model consists of two long adjacent vertical fins attached to a base with constant temperature $T_s > T_\infty$. The mathematical model of two-dimensional Navier-Stokes equations in vorticity-stream function form (S. 7.1.2.3 and 7.1.2.4) and full energy equation for the fluid, one-dimensional conduction equation for the fins and conjugate boundary conditions (S. 1.1) governed the problem. The radiation heat transfer fluxes are included in the energy equation as the sources. Since solution of Navier-Stokes and full energy equations requires the closed domains (S. 7.2 and 1.1), the configuration of two fins on the base is considered as a closure with the open top, front, and rear sides regarded as imaginary surfaces. The radiation heat fluxes are calculated as sums of heat exchanges between all six surfaces: two fins, base, and three open imaginary sides. The system of governing equations is solved numerically. Some of basic results are as follows:

- Calculation of average Nusselt numbers for a four-fin array agree with experimental data presented in [318]. Analysis of these results show that the contributions of the fins, base, and end fins to total heat transfer are 36, 13.5, and 50.5%, which agrees with the observation in [318]. The effect of fins spacing on heat fluxes is studied for arrays with different number of fins over a fixed base. As the number of fins increases from 4 to 16 and the value of spacing decreases from 20 to 2.8 mm, the heat fluxes from fin and from base decrease from 149 to 44 W/m^2 and from 379 to 148 W/m^2. Despite increased numbers of fins, the heat transfer rate and effectiveness remain almost the same, but average heat transfer coefficient lessens remrkable om 5.29 to 1.48 W/Km^2.

- The effect of the base temperature indicate that the total heat transfer rate increases as the base temperature grows for any studied values of spacing and heights. The effectiveness increases as well for all heights, but it is found that for small values of spacing, effectiveness decreases as the base temperature grows.
- The results obtained for different fins thicknesses indicate that in the case of low heights and high thermal conductivities, the heat flux from the fin does not depend on thickness. For instance, as the thickness increases from 1.5 to 6 mm, the heat flux changes only from 146.6 to 150.6 W/m^2. The role of thermal conductivity and emissivity of the fin is also studied. It is observed that thermal conductivity decreasing leads to reduction of the fin heat flux, and increasing in emissivity yields growing heat flux due to increasing the radiation component.
- The temperature profiles obtained for two different spacings show that the temperature far away from the fins is lower for higher spacing than that for smaller spacing. At the same time, the velocity profiles indicate that at larger spacings, the greater recirculation results in higher velocities near the wall and lower velocities farther from the wall. The isotherms and streamlines for the same two enclosures indicate that the air temperature is high in the middle of the enclosure with smaller spacing, whereas in the other enclosure with the larger spacing, the heating is confined to the air near the fins and the base. It can be seen from the streamlined distribution that the streamlines travel upward along the fins, where the temperature is high compared to that of the enclosure.

■ Example 7.16: Stability of Fluid Between Two Horizontal Plates

The stability of fluid located between two long parallel heated plats depends on the temperature gradient. If the temperature of the upper plate is higher than that of the lower one, the temperature gradient is directed along the gravitational force. In such a case, the fluid density increases in gravitational force direction so that lighter fluid layers are located above heavier part of fluid. This situation is stable, and heat is transformed in gravitational force direction from upper to lower plate by conduction. In the opposite case, the density decreases in direction of gravitational force, resulting in situation when the buoyancy force rises the lighter fluid layers cooling those, whereas the heavier layers are descended by the gravitational force being warmed. This situation is unstable resulting in circulation pattern. In this case, the Rayleigh-Benard cells are forming if the Rayleigh number $\mathrm{Ra}_L = g\beta(T_u - T_b)L^3/\nu\alpha$ (indices u and b refer to upper and bottom plats, L is the height between plates) exceeds it critical value. This effect was first observed by Benard experimentally in 1900, and was first studied by Rayleigh in 1916.

Comment 7.12 Natural convection has many applications. In meteorology, free convection processes are relevant since the stable and unstable phenomena similar to that between two parallel plates occur in the atmosphere and ocean significantly affecting the weather. In the Earth's mantle convection results due to the temperature difference between the warmer inside and surface. Small heat transfer rate are required for cooling in many engineering systems: electronic devices, thermal pipes, refrigerators, and room radiators.

8

Turbulent Fluid Flow and Heat Transfer

8.1 Transition from Laminar to Turbulent Flow

Laminar flow exists only at relatively small Reynolds numbers. As the Reynolds number increases, the laminar regime of flow transients in turbulent flow. The laminar flow is well organized so that it looks like thin parallel layers (lamina in Greek means plate or layer) of fluid move unmixed along a pipe or a plate. In contrast to that, the mixing process inside a fluid leading to homogeneous disturbed medium is one of the basic characteristics of turbulent flow. The patterns of these two regimes were first observed by Reynolds in the nineteenth century, who put the dye inside the flow to make it visible. He was also the first to understand that there exists a universal dimensionless number (now known as critical Reynolds number) at which the transition occurs. The critical Reynolds numbers experimentally determined for flows in a circular section pipe and past a plate are: $\mathrm{Re}_{cr} = \hat{u}D/v = 2300$ and $\mathrm{Re}_{cr} = Ux/v = 3.5 \cdot 10^5 \div 10^6$, where $\hat{u}$ is an average velocity in a pipe. The value of critical Reynolds number depends on the conditions outside of a pipe or a body and increases as the level of disturbances in the inlet flow decreases. The just indicated critical Reynolds numbers correspond to usually disturbed environment, whereas in the experiments when the disturbance in the inlet flow was reduced, the flow in a pipe remained laminar up to Reynolds number 40000 [338]. At the same time at Reynolds numbers less than 2000, the flow in a pipe remains laminar independent of the level of inlet disturbances because these are dissipated by viscosity in flows with smaller than critical Reynolds numbers.

The transition occurs not at one specific Reynolds number, rather it develops in a transition interval from critical to some greater Reynolds numbers. In this interval, the flow alternately becomes laminar or turbulent being fully laminar at the beginning and fully turbulent at the end of the transition interval. This phenomenon called intermittency is characterized by the intermittency factor γ, which is defined as a fraction of the interval when the flow is turbulent. Thus, the intermittency factor is a function of the point coordinates and of the Reynolds numbers being equal $\gamma = 0$ and $\gamma = 1$ at the beginning and at the end of the transition interval. The experiments indicate that intermittency factor in pipe at fixed Reynolds number increases with

Applications of Mathematical Heat Transfer and Fluid Flow Models in Engineering and Medicine,
First Edition. Abram S. Dorfman.
© 2017 John Wiley & Sons Ltd. Published 2017 by John Wiley & Sons Ltd.

distance from leading edge and in direction from the axis to the wall [338]. In the boundary layer in the transition interval, the intermittence factor also increases toward the wall becoming constant close to surface on a distance of about 20% of boundary layer thickness [145].

As the transition occurs, the flow characteristics change: (i) the local parameters, velocity and pressure, at each point become unstable randomly fluctuating, (ii) a velocity distribution across the cross-section in pipe and in boundary layer becomes more uniform due the mixing process in turbulent flow, (iii) the boundary layer becomes thicker being proportional to $x^{4/5}$ instead of $x^{1/2}$ in laminar flow, (iv) in conformity with this the skin friction on the flat plate increases becoming proportional to the free stream velocity of power 1.8 instead of power 1.5, and similarly, the proportionality of the skin friction to average velocity for the laminar flow in pipe changes to square dependency of average velocity for turbulent flow, (v) analogous to the case of flat plate, changes occur in the laminar flows past thin bodies streamlined without or with late separation (e.g., past aerofoil), (vi) in contrast, in the case of thick bodies (like sphere or cylinder), the resistance decreases after transition because in this case, the laminar flow separates far from the trailing edge creating significant wake behind body, whereas the turbulent flow separates downstream later resulting in smaller wake and reducing energy loss.

Comment 8.1 The basic properties of laminar and turbulent flows are important to know to distinguish these two regimes as well as to find the way to reduce the energy losses. For example, in the case of a thin body such as airfoil, the energy losses are reduced by using special body form (laminar airfoils) providing the separation downstream as far as possible. An opposite technique is employed for reducing energy losses in flow past thick body, providing the transition as early as possible because turbulent flow separates later downstream than laminar flow. Prandtl used the wire on a sphere as a turbulator to get an early transition, showing the remarkable decreasing of the resistance ([338], p. 41).

8.2   Reynolds Averaged Navier-Stokes Equation (RANS)

8.2.1   Some Physical Aspects

The turbulent flow is extremely complicated phenomenon. In that case, the velocity and pressure at each point of flow fluctuate randomly depending on coordinates and time. Despite the range of the scales of those fluctuations is extremely wide, the unsteady three-dimensional Navier-Stokes equation, in principle, describes the real flow patterns. However, it is impossible, at least at present, to get a solution of such a problem due to enormous computer memory and speed required for this procedure. Because of that for a long time, only semi-empirical, based on hypothesis, or statistically grounded approaches are available for studying turbulent flows. Only in the last few decades, the essential new methods of direct numerical simulation of turbulence were developed and used to solve the practically important problems. These methods and current situation in this area are outlined in the Chapters 5 and 6, whereas the classical methods are considered in this chapter as well as other helpful special information in Part III.

The turbulent flow behavior is characterized by interaction between fluctuations of quite different scales of length and time, ranging in length from a largest of the order of the boundary layer thickness to the extremely small sizes. This process of interaction is usually described in terms of turbulent eddies-small swirling flows of size, corresponding to fluctuation scales.

Such eddies provide the energy transport across the boundary layer by cascading manner: the large eddies take the energy from the mean flow transferring it to smaller ones, whereas the last transforms the energy to miner eddies and so on until the smallest eddies are achieved, which energy is dissipated in the heat by viscous effects. This energy transferring process leads to great transfer of momentum and mass and yields huge waste of energy and additional turbulent stresses several order larger than laminar stresses. More detail analysis of a cascading process may be found, for example, in [422].

8.2.2 Reynolds Averaging

To apply averaging following Reynolds, the instantaneous turbulence parameters should be expressed as a sum of a mean $U_i(x)$ and the fluctuating $u_i'(x, t)$ components

$$u_i(x, t) = U_i(x) + u_i'(x, t), \quad U_i(x) = \frac{1}{t_2} \int_{t_1}^{t_1+t_2} u_i(x, t)dt, \quad \overline{u_i'} = \frac{1}{t_2} \int_{t_1}^{t_1+t_2} [u_i(x, t) - U_i(x)]dt = 0$$

(8.1)

These relations are obtained under an assumption that the averaging time is sufficiently large compared to time of scale $t_2 \gg t_1$. This makes sure that the mean velocity $U_i(x)$ is independent of time and hence, according to the first and the last relations (8.1), provides the average fluctuation $\overline{u_i'}$ to be zero. This becomes evident after substitution $U_i(x)$ from the first equation (8.1) into third one. Comparing the time of averaging with the scale of time shows also that the averaging value of time derivative of a turbulence velocity is equal to derivative of the mean flow velocity according to expression

$$\frac{\overline{\partial u_i}}{\partial t} = \frac{1}{t_2} \int_{t_1}^{t_1+t_2} \frac{\partial}{\partial t}(U_i + u_i') \, dt = \frac{U_i(x, t_1 + t_2) - U_i(x, t_1)}{t_2} + \frac{u_i'(x, t_1 + t_2) - u_i'(x, t_1)}{t_2} = \frac{\partial U_i}{\partial t}$$

(8.2)

This relation is true because in the right-hand side: (i) the scale t_1 of mean flow in the first term is large in comparison with averaging time t_2 so that in a limit at $t_2 \to 0$ this term corresponds to derivative $\partial U_i/\partial t$, whereas (ii) the second term of sum (8.2) ceases because the same time of averaging t_2 in the denominator in this term is large comparing to a very small scale of fluctuation.

In contrast to zero result (8.1) for average single fluctuation $\overline{u'}_i$, the average product of two fluctuating values $\overline{u_i u_j}$ differs from the corresponding mean product $U_i U_j$ by average product of fluctuation components $\overline{u_i' u_j'}$

$$\overline{u_i u_j} = \overline{\left(U_i + u_i'\right)\left(U_j + u_j'\right)} = \overline{U_i U_j + U_j u_i' + U_i u_j' + u_i' u_j'} = U_i U_j + \overline{u_i' u_j'}$$

(8.3)

This follows from the central part of this equation because the second and third terms vanish due to the last equation (8.1). That is because U does not depend on time, and therefore, the integral of a product Uu' is equal to the product of U and the last integral (8.1) for $\overline{u'}_i$, which is zero. Two quantities for which product is not zero $\overline{u_i u_j} \neq 0$ are considered as correlated quantities, otherwise these are uncorrelated.

To understand physically that the turbulent fluctuation velocities are correlated, consider a simple example of two adjusted flow layers with longitudinal mean velocities $\bar{u}_1$ and $\bar{u}_2$ [338]. Let the velocity of upper layer is $\bar{u}_1 > \bar{u}_2$. If a particle from lower layer enters the upper layer having transverse velocity v', which we count to be positive, it provides due to inertia a negative change u' of velocity $\bar{u}_1$ because the velocity of coming particle is $\bar{u}_2 < \bar{u}_1$. This results in a negative averaged product $(-\overline{u'v'})$ since v' > 0 and u' < 0. Similarly, the particle with negative transverse velocity v', which enters the lower layer from upper layer, gives a positive rise u' to velocity $\bar{u}_2$ leading again to a negative averaged product $(-\overline{u'v'})$ because in this case v' < 0 and u' > 0.

This simple example illustrates that the turbulent fluctuation velocities are inherently correlated. The considered here principles of turbulent flow averaging were first formulated by Reynolds in 1895.

8.2.3   Reynolds Equations and Reynolds Stresses

For averaging, the Navier-Stokes equations in the Einstein notation (S.7.1.2.2)

$$\frac{\partial u_i}{\partial x_i} = 0, \qquad \rho\left(\frac{\partial u_i}{\partial t} + u_j\frac{\partial u_i}{\partial x_j}\right) = -\frac{\partial p}{\partial x_i} + \mu\frac{\partial^2 u_i}{\partial x_j\partial x_j} \tag{8.4}$$

is convenient to present in the form similar to equation (7.55). To do this, the convection terms of equation (8.4) are modified by adding and subtracting the second term to obtain

$$u_j\frac{\partial u_i}{\partial x_j} = \frac{\partial(u_i u_j)}{\partial x_j} - u_i\frac{\partial u_j}{\partial x_j} = \frac{\partial(u_i u_j)}{\partial x_j}, \qquad \rho\frac{\partial u_i}{\partial t} + \rho\frac{\partial(u_i u_j)}{\partial x_j} = -\frac{\partial p}{\partial x_i} + \mu\frac{\partial^2 u_i}{\partial x_j\partial x_j} \tag{8.5}$$

where finally this term is omitted taken into account that according to first equation (8.4) it equals zero. Replacing the modified convection terms gives the Navier-Stokes equation in the form (8.5). Employing equations (8.1)–(8.3) yields

$$\frac{\partial U_i}{\partial x_i} = 0, \qquad \frac{\partial u_i'}{\partial x_i} = 0, \qquad \rho\frac{\partial U_i}{\partial t} + \rho\frac{\partial}{\partial x_j}(U_i U_j + \overline{u_i' u_j'}) = -\frac{\partial P}{\partial x_i} + \mu\frac{\partial^2 U_i}{\partial x_j\partial x_j} \tag{8.6}$$

First equation (8.6) is found from continuity equation (8.4) since averaging fluctuation components are zero, whereas the second equation (8.6) for fluctuation velocities (not averaging) is a result of subtracting the first equation (8.6) from continuity equation (8.4). A third equation (8.6) is the last equation (8.5) with convection term changed by relation (8.3). Returning to initial equations form (8.4) gives the Reynolds averaging equations

$$\frac{\partial U_i}{\partial x_i} = 0, \qquad \rho\frac{\partial U_i}{\partial t} + \rho U_j\frac{\partial U_i}{\partial x_j} = -\frac{\partial P}{\partial x_i} + \frac{\partial}{\partial x_j}\left(\mu\frac{\partial U_i}{\partial x_j} - \overline{\rho u_i' u_j'}\right) \tag{8.7}$$

It is seen that the Reynolds averaging equations (8.7) are Navier-Stokes equations in which the instantaneous parameters are replaced by mean values, but the additional terms $(-\overline{\rho u_i' u_j'})$ known as Reynolds stresses fundamentally change the situation. The Reynolds stresses form a symmetric $(\tau_{ij} = \tau_{ji})$ tensor $\tau_{ij} = -\overline{\rho u_i' u_j'}$ defined by six unknown independent components. This means that the averaging process creates six extra unknown quantities without producing additional equations, which results in unclosed system of four equations containing ten instead of four unknown. To close the system, the deficient equations or other means defining the unknown Reynolds stresses must be created. In classical turbulent theory, the closure problem is solved by semi-empirical or statistical models.

According to current terminology, the turbulence models are classified, depending on a number of differential equations used in addition to continuity and Navier-Stokes equations. In conformity with that, the semi-empirical models based on algebraic (without differential) equations are zero-equation models. The models grounded on one or two differential equations are called as one- or two-equations models.

Comment 8.2 Attempts to create additional equations for estimating Reynolds stresses by employing, for example, method of moments (S. 9.6) failed. The new equations are produced in moment method by multiplying an existing equation by some parameter in first, second, third, and higher powers. Such procedure results in moment equations of the first, and higher orders. However, even the first order moment equation obtained by averaging Navier-Stokes equation contains due to its nonlinearity 22 new unknown [422].

8.3   Algebraic Models

Boussinesq was the first who in 1877 tried to estimate the turbulent stresses. Proceeding from analogy, he proposed a relation $\mu_{tb}(\partial U/\partial y)$ similar to that for shear stress in laminar flow $\mu(\partial u/\partial y)$. These two very similarly looking expressions are in fact different in essence. First of all, the viscosity coefficient μ is physical property of fluid and usually is known, whereas the eddy viscosity coefficient μ_{tb} is a flow characteristic, depending practically on the same parameters as the flow itself. In particular, in contrast to viscosity coefficient, the eddy viscosity should depend on velocity characteristics because the viscous forces in laminar flow are proportional to velocity, whereas in turbulent flow these depend on square of mean velocity. The other essential difference between both flows is the mechanism of transport processes responsible for stresses generation. Whereas in a laminar flows, these transport processes are of the molecular nature, in turbulent, they are produced by eddies motion, which is of macroscopic type because even smaller eddies are many orders of magnitude larger than molecules. So, it is clear that an analogy between the Boussnesq and laminar shear stress formulae does not help for calculation of turbulent flow characteristics until the relations between eddy viscosity coefficient and flow parameters are established.

Today, the problem of turbulent flow prediction is still far from its complete solution. Nevertheless, the turbulence models enable us to solve more-or-less accurate the turbulent flow problems. Our prediction methods and understanding of turbulence nature improves remarkable due to modern experimental technique and the recently developed direct numerical simulation approaches considered in Chapters 5 and 6.

8.3.1   Prandtl's Mixing-Length Hypothesis

In 1925, Prandtl gave the first method for estimating the eddy-viscosity coefficient, introducing the mixing-length hypothesis. He considered a simple model assuming that in turbulent flow particles coalesce forming lumps, which moves remaining value of their momentum some distance l that he named mixing length.

The basic Prandtl's idea may be explained considering simple model of three parallel layers flow in x-direction. If the particles at a middle layer with coordinate y have mean velocity $U(y)$, then, particles at the layers above and below with coordinates $y + l$ and $y - l$ have the velocities $U(y + l)$ and $U(y - l)$, where l is the mixing length. These particles arriving into the middle layer due to transverse fluctuation $v' < 0$ and $v' > 0$ from above and below, respectively, change the velocity $U(y)$ at the middle layer by $\Delta U_a = U(y + l) - U(y)$ and by $\Delta U_b = U(y) - U(y - l)$. Expansion of the functions $U(y + l)$ and $U(y - l)$ in Taylor series in the vicinity of $U(y)$ and taking into account only first two terms $U(y) \pm l(dU/dy)$ shows that both changes equal approximately each other $\Delta U_a = \Delta U_b = l(dU/dy)$. The quantities ΔU might be considered as the turbulent fluctuations since they estimate the small changes of the mean velocity $U(y)$. Averaging the absolute values of these fluctuations arriving to the middle layer from adjacent two layers give the estimation of absolute value $\overline{|u'|}$ of x- component fluctuation in the layer with velocity $U(y)$. Then, employing another Prandtl's assumption that both fluctuation components are proportional to each other yields basic relations of mixing-length model

$$\overline{|u'|} = \frac{|\Delta U_a| + |\Delta U_b|}{2} = l\left|\frac{dU}{dy}\right|, \quad \overline{|v'|} = const. \cdot l\left|\frac{dU}{dy}\right|, \quad \tau_{tb} = \mu_{tb}\frac{dU}{dy}, \quad \mu_{tb} = \rho l^2 \left|\frac{dU}{dy}\right|$$

(8.8)

The last expression is obtained comparing a Boussinesq relation (third one) with relation for turbulent stresses gained via the first two relations (8.8) $\tau_{tb} = -\rho u'v' = \rho|l(dU/dy)|^2$.

It follows from the preceding discussion that the mixing-length l is a distance that a particle lump travels in transverse direction before this lump affects the mean velocities of adjacent layers of flow. Mixing-length concept is similar to the free path notion in the kinetic theory of gases in which the molecules motion is replaced by flow of macroscopic particles lumps. Prandtl postulated that close to the surface mixing-length is proportional to a distance from the surface $l = \kappa y$, where $\kappa = 0.41$ is the Karman constant.

The Reynolds stresses determined by Boussinesq formula (8.8) through mixing-length vanish at the points with $dU/dy = 0$, which are points of maximum or minimum of velocity. This is against the fact that the turbulent mixing exists all over the turbulent flow including as well these points. To fix this shortage, Prandtl suggested another relation $\mu_{tb} = \kappa_1 \rho \delta(x)(U_{max} - U_{min})$ according to which the eddy viscosity is proportional to a maximum velocity difference and width of mixing zone $\delta(x)$ via empirical constant κ_1. Time showed that this relation yields fair results basically for free turbulent flows.

Comment 8.3 Prandtl himself counted his mixing-length model as a first approximation. Later, detail analysis based on comparison with molecular transport process indicated some theoretical shortages of mixing-length model [422]. Nevertheless, for many years, until the computer came to use, Prandtl's model was widely used for studying turbulent flows inside the pipes, around bodies, and in free streams showing satisfactory agreement with experimental data in a number of cases.

8.3.2 Modern Structure of Velocity Profile in Turbulent Boundary Layer

Experimental and theoretical results presented, in particular in [76, 79, 327, 392 and 422], lead to detailed analysis of special type of turbulent flows known as equilibrium boundary layer flows. Despite strictly speaking turbulence is never in equilibrium, this term becomes common defining in fact the turbulent flows with small changes characteristics [422]. As indicated in Section 1.7, such a type of boundary layers are characterized by constant dimensionless pressure gradient parameter (1.48) analogous to the parameter β for self-similar laminar boundary layers (S. 7.5.2). The equilibrium turbulent flows as well as self-similar laminar flows possess of the similarity property (S.7.4.4.3), significantly simplifying the studying the turbulence processes.

Modern models adopted the equilibrium velocity profile and corresponding eddy viscosity distribution across the boundary layer as a basis of the models. Such typical velocity profile consists of three parts: (i) the viscous sublayer (inner part), the relatively small laminar region near the surface, where the law of the wall holds, (ii) the defect layer (outer part), the major region of the boundary layer with dominant turbulent Reynolds stresses and Clauser's law located between the viscous sublayer and the free stream region, and (iii) the overlap region of inner and outer layers covered by the log layer where both laws are asymptotically valid. Thus, the equilibrium turbulent velocity profile is basically determined by the wall and defect laws

$$U^+ = f_1(y^+), \qquad \frac{U_e - U}{u_\tau} = f_2\left(\frac{y}{\delta_\tau}\right), \qquad \delta_\tau = \frac{U_e \delta_1}{u_\tau}, \qquad \mu_{tb} = \kappa_2 \rho U_e \delta_1 \qquad (8.9)$$

where $U^+ = U/u_\tau$, $y^+ = y u_\tau / \nu$, $u_\tau = \sqrt{\tau_w / \rho}$, δ_1 is the displacement thickness defined by equation (7.48), U_e is the velocity on the outer edge in turbulent flow (in this case U is used for mean flow velocity), u_τ is the friction velocity so called because it is defined via shear stress on the surface τ_w, $f_1(y^+)$ and $f_2(y/\delta_\tau)$ are universal functions found from a number of experimental data, (e.g.., Fig. 3.7 in [442]) with a linear part $U^+ = y^+$ of $f_1(y^+)$ at the surface.

The velocity log layer profile is found by considering log layer region as limiting case of the inner and outer layers. Two assumptions are employed as properties of the log layer structure: (i) the inertia terms are small compared to viscous terms, and (ii) the turbulent stresses are dominant. While the first affirmation is based on the fact that inner layer lies sufficiently close to the surface, where the velocities are small, the second assumption is true because the outer layer is located enough far from the body, where the laminar stresses are negligible. It may be shown that flow with such properties has the logarithmic velocity profile. This follows from Reynolds equation (8.7), which shows that in the case of a steady-state at zero pressure gradient and negligible convective terms (see (i)), the derivative of a sum of laminar and turbulent stresses (last term in (8.7)) is zero.

$$\frac{\partial}{\partial y}\left[(\mu + \mu_{tb})\frac{\partial U}{\partial y}\right] = 0, \qquad \mu\frac{\partial U}{\partial y} + \mu_{tb}\frac{\partial U}{\partial y} = const., \qquad \mu\frac{\partial U}{\partial y} + \rho\left(\kappa y \frac{\partial U}{\partial y}\right)^2 = const.$$

$$(8.10)$$

Integrating this relation after using Prandtl's formulae (8.8) for turbulent stress and for mixing length $l = \kappa y$ leads to the second and to the last expressions (8.10). At the surface ($y = 0$), the second term in the last equation vanishes telling us that a constant in this equation is equal to stress $\tau_w = \mu(\partial U/\partial y)_w$ on the surface. On the other hand, from the assuming (ii) it follows that far from surface the first term in this relation (laminar stress) may be omitted. Taken this into account and the just gained fact that a constant in the last relation (8.10) is τ_w, we get an equation, which integration leads to logarithmic profile.

$$\rho\left(\kappa y \frac{\partial U}{\partial y}\right)^2 \approx \tau_w \qquad \kappa y \frac{\partial U}{\partial y} \approx \sqrt{\frac{\tau_w}{\rho}} = u_\tau, \qquad U = \frac{u_\tau}{\kappa}\ln y + C, \qquad U^+ = 2.5 \ln y^+ + 5$$

$$(8.11)$$

In the last equation the variables (8.9) y^+, U^+ are used. According to experimental data, a log layer is located between $y^+ = 30$ and $y = 0.1\delta$, so that the sublayer and defect layer are disposed from $y^+ = 0$ to $y^+ = 30$ and from $y = 0.1\delta$ to $y = \delta$, respectively.

Wilcox showed that the features discussed above of equilibrium flows found on empirical basis may be obtained theoretically applying the perturbation approach [422].

In what follows, we consider three algebraic models based on modern velocity profiles: one of the earliest Mellor-Gibson model that the author used in applications (S. 1.7, 2.1.2.3, 2.1.2.4, 2.3) as well as the Cebecy-Smith and Baldwin-Lomax models published about ten years after the Mellor-Gibson model and taking into account some additional effects on eddy viscosity.

8.3.3   Mellor-Gibson Model [260, 261]

In this model, the eddy-viscosity function is constructed using: the Laufer [216] measured data for the inner part, the Clauser law (8.9) for the outer part and the Prandtl mixing-length relations (8.8) for the log layer. The Laufer data for inner part is presented as dimensionless total viscosity $\tilde{v}_\Sigma = v_\Sigma/v = 1 + v_{tb}/v$

$$\left.\begin{array}{lll} 0 < r < 2, & \omega = r^2 - r, & \tilde{v}_\Sigma(\omega) = 1 \\[4pt] 2 < r < 4.5, & \omega = 2.75r - 1.5, & \tilde{v}_\Sigma(\omega) = \dfrac{r^2}{\omega} \\[4pt] 4.5 < r < 11, & \omega = 11, & \tilde{v}_\Sigma(\omega) = \dfrac{r^2}{11} \end{array}\right\} \quad v_{tb} = 0.016\,U_e\delta_1, \quad v_{tb} = \kappa^2 y^2 \left|\dfrac{dU}{dy}\right|$$

$$(8.12)$$

where the function $\tilde{v}_\Sigma(\omega)$ is composed using two other functions $\omega = \kappa^2 y^{+2}(dU^+/dy^+)$, $r = \kappa y^+(1 + \beta_v y^+)^{1/2}$, and three pieces for the experimental curve approximation (more details may be found in [111] or [119]). Parameter β_v in function r is connected with parameter (1.48)β as $\beta_v = \beta/\mathrm{Re}_{\delta_1}\sqrt{c_f/2}$. The eddy viscosity in the outer part of boundary layer and in the log layer is presented by second (Clauser) and third (Prandtl) right-hand formulae (8.12).

The velocity profiles are obtained by integrating the steady-state boundary layer equation (8.7) using relations (8.12). For the inner layer this gives: [260]

$$U^+ = u_v^+ + \frac{2}{\kappa}\left[(1 + \beta_v y^+)^{1/2} - 1\right] + \frac{1}{\kappa}\ln\left[\frac{4}{\beta_v}\frac{(1 + \beta_v y^+)^{1/2} - 1}{(1 + \beta_v y^+)^{1/2} + 1}\right],$$

$$u_v^+ = \lim_{\zeta \to 0}\left[\int_\zeta^{y^+} \frac{\sigma(r)}{\kappa^2 y^{+2}}dy^+ - \frac{\ln\zeta}{\kappa}\right], \quad \sigma(r) = \omega - r \qquad (8.13)$$

Far from the wall, at $\omega \geq 11$ function $\sigma(r)$ vanishes, and the first term in equation (8.13) becomes a constant $u_v^+ = D^+(\beta_v)$. Equation (8.13) unlike a common logarithmic profile takes into account the effect of the pressure gradient through parameter β_v. In the case of zero pressure gradient ($\beta_v = 0$) close to the surface when $y^+ \ll 1/\beta_v$, an equation (8.13) has a limit taking usual simple form $U^+ = (1/\kappa)\ln y^+ + D^+$.

Comment 8.4 The notes "far from the wall" in the first sentence after equation (8.13) and "close to the surface" in the last sentence means in fact the same area of the log layer. Here, as well as in the case of analyzing equations (8.10), the log layer overlaps the inner and outer regions, and words "relatively close" and "relatively far" are omitted.

For positive pressure gradient, especially close to the point of separation, friction coefficient $c_f \to 0$, and both parameters $\beta \sim \beta_v \to \infty$, so that relations (8.13) become not applicable. For this case, Mellor [260] introduced new variables suitable for large values of β and obtained the velocity profile for the inner part in the other form

$$u^{++} = u_v^{++} + \frac{2}{\kappa}\left[\left(\beta_v^{-2/3} + y^{++}\right)^{1/2} - \beta_v^{-1/3}\right] + \frac{1}{\kappa\beta_v^{-1/3}} \ln \frac{4}{\beta_v} \left[\frac{\left(\beta_v^{-2/3} + y^{++}\right)^{1/2} - \beta_v^{-1/3}}{\left(\beta_v^{-2/3} + y^{++}\right)^{1/2} + \beta_v^{-1/3}}\right]$$

$$u_v^{++} = \lim_{\zeta \to 0}\left[\int_\zeta^{y^{++}} \frac{\sigma(r)}{\kappa^2 y^{++2}}dy^{++} - \frac{1}{\kappa\beta_v^{1/3}} \ln\frac{\zeta}{\beta_v^{1/3}}\right], \quad r = \kappa\left(\beta_v^{-2/3} + y^{++}\right)^{1/2} y^{++} \quad (8.14)$$

Here, $u^{++} = u/u_p$, $y^{++} = u_p y/v$, $u_p = [(v/\rho)(dp/dx)]^{1/3}$ is the pressure velocity so called because it determines velocity via pressure similar to u_τ defined friction velocity through skin friction. Equation (8.14) as well as (8.13) close to the wall has the limit $(2/\kappa)y^{++1/2} + (1/\kappa\beta_v^{1/3})[\ln(4/\beta_v) - 2] + D^+/\beta_v^{1/3}$ when $1/\beta_v \to 0$, $y^{++} \gg 1/\beta_v$ and $\omega \geq 11$. Coefficients D^+ in this and above present limiting formulae as well as pertinent logarithmic formulae for friction may be found in original papers or in [119].

The outer defect layer profiles are found by integrating boundary layer equations for entire range $-0.5 \leq \beta \leq \infty$ and $\mathrm{Re}_{\delta_1} = U\delta_1/v = 10^3, 10^5, 10^9$. For the not large values of β corresponding to inner profiles (8.13), the calculated defect profiles are tabulated in Clauser variables (8.9) as $f'(\eta)$ and independent variable η

$$\frac{U-u}{u_\tau} = f'(\eta), \quad \eta = \frac{y}{\delta_\tau}, \quad \delta_\tau = \frac{\delta_1 U}{u_\tau}, \quad \frac{U-u}{u_p} = F'(\xi), \quad \xi = \eta\beta^{1/2}, \quad u_p = u_\tau\beta^{1/2}$$

$$(8.15)$$

For the large β pertaining to inner profile (8.14), the outer defect profiles are tabulated in different suitable variables $F'(\xi)$ and ξ. The results show that the data in the range $10^3 < \mathrm{Re}_{\delta_1} < 10^9$ differ by less than 2% from corresponding data for $\mathrm{Re}_{\delta_1} = 10^5$. Due to that, only the last numerical results are tabulated in [261]. The velocity distribution across the entire boundary layer is obtained by matching equations (8.13) or (8.14) for inner and tabulated data from [261] for outer parts. It is shown that inner and outer profiles coincide in overlap area crating the log layer region.

8.3.4   Cebeci-Smith Model [59]

The eddy viscosity is build as well of the inner μ_{in} and outer μ_{ou} parts

$$\mu_{in} = \rho l^2\left[\left(\frac{\partial U}{\partial y}\right)^2 + \left(\frac{\partial V}{\partial x}\right)^2\right]^{1/2}, \quad l = \kappa y\left[1 - \exp\left(-\frac{y^+}{A^+}\right)\right], \quad \mu_{ou} = 0.0168\,\rho U_e \delta_1 F_K$$

$$(8.16)$$

The first two relations are applicable for $y \leq y_m$, and the third equation is used for $y > y_m$, where y_m is a coordinate of the matching point. The value of y_m may be found from equation $\mu_{in} = \mu_{ou}$. Since the matching point lies in log layer, the μ_{in} is found from first equation (8.16). Neglecting the second term in this equation, one obtains as above a logarithmical formula (8.11) $U = (u_\tau/\kappa) \ln y + C$, and after it is differentiated gets a simplified first relation (8.16) $\mu_{in} = \rho\kappa^2 y^2 u_\tau/\kappa y = \rho\kappa u_\tau y$. Equating this result to the last relation (8.16), defining μ_{ou} (taking $F_K = 1$) gives $y^+ \approx 0.04\mathrm{Re}_{\delta_1}$.

The second equation (8.16) is the Van Driest's formula, which takes into account the damping effect near the surface. This effect occurs because in vicinity of the wall, the laminar stress is so weak that the turbulent fluctuations contribution becomes significant. Assuming that at the surface $u' \sim y$, we obtain from continuity equation that $v' \sim y^2$, and hence, near a surface the turbulent stresses $(-u'v')$ are proportional to y^3. Nevertheless, some authors think that the term with y^3 is small so that actually the turbulence decays as y^4 [274]. The same result follows from formulae (8.16) because according to second relation (8.16) a mixing length is proportional to y^2 (the first term of Taylor series of difference in brackets is of order y). Consequently, μ_{in} is of order y^4. The other two functions in relations (8.16) defining A^+ and F_K are

$$A^+ = 26\left[1 + y\frac{dp/dx}{\rho u_\tau^2}\right]^{-1/2}, \quad F_K = \left[1 + 5.5\left(\frac{y}{\delta}\right)^6\right]^{-1} \tag{8.17}$$

The first function differs from the initial constant $A^+ = 26$, given in [403], by expression in brackets, which takes into account the effect of pressure gradient. Another function (8.17) F_K accounts for the intermittence influence. It was experimentally observed that the intermittence phenomenon of alternative laminar and turbulent flows well-known in transition regime in a tube (S. 8.1) takes place also close to the border between boundary layer and external potential flow, and function F_K was introduced in [206] to take into account the eddy viscosity decreasing in outer boundary layer caused by this effect.

8.3.5   Baldwin-Lomax Model [25]

This model is constructed as well using two layers for defining the eddy viscosity

$$\mu_{in} = \rho l^2 |\omega|, \quad |\omega| = \left|\frac{\partial V}{\partial x} - \frac{\partial U}{\partial y}\right|, \quad \mu_{ou} = 0.027 F_{wake}F_K(y, y_{max}/0.3)$$

$$F_{wake} = \min\left[y_{max}F_{max}, y_{max}U_{dif}^2/F_{max}\right], \quad F_{max} = (1/\kappa)[\max(l|\omega|)] \tag{8.18}$$

where the mixing length l is defined by Van Driest's formula (8.16) with $A^+ = 26$, y_{max} is the ordinate at which the product $(l|\omega|)$ achieved the maximum value, $|\omega|$ is the magnitude of vorticity (S.7.1.2.3), $F_K(y, y_{max}/0.3)$ is function (8.17) with $y_{max}/0.3$ replaced for δ in Clauser formula (last equation (8.18)), and U_{dif} is the maximum of U. This model was formulated specially for the cases when the usual boundary layer properties like thicknesses are difficult to estimate, for example, for separated flows. Because of that δ in function μ_{ou} and some quantities in other relations are replaced by unusual characteristics. Nevertheless, the results obtained for separated flows by this as well as by other algebraic models are unsuccessful [422].

8.3.6   Application of the Algebraic Models

8.3.6.1   Flows in Channel and Pipe

In a channel or in a pipe with fully developed flow, the velocity is independent on x, and because of that from continuity equation one gets $\partial V/\partial y = 0$. This gives $V = cont.$ and $V = 0$ since the velocity is zero at a surface. In such a case, the inertia terms vanish simplifying boundary layer equation (8.7) to the form, which integrating leads to solution

$$\frac{dp}{dx} = \frac{1}{r^j} \frac{\partial}{\partial r} \left[r^j \left(\mu \frac{\partial U}{\partial r} + \tau_{tb} \right) \right], \quad (\mu + \mu_{tb}) \frac{\partial U}{\partial y} = \rho u_\tau^2 \left(1 - \frac{y}{R} \right) \tag{8.19}$$

Here, r (or y) and R are transverse coordinate and half of channel height (for $j = 0$) or radius of pipe (for $j = 1$). The solution (8.19) is obtained as follows: (i) both sides of equation (8.19) are multiplied by r^j and then integrated, knowing that in the fully developed flow $dp/dx = const.$; this yields: $r(dp/dx) = (j + 1)[\mu(\partial U/\partial r) + \tau_{tb}]$, (ii) for the surface, this relation becomes $dp/dx = \tau_w(j + 1)/R$ because the expression in brackets is a surface shear stress, and (iii) substituting the last relation in the previous one, using formula Boussinesq for τ_{tb}, and y instead of $r = R - y$ lead to solution (8.19).

The numerical results are attained using the Cebeci-Smith and Baldwin-Lomax models for eddy viscosity estimation. Since U_e and δ_1 in the first as well as U_{dif} and y_{max} in the second models are unknown in advance, the relaxation iterative method (Com. 4.12) is employed. The agreement with direct numerical simulation (S. 5.6) and experimental results is gained with difference (1–8) % for velocity profiles and skin friction (Fig. 3.11 and 3.12 in [422]).

Comment 8.5 This problem in fact is the same as the Hagen-Poiseuille laminar flow in a tube (S. 7.3.2). Comparison shows how much complicated is the turbulent case even for such relatively simple problem. Whereas the Hagen-Poiseuille problem has an exact solution, the analogous turbulent task is solved numerically using at least one empirical constant. Because of that, along with semi-empirical turbulence models, in practical calculations widely are employed empirical correlations. For example, the well-known simple relation for velocity profiles in a tube $u/U = (y/R)^{1/n}$, where U is the maximum velocity and n is a function of $Re = \bar{u}D/\nu$, or formula $c_f^{-1/2} = 4\log(2Re\sqrt{c_f}) - 1.6$ for friction coefficient. Such correlations for different cases usually are given in engineering fluid flow courses, some examples with references may be found also in [338].

8.3.6.2   The Boundary Layer Flows

We consider examples of prediction of boundary layer characteristics by Cebeci-Smith and Baldwin-Lomax models. Comparison of computed and measured results including data from AFOSR (Com. 8.6) shows that predictions of both models are accurate: (i) satisfactory for flows with zero and favorable (negative) pressure gradients, (ii) reasonable for flows with mild adverse (positive) pressure gradients, and (iii) unsatisfactory for flows with strongly adverse pressure gradients. Whereas the differences from measured data for velocity profiles are small in the first two cases, these for integral characteristics, such as friction coefficient and shape factor $H = \delta_2/\delta_1$, are usually of 8-10%. In the third case, the results for c_f are significant higher than the measured data. A special case is the separated flows. Because in this case, a bubble forms between the flow and surface, and there is no longer equilibrium, the algebraic models could not describe properly the flow separation.

Comment 8.6 The AFOSR (Air Force Office of Scientific Research) Conference on Computational of Turbulent Boundary Layer was held in Stanford University in 1968. There were 75 invited famous scholars, and their presented results became standard examples for checking the accuracy of turbulent boundary layer researches.

8.3.6.3 Heat Transfer From an Isothermal and Nonisothermal Surface

The investigation of heat transfer in turbulent boundary layer obtained using Mellor-Gibson model were performed in the 1970s and presented in articles [102, 106], shortly after this turbulence model was published. We discussed the basic parts of these results for arbitrary nonisothermal surfaces in the form of universal functions in Chapter 1 and for isothermal surfaces including comparison with experimental data in Chapter 2. Here, we present some details that may be useful for interested reader.

The thermal boundary layer equation and boundary conditions for the turbulent flow as well as for laminar flow are used in Prantl-Mizes-Görtler's form (7.40)

$$2\Phi\frac{\partial\theta}{\partial\Phi} - \varphi\frac{\partial\theta}{\partial\varphi} - \frac{1}{\Pr}\frac{\partial}{\partial\varphi}\left(\frac{u}{U}\varepsilon_a\frac{\partial\theta}{\partial\varphi}\right) = 0, \quad \varepsilon_a = \frac{\alpha_e}{\nu\Phi} = \frac{1}{\Phi}\left(\frac{1}{\Pr} + \frac{\varepsilon-1}{\Pr_{tb}}\right) \tag{8.20}$$

where $\varphi = \psi/\nu\Phi$ slightly differs from similar variable (7.38) for the case of laminar boundary laer, $\varepsilon = \nu_{tb}/\nu$ (S. 8.3.3) and $\alpha_e = \alpha + \alpha_{tb}$ (S. 1.7). Solution of this equation in the form of universal function (1.39) after substitution into equation (8.20) yields ordinary differential equations, which were solved numerically defining coefficients g_k as well as for laminar and other universal functions considered in Chapter 1. Then, the exponents C for the integral form of universal function (1.40) are calculated using known coefficients g_k as explained in Section 1.6. More detailed description of this procedure may be found in original papers or in [119].

8.3.7 The 1/2 Equation Model

This model proposed by Johonson and King [185] takes an intermediate position between the algebraic and energy equations models. In fact, it is an algebraic model, like the Cebeci-Smith or the Baldwin-Lomax model, which applies in addition one ordinary differential equation. That is maybe the reason way this model is called "1/2 equation" in contrast to one- or two-equations models based on the partial differential equations. The eddy viscosity is defined similar to other models using the inner and outer layers

$$\mu_{in} = \rho\left[1 - \exp\left(-\frac{u_D y}{\nu A^+}\right)\right]^2 \kappa u_s y, \quad \mu_{ou} = 0.0168\rho U_e \delta_1 F_K(y, \delta)\sigma(x) \tag{8.21}$$

but the relation for eddy viscosity contains both components $\mu_{tb} = \mu_{ou}\tanh(\mu_{in}/\mu_{ou})$. Here, $u_D = \max(u_m, u_\tau)$ and $u_m = \sqrt{\tau_m/\rho_m}$ are velocity scales, index m denotes the coordinate y_m at which the Reynolds stress achieves the maximum value τ_m, the scale u_s is given through several quantities: $u_\tau, u_m, y_m/\delta, \rho_w, \rho_m, L_m$, where ρ_w is a density on the surface,

$L_m = \kappa y_m$ if $y_m/\delta \le 0.09/\kappa$, and $L_m = 0.09\delta$ if $y_m/\delta > 0.09/\kappa$. The additional scales are also employed to improve the prediction of separated flows.

The last relation (8.21) is the equation (8.16) from the Cebeci-Smith model with an additional function $\sigma(x)$ that provides departure from equilibrium, which corresponds to $\sigma(x) = 1$. The maximum Reynolds stress is defined by ordinary differential equation

$$U_m \frac{d}{dx}\left(\frac{\tau_m}{\rho_m}\right) = 0.25\frac{(u_m)_{eq} - u_m}{L_m}\left(\frac{\tau_m}{\rho_m}\right) - c\frac{(\tau_m/\rho_m)^{3/2}}{0.7\delta - y_m}\left[1 - \sqrt{\sigma(x)}\right] \qquad (8.22)$$

where $\tau_m = (\mu_{tb})_m(\partial U/\partial y + \partial V/\partial x)_m$, $c = 0.5$ if $\sigma(x) \ge 1$, otherwise $c = 0$, and $(u_m)_{eq}$ corresponds to the equilibrium case with $\sigma(x) = 1$. The function $\sigma(x)$ is found iteratively so that a maximum Reynolds stress τ_m obtained from equation (8.22) will be finally equal to the maximum stress defined by relation given just after equation (8.22). Satisfying of this equality ensures that eddy viscosity μ_{tb} and the maximum Reynolds stress τ_m gained separately are adjusted to each other. Examples show that this more complicated model predicts better the attached and separated flows than simple algebraic models.

8.3.8   Applicability of the Algebraic Models

The algebraic models are the simples and easies in performance among turbulence models. Therefore, it is reasonable to start with such relatively simple approach and then use more complicated models if alternatives are significantly better in accuracy or in some specific features. For the flows with zero, favorable, and not very strong adverse pressure gradients, algebraic modal predictions are usually satisfactory. At the same time, the algebraic models are incomplete, since the mixing length in those models is not specified. Because of that, the algebraic models consist of free constants, which are defined by meeting experimental data leading to significantly restricted range of models applicability. Nevertheless, algebraic models are used in solving turbulence problems including complex flows without separation. For example, the results for isothermal (S. 2.1.2.3) and nonisothermal (S. 1.7) heat transfer characteristics obtained using Mellor-Gibson model well fit the experimental correlation for different velocity and temperature distribution for wide range of Prandtl and Reynolds numbers (S. 2.1.2.2). Other satisfactory applications of algebraic models may be seen, for example, in [422]. At the same time, the results providing by the 1/2 equation model better correlate with measured data of separation flows than algebraic models predictions.

8.4   One-Equation and Two-Equations Models

Since the 1960s, after the advent of computers, the one- and two-equations models become a basic powerful tool of turbulence flows investigation. These types of models, which are grounded on kinetic energy equation and other differential equations, simulate the actual physical patterns of turbulence much closer than the algebraic models. Whereas the one-equation models are still incomplete because they adapt the length scale from some typical flows, the two-equations models are complete, determining the turbulence length scale or some equivalent parameter by special differential equation. The improved turbulence models are more complicated and required additional closure coefficients. However, using such models is worth because modern models enhance the computation results and take into account nonlocal and flow history effects that help to develop our insight into turbulence nature.

8.4.1 Turbulence Kinetic Energy Equation

Prandtl defined the kinetic energy per unit mass as a sum of square of fluctuation velocity components. Then, using the dimensional analysis of units, he defined eddy viscosity and Reynolds stress and in terms of turbulence energy and density as

$$k = (1/2)\left(\overline{u'^2} + \overline{v'^2} + \overline{w^2}\right), \quad \mu_{tb} = const.\rho k^{1/2}l, \quad \tau_{ii} = -\rho\overline{u_i'u_i'} = -2\rho k \qquad (8.23)$$

The differential equation determining the turbulence energy, which is called the transport equation, is similar to average Reynolds equation (8.7) but consist of second and third specific terms in the right-hand side instead of pressure gradient

$$\rho\frac{\partial k}{\partial t} + \rho U_j\frac{\partial k}{\partial x_j} = \tau_{ij}\frac{\partial U_i}{\partial x_j} - \rho\varepsilon + \frac{\partial}{\partial x_j}\left[(\mu + \sigma_k\mu_{tb})\frac{\partial k}{\partial x_j}\right], \quad \varepsilon = C_D\frac{k^{3/2}}{l} \qquad (8.24)$$

Here, σ_k and C_D are closure coefficients. The terms in equation (8.24) physically signify the following. The two terms at the left-hand side define the substantial derivative (Exam. 7.2) of the turbulence energy. The first and the second terms at the right-hand side called production and dissipation give the rate of energy that turbulence takes from mean flow and the rate of the loss of turbulence energy transforming into thermal energy. The last term in equation (8.24) consists of two parts. The first one, which is similar to terms in other analogous equations, takes into account the molecular diffusion, whereas the second part of this term, also looking similar to analogous terms, here represents two complex processes. The turbulent transport provided by turbulent fluctuations, and the pressure diffusion occurring due to correlation of the pressure and velocity fluctuations. The two terms on the left-hand side and the molecular diffusion term are exact items, whereas the others in equation (8.24) are approximated via turbulence energy k using postulated by Prandtl relations: for the Reynolds stress $\tau_{ij} = -(2/3)\rho k\delta_{ij}$ and the last equation (8.24) for the dissipation term ε, respectively. The second part of the last term of equation (8.24), which represents the turbulent transport and the pressure diffusion is defined though Boussinesq approximation as a product of eddy viscosity μ_{tb} and gradient of turbulent energy k. Thus, three approximate relations and two closure coefficients σ_k and C_D are needed to close the equation for turbulence energy.

8.4.2 One-Equation Models

Two equations (8.24) for turbulence kinetic energy and for dissipation with equation (8.23) for Reynolds stresses in addition form the basic set of equations for one-equation model. To complete the model, the length scale should be specified. Prandtl used in his model the mixing length for this purpose. It can be seen applying equations (8.23) and (8.24) together with two last relations (8.8) that this assumption leads to proportionality between production $\tau(\partial U/\partial y) = \mu_{tb}(\partial U/\partial y)^2$ and dissipation $\rho\varepsilon$ given through relation (8.24). To see this note that according to (8.8) and (8.23) we have $\tau = \rho l^2(\partial U/\partial y)^2 = 2\rho k$, and hence, $(\partial U/\partial y)^2 = 2k/l^2$ giving for production $\tau(\partial U/\partial y) = \mu_{tb}(\partial U/\partial y)^2 = const.\rho k^{1/2}l \cdot 2k/l^2 \sim \rho k^{3/2}/l$ an expression proportional to dissipation defined by (8.24). Due to that in this case, the constant in formula (8.23) for eddy viscosity may be taken unity leading to formula $\mu_{tb} = \rho k^{1/2} l$, and both mixing lengths in obtained independently equations (8.8) and (8.24) are proportional. The other one-equation models using Prandtl's turbulence energy equation (8.24) differ from initial Prandtl model basically by closure coefficients and length scale functions [422].

A different type of one-equation model was introduced by Bradshaw, Ferriss, and Atwell [48]. They employed the same equation (8.24), but instead of using the Boussinesq approximation in the last term, they used an experimental result, which according to the Reynolds stress in the boundary layer is approximately proportional to turbulence kinetic energy with constant factor resulting in $\tau_{tb} \approx 0.3k$. This modification changes the type of partial differential equation (8.24) from parabolic to hyperbolic type (S. 1.1 and Com. 8.7) and is remarkable at simplifying the solution of this equation. This model shows well predictions, in particular, for the adverse pressure gradients and provides the best results among others models tested at AFOSR 1968 (Com. 8.6).

Comment 8.7 Hyperbolic equation has real characteristics: two lines crossing at each point of solution domain, which are used to build a method of solution. See also Com. 6.1.

There are also one-equation models based on equations other than turbulence energy equation (8.24). Several models used equations similar to (8.24) but are written for kinematics eddy viscosity $v_{tb} = \mu_{tb}/\rho$. Models using other than (8.24) energy equations contain more closure coefficients (up to eight) and additional empirical function including definition for the length scale. Two models of this type, Baldwin-Barth and Spalart-Allmaras are analyzed in [422]. Comparing obtained computational results with measurements shows that the predictions of the Spalart-Allmaras model are satisfactory for special problems, such as airfoil and wings, for which this model was calibrated. The results obtained by Baldwin-Barth model differ more from the measured data even than simpler algebraic models predictions. In general, the one-equation models predictions are close to those of the algebraic models. If one-equation model is specified giving satisfactory results for some separated flows than it could not predict the other flows, like wake or mixing layers, even with accuracy of algebraic models.

8.4.3   Two-Equation Models

The modern two-equation models are currently the basic tools for solving complex engineering and scientific problems. The popularity of these models came due to the fact that two-equation models are the simplest complete models. This means that problem solution is found by the model without using the experimental data. Because of importance of the two-equation models, at the Conference of AFOSR 1980-81 basically these models are tested. Considering the one-equation models, we have seen the difficulties associated with length scale definition. In contrast to that, the two-equation models provide along with turbulence energy equation the analogous equation for length scale or for some equivalent parameter.

Whereas any model consists of the turbulence energy equation, there is no universal parameter for second equation. Kolmogorov first used for the turbulence model a second partial differential equation and formulated it for specific dissipation energy rate per unit volume and time ω. Since this quantity has the dimension (1/s), it follows from dimensional analysis that the eddy viscosity, length scale, and the dissipation are defined as $\mu_{tb} \sim \rho k/\omega$, $l \sim k^{1/2}/\omega$, $\varepsilon \sim k\omega$. Some other developers suggested the second equation for time $t \sim 1/\omega$, length scale, dissipation ε, or product kl. As was pointed at the Conference of AFOSR 1980-81, the uncertainty about the two-equation models is the vague choice of variable for second equation. Whereas during the last time some clearness was gained, there still is no full answer what variable is most suitable [422].

8.4.3.1 The k − ω Model

The first two-equation model was suggested in 1942 by Kolmogorov. His second equation in terms of dissipation per unit volume and time ω was formulated in the form similar to the equation for turbulence kinetic energy (8.24)

$$\rho\frac{\partial\omega}{\partial t} + \rho U_j\frac{\partial\omega}{\partial x_j} = -\beta_\omega\rho\omega^2 + \frac{\partial}{\partial x_j}\left(\sigma\mu_{tb}\frac{\partial\omega}{\partial x_j}\right), \quad \omega = C\frac{k^{1/2}}{l} \tag{8.25}$$

This equation as compared to equation (8.24) for k has no production and molecular diffusion terms. In the early days when it was difficult to use complex turbulent models, Kolmogorov did not develop the model complete, rather he published an idea including in the equation for specific dissipation ω only the most important terms. The production term was not taken into account because the smallest eddies are responsible for dissipation for which production is not as important. Under the same reason, the molecular diffusion term was omitted, which is not very important for the case of high Reynolds numbers considered by Kolmogorov.

In the 1960s when the interest in turbulence models was rising, the followers added to Kolmogorov's model missing terms and improved it. Further improvements were achieved by testing models comparing the computed and measured results for estimating the closure coefficients. Currently, the several versions of $k - \omega$ models exist, in particular, based on different variables for ω−equation mentioned above. Here we present the model developed by Wilcox as it is given in his first edition (1994) book [422]

$$\rho\frac{\partial k}{\partial t} + \rho U_j\frac{\partial k}{\partial x_j} = \tau_{ij}\frac{\partial U_i}{\partial x_j} - \beta_k\rho k\omega + \frac{\partial}{\partial x_j}\left[(\mu + \sigma_k\mu_{tb})\frac{\partial k}{\partial x_j}\right], \quad l = \frac{k^{1/2}}{\omega} \tag{8.26}$$

$$\rho\frac{\partial\omega}{\partial t} + \rho U_j\frac{\partial\omega}{\partial x_j} = \alpha_\omega\frac{\omega}{k}\tau_{ij}\frac{\partial U_i}{\partial x_j} - \beta_\omega\rho\omega^2 + \frac{\partial}{\partial x_j}\left[(\mu + \sigma_\omega\mu_{tb})\frac{\partial\omega}{\partial x_j}\right], \quad \varepsilon = \beta_k\omega k, \tag{8.27}$$

where $\mu_{tb} = \rho k/\omega$. Closure coefficients are: $\alpha_\omega = 5/9$, $\beta_k = 0.09$, $\beta_\omega = 3/40$, $\sigma = 1/2$.

Comment 8.8 A more complicated system of equations determines the $k - \omega$ model in the third edition (2006) of this book, which in addition consists of three closure coefficients: two tensors and equation for dissipation with one term more.

8.4.3.2 The k − ε Model

The first version of this model was suggested in 1945 by Chou. The improved version, which is known in the turbulence modeling as a standard $k - \varepsilon$ model was developed in several papers published in the middle of 1970s by Launder with co-authors. The model is as follows:

$$\rho\frac{\partial k}{\partial t} + \rho U_j\frac{\partial k}{\partial x_j} = \tau_{ij}\frac{\partial U_i}{\partial x_j} - \rho\varepsilon + \frac{\partial}{\partial x_j}\left[\left(\mu + \frac{\mu_{tb}}{\sigma_k}\right)\frac{\partial k}{\partial x_j}\right], \quad l = C_\mu\frac{k^{3/2}}{\varepsilon} \tag{8.28}$$

$$\rho\frac{\partial\varepsilon}{\partial t} + \rho U_j\frac{\partial\varepsilon}{\partial x_j} = C_{\varepsilon 1}\frac{\varepsilon}{k}\tau_{ij}\frac{\partial U_i}{\partial x_j} - C_{\varepsilon 2}\rho\frac{\varepsilon^2}{k} + \frac{\partial}{\partial x_j}\left[\left(\mu + \frac{\mu_{tb}}{\sigma_\varepsilon}\right)\frac{\partial\varepsilon}{\partial x_j}\right], \quad \omega = \varepsilon/C_\mu k, \tag{8.29}$$

with $\mu_{tb} = C_\mu\rho k^2/\varepsilon$, closure coefficients $C_{\varepsilon 1} = 1.44$, $C_{\varepsilon 2} = 1.92$, $C_\mu = 0.09$, $\sigma_k = 1$, $\sigma_\varepsilon = 1.3$, and additional correction function for low Reynolds numbers. Later a more complicated version was published [422, 2006] for the $k - \omega$ model.

Comment 8.9 The other two-equation models applying different second differential equations, for example, $k - \tau$ or $k - kl$ (with the second equation for time τ or for product kl) are used and tested much less than $k - \varepsilon$ and $k - \omega$ models.

8.4.4 Applicability of the One-Equation and Two-Equation Models

The one-equation models as well as algebraic models are uncompleted, and due to that, the accuracy of those methods often are comparable with the algebraic models predictions. In contrast, the completed two-equation models are superior to both one-equation and algebraic models because those consist of the whole required for solution information inside the model. The turbulent flows with strong adverse pressure gradients, separated or/and reattachment flows, compressible flows with high Mach numbers, and other complications may be studied with reasonable accuracy only by two-equation models, since applications of more accurate and reliable new methods of direct numerical simulation at present are relatively restricted (Chaps. 5 and 6). At the same time, the algebraic models are preferable for solution of problems at zero and favorable pressure gradients, especially when a close analytic solution may be obtained to get insight into physics.

Comment 8.10 The Karman-Pohlhausen integral methods described in Section 7.6 for laminar boundary layer are applicable to turbulent flows as well. Those methods were developed and also widely used before computer era. Because the polynomial velocity profiles usually applied for laminar boundary layer problems are not proper for turbulent flows, in this case for velocity distribution instead of polynomial profiles are used other simple functions reviewed, for example, in [113], [202], and [304].

Even now, there are some problems for which the solution of a plain integral method is reasonable to use, in particular, when simple analytical formula is required. We encountered such cases, considering different relations for influence functions (S. 1.3.2, Exam. 7.13).

9

Analytical and Numerical Methods in Fluid Flow and Heat Transfer

This chapter presents the mathematical methods frequently applied in applications. Although the reviewed methods are general, the following examples are mainly the problems of heat transfer in solids. That is because the other two topics laminar and turbulent fluid flow and heat transfer important for studying the basic text are considered with examples in previous chapters of this part of the book. As well as in previous chapters, here, the information from basic text is not repeated, rather at advising the additional explanations a reader is referred to the relevant basic text sections.

Analytical Methods

9.1   Solutions Using Error Functions

Error integral is usually used in two tabulated functions $erf(z)$ and $erfc(z)$

$$erf(z) = \frac{2}{\sqrt{\pi}} \int_0^z \exp(-\xi^2)d\xi, \quad erf(-z) = -erf(z), \quad erfc(z) = 1 - erf(z) \qquad (9.1)$$

with different limiting properties $erf(0) = 0$, $erf(\infty) = 1$ and $erfc(0) = 1$, $erfc(\infty) = 0$.

It can be shown that functions (9.1) satisfied the unsteady one-dimensional (without source) conduction equation (1.1) and boundary conditions of two types of one-dimensional problems: (i) two-dimensional solid with infinite transverse size and infinite or semi-infinite length in longitudinal direction and (ii) thin rod with insulated literal surface. To see that these problems are inherent one-dimensional, note that in both cases, the transverse resistance is infinite, and due to that heat flows only in the longitudinaldirection. Therefore, in such a problem, similarity variable should be exists since in this case with $l \rightarrow \infty$, the usual dimensionless variable x/l could not be used. Knowing that one-dimensional equation (1.1) contains three variables: α, x and t with units m²/s, m, and s, one finds a combination $x/\sqrt{\alpha t}$ giving a desired dimensionless

Applications of Mathematical Heat Transfer and Fluid Flow Models in Engineering and Medicine,
First Edition. Abram S. Dorfman.
© 2017 John Wiley & Sons Ltd. Published 2017 by John Wiley & Sons Ltd.

variable. Then, the equation (1.1) and its solution become

$$z\frac{d\theta}{dz} + c^2\frac{d^2\theta}{dz^2} = 0, \quad \theta = c_1\int_0^z \exp(-\xi^2)d\xi + c_2 = c_1 erf(z) + c_2, \quad z = \frac{x}{2\sqrt{\alpha t}} \quad (9.2)$$

where $\theta = (T - T_i)/(T_w - T_i)$, T_i and T_w are initial and surface temperatures. To adjust the solution to function (9.1) $erf(z)$, the constant in differential equation (9.2) is taken as $c = \sqrt{2}$, which yields the variable $z = x/2\sqrt{\alpha t}$ instead of initial $x/\sqrt{\alpha t}$. The constants c_1 and c_2 in solution (9.2) are found from boundary conditions.

■ **Example 9.1: A solid or thin rod with constant initial T_i and surface T_w temperatures**

From boundary conditions we have: $t = 0$, $z \to \infty$, $T = T_i$, $\theta = 0$ and $x = 0$, $z = 0$, $T = T_w$, $\theta = 1$. Then, according to (9.1) and (9.2) the constants are: $z = 0$, $erf(z) = 0$, $c_2 = 1$ and $z \to \infty$, $erf(z) = 1$, $c_1 = -1$ resulting in solution (9.2)

$$\frac{T(x,t) - T_i}{T_w - T_i} = 1 - erf\left(\frac{x}{2\sqrt{\alpha t}}\right) = ertc\left(\frac{x}{2\sqrt{\alpha t}}\right) = \frac{2}{\sqrt{\pi}}\int_{x/2\sqrt{\alpha t}}^{\infty} \exp(-\xi^2)d\xi \quad (9.3)$$

Other examples of solutions in terms of error functions may be found in [58].

■ **Example 9.2: A solid or thin rod with zero initial and function $\phi(t)$ surface temperatures**

Solution of this problem for dependent of time temperature $\phi(t)$ may be found by Duhamel's integral (1.21) employing (9.3) as a simple solution $f(x,t)$ and function $\phi(t)$ as $F(t)$ (S. 1.3.1). In this case, another form of Duhamel integral that is obtained after integrating (1.21) by parts is convenient to use. Putting $u = f(x,t-\tau)$, $dv = F'(\tau)d\tau$, one transforms (1.21) by parts, gets the other form of Duhamel integral, and then find solution of considering problem with initial temperature $T_i = 0$

$$T(x,t) = f(x,0)F(t) + \int_0^t f'(x,t-\tau)F(\tau)d\tau \quad (9.4)$$

$$T(x,t) = \int_0^t \phi(\tau)\frac{d}{dt}f(x,t-\tau)d\tau = \frac{x}{2\sqrt{\pi\alpha}}\int_0^t \frac{\phi(\tau)}{(t-\tau)^{3/2}}\exp\left[-\frac{x^2}{4\alpha(t-\tau)}\right]d\tau \quad (9.5)$$

Comment 9.1 A solution of this problem for the case of spatial variable $T_i(x)$ initial temperature and time variable $\phi(t)$ surface temperature may be gained using principle of superposition as a sum of solutions (9.3) and (9.5) for body at temperatures $T_i(x)$ and at $\phi(t)$, respectively.

9.2   Method of Separation Variables

The idea of separation variable is to present a solution of a partial differential equation as a product of several parts in which each part depends only on one variable. Such procedure reduces partial differential equation to a set of ordinary equations.

9.2.1 General Approach, Homogeneous, and Inhomogeneous Problems

The problem is homogeneous if the right-hand sides of equation and boundary conditions are zero; otherwise, the equation or boundary condition is inhomogeneous. For example, both equations (1.1) having sources in the right hand sides are inhomogeneous. Consider three cases when separation of variable is possible for conduction equation:

(i) Homogeneous equations (1.1) and boundary conditions when substitution of a product of two functions $T = f_1(t)f_2(x_i)$, one depending on time $f_1(t)$ and another on coordinates $f_2(x_i)$, reduces the problem to following system

$$\frac{\partial f_1(t)}{\partial t}f_2(x_i) = \alpha f_1(t)\nabla^2 f_2(x_i), \quad T(x_i,0) = F(t) \quad \lambda_{wi}\frac{\partial f_1(x_i)}{\partial n_i} + h_i f_1(x_i) = 0 \quad (9.6)$$

Here, T is temperature excess beyond surrounding, the function $F(t)$ specifies initial conditions at $t = 0, n$ is the outward normal, and the homogeneous general form of boundary condition of the third kind becomes boundary condition of the first kind at $\lambda_{wi} = 0$ and of the second kind at $h_i = 0$. Separation of variables in first equation (9.6) transforms the solution of partial differential equations (1.1) to the simplified equations for the three- (or two-) and one-dimensional conduction problems, respectively, as

$$\frac{1}{\alpha f_1(t)}\frac{df_1(t)}{dt} = \frac{1}{f_2(x_i)}\nabla^2 f_2(x_i) = -\mu^2 \qquad \frac{1}{\alpha f_1(t)}\frac{df_1(t)}{dt} = \frac{1}{f_2(x)}\frac{d^2 f_2(x)}{dx^2} = -\mu^2 \quad (9.7)$$

These equations are in separated form because the left hand side of each relation depends only on time, whereas expressions of the right hand side of both equations (9.7) depend only on coordinates. Because of that, two parts of each equation (9.7) may be equal to each other only if each of these parts is constant and are equal to some negative value $(-\mu^2)$. Integrating the left-hand side of equations (9.7) gives $f_1(t) = C\exp(-\mu^2 t)$. This result tells us that μ^2 should be negative: otherwise, the solution will be infinite at $t \to \infty$, which contradicts the physics. The right-hand part of each equation (9.7) constitutes the second separated equation—an ordinary or a partial differential equation in the cases of one-dimensional and of two- or three-dimensional conduction problems, respectively

$$\frac{d^2 f_2(x)}{dx^2} + \mu^2 f_2(x) = 0 \quad \nabla^2 f_2(x_i) + \mu^2 f_2(x_i) = 0 \quad (9.8)$$

The partial differential equation (9.8) known as Helmholtz equation may be separated further to ordinary differential equations, if the solution of this equation may be presented as a product of two or three functions each depending only on one of coordinates.
(ii) Problems with time-independent inhomogeneous boundary conditions may be reduced to homogeneous problems of type considered in (i), if suitable for separation new variables are possible.
(iii) Solution of an unsteady inhomogeneous problem with time-dependent sources or/and boundary conditions may be reduced by Duhamel's integral to a simpler problem with independent of time sources and boundary conditions [295]. Examples are considered next.

9.2.2   One-Dimensional Unsteady Problems

■ **Example 9.3: Thin laterally insulated rod of length L initially at the temperature $T_i(x)$ and constant ends temperatures T_0 at $x = 0$ and T_L at $x = L$**

Because both boundary conditions are inhomogeneous (not zero), a new variables $\bar{x}$ and ϑ are used, transforming boundary conditions in homogeneous form

$$\bar{x} = x/L, \quad \vartheta = T - T_0 - (T_L - T_0)\bar{x}, \quad \vartheta = 0 \text{ at } \bar{x} = 0, \quad \vartheta = 0 \text{ at } \bar{x} = 1 \quad (9.9)$$

Because the conduction equation in these variables is also homogeneous, the separation variable approach is possible resulting in two ordinary differential equations

$$\frac{\partial \vartheta}{\partial \text{Fo}} - \frac{\partial^2 \vartheta}{\partial \bar{x}^2} = 0, \quad \vartheta = f_1(\text{Fo})f_2(\bar{x}), \quad \frac{df_1}{d\text{Fo}} + \mu^2 = 0, \quad \frac{d^2 f_2}{d\bar{x}^2} + \mu^2 f_2 = 0 \quad (9.10)$$

where the Fourier number $\text{Fo} = \alpha t/L^2$ is used instead of time. As we know, integrating the first equation gives $f_1(\text{Fo}) = C \exp(-\mu^2 \text{Fo})$ (see (9.7)). Integrating a second equation is also a simple task leading to $f_2(\bar{x}) = C_1 \sin \mu\bar{x} + C_2 \cos \mu\bar{x}$. Thus, the solution in the form of product (9.10) should satisfy two boundary conditions (9.9)

$$\vartheta = (C_1 \sin \mu\bar{x} + C_2 \cos \mu\bar{x}) \exp(-\mu^2 \text{Fo}), \quad \vartheta(0, \text{Fo}) = \vartheta(1, \text{Fo}) = 0 \quad (9.11)$$

To satisfy the first condition at $\bar{x} = 0$, it is necessary to take $C_2 = 0$. Then, the second condition at $\bar{x} = 1$ can be satisfied only by putting $\mu = n\pi$ as the angle of a sine, which gives $\vartheta = C_1 \exp(-n^2\pi^2 \text{Fo}) \sin \pi n\bar{x}$. Since here, constant C_1 remains free, it is replaced by C_n forming a family of partial solutions of conduction differential equation (9.10)

$$\vartheta = \sum_{n=1}^{\infty} C_n \sin(n\pi\bar{x}) \exp(-n^2\pi^2 \text{Fo}) \quad \vartheta_i(\bar{x}, 0) = \sum_{n=1}^{\infty} C_n \sin(n\pi\bar{x}) \quad (9.12)$$

This solution satisfies all assigned conditions, except the initial temperature $T_i(x)$ at $\text{Fo} = 0$. Setting in this equation $\text{Fo} = 0$, we find that initial condition is presented by the Fourier series (S. 9.3.1) given by the second expression (9.12).

The coefficients C_n of Fourier series are defined by the Euler formula (S. 9.2.3)

$$C_n = 2 \int_0^1 \vartheta_i(\bar{x}) \sin(n\pi\bar{x}) d\bar{x}, \quad C_n = 2 \int_0^1 T_i(\bar{x}) \sin(n\pi\bar{x}) d\bar{x} - \frac{2}{n\pi} T_0 \quad (9.13)$$

which for initial temperature (9.9) $\vartheta_i(\bar{x}) = T_i(\bar{x}) - T_0 - (T_L - T_0)\bar{x}\, \vartheta_i$ leads to the second relation (9.13). Substituting this result into first relation (9.12) gives the problem solution

$$T(\bar{x}, \text{Fo}) = T_0 + (T_L - T_0)\bar{x} + 2 \sum_{n=1}^{\infty} \left[\int_0^1 T_i(\bar{x}) \sin(n\pi\bar{x}) d\bar{x} - \frac{T_0}{n\pi} \right] \sin(n\pi\bar{x}) \exp(-n^2\pi^2 \text{Fo})$$

$$(9.14)$$

In the case of constant initial temperature, solution (9.14) simplifies to

$$\frac{T(\bar{x}, \text{Fo}) - T_0}{T_L - T_0} = \bar{x} + \frac{T_i - T_0}{T_L - T_0} \frac{2}{\pi} \sum_{n=1}^{\infty} \frac{\sin(n\pi\bar{x})}{n} \exp(-n^2\pi^2\text{Fo}) \tag{9.15}$$

■ **Example 9.4: Thin rod laterally insulated of length L initially at temperature $T_i(x)$, at temperatures T_0 at the end $x = 0$ and at insulated $(\partial T/\partial x = 0)$ end $x = L$**

Variables similar to (9.9) transform the boundary condition to homogeneous form

$$\vartheta(\bar{x}, \text{Fo}) = T(\bar{x}, \text{Fo}) - T_0 \qquad \vartheta = 0 \text{ at } \bar{x} = 0 \qquad \partial\vartheta/\partial\bar{x} = 0 \text{ at } \bar{x} = 1 \tag{9.16}$$

Similarly, solution (9.11) subjected to the second and then to the first conditions yields

$$\vartheta = \sum_{n=0}^{\infty} C_n \cos\left[\left(n + \frac{1}{2}\right)\pi\bar{x}\right] \exp\left[-\left(n + \frac{1}{2}\right)^2 \pi^2 \text{Fo}\right] \qquad \vartheta_i = \sum_{n=1}^{\infty} C_n \cos\left[\left(n + \frac{1}{2}\right)\pi\bar{x}\right] \tag{9.17}$$

Determining coefficients C_n as in previous example, one obtains the problem solution

$$C_n = 2 \int_0^1 T_i(\bar{x}) \cos\left[\left(n + \frac{1}{2}\right)\pi\bar{x}\right] d\bar{x} - \frac{2(-1)^n}{(n + 1/2)\pi} T_0 \tag{9.18}$$

$$T(\bar{x}, \text{Fo}) = T_0 + 2 \sum_{n=0}^{\infty} \left\{ \int_0^1 T_i(\bar{x}) \cos\left[\left(n + \frac{1}{2}\right)\pi\bar{x}\right] d\bar{x} - \frac{2(-1)^n T_0}{(n + 1/2)\pi} \right.$$

$$\left. \times \cos\left[\left(n + \frac{1}{2}\right)\pi\bar{x}\right] \exp\left[-\left(n + \frac{1}{2}\right)^2 \pi^2 \text{Fo}\right] \right. \tag{9.19}$$

In the case of constant initial temperature, solution (9.19) becomes

$$\frac{T(\bar{x}, \text{Fo}) - T_0}{T_i - T_0} = \frac{2}{\pi} \sum_{n=0}^{\infty} \frac{(-1)^n}{(n + 1/2)} \cos\left[\left(n + \frac{1}{2}\right)\pi\bar{x}\right] \exp\left[-\left(n + \frac{1}{2}\right)^2 \pi^2 \text{Fo}\right] \tag{9.20}$$

9.2.3 Orthogonal Eigenfunctions

In solutions just considered, the coefficients C_n were determined using Fourier's trigonometric series. However, there are problems that require for such procedure other types of series. The required properties of such a set of functions gives a solution of the Sturm-Liouville problem named after pioneers first studied this subject.

Consider the Sturm-Liouville problem in interval $a < x < b$ with boundary conditions at $x = a$ and $x = b$, respectively

$$\frac{d}{dx}\left[p(x)\frac{dy(x, \mu)}{dx}\right] + q(x)y - \mu w(x)y = 0, \quad C_1\frac{dy}{dx} + C_2y = 0, \quad C_3\frac{dy}{dx} + C_4y = 0 \tag{9.21}$$

This homogeneous problem has solution only for a set of values $\mu_1 < \mu_2 < \mu_3 \ldots$ which are called eigenvalues of problem (9.21). The corresponding solutions $y_n(x)$ of equation (9.21) form a set of functions that are called orthogonal eigenfunction with respect to weighting function $w(x)$ if they satisfy the following integral conditions

$$I = \int_a^b y_n(x)y_m(x)w(x)dx = 0 \quad \text{for any } \mu_n \neq \mu_m, \quad I \neq 0 \text{ for } \mu_n = \mu_m \tag{9.22}$$

Or in other words: a set of functions is orthogonal if the integral (9.22) equals zero for each pair of eigenfunctions $y_n(x)$ and $y_m(x)$ corresponding to eigenvalues μ_n and μ_m except the case of their equalities when $\mu_n = \mu_m$ and $y_n(x) = y_m(x)$.

The orthogonal property ensures that a given set of functions is presentable as a series of eigenfunctions similar to Fourier's expansion. To see that this is true, multiply both sides of series $f(x) = \sum_{n=0}^{\infty} C_n y_n$ by eigen $y_n(x)$ and weighting $w(x)$ functions to get

$$\int_a^b f(x)y_n(x)w(x)dx = C_1 \int_a^b y_1(x)y_n(x)w(x)dx + C_2 \int_a^b y_2(x)y_n(x)w(x)dx + \dots +$$

$$C_n \int_a^b y_n(x)y_n(x)w(x)dx \tag{9.23}$$

Due to orthogonal conditions (9.22), all integrals except the last one containing $y_n^2(x)$ become zero, and then, formula-defining coefficients C_n follows from expression (9.23)

$$C_n = \frac{\int_a^b f(x)y_n(x)w(x)dx}{\int_a^b y_n^2(x)w(x)dx} \qquad f(x) = \sum_{n=1}^{\infty} C_n y_n(x) \tag{9.24}$$

It is easy to check that Fourier series are orthogonal, and that the Euler formula (9.13) is a particular case of this equation at $a = 0$, $b = 1$, $f(x) = \vartheta_i(\overline{x})$, $y_n(x) = \sin n\pi\overline{x}$ and $w(x) = 1$.

■ Example 9.5: Thin laterally insulated rod of length L initially at $T_i(x)$, with T_0 at $x = 0$ and heat transfer into surrounding at $x = L$ according to boundary condition of the third kind $\lambda_w(\partial T/\partial x) = h(T - T_\infty)$

Both given boundary condition are inhomogeneous since these contain specific values of temperatures. However, in this case, it is not as easy as before to modify given conditions to the homogeneous forms. We will show one possible way to solve this problem and to find a proper variable. Consider an expression relating the prescribed temperatures T_0 and T_∞ to a new variable ϑ through two linear functions with undetermined coefficients $T = \vartheta + (a_1 + b_1\overline{x})T_0 + (a_2 + b_2\overline{x})T_\infty$, where $\overline{x} = x/L$. It is easy to understand that to modify the given conditions at $x = 0$ to homogeneous form we should take $a_1 = 1$ and $a_2 = 0$. Then, substituting the reduced relation in the second prescribed boundary condition at $\overline{x} = 1$ leads to following expression

$$\frac{\partial \vartheta}{\partial \overline{x}} + b_1 T_0 + b_2 T_\infty = \text{Bi}[\vartheta + (1 + b_1)T_0 + (b_2 - 1)T_\infty, \quad \text{Bi} = \frac{hL}{\lambda_w} \tag{9.25}$$

It is seen that to have homogeneous boundary condition $\partial \vartheta / \partial \bar{x} - Bi \vartheta = 0$, it is necessary to satisfy two conditions $[b_1 - Bi(1 + b_1)]T_0 = 0$ and $[b_2 - Bi(b_2 - 1)]T_\infty = 0$ which after solution for coefficients give $b_1 = -b_2 = Bi/(1 - Bi)$ resulting in proper new variable and homogeneous form of both prescribed boundary conditions

$$\vartheta = T - T_0 - \frac{Bi}{1 - Bi}\bar{x}(T_0 - T_\infty), \qquad \vartheta = 0 \quad \text{at} \quad \bar{x} = 0, \qquad \frac{\partial \vartheta}{\partial \bar{x}} - Bi\vartheta = 0 \quad \text{at} \quad \bar{x} = 1 \quad (9.26)$$

The conduction equation in a new variable ϑ remains homogeneous because the second derivative of additional term, appearing in transformed equation due to the new variable, which is proportional to $\bar{x}$, is zero. The solution of transformed equation is the same relation (9.11). Satisfying the first condition (9.26) at $\bar{x} = 0$ requires to take $C_2 = 0$. Then, we get as before $\vartheta = C_1 \sin \mu \bar{x} \exp(-\mu^2 Fo)$. Substituting this result in another condition (9.26) at $\bar{x} = 1$ gives the second expression, which should be satisfied

$$\mu \cos \mu = Bi \sin \mu \quad \text{at} \quad \bar{x} = 1 \quad \text{or} \quad \tan \mu = \mu/Bi \qquad (9.27)$$

Numerical solution shows that this equation has infinite number of roots $\mu_1, \mu_2, \mu_3, \ldots,$ and hence, the problem solution may be presented by series similar to (9.12)

$$\vartheta = \sum_{n=1}^{\infty} C_n \sin(\mu_n \bar{x}) \exp(-\mu_n^2 Fo) \qquad \vartheta_i = \sum_{n=1}^{\infty} C_n \sin(\mu_n \bar{x}) \qquad (9.28)$$

Although both series are looking identical, they differ in essence. Comparison indicates that a set of roots of equation (9.27), in contrast to usual evenly spaced roots defined by trigonometric function in series (9.12), are spaced not evenly so that the interval between roots grows with n increasing and reaches π as $n \to \infty$. Therefore, the series (9.28) with roots μ_n of equation (9.27) are not usual Fourier series. In such a case, coefficients C_n of series (9.28) can be determined applying formula (9.24) only if the roots μ_n are eigenvalues of Sturm-Liouville problem.

To check if in some particular case roots μ_n are eigenvalues and formula (9.24) is applicable, one should compare the problem in question with Sturm-Liouville standard model (9.21). In this case, relation (9.27) is a solution of ordinary differential equation (9.10) $f'' + \mu^2 f = 0$ obtained for function f_2 after separation of variables. For the problem in question, this differential equation was solved under two boundary conditions (9.26), which in terms of function f are: $f(0) = 0, f'(1) + Bif(1) = 0$. Comparing this system of equation and boundary condition with Sturm-Liouville mathematical model (9.21) shows that the problem in question is a particular case of standard model (9.21) for interval $0 < x < 1$ with $p(x) = w(x) = 1, q(x) = 0,$ $C_1 = 0, C_2 = C_3 = 1,$ and $C_4 = Bi$. This implies that μ_n defined by equation (9.27) and functions $\sin(\mu_n \bar{x})$ in series (9.28) are eigenvalues and eigenfunctions, respectively, and hence, the coefficients C_n may be defined by formula (9.24). Taken into account equations (9.26) and carry out integration, one gets coefficients C_n and then, via first equation (9.28) obtains the problem solution

$$C_n = \frac{\int_0^1 \vartheta_i(\bar{x}) \sin(\mu_n \bar{x}) d\bar{x}}{\int_0^1 \sin^2(\mu_n \bar{x}) d\bar{x}} = \frac{\int_0^1 [T_i(\bar{x}) - T_0 - B\bar{x}(T_0 - T_\infty)] \sin(\mu_n \bar{x}) d\bar{x}}{\int_0^1 \sin^2(\mu_n \bar{x}) d\bar{x}}, \qquad B = \frac{Bi}{1 - Bi} \qquad (9.29)$$

$$T(x, \text{Fo}) = T_0 + B\bar{x}(T_0 - T_\infty) + 2\sum_{n=1}^{\infty} \frac{1}{\mu_n(\mu_n - \sin\mu_n \cos\mu_n)} \left[\mu_n^2 \int_0^1 T_i(\bar{x})\sin(\mu_n\bar{x})d\bar{x} \right.$$

$$\left. -T_0\mu_n\left(1 - \cos\mu_n\right) - B(\sin\mu - \mu\cos\mu)(T_0 - T_\infty) \right] \sin(\mu_n\bar{x})\exp(-\mu_n^2\text{Fo}) \qquad (9.30)$$

Some more simple cases follow from this solution. For the case of heat transfer from both ends of a rod into surrounding at T_∞, we obtain since in this case $T_0 = T_\infty$ at $x = 0$

$$T(x, \text{Fo}) = T_\infty + 2\sum_{n=1}^{\infty} \frac{\mu_n \int_0^1 T_i(\bar{x})\sin(\mu_n\bar{x})d\bar{x} - T_\infty(1 - \cos\mu_n)}{\mu_n - \sin\mu_n \cos\mu_n} \sin(\mu_n\bar{x})\exp(-\mu_n^2\text{Fo}) \qquad (9.31)$$

If in addition, the initial temperature is constant, this expression becomes farther simpler

$$\frac{T(x, \text{Fo}) - T_\infty}{T_i - T_\infty} = 2\sum_{n=1}^{\infty} \frac{1 - \cos\mu_n}{\mu_n - \sin\mu_n \cos\mu_n} \sin(\mu_n\bar{x})\exp(-\mu_n^2\text{Fo}) \qquad (9.32)$$

Comment 9.2 Nonstandard series satisfying Sturm-Liouville requirements such as a problem considered in the last example are called generalized Fourier series.

9.2.4   Two-Dimensional Steady Problems

The two-dimensional steady heat transfer problems are governed by homogeneous Laplace's or by inhomogeneous Poisson's equation (1.2). Since these equations are of elliptic type, the boundary conditions should be specified on each side of the computation domain (S. 1.1). Two types of problems are usually considered: the Dirichlet problem when the boundary condition of the first kind is used, specifying the temperature on the boundaries of domain, and the Neumann problem in which the second kind of boundary condition in the form of normal temperature derivative on domain sides is specified. The Neumann problem is ill-posted, which means that a solution of the problem requires the thermal equilibrium when the integral of total heat flow inside the object is zero (S. 1.1).

■ **Example 9.6: Two-dimensional rectangular sheet in xy plane of length a, height b and prescribed sides temperatures: left $T(0,y) = \varphi_1(y)$, right $T(a,y) = \varphi_2(y)$, lower $T(x,0) = \varphi_3(x)$, and upper $T(x,b) = \varphi_4(x)$ (Dirichlet problem)**

Substitution of the product $f_1(x)f_2(y)$ into Laplace equation (1.2) and separation of variables yields as in above examples two differential equations and their solutions

$$\frac{f_1''(x)}{f_1(x)} = -\frac{f_2''(y)}{f_2(y)} = -\mu^2 \quad f_1''(x) + \mu^2 f_1(x) = 0 \quad f_2''(y) - \mu^2 f_2(y) = 0 \qquad (9.33)$$

$$T(x, y) = [C_1\sin(\mu x) + C_2\cos(\mu x)][C_3 sh(\mu y) - C_4 ch(\mu y)] \qquad (9.34)$$

Assuming that the temperatures along the vertical sides are zero $\varphi_1(y) = \varphi_2(y) = 0$, we satisfy zero conditions at $x = 0$ and $x = a$ by taking $C_2 = 0$ and $\mu = n\pi/a$, respectively. It should be also taken $C_4 = 0$ because in the case of zero temperatures on the vertical sides, the temperatures at $y = 0$ and $y = b$ are zero as well. As a result, we obtain from (9.34) the solution $C_{1n} \sin(n\pi x/a) sh(n\pi y/a)$ which meets three conditions at $x = 0$, $x = a$, and $y = 0$. To satisfy the forth condition of zero at $y = b$, the superposition principle is used by adding a similar relation $C_{2n} \sin(n\pi x/a) sh[(n\pi/a)(b-y)]$. A sum of these two results gives the problem solution with two unknown coefficients C_{n1} and C_{n2}

$$T(x,y) = \sum_{n=1}^{\infty} \left[C_{1n} sh\frac{n\pi}{a}y + C_{2n} sh\frac{n\pi}{a}(b-y) \right] \sin\frac{n\pi}{a}x \qquad (9.35)$$

These coefficients are determined applying giving boundary conditions at $y = 0$, $\varphi_3(x)$ and at $y = b$, $\varphi_4(x)$. Corresponding equations are obtained by putting $y = 0$ or $y = b$ into solution (9.35) and employing formula (9.24), which yields two expressions

$$\varphi(x) = \sum_{n=1}^{\infty} C_n sh\frac{n\pi b}{a} \sin\frac{n\pi}{a}x \qquad C_n = \frac{2}{ash\frac{n\pi b}{a}} \int_0^a \varphi(x) \sin\frac{n\pi}{a}x dx \qquad (9.36)$$

defining C_{1n} and C_{2n} in both considering cases at $\varphi(x) = \varphi_3(x)$ and $\varphi_4(x)$, respectively. Substituting these constants into (9.35) completed the solution of this part of the problem.

The other part of the solution is considered in a similar way, assuming two other boundary conditions being zero: $\varphi_3(x) = \varphi_4(x) = 0$. The full solution is established as a sum of results obtained for horizontal and vertical sides considering cases.

■ **Example 9.7: The same problem at mixed type of boundary conditions: left $T(0,y) = 0$, right $T(a,y) = 0$, lower $T(x,0) - (\partial T/\partial y)_{y=0} = 0$ and upper $T(x,b) = \varphi(x)$**

The partial solution of this problem is the same expression (9.34) that after satisfying conditions $T = 0$ at $x = 0$ and $x = a$ by taking $C_2 = 0$ becomes

$$T(x,y) = C_1 \sin\frac{n\pi}{a}x \left(C_3 sh\frac{n\pi}{a}y - C_4 ch\frac{n\pi}{a}y \right) \qquad (9.37)$$

Meeting the condition at the lower side by applying this relation leads to expression

$$C_1 \sin\frac{n\pi}{a}x \left[C_3 sh\frac{n\pi}{a}y - C_4 ch\frac{n\pi}{a}y - \frac{n\pi}{a}\left(C_3 ch\frac{n\pi}{a}y - C_4 sh\frac{n\pi}{a}y \right) \right]_{y=0} = 0 \qquad (9.38)$$

Putting here $y = 0$ gives condition $C_4 + (n\pi/a)C_3 = 0$, which substituting into equation (9.37) and changing C_3 to C_n yields the problem solution

$$T(x,y) = \sum_{n=1}^{\infty} C_n \sin\frac{n\pi}{a}x \left(sh\frac{n\pi}{a}y + \frac{n\pi}{a}ch\frac{n\pi}{a}y \right) \qquad (9.39)$$

Then, employing upper side condition results in relation giving via (9.24) coefficients C_n

$$\varphi(x) = \sum_{n=1}^{\infty} C_n \sin\frac{n\pi}{a}x \left(sh\frac{n\pi}{a}b + \frac{n\pi}{a}ch\frac{n\pi}{a}b \right) \qquad C_n = \frac{2\int_0^a \phi(x)\sin\frac{n\pi}{a}x dx}{a\left(sh\frac{n\pi}{a}b + \frac{n\pi}{a}ch\frac{n\pi}{a}b \right)} \qquad (9.40)$$

that along with expression (9.39) issued the complete solution of the problem in question.

9.3   Integral Transforms

The integral transform technique significantly simplifies differential equations solutions reducing the ordinary differential equations to algebraic relations and modifying partial differential equations into ordinary differential equations. We consider briefly the Fourier and Laplace transforms, that most often are used in applications. A systematical usage of integral transforms to heat conduction problems one may find in [295].

Although the transformed equations are simpler than the originals and usually can be solved readily, the most difficult procedure consists of inversion of the solution from the subsidiary space in physical variables. Therefore, it is common to use transforms tablets presenting the inverse solution. Such relatively short tables are given in advanced mathematic courses. More complete tables may by found in special books [129, 310].

9.3.1   Fourier Transform

The expansion in Fourier series, that we use above, presents an arbitrary functions as a sum of harmonic oscillations with finite frequencies $n\pi/L = \pi/L, 2\pi/L, 3\pi/L \ldots$ As L increases, the distance between frequencies π/L decreases so that the numbers of terms in the series increases. Therefore, in the case of infinite or semi-infinite domain, the difference between frequencies goes to zero, whereas the number of terms becomes infinite, and in the limit the Fourier series converts into integral

$$f(x) = \frac{1}{2\pi} \int_{-\infty}^{\infty} C_n \exp(i\omega x)d\omega \qquad C_n = \int_{-\infty}^{\infty} f(x) \exp(-i\omega x)dx \qquad (9.41)$$

with continuous spectrum of frequency ω. The last integral denoting as $\hat{f}(\omega)$ gives the Fourier transform of function $f(x)$. The first integral in which the integrand is replaced by the Fourier transform $\hat{f}(x)$ instead of coefficients C_n presents the inverse formula that returns the initial function $f(x)$. These two integrals form the Fourier transform

$$\hat{f}(\omega) = \int_{-\infty}^{\infty} f(x) \exp(-i\omega x)dx \qquad f(x) = \frac{1}{2\pi} \int_{-\infty}^{\infty} \hat{f}(\omega) \exp(i\omega x)d\omega \qquad (9.42)$$

Relations (9.42) become simpler for even and odd function $f(x)$ resulting in two pairs of expressions known as cosine and sine Fourier transforms:

$$\hat{f}_C(\omega) = \int_0^{\infty} f(x) \cos \omega x dx \qquad f(x) = \frac{2}{\pi} \int_0^{\infty} \hat{f}_C(\omega) \cos \omega x d\omega \qquad (9.43)$$

$$\hat{f}_S(\omega) = \int_0^{\infty} f(x) \sin \omega x dx \qquad f(x) = \frac{2}{\pi} \int_0^{\infty} \hat{f}_S(\omega) \sin \omega x d\omega \qquad (9.44)$$

These equations follow from relations (9.42) after using formula $\exp(\pm i\omega x) = \cos \omega x \pm i \sin \omega x$. Putting this expression, for example, in the first integral (9.42), we get

$$\hat{f}(\omega) = \int_{-\infty}^{\infty} f(x)(\cos \omega x - i \sin \omega x)dx = 2 \int_0^{\infty} f(x) \cos \omega x dx \qquad (9.45)$$

This result is obtained by taking into account that cosine is an even function but sine is odd one. Due to that, the integral in equation (9.42) with limits $\mp\infty$ containing cosine doubles, whereas the similar integral with sine vanishes. Analogous procedure leads to the other integral (9.43).

■ **Example 9.8: Infinite solid or thin laterally insulated rod initially at $T_i(x)$**

Fourier transform of one-dimensional equation (1.1) according to (9.42) yields

$$\alpha \int\limits_{-\infty}^{\infty} \frac{\partial^2 T}{\partial x^2} \exp(-i\omega x)dx = \int\limits_{-\infty}^{\infty} \frac{\partial T}{\partial t} \exp(-i\omega x)dx \qquad \frac{d\hat{T}}{dt} + \alpha\omega^2\hat{T} = 0 \qquad (9.46)$$

The first term in the last equation is obtained using in the right hand side of integral (9.46) the Leibniz rule of interchanging the integration and differentiation, whereas the second term is derived from the left hand side of this relation applying double integration by parts with $u = \exp(-i\omega x)$ and $dv = (\partial^2 T/\partial x^2)dx$ which gives after first integration

$$\int\limits_{-\infty}^{\infty} \frac{\partial^2 T}{\partial x^2} \exp(-i\omega x)dx = \frac{\partial T}{\partial x} \exp(-i\omega x)\Big|_{-\infty}^{\infty} + i\omega \int\limits_{-\infty}^{\infty} \frac{\partial T}{\partial x} \exp(-i\omega x)dx \qquad (9.47)$$

In the right-hand side of this expression, the first term vanishes due to usual assumption that the temperature and its derivatives go to zero as $x \to \pm\infty$. Then, repeating integration by part of the last integral (9.47) with $u = \exp(-i\omega x)$ and $dv = (\partial T/\partial x)dx$ leads to the second term of equation (9.46). In this equation, the frequency ω is a constant parameter, and hence, this equation is a simple ordinary differential equation which solution is

$$\hat{T} = C\exp(-\alpha\omega^2 t), \qquad C = \hat{T}|_{t=0} = \hat{T}_i(\omega), \qquad \hat{T}(\omega, t) = \hat{T}_i(\omega)\exp(-\alpha\omega^2 t) \qquad (9.48)$$

Determining the constant C of this solution using a transformed initial temperature $\hat{T}_i(\omega)$ by second relation leads to the problem solution (9.48) in Fourier space.

To return to physical variables, the so-called convolution theorem is employed. According to this theorem, the inverse of a product of two transformed functions is given by integral of a product of the inverted functions gained for each of these functions. Thus, for solution (9.48) of two transformed functions $\hat{T}_i(\omega)$ and $\exp(-\alpha\omega^2 t)$, we have

$$T(x, t) = \int\limits_{-\infty}^{\infty} T_i(\xi)I(x - \xi, t)d\xi, \qquad I(x, t) = \frac{1}{\pi} \int\limits_{0}^{\infty} \exp(-\alpha\omega^2 t)\cos\omega x\, d\omega \qquad (9.49)$$

where $I(x, t)$ is an inverted function obtained for $\exp(-\alpha\omega^2 t)$ using cosine transform (9.43) since the exponential function is even with respect to ω. A half of this function is taken into account only because of the relationship (9.45) between Fourier integrals with infinite and semi-infinite limits. As the next step, the integral (9.49) should be inversed in the physical space. As mentioned at the beginning of Section 9.3, in contrast to other standard steps in Fourier transform, the inverse procedure does not have a standard technique. In this case, the inverse expression may be found by artificial approach based on the fact that the derivative of integral (9.49) with respect to x after integration by parts ($u = \sin(\omega x)$ and $dv = \exp(-\alpha\omega^2 t)\omega d\omega$) yields a relation proportional to the self integral (9.49)

$$\frac{dI}{dx} = -\frac{1}{\pi}\int_0^\infty \exp(-\alpha\omega^2 t)\sin(\omega x)\omega d\omega = -\frac{x}{2\pi\alpha t}\int_0^\infty \exp(-\alpha\omega^2 t)\cos\omega x dx = -\frac{x}{2\alpha t}I \quad (9.50)$$

The first and the last terms of that expression comprise an ordinary differential equation $dI/dx = -(x/2\alpha t)I$, which solution gives the integral (9.49) in the physical variables $I = C_1 \exp(-x^2/4\alpha t)$. The constant C_1 is defined as a value of this integral at $x \to 0$ when $\cos \omega x \to 1$, and integral (9.49) becomes error function (9.1) giving $C_1 = 1/2\sqrt{\pi\alpha t}$. Substituting these results into the first equation (9.49) results in the solution

$$T(x,t) = \frac{1}{2\sqrt{\pi\alpha t}}\int_{-\infty}^\infty T_i(\xi)\exp[-(x-\xi)^2/4\alpha t]d\xi \quad (9.51)$$

It may be shown that for the case of the constant initial temperature, this result coincides with relation (9.3) obtained using the error function directly.

Comment 9.3 The inversed integral (9.49) may be found much easier using the table of Fourier transforms. We present this example to show one of artificial inverse means.

■ Example 9.9: Two-dimensional infinite sheet in xy plane of semi-infinite height (half plane) initially at $T_i(x)$

The mathematical model of this problem consists of the homogeneous two-dimensional Laplace equation (1.2), which after Fourier transform yields similar to (9.46) simple ordinary differential equation

$$\int_{-\infty}^\infty \frac{\partial^2 T}{\partial x^2}\exp(-i\omega x)dx + \int_{-\infty}^\infty \frac{\partial^2 T}{\partial y^2}\exp(-i\omega x)dx = 0 \qquad \frac{d^2\hat{T}}{dy^2} - \omega^2\hat{T} = 0 \quad (9.52)$$

This differential equation is derived in the same way as equation (9.46). The first term is obtained by changing the integration and differentiation in the second integral (9.52) as well as the first term of equation (9.46) is gained. Such procedure is authorized because in both cases the operations of differentiation and integration are designated with respect to different variables, t and x in the first, and, y and x in the second integrals, respectively.

The second term of the last equation (9.52) is derived by double integration by parts of the first integral (9.52) also as the same term of equation (9.46) resulting in both cases after first integration in expression (9.47) and in desired term after second integration.

The solution of ordinary differential equation (9.52) satisfying the transformed initial condition $\hat{T}|_{y=0} = \hat{T}_i(\omega)$ is $\hat{T} = \hat{T}(\omega)\exp(-|\omega|y)$ (the absolute value is applied to avoid the case $\hat{T} \to \infty$). To return to physical variables, the convolution theorem and sine inverse formula (9.44) are used in the way similar to that in the previous example

$$T(x,y) = \int_{-\infty}^\infty T_i(\xi)I(x-\xi,y)d\xi, \qquad I(x,y) = \frac{1}{\pi}\int_0^\infty \exp(-|\omega|y)\cos\omega x d\omega \quad (9.53)$$

The inverted integral (9.53) in contrast to resemble integral (9.49) in the former problem may be found performing integration. Because the operation variable in (9.53) is ω, the coordinates x and y are considered as parameters, and calculation leads to the expression

$$I(x, y) = \frac{1}{\pi} \int_0^\infty \exp(-|\omega|y) \cos \omega x d\omega = \frac{\exp(-|\omega y|)}{\pi(x^2 + y^2)} (-y \cos \omega x + x \sin \omega x) \Big|_0^\infty = \frac{y}{\pi(x^2 + y^2)}$$

(9.54)

The last result follows from limiting values estimation giving zero for upper limit since $\exp(-|\omega y|) \to 0$ and y for lower limit because $x \sin \omega x = 0$. Substituting the result (9.54) in the first equation (9.53) completes the problem solution

$$T(x, y) = \frac{y}{\pi} \int_{-\infty}^\infty \frac{T_i(\xi)}{(x - \xi)^2 + y^2} d\xi$$

(9.55)

9.3.2   Laplace Transform

Laplace transform is another widely used integral transform. Whereas the Fourier transform is usually used for infinite variable domains, the Laplace transform is suitable to problems with domains restricted to semi-infinite positive part of numerical axis. Accordingly, in this case, the basic equations (9.42) contains a variable $(-st)$ with $s > 0$ instead of variable $(-i\omega x)$ in the Fourier integral, and the transform expressions are

$$\hat{f}(s) = \int_0^\infty f(t) \exp(-st) dt \qquad f(t) = \frac{1}{2\pi i} \int_{\gamma-\infty}^{\gamma+\infty} \hat{f}(s) \exp(st) ds$$

(9.56)

where functions $f(t)$ and $\hat{f}(s)$ substitute functions $f(x)$ and $\hat{f}(\omega)$ in Fourier integrals.

It is common to use in Laplace transforms the variable t instead of x because time is often an independent variable in the relevant applications. In fact, for dummy variables in integrals (9.56) it does not matter. Laplace transform is most often applied integral transform due to better integrals convergence with real kernels in comparison with complex ones, which sometimes results in divergent trigonometric functions as $x \to \pm\infty$. At the same time, the more complex inverse procedure in Laplace transform via second formula (9.56) is usually overcome employing tables of transforms.

■ **Example 9.10: Semi-infinite solid or thin insulated rod initially at zero temperature and time-dependent surface temperature** $T(0, t) = \phi(t)$

Because the domain is semi-infinite and temperature is a function of time, the Laplace transform is appropriate. For the same one-dimensional Laplace equation one gets

$$\alpha \int_0^\infty \frac{\partial^2 T}{\partial x^2} \exp(-st) dt = \int_0^\infty \frac{\partial T}{\partial t} \exp(-st) dt \qquad \alpha \frac{d^2 \hat{T}}{dx^2} - s\hat{T} = 0$$

(9.57)

The first term in the last equation is obtained by interchanging the integration and differentiation as in prior examples. The second term is the Laplace transform of derivative $\partial T / \partial t$, which is found by integrating by parts the second integral (9.57).

Putting $u = \exp(-st)$, $dv = (\partial T/\partial t)dt$ and knowing that $T(0) = 0$, we obtain

$$\int_0^\infty \frac{\partial T}{\partial t} \exp(-st)dt = \exp(-st)T(t)|_0^\infty + s\int_0^\infty T(t)\exp(-st)dt = s\hat{T} - T(0) = s\hat{T} \qquad (9.58)$$

Solving the ordinary differential equation (9.57) and using the common condition for the limited final result $\lim_{x\to\infty} \hat{T}(x, s) = 0$ leads to the solution in the Laplace space

$$\hat{T}(x, s) = C_1 \exp(\sqrt{s/\alpha}x) + C_2 \exp(-\sqrt{s/\alpha}x) \qquad \hat{T}(x, s) = \hat{\phi}(s)\exp(-x\sqrt{s/\alpha}) \qquad (9.59)$$

in which the constant $C_1 = 0$ and C_2 are defined applying the boundary condition in the transformed space $\hat{T}(0, s) = \hat{\phi}(s)$. To inverse the solution (9.59), the convolution theorem in the same way as in the derivation of equation (9.49) is used. First, applying the table of transforms, one gets the inverse of the function $\hat{f}(x, s) = \exp(-x\sqrt{s/\alpha})$

$$f(x, t) = \frac{x\exp(-x^2/4\alpha t)}{2\sqrt{\pi\alpha}\, t^{3/2}}, \qquad T(x, t) = \int_0^t \phi(\tau)f(t - \tau, x)d\tau \qquad (9.60)$$

and then, obtains according to the convolution theorem the second expression (9.60) for temperature in the physical space. Combination of two relation (9.60) gives the solution

$$T(x, t) = \frac{x}{2\sqrt{\pi\alpha}}\int_0^t \frac{\phi(\tau)}{(t - \tau)^{3/2}} \exp\left[-\frac{x^2}{4\alpha(t - \tau)}\right]d\tau \qquad (9.61)$$

This outcome agrees with solution (9.5) obtained by Duhamel integral and error function.

■ **Example 9.11: Thin rod literally insulated of length L initially at zero temperature with insulated end at $x = 0$ and constant temperature T_L at $x = L$**

We considered analogous problem in example 9.4. The series obtained in that example as well as others of such type convergence slowly at small values of Fourier number, close to $t = 0$. Here, we present the solution in series obtained by Laplace transform that converges fast at small times. In Laplace space, the problem is governed by the same equation (9.57) as in previous example and following boundary conditions

$$\alpha\frac{d^2\hat{T}}{dx^2} - s\hat{T} = 0, \qquad \frac{d\hat{T}}{dx} = 0 \text{ at } x = 0, \qquad \int_0^\infty T_L \exp(-st)dt = \frac{\hat{T}_L}{s} \text{ at } x = L \qquad (9.62)$$

The solution of this simple differential equation is

$$\hat{T}(x, s) = C_1 sh\sqrt{\frac{s}{\alpha}}x + C_2 ch\sqrt{\frac{s}{\alpha}}x, \qquad C_2 ch\sqrt{\frac{s}{\alpha}}L = \frac{\hat{T}_L}{s}, \qquad \hat{T}(x, s) = \frac{\hat{T}_L ch\sqrt{s/\alpha}x}{s\, ch\sqrt{s/\alpha}L} \qquad (9.63)$$

The constants here are found satisfying the boundary conditions: first at $x = 0$ by taking $C_1 = 0$, which gives $d\hat{T}/dx = C_2 sh\sqrt{s/\alpha}$, and then, using the other condition at $x = L$ via the second equation (9.63) determining the constant C_2.

One way to inverse the solution (9.63) gained after substitution of the constants in the first equation (9.63) is to express hyperbolic functions in Taylor series to get [58]

$$\hat{T}(x,s) = \frac{\hat{T}_L}{s}(e^{ax} + e^{-ax})e^{-aL}(1 + e^{-2aL})^{-1} = \frac{\hat{T}_L}{s}[e^{-a(L-x)} + e^{-a(L+x)}]\sum_{n=0}^{\infty}(-1)^n e^{-2naL}$$

$$= \frac{\hat{T}_L}{s}\sum_{n=0}^{\infty}(-1)^n e^{-a[(2n+1)L-x]} + \frac{\hat{T}_L}{s}\sum_{n=0}^{\infty}(-1)^n e^{-a[(2n+1)L+x]}, \qquad a = \sqrt{\frac{s}{\alpha}} \qquad (9.64)$$

Then, using the table of transforms for e^{-ax}/s, results in the problem solution

$$\frac{T(x,t)}{T_L} = \sum_{n=0}^{\infty}(-1)^n erfc\frac{(2n+1)L-x}{2\sqrt{\alpha t}} + \sum_{n=0}^{\infty}(-1)^n erfc\frac{(2n+1)L+x}{2\sqrt{\alpha t}} \qquad (9.65)$$

This solution is reasonable to use along with solution of the same problem in series (9.20), which converges fast at large time.

Comment 9.4 There are two forms of hyperbolic function expressed by notations similar to trigonometric function and via exponential function like in series (9.64).

Comment 9.5 Integral transforms are widely used for solving differential equations in different areas. Advanced Engineering Mathematics courses usually consider Fourier and Laplace transforms and offer problems for exercises as well as other analytical methods including drills for practice shortly reviewed in this part of Chapter 9.

9.4   Green's Function Method

The idea of Green's function is similar to Duhamel's principle. This method presents a solution of a given problem in terms of a simple problem of the same type. In creating Green's function, the simplicity is achieved due to applying homogeneous boundary conditions and Dirac delta function instead of inhomogeneous conditions and space-time dependent sources, respectively. Dirac delta function is defined as a zero value for all x except one $x = x_0$ for which an infinite value is assigned so that integral of this function is equal to unity

$$\delta(x - x_0) = \begin{cases} 0 & x \neq x_0 \\ \infty & x = x_0 \end{cases}, \quad \int_{-\infty}^{\infty}\delta(x - x_0)dx = 1, \quad \int_{-\infty}^{\infty}\delta(x - x_0)f(x)dx = \begin{cases} f(x_0) & x = x_0 \\ 0 & x \neq x_0 \end{cases}$$

$$(9.66)$$

Here, the third relation presents one of Dirac function property called general sampling, which specifies the value $f(x_0)$ of any function $f(x)$ that is continuous at origin point $x = x_0$ (Com. 9.4). In formulating Green's function, the space-time dependent heat source is substituted by the product of delta functions $\delta(x - \xi)\delta(t - \tau)$. Physically, it means that such a product of delta functions determines the temperature at the location x and time t produced by an instantaneous source of strength of unity at point with coordinate ξ at time τ.

The solution of one-dimensional conduction problem for finite domain is given in terms of Green's function $G_{\tau=0}(x,t\,|\,\xi,\tau)$ by the following expression [295]

$$T(x,t) = \int_0^L G|_{\tau=0}T_i(\xi)d\xi + \frac{\alpha}{\lambda_w}\int_0^t d\tau \int_0^L q_v(\xi,\tau)Gd\xi + \frac{\alpha}{\lambda_w}\int_0^t d\tau \int_0^L (Gf|_{\xi=0} + Gf|_{\xi=L})d\xi$$

(9.67)

This equation is written for Green's function satisfying the boundary conditions $G = 0$ at $t < \tau$ and the general boundary condition $hG - \lambda_w(\partial G/\partial n) = f(x,t)$ at $t > \tau$. The first term in equation (9.67) takes into account the initial temperature distribution, the second determines the effect of source $q_v(\xi,\tau)$, and the third one defines the contribution of the boundary conditions at $x = 0$ and $x = L$ via functions $Gf|_{\xi=0}$ and $Gf|_{\xi=L}$, respectively. If the problem in question consists of a boundary condition of the first kind, the last integral in equation (9.67) should be replaced by $(-1/h)(\partial G/\partial n)$.

■ **Example 9.12: One-Dimensional Solid at Initial Temperature $T_i(x)$ and Boundary Conditions: $\partial T/\partial x = 0$ at $x = 0$, Temperature $\phi(t)$ at $x = L$ and Source $q_v(x,t)$**

The mathematical model includes equation (1.1), two boundary conditions

$$\alpha\frac{\partial^2 T}{\partial x^2} + \frac{\alpha}{\lambda_w}q_v - \frac{\partial T}{\partial t} = 0, \qquad \frac{\partial T}{\partial x} = 0 \text{ at } x = 0, \qquad T = \phi(t) \text{ at } x = L$$

(9.68)

and temperature $T = T_i(x)$ at $t = 0$. The Green's function is formulated as it described above using homogeneous boundary conditions and a product of delta functions instead of source $q_v(x,t)$. This results in a system of equation and boundary conditions similar to initial system (9.68), but significantly simpler

$$\alpha\frac{\partial^2 G}{\partial x^2} + \frac{\alpha}{\lambda_w}\delta(x-\xi)\delta(t-\tau) = \frac{\partial G}{\partial t}, \qquad G = 0 \ t < \tau, \qquad \frac{\partial G}{\partial x}\bigg|_{x=0} = 0, \qquad G|_{x=L} = 0 \quad (9.69)$$

To satisfy the homogeneous (without delta functions) problem, the separation of variables is employed leading according to expressions (9.11) and (9.69) to the following relations

$$(C_1 \sin \mu_n x + C_2 \cos \mu_n x)\exp(-\alpha\mu_n^2 t) = 0, \qquad C_1 = 0, \qquad \mu_n = \frac{(2n+1)\pi}{2L}, \quad n = 0, 1, 2 \ldots$$

(9.70)

In (9.70), the first constant is zero due to the condition (9.68) at $x = 0$, and the value of μ_n follows from the other condition at $x = L$, which requires $C_2 \cos \mu_n x = 0$. To find the Green's function associated with equation (9.69), the cosine Fourier transform (9.43) with respect to x is applied taken into account that the cosine is an even function

$$\alpha\int_0^L \frac{\partial^2 G}{\partial x^2}\cos \mu_n x dx + \frac{\alpha}{\lambda_w}\int_0^L \delta(x-\xi)\delta(t-\tau)\cos \mu_n x dx - \int_0^L \frac{\partial G}{\partial t}\cos \mu_n x dx = 0 \quad (9.71)$$

The integrals in (9.71) are evaluated employing: the Green's identity

$$\int_0^L \left(u\frac{\partial^2 v}{\partial x^2} - v\frac{\partial^2 u}{\partial x^2} \right) dx = u\frac{\partial v}{\partial x}\Big|_0^L - v\frac{\partial u}{\partial x}\Big|_0^L, \tag{9.72}$$

for the first integral, the property of delta function for the second, and the interchanging integration and differentiating for the third integral. Setting in relation (9.72) $u = \cos\mu_n x, v = G$, and solving the obtained equation for the first integral, one gets

$$\int_0^L \frac{\partial^2 G}{\partial x^2}\cos\mu_n x\, dx = -\mu_n^2 \int_0^L G\cos\mu x\, dx + \cos\mu_n x\frac{\partial G}{\partial x}\Big|_0^L + G\sin\mu_n x\Big|_0^L = -\mu_n^2\hat{G} \tag{9.73}$$

The second and third terms in the right-hand side of this expression vanish due to the two last conditions (9.69) and to value (9.70) of μ_n giving the Green function in Fourier space. For the second integral from expression (9.71) we have

$$\int_0^L \delta(x-\xi)\delta(t-\tau)\cos\mu_n x\, dx = \delta(t-\tau)\int_0^L \delta(x-\xi)\cos\mu_n x\, dx = \delta(t-\tau)\cos\mu_n\xi \tag{9.74}$$

This result is obtained considering that: (i) the integration variable is x, so that time-dependent function $\delta(t-\tau)$ is placed out of integral and (ii) a reduced integral (9.74) according to sampling property of Dirac function (third relation (9.66)) equals $\cos\mu_n\xi$. The evaluation of the last integral (9.71) through change of integration and differentiating with regard to the value (9.70) of μ_n results in derivative of the Green function in Fourier space $d\hat{G}/dt$. As a consequence, these three transformed integrals comprise a linear first order ordinary differential equation in Fourier space and which solution is found by standard procedure or using well-known formula as follows

$$\frac{d\hat{G}}{dt} + \alpha\mu_n^2\hat{G} = \delta(t-\tau)\cos\mu_n\xi, \quad \hat{G} = \cos\mu_n\xi\exp(-\alpha\mu_n^2 t)\left[C + \int_0^t \exp\left(\alpha\mu_n^2\tau\right)\delta(t-\tau)d\tau \right] \tag{9.75}$$

After taking $C = 0$ according to initial condition (9.69) and using sampling property of delta function (9.66), relation (9.75) of Green function in Fourier space $\hat{G}$ simplifies to

$$\hat{G} = \cos\mu_n\xi\exp[-\alpha\mu_n^2(t-\tau)], \quad G(x,t) = \frac{2}{L}\sum_{n=1}^{\infty}\exp[-\alpha\mu_n^2(t-\tau)]\cos\mu_n\xi\cos\mu_n x \tag{9.76}$$

Because the space domain is finite, to get the Green function in physical space, we apply Fourier series instead of the first integral (9.43) with infinite limit and obtain the last equation (9.76). Finally, substitution of the Green function (9.76) into expression (9.67) leads to the problem solution

$$
T(x,t) = \frac{2}{L} \sum_{n=1}^{\infty} \exp(-\alpha \mu_n^2 t) \cos \mu_n x \left\{ \int_0^L T_i(\xi) \cos \mu_n \xi d\xi + \int_0^t \exp(\alpha \mu_n^2 \tau) \right.
$$

$$
\left. \times \left[\frac{\alpha}{\lambda_w} \int_0^L q_v(\xi, \tau) \cos \mu_n \xi d\xi + (-1)^n \alpha \mu_n \phi(\tau) \right] d\tau \right\} \tag{9.77}
$$

This result is gained taking into account that: (i) in considering case the Green function is $\cos \mu_n x$ with μ_n defined by (9.70), and (ii) the last term in equation (9.77) corresponds to the boundary condition of the first kind $T = \phi(t)$ at $x = L$, and therefore, in conformity with notes concerning the Green function (9.67), the derivative $\partial G / \partial x$ along with given temperature $\phi(t)$ at $x = L$ is replaced for the last integral in Green function (9.67).

Comment 9.6 Different methods may be used to solve the same problem. For example, we considered solutions of similar problems as the last one applying the Duhamel's integral (Exam. 9.2) and Laplace transform (Exam 9.10). Despite the fact that these solution should follow one from another, it is not easy sometimes to show this, as it is, in particular, in the case of comparing solutions (9.61) and (9.5) with the last result (9.77).

Numerical Methods

9.5   What Method is Proper?

Finite-difference methods were developed and used long before the numerical methods became a powerful common tool for solving differential equations due to computers [301]. Because the understanding and technique of the finite-difference methods seems to be simpler than that of analytical methods, it was believed that the time of analytical methods was over. Although the techniques of analytical and numerical approaches indeed are different, both methods are based on the same fundamental principles. The only distinction between both approaches is that these basic principles are applied in the former case to infinite-small differences, whereas in the latter one they are used for small but finite size values. For example, both derivatives analytical and numerical are determined by the same principle. However, to calculate an analytical derivative, one needs to have some knowledge, and nevertheless, sometimes that might be not easy. At the same time, to obtain the finite-difference derivative using the difference between function values at two grid points is not at all a problem. This feature of numerical methods gives an impression that numerical approach is much simpler then analytical methods

In the early 1960s, this seeming simplicity leads to many unsuccessful attempts to solve numerically complex contemporary problems, showing that only a deep physics understanding, together with careful testing software, may yield the proper solution. It becomes clear that only an investigator who adopted a corresponding part of current knowledge can possess the complex technique of a numerical solution, which just seems to be simple, and then interpret the obtained results. Otherwise, an insufficiently considered and prepared program can give an unrealistic outcome.

After the applications of computers were expanded, analytical methods not only retained their importance, but gained new functions as well. In particular, despite the fact that there are many recommendations and rules for preparing and checking numerical programs, one of the best ways to test and control the accuracy is to compare the result obtained by software with the available proper analytical solution [306]. We mention below some cases where analytical solutions are especially useful in preparing and testing the software:

- The formulae for the finite-difference derivatives are usually obtained using a Taylor series. For a grid point i located midway between points $i - 1$ and $i + 1$, one obtains the following two expressions, using the first three terms of the Taylor series

$$f_{i-1} = f_i - (x_i - x_{i-1}) \left(\frac{df}{dx} \right)_i + \frac{1}{2}(x_i - x_{i-1})^2 \left(\frac{d^2f}{dx^2} \right)_i + \dots \tag{9.78}$$

$$f_{i+1} = f_i + (x_i - x_{i-1}) \left(\frac{df}{dx} \right)_i + \frac{1}{2}(x_i - x_{i-1})^2 \left(\frac{d^2f}{dx^2} \right)_i - \dots \tag{9.79}$$

Adding and subtracting these equations give formulae for the first two derivatives

$$\left(\frac{df}{dx} \right)_i = \frac{f_{i+1} - f_{i-1}}{x_i - x_{i-1}}, \qquad \left(\frac{d^2f}{dx^2} \right)_i = \frac{f_{i-1} - 2f_i + f_{i+1}}{(x_i - x_{i-1})^2} \tag{9.80}$$

These formulae may be used only if the function in question is analytic. If at some grid points the function is singular, such as, for example, at these points one or more derivatives become infinite, equations (9.80) cannot be applied. We encountered such a case in Section 3.1.1 showing that a wall temperature of a thermally thin plate at leading edge ($x = 0$) is not an analytic function of the coordinate x. It is rather presented as a series in integer powers of variable $x^{1/s}$, where s is the denominator of the exponent in the relation (3.1) for an isothermal heat transfer coefficient. For laminar or turbulent flow, this variable is $x^{1/2}$ or $x^{1/5}$, respectively. It is clear, that in this case the derivative with respect to x is proportional to $x^{-1/2}$, or to $x^{-4/5}$ for laminar, or turbulent flow, or to $x^{(1/s)-1}$ for any other value of s. Hence, this derivative becomes infinite at $x = 0$ for laminar, turbulent flows or for other cases in which an exponent r/s in the relation (3.1) for an isothermal heat transfer coefficient is less than unity. However, if one introduces a new variable $z = x^{1/s}$, the temperature distribution turns into analytical function, the derivative with respect to z becomes finite, and equations (9.80) can be used.

A similar situation is observed for the boundary layer equations in Prandtl-Mises form (7.36). These equations have a singularity at the surface where the stream function is zero because the velocity and temperature near the surface are presented in series of variable $\psi^{1/2}$ (see (7.37)). Therefore, near the surface, both derivatives of the velocity and of the temperature with respect to variable ψ become infinite, but introducing a new variable $z = \psi^{1/2}$ solves this problem again. Thus, analytical analysis gives us knowledge what variable should be used to overcome the singularities in general and particularly in the case of numerical approach.

- The other difficulty that usually arises in preparing a program for numerical solution is an attaining the proper distribution of the grid points inside a considering domain. This distribution should correspond to the studied function gradient distribution. In resolving this problem, significantly may assist an analyzing a field of analytical solution of similar problem. For example, it is known from exact solutions of the boundary layer problems that at the vicinity of the wall, the values of the velocity, and temperature gradients are maximum gradually decreasing as the distance from the wall increases, becoming zero at $y \to \infty$.
- Analytical solutions are useful also in approximating computation mesh pattern between the grid points. In particular, for the case of boundary layer equations, it is reasonable to apply the polynomial profiles that are usually used in integral methods (S. 7.6) or some other well-known distribution, for example, self-similar profiles (S. 7.5.2).
- In numerical applications, apparently the most important function of the analytic solution is that these can be applied as references in checking and testing the software. Comparing computation results with the exact solution estimates not only the usefulness of the tested program but gives the expected deviation from the exact data as well.

These examples demonstrate the role of the analytical solutions in creating the software. At the same time, there is no doubt of the significance of the numerical data without which the analytic solutions are practically unproductive. Therefore, it is meaningless to oppose the analytical and numerical methods. Rather, it is reasonable to consider both approaches as a united, combined method for investigation and solution of the contemporary problems. In fact, the numerical and analytical methods are means supplementing each other. Whereas the former is the a powerful technique for approximate solution of almost any complex problem using known mathematical models, the latter gives a possibility to find exact solutions of relatively simple problems, investigate general properties of a particular phenomena, and think of a new models on the basis of this data. Because the solution of any contemporary problem is a challenge, only employing both methods in combination gives hopes for getting the adequate results.

9.6 Approximate Methods for Solving Differential Equations

The approximate methods for solving differential equations were developed and widely used many years before they became a basis of modern numerical methods. However, before computers, these methods were usually used for entire computation domain as analytical means. The use of computers makes it possible to divide the computation domain on small subdomains and apply the same approximate methods for each small mesh. This modification vastly increases the calculation accuracy and converts these simple analytic approaches into the contemporary numerical methods.

Numerical methods differ from each other by means of discretization of the computation domain and by just-mentioned analytical methods for problem solutions in the small subdomains. Depending on the first procedure, the numerical methods can be classified in three basic groups: the finite-difference (FDM), finite-element (FEM), and boundary element (BEM) methods. The old one, the finite-difference method, usually uses for discretization the uniform grids and calculates in the points of these grids the derivatives by formulae (9.80). The two other techniques compute the values of the studied functions in each of subdomains, usually irregular distributed, employing the approximate analytical methods. The distinction between these two approaches results from various numbers of subdomains that are needed for a solution. Whereas the finite-element method requires the subdomains of the whole field of a function of interest, the boundary-element means uses only subdomins located on boundaries (see details below).

The finite-difference method is employed also in two modern modifications based on control-volume (CV) approach: the finite-difference method (CVFDM), and finite-element method (CVFEM). In the control-volume formulation, discretization equations are obtained as a result of integrating the applying differential equations over each of control volume. Here, the basic idea is that such an equation expresses the conservation laws for small finite volume just as the differential equation expresses these laws for an infinitesimal volume. The modified finite-difference method (CVFDM) for simplicity is called as previously the finite-difference method (FDM), but the modified finite-element method (FEM) is in the essence the finite-volume method (FVM) since the finite volume coincides with considering element. Thus, there are three commonly used types of finite-difference approaches: FDM, FVM, and BEM.

The distinction between different analytical approximate methods employed in numerical methods is convenient to describe using the weighted residual approach (see, e.g., [306], [49], or [50]). The weighted residual method is a generalized well-known approach of moments, which was widely used before computers became common, in particular, in the integral methods (S. 7.6). The concept of weighted residual method can be explained as follows.

Let's say we need to find an approximate solution of differential equation $F(u) = 0$ subjected to a given boundary condition. First, the given boundary condition is converted to a homogeneous form (S. 9.2.1). Then, some function $\tilde{u} = f(x)$ is chosen that exactly satisfies the boundary condition but contains one or more unknown parameters; for instance, a polynomial with undefined coefficients. Substituting this function into the just-mentioned differential equation yields a residual $R = F(\tilde{u})$ because $\tilde{u}$ is an approximate solution and, hence, does not satisfy the equation under consideration. Multiplying this residual by some weighted function w and integrating the result over the considering domain S, one tries to minimize an average error by equating this integral to zero

$$\int_S wR dx = \int_S wF(\tilde{u})dx = 0 \tag{9.81}$$

Applying a set of weighted functions gives as many algebraic equations as are required to determine the unknown parameters. Solving gained in this set of equations and substituting evaluated parameters into function $\tilde{u} = f(x)$ completed the desired solution of the considering equation $F(u) = 0$.

Various approximate methods applied to create a set of algebraic equation differ from each other by classes of weighted functions. For instance, the method of moments results in a set of weighted function: $1, x, x^2, \ldots$ The above-mentioned integral method is a case when the first moment is used only, that is $w = 1$. Using other approximate method leads us to different set of weighted functions and slightly different approximate solution (see examples below).

■ **Example 9.13: Consider a simple conduction problem for a plane plate governed by one-dimensional equation and simple boundary conditions**

$$\lambda \frac{d^2T}{dx^2} + q = 0, \quad x = 0 \quad T = T_0, \quad x = L \quad T = T_L \tag{9.82}$$

where T_0 and T_L are temperatures of edges of a plate and q is uniform heat source. To solve the problem using the method of moments, a new variables ϑ and $\xi = x/L$ are introduced, which change the problem to homogeneous form (S. 9.2.2)

$$\vartheta = T - T_0(1 - \xi) - T_L\xi, \qquad \frac{d^2\vartheta}{d\xi^2} + \overline{q} = 0, \qquad \xi = 0, \ \xi = 1 \ \vartheta = 0, \qquad \overline{q} = \frac{qL^2}{\lambda} \quad (9.83)$$

For using first two moments, one should apply function with two parameters that satisfies-boundary conditions, for instance, a relation $\vartheta = a_1\xi(\xi - 1) + a_2\xi^2(\xi^2 - 1)$. Substituting this relation into equation (9.83) gives the residual $R = 2a_1 + (12\xi^2 - 2) + \overline{q}$. Then, using weighted functions $w = 1$ and ξ for first two moments, we get from equation (9.81)

$$\int_0^1 [2a_1 + (12\xi^2 - 2)a_2 + \overline{q}]\,d\xi = 0 \qquad \int_0^1 [2a_1 + (12\xi^2 - 2)a_2 + \overline{q}]\xi\,d\xi = 0 \qquad (9.84)$$

These two equations determine coefficients $a_1 = \overline{q}/2$ and $a_2 = 0$ resulting in solution

$$T = \frac{qL^2}{\lambda}\frac{x}{L}\left(\frac{x}{L} - 1\right) + T_0\left(1 - \frac{x}{L}\right) + T_L\frac{x}{L} \qquad (9.85)$$

In this particular case, approximate method yields exact solution.

Similar solutions are obtained employing other methods. Some differences arise due to other sets of weighted functions. Thus, in Galerkin's method, the weighted functions are the same as the functions satisfying the boundary conditions. Therefore, in this case, the equations (9.84) should be constructed by multiplying the same residual $R = 2a_1 + (12\xi^2 - 2) + \overline{q}$ by parts of function $\vartheta = a_1\xi(\xi - 1) + a_2\xi^2(\xi^2 - 1)$ to find from equation (9.81) the following two expressions

$$\int_0^1 [2a_1 + (12\xi^2 - 2)a_2 + \overline{q}]\xi(\xi - 1)\,d\xi = 0, \qquad \int_0^1 [2a_1 + (12\xi^2 - 2)a_2 + \overline{q}]\xi^2(\xi^2 - 1)\,d\xi = 0$$
$$(9.86)$$

Solution of these equations gives the same results (9.85).

We show that analogous results yields also point collocation method with Dirac delta weighted function (9.66) and subdomain collocation method. In the last case, instead of multiplying a residual by weighted functions, the domain is divided into some subdomains. For instance, for two domains, instead of equations (9.84) one gets a system

$$\int_0^{1/2} [2a_1 + (12\xi^2 - 2)a_2 + \overline{q}]\,d\xi \qquad \int_{1/2}^1 [2a_1 + (12\xi^2 - 2)a_2 + \overline{q}]\,d\xi \qquad (9.87)$$

which again leads to the same solution (9.85).

Comment 9.7 The point collocation method mentioned above consists of satisfying the boundary conditions functions with some undefined parameters that are evaluated by fitting the solution in corresponding number of grid points.

Comment 9.8 The relation $\vartheta = a_1\xi(\xi - 1) + a_2\xi^2(\xi^2 - 1)$ that was used in the considered examples is not unique; there are many others with free coefficients satisfying the same boundary conditions that result in slightly different approximate solutions. We consider several methods with the same relation for ϑ in order to show that the basic distinction between those methods lies in the different weighted functions.

A special case is the subdomains method when the weighted function is taken as $w = 1$ for one of subdomains and $w = 0$ for all others at a time. Physically, this implies that the average residual error is zero over the each small domain. In particular, the control-volume formulation pertains to this type of methods. Applying this approach to the same simple one-dimensional equation (9.82), one gets after its integration

$$\lambda\left[\left(\frac{dT}{dx}\right)_b - \left(\frac{dT}{dx}\right)_a\right] + \int_a^b qdx = 0, \qquad \frac{T_{i+1} - T_i}{x_{i+1} - x_i} - \frac{T_i - T_{i-1}}{x_i - x_{i-1}} + \bar{q}(x_b - x_a) \qquad (9.88)$$

Here, a and b denote the midways points between x_{i-1}, x_i and x_{i+1}. In deriving the last equation (9.88) from the first one, it is assumed that the temperature between grid points changes linearly. This example shows the usual way of using a control-volume approach for determining derivatives in the form of first part of the second equation (9.88).

Comment 9.9 This way of defining the finite-difference derivatives differs from relations (9.80) certifying the derivatives via the Taylor series and is more grounded because the control-volume method provides satisfaction of conservation laws over each subdomain.

Another specific case, which is called weak formulation, is employed in the finite-element and boundary-element numerical methods. While in finite-difference method, an approximate solution is obtained by satisfying the differential equation at the grid points, in FEM and BEM, the solution is found by distributing the solution error over the each subdomain. To introduce the basic concepts of boundary-element approaches, we start again from the simple one-dimensional equation (9.82) for domain (0, 1). Multiplying this equation by some weighted function w and transforming the result by double integration by parts (setting: $u = w$, $dv = (d^2T/dx^2)dx$ and $u = (dw/dx)dx$, $dv = (dT/dx)dx)$), one obtains the following two relation

$$\int_0^1 \left(\frac{d^2T}{dx^2}w + \frac{q}{\lambda}w\right) dx = -\int_0^1 \left(\frac{dT}{dx}\frac{dw}{dx} + \frac{q}{\lambda}w\right) dx + \left[\frac{dT}{dx}w\right]_0^1 = 0 \qquad (9.89)$$

$$\int_0^1 \left(-\frac{dT}{dx}\frac{dw}{dx} + \frac{q}{\lambda}w\right) dx + \left[\frac{dT}{dx}w\right]_0^1 = \int_0^1 \left(T\frac{d^2w}{dx^2} + \frac{q}{\lambda}w\right) dx + \left[\frac{dT}{dx}w\right]_0^1 - \left[T\frac{dw}{dx}\right]_0^1 = 0$$

$$(9.90)$$

These two final weak expressions are starting statements for both element methods: the first expression for FEM and the second for BEM.

Comment 9.10 A weak solution of differential equation means that despite such solution is not differentiable, it may be used due to reduction of the continuity requirements (see S. 7.1.2.5) for weak defined expressions.

The basic idea of employing the two last equations is that those relations give a possibility to substitute the searching of some approximate solution by using a proper weighted function such that only boundary values would be needed to obtain the final result instead of the data of the whole computation domain usually required for that. This can be achieved in two ways [49, 50]:

- Selecting a weighted function that satisfies the homogenous form of the governing equation. Considering again the simple problem (9.82), we have the system of the following homogeneous differential equation with boundary conditions (9.82) and the weighted function satisfying this system.

$$d^2T/dx^2 = 0, \qquad T_0 = T_L = 0, \qquad w = a_1 x + a_2 \qquad (9.91)$$

Substituting this result into last expression (9.90) yields

$$\int_0^1 \frac{q}{\lambda} w \, dx + \left[\frac{dT}{dx} w\right]_0^1 - \left[T\frac{dw}{dx}\right]_0^1 = \int_0^1 \frac{q}{\lambda}(a_1 x + a_2) dx + \left(\frac{dT}{dx}\right)_1 (a_1 + a_2) - \left(\frac{dT}{dx}\right)_0 a_2 = 0$$
$$(9.92)$$

This equation should be satisfied for arbitrary a_1 and a_2. Therefore, collecting terms containing these coefficients leads to equations determining the derivatives at $x = 0$ and $x = 1$ and then gives the solution of the problem knowing only a weighted function

$$\int_0^1 \frac{q}{\lambda} x \, dx + \left(\frac{dT}{dx}\right)_1 = 0, \qquad \int_0^1 \frac{q}{\lambda} dx + \left(\frac{dT}{dx}\right)_1 - \left(\frac{dT}{dx}\right)_0 = 0, \qquad (9.93)$$

$$\left(\frac{dT}{dx}\right)_1 = -\frac{q}{2\lambda}, \qquad \left(\frac{dT}{dx}\right)_0 = \frac{q}{2\lambda}, \qquad T = \frac{q}{2\lambda} x(1-x) \qquad (9.94)$$

- Using a function (usually Dirac function) satisfying a homogeneous governing equation. Assuming that delta function meets the homogeneous equation (9.82), we find the corresponding weighted function

$$\frac{d^2w}{dx^2} = -\delta_i, \qquad \delta_i = \begin{cases} 1, & x = x_i \\ 0, & x \neq x_i \end{cases} \qquad w = \begin{cases} x \text{ at } x \leq x_i \\ x_i \text{ at } x > x_i \end{cases}, \qquad (9.95)$$

After that, applying relation (9.90) and first equation (9.95), we obtain

$$-\int_0^1 T\delta(x) dx + \int_0^1 \frac{q}{\lambda} w \, dx + \left(\frac{dT}{dx}\right)_1 w_1 - \left(\frac{dT}{dx}\right)_0 w_0 = 0 \qquad (9.96)$$

The first integral equals T_i for $\delta = \delta_i$ because the integral of delta function equals unit (see (9.66)). Solving equation (9.96) for this integral and taking in conformity with relation (9.91) that $w_0 = 0$, we determine temperature T_i as follows

$$T_i = \frac{q}{\lambda}\left(\int_0^{x_i} x\,dx + \int_{x_i}^1 x_i\,dx\right) + \left(\frac{dT}{dx}\right)_1 w_1 = \frac{q}{\lambda}\left(\frac{x_i^2}{2} + x_i - x_i^2\right) + \left(\frac{dT}{dx}\right)_1 x_i \qquad (9.97)$$

Using the boundary condition $T_i = 0$ at $x_i = 1$ yields unknown derivative $(dT/dx)_1$ at $x_i = 1$ and then gives the same solution (9.94) after substitution x for x_i in (9.97)

$$\left(\frac{dT}{dx}\right)_1 = -\frac{q}{2\lambda}, \qquad T = \frac{q}{\lambda}\left(x - \frac{x^2}{2}\right) - \frac{q}{2\lambda}x = \frac{q}{2\lambda}x(1-x) \qquad (9.98)$$

Transforming equation (9.90) applying delta function as it just was shown is a routine procedure in boundary element method. Modifying the last term of this equation using delta function (9.95) leads to the usual form of equation in the boundary element approach

$$T_i = \int_0^1 \frac{q}{\lambda}w\,dx + \left[\frac{dT}{dx}w\right]_0^1 - \left[T\frac{dw}{dx}\right]_0^1 \qquad (9.99)$$

This relation indicates that the unknown function is defined only by boundary conditions data. In contrast to that, the finite element method is based on equation (9.89), which reveals that the unknown function is defined by information of the whole domain. This may be seen by applying a similar modifying procedure to the last equation (9.89).

Comment 9.11 The procedures and features described here for a simple one-dimensional equations are valid for two- and three-dimensional problems [49, 50].

9.7   Computing Flow and Heat Transfer Characteristics

In practice, computing flow and heat transfer characteristics is associated with some specific difficulties [6, 306].

9.7.1   Control-Volume Finite-Difference Method

9.7.1.1   Computing Pressure and Velocity

The main difficulty in solving the Navier-Stokes equations is that the pressure is unknown. Although there is no a special equation for pressure, it is indirectly controlled by the continuity equation. This is achieved due to connection between the velocity field calculation and satisfying the continuity equation. As it is clear from Navier-Stokes equations (1.5) and (1.6), the velocity field can be calculated only if the pressure is known. At the same time, the continuity equation can be satisfied only when the velocities are computed using a proper pressure. Thus, to calculate the velocities and satisfy the continuity equation, the pressure should be known.

One well-known simple method to overcome this difficulty is to use Navier-Stokes equation in the stream function form that is obtained by cross-differentiating equations (1.5) and (1.6) to eliminate the pressure (S. 7.1.2.3). However, this approach is not applicable for three-dimensional case in which the stream function does not exist.

As shown in [306], the direst computing pressure procedure for standard three-point control volume approach with midway located grid point fails resulting in zero pressure. A real pressure distribution can be obtained by using special so-called staggered control volumes. In staggered control volume, the velocity components and pressure are calculated for main points located on the control volume faces that are set midway between two adjacent points in direction normal to corresponding velocity component. This means that x-velocity component u is calculated at the y-directed feces, and vice versa, the y-directed component v is calculated at the x-directed faces. Such method results in the u-components located on the left and on the right faces from the main point and in the v-components located above and below the main point. The location of the main point with respect to two adjacent points is not important. It is significant only that the main point should be on the control volume face.

Using the staggered grid eliminates the difficulties of calculating the pressure field, but a corresponding computer program becomes more complicated because it must record all information about the location of the velocity components and must perform tiresome interpolations. The software SIMPLE (Semi-Implicit Method for Pressure-Linked Equations) takes into account just discussed peculiarities of computing the velocity components and pressure [306]. This iterative procedure starts from guessing of the pressure field. Then, using the finite-difference technique and guessed pressure field, the Navier- Stokes equations in the form (1.5) and (1.6) for velocity components are solved. The second iteration consists of applying just found velocity components to calculate the pressure difference between adjacent grid points. This gives the new pressure field that is used to get new velocity components. These iterations are carried on until the continuity equation is satisfied. To control the process of iterations, the special equation is derived.

Later, to improve the process of convergence, revised versions were developed. The most used among others are: SIMPLER (SIMPLE Revised) [306], SIMPLEC (SIMPLE Consistent) [402], and SIMPLEM (SIMPLE Modified) [276]. A nonitertive procedure PISO (Pressure-Implicit with Splitting of Operators) also is developed [175].

Available studies reveal that family SIMPLE is a reliable, practical computing tool [175].

9.7.1.2   Computing Convection-Diffusion Equations

Consider a steady one-dimensional equation with only the convection and diffusion (or conduction, S. 7.1.1) terms, in which integration over a control volume yields

$$\frac{d}{dx}(\rho c u T) = \frac{d}{dx}\left(\lambda \frac{dT}{dx}\right), \quad (\rho c u T)_{i+1/2} - (\rho c u T)_{i-1/2} = \left(\lambda \frac{dT}{dx}\right)_{i+1/2} - \left(\lambda \frac{dT}{dx}\right)_{i-1/2} \tag{9.100}$$

Here, $i + 1/2$ and $i - 1/2$ denote a midway position between points $i + 1$, i and i, $i - 1$, respectively. Using the lineal approximation between grid points and the same simple equation (9.80) for the first derivative, we obtain from the last equation

$$\frac{1}{2}(\rho u c)_{i+1/2}(T_{i+1} + T_i) - \frac{1}{2}(\rho u c)_{i-1/2}(T_i + T_{i-1}) = \frac{\lambda_{i+1/2}(T_{i+1} - T_i)}{x_{i+1} - x_i} - \frac{\lambda_{i-1/2}(T_i - T_{i-1})}{x_i - x_{i-1}} \tag{9.101}$$

It can be shown that this fine-looking central-difference scheme leads to unrealistic results [306]. To see this, rearrange the last equation by collecting terms with the same T_n

$$T_i \left[\frac{(\rho cu)_{i+1/2}}{2} - \frac{(\rho cu)_{i-1/2}}{2} + \frac{\lambda_{i+1/2}}{x_{i+1} - x_i} + \frac{\lambda_{i-1/2}}{x_i - x_{i-1}} \right] = T_{i+1} \left[\frac{\lambda_{i+1/2}}{x_{i+1} - x_i} - \frac{(\rho cu)_{i+1/2}}{2} \right]$$

$$+ T_{i-1} \left[\frac{\lambda_{i-1/2}}{x_i - x_{i-1}} + \frac{(\rho cu)_{i-1/2}}{2} \right] \quad a_i = a_{i+1} + a_{i-1} + (\rho cu)_{i+1} - (\rho cu)_{i-1} \quad (9.102)$$

In the last equation, notation a stands for relations in the brackets. The two additional terms do not change this equation because according to continuity law they are equal: $(\rho u)_{i+1} = (\rho u)_{i-1}$. Consider now a simple example. Let, for instance, we have: $(\rho cu)_{i+1} = (\rho cu)_{i-1} = 3/2$, $\lambda_{i+1}/(x_{i+1} - x_i) = \lambda_{i-1}/(x_i - x_{i-1}) = 1/2$, $T_{i+1} = 200$, $T_{i-1} = 100$. Then, from (9.102) it follows that: $T_i = 75$, but if $T_{i+1} = 100$, $T_{i-1} = 200$, we get from (9.102) that $T_i = 225$. These results are unrealistic because T_i cannot fall outside its neighbors T_{i+1} and T_{i-1}. It is obvious that such unrealistic results are possible in any case if $|\rho cu|$ exceeds $2\lambda/\Delta x$.

There are some possibilities to overcome this difficulty. The simplest way is to apply the upwind scheme where the midway between i and $i \pm 1$ points are determined as: $T_{i\pm1/2} = T_i$ if $(\rho cu)_{i\pm1/2} > 0$ and $T_{i\pm1/2} = T_{i\pm1}$ if $(\rho cu)_{i\pm1/2} < 0$ and with unchanged diffusion terms. The exact solution of the first equation (9.100) subjected to boundary conditions $x = 0$, $T = T_0$, $x = L$, $T = T_L$ and presented in the form

$$\frac{T - T_0}{T_L - T_0} = \frac{\exp \text{Pe}(x/L) - 1}{\exp \text{Pe} - 1}, \quad \text{Pe} = \frac{\rho cuL}{\lambda} \quad (9.103)$$

gives an understanding of the applicability of upwind scheme. Analysis shows that for large values of $|\text{Pe}|$, the temperature at the middle is nearly equal to that at upwind boundary, and this is the assumption used in upwind scheme. However, in this scheme, the boundary temperature remains the same for all values of Peclet number. The other imperfection is that for large $|\text{Pe}|$ at the middle, the derivative dT/dx is almost zero so that the diffusion is almost absent. At the same time, in upwind scheme the diffusion is calculated always applying linear profile, which overestimates it for large Peclet numbers.

Two schemes with qualitative behavior close to solution (9.103) are usually used instead of central-difference scheme. One is the hybrid scheme that approximated fairly such behavior working alike the central-difference scheme for the range $-2 \le \text{Pe} \le 2$ and reducing the diffusion to zero similar to upwind scheme outside of this range. That is the reason why this scheme is named hybrid scheme. A power-law scheme constructed of four ranges of Peclet number, $\text{Pe} < -10$, $-10 \le \text{Pe} < 0$, $0 \le \text{Pe} \le 10$, and $\text{Pe} > 10$, better approximate the exact solution (9.103).

9.7.1.3 False Diffusion

There are two types of false diffusion. The first type is a usual misunderstanding when by comparison of central-difference and upwind schemes using the Taylor series, one concludes that the upwind scheme produces the false diffusion. Indeed, the central-difference scheme is better than upwind one only in the case of small Peclet numbers when the Taylor expansion is applicable. For large Peclet numbers, the truncating Taylor series cannot be used for analyzing convention-diffusion dependence, which in this case is of exponential type. At the same time, as is discussed above a central-difference scheme leads to unrealistic results in the case of large Peclet numbers.

The other type is the real false diffusion that arises in situation when the calculations show the present of diffusion despite the diffusion coefficient is zero. For example, if two parallel two-dimensional streams of equal velocities and different temperatures come in contact, the diffusion process forms a mixed layer only in the case of nonzero diffusion coefficient. However, when the diffusion coefficient is zero, such two streams remains separated with temperature discontinuity at the interface. Thus, if in such situation, the computer program shows smeared profile in the cross section, it is obvious that a real false diffusion is taking place.

In general, false diffusion arises when the flow is oblique to the grid lines and there is a gradient in the direction normal to the flow. Since the false diffusion is the most severe when the flow direction makes an angle 45° with the grid lines [306], the intensity of false diffusion can be reduced by adjusting the flow along the grid lines. The other way to reduce the false diffusion is to use a small Δx and Δy, knowing that this results in small Peclet numbers when the central-difference scheme works perfectly.

9.7.2   Control-Volume Finite-Element Method

The control-volume finite-element method is basically very close to the control-volume finite-difference method described above. The only advantage that distinguishes the former from the latter is the ability to use the irregular grids, since in the finite-difference approach, mainly uniform grids are employed. Such irregular, for example, triangular grids are more flexible and allow providing local grid refinement.

The difficulties just discussed are inherent in the finite-element method as well. These difficulties have been resolved leading to a similar control-volume finite-element method with the following basic features [6, 26]:

- For the triangular grids, the values of dependent variables are calculated for the grid points that lie at the vertices of the triangles, which plays a role of main points. The lines joined the centroid of each triangular element with midpoints of its sides divide each element in three equal areas, regardless of the form of triangle element. These areas collectively construct the nonoverlapping contiguous polygonal volume elements that are similar to these in finite-difference approach.
- Many CFD (computational fluid dynamics) codes used the staggered grids that do not have the problems of central-difference schemes. However, the staggered grids cannot be used for nonorthogonal grids and unstructured meshes, which are typical for the finite-element approach. Therefore, in the finite-element method, instead of the staggered grids, two other approaches are used. One consists of unequal-order formulations for pressure and velocity components in which for the former are used a sparser grid and a lower-order interpolation than that for the latter. In the second approach, the equivalent of the co-located momentum-interpolation scheme is applied. However, the first approach requires two sets control volumes that makes the calculation awkward and an excessively fine grids in the case of high Reynolds numbers or pressure gradients.

- As a result, the co-located momentum-interpolation scheme has been adopted in computation practice. In these schemes, both the velocities and pressure are calculated at the same set of nodes located at the centre of the control volume in contrast to the case of staggered grids when the pressure and the velocities are determined at different faces of the control volume. The formulae for velocities and pressure in co-located methods are derived employing the discretization of the momentum and continuity equations for the staggered grids and due to that do not have problems of the central-difference as well.

10

Conclusion

How Close should be a Model to Nature?

> For every complex problem there is an answer that is clear, simple, and wrong.
>
> Henry Mencken

This book presents applications of modern mathematical models of heat transfer and fluid flow developed and widely used during the last fifty years after computers became common. These models and corresponding methods of problem solutions in comparison with previous approaches are physically and mathematically much stronger grounded, comprising the powerful tools for investigating natural and engineering systems. Although the contemporary models are formulated relatively recent, they are obtained as a result of numerous improvements of the first simple models, gradually approaching the current stage of development. Because of that, the proximity of the model to the natural prototype represents on each step the level of knowledge in a specific field.

For example, the contemporary direct numerical simulation methods were created as a result of enormous studies for the past 200 years. Starting from Navier and other scholars' attempts as well as equations given by Stokes in 1845, the developers greatly improved the initial simple models of turbulent flow in many steps, including such keystone contributions as proposed by Reynolds' dimensionless number, his averaging of Navier-Stokes equations, the first simple Prandtl mixing-length model, the Kolmogorov turbulent energy equation, the Smogorinsky LES procedure, and finally DES models.

Examples considered in this text show that despite the current restricted computer resources, the results obtained by the last versions of DES differ in principle from former solutions. In particular, the presented numerical data and contour patterns demonstrate successful simulation of flow separation around the bluff bodies and models of aircraft at high Reynolds numbers. In contrast to other methods, the direct simulation approaches also provide the numerical information of the instantaneous parameters such as velocity components fluctuation, stresses, correlations of parameters, and so on, which before could only be attained experimentally. Due to that development, these results are viewed as experimental data, albeit gained computationally.

Applications of Mathematical Heat Transfer and Fluid Flow Models in Engineering and Medicine,
First Edition. Abram S. Dorfman.
© 2017 John Wiley & Sons Ltd. Published 2017 by John Wiley & Sons Ltd.

As the computer power increases, new feasibilities will be achieved, resulting in fresh, original thoughts, ideas, and improved models.

The leading role of mathematical models in research is clearly seen as well from another part of the book, which presents applications of modern methods in heat transfer. Comparing the boundary conditions of third and forth kinds reveals that using conjugate models leads to a better physical understanding of heat transfer features as well as to diverse new applications. Such results as essentially different effects of positive and negative temperature head gradients on heat transfer coefficient, the analogy of this effect with a well-known effect of pressure gradients on the friction coefficient, as well as the data of Reynolds and Prandtl numbers influence on the temperature head effects could not be observed within the frame of boundary conditions of the third kind. The same merit of conjugate models showed applications in various areas from modeling industrial and technology processes to cooling, drying, thermal treatment, and production of goods and food.

A special case is the usage of mathematical models in biology and medicine. The employing of mathematical models cardinally changed the theoretical methods in these fields of science from qualitative considerations to the quantitative approaches presenting the results in analytical, numerical, or graphical dependences between parameters. Even the first results of possible bacteria trajectories in the ureter obtained by a simple linear model showed just how reliable and informative the modeling of fluid flow in human organs is. The further advanced studies applying $k - \omega$ and $k - \varepsilon$ turbulent models and peristaltic flow simulations gave new insight into processes in healthy and pathologic body organs. This was achieved due to gained patterns of flow in normal and disordered blood vessels, urine and bile channels with or without stones, simulation data of thermal, magnetic, and electromagnetic effects during medical procedures, in particular, direct drag infusion, delivery antibodies into tumors, endoscopy, and hyperthermia for cancer treatment.

In fact, any research starts from choosing or creating a model, and the question of proximity of the employing model to nature is one of the first to answer. A simple response to this question is unrealistic because creating a model is a challenge. The two conflicting essential parts—the system of equations describing a problem and algorithm specifying the solution—constitute this issue. Usually, the more complete and hence, more complex, is the system of governing system of equations, the more involved is the algorithm for solving it. Although it is obvious that a complete model is desired, the realization of a complex pattern requires high computer memory and computing time, which leads to excessive costs. Therefore, a balance between the simplicity and usefulness is one of the basic model characteristics that have to be provided.

The complexity of the model largely depends particularly on the problem in question and on the details that we are supposed to get. For example, in the case of a high Reynolds number, if only friction, heat, and/or mass transfer coefficients have to be estimated, the model may be based on boundary layer equations. In such a case, there is no reason to apply software; rather it is much easier to use the analytical or correlative relations. A similar situation takes place if the flow is characterized by a low Reynolds number, and if the more simple creeping relations are applicable, instead of Navie-Stokes and full energy equations. On the other hand, in the case of moderate Reynolds numbers as well as regardless of Reynolds number values when the parameter profiles or flow patterns are required, the Navie-Stokes and full energy equations should be solved. Such a compromise between possibilities and specific requirements one should achieve in all other aspects of the model, including initial and boundary conditions, type of numerical methods and grid discretization, accuracy and applicability of results, computer memory, time and costs of calculations.

The great investigative experience represented by examples reviewed in the basic text along with conclusions outlined in this short analysis show that **the model should be as simple as possible while describing the essential physical properties of studied phenomenon with satisfactory accuracy.**

The modern methods introduced by this book have a long-term bright future. Nevertheless, the detailed forecast for each means widely depends on the current degree of it evolution. Thus, to predict the future of conjugate heat transfer approach, we take into account that this subject has been studied since the time of Newton, and that the current methods are highly developed. Proceeding from these facts, it may be expected that in the future, the existing conjugate methods will be widely used for solving more complicated, realistic problems, and the conjugate principles will be extended to study the other phenomena where interface effects are important, such as, for instance, in combustion or biology structures.

From quite a different situation one proceeds to predict the future of modeling in biology or medicine in which the mathematical methods are only first applied to obtain the quantitative data of processes in nature and in human organs. The review of examples shows that these first mathematical models are simple, based on weighty assumptions, and are mainly semi-conjugate considering only the impact of the imposed propagation wave on fluid flow, ignoring the backward effects of fluid flow on the organ wall's motion. In such a case, it is reasonable to anticipate that the coming years will bring the more physically grounded models, and wide, profound investigations will be accomplished, providing a deep insight and resulting in new methods of treatment.

The more assured forecast may be devised for the turbulence direct simulation. We have seen that even now, under limited computer means, the new modern methods made great progress, especially because the last models provided the first successful simulation of separation flows around real objects at rather high Reynolds numbers. Therefore, there is no doubt that growing computer resources will lead to further essential improvements of models and simulation techniques, opening new horizons in understanding the turbulence nature and in practical applications. However, estimations show that direct numerical simulation of flow past airborne or ground vehicle at real Reynolds numbers requires computer efficiency that may be possible approximately in 2045 (Spalart) [368]. When it came true, the turbulence "the most important unsolved problem of classical physics", as it was characterized by Richard Feynman, would be finally solved almost two centuries after its first publication.

Finally, it is worth to mention that although studying the new efficient methods requires much time and energy, this gives a person a powerful tool for solving complicated contemporary problems, which provides physical understanding and highly accurate data of processes of interest.

References

1. Abbasi, F.M., Hayat, T., Ahmad, B. and Chen, G.Q. (2014) Peristaltic motion of a non-Newtonian nanofluid in an asymmetric channel. *Zeitschrift für Naturforschung Section, A-A J. Physical Science*, **69**, 451–461.
2. Abd-Alla, A.M., Abo-Dahab, S.M. and El-Semiry, R.D. (2014) Peristaltic flow in cylindrical tubes with an endoscope subjected to effect of rotation and magnetic field. *J. Comput. Theor. Nanoscin.*, **11** (4), 1040–1048.
3. Abd-Alla, A.M. and Abo-Dahab, S.M. (2015) Magnetic field and rotation effects on peristaltic transport of a Jeffrey fluid in an asymmetric channel. *J. Magnetism Magnetism Materials*, **374**, 680–689.
4. Abdoli, A., Dulikravich, G.S., Bajaj, C. and Jahania, M.S. (2014) Human heart conjugate simulation: unsteady thermo-fluid-stress analysis. *Int. J. Num. Meth. Biomed. Engin*, **30** (11), 1372–1386.
5. Abramzon, B.M. and Borde, I. (1980) Conjugate unsteady heat transfer from a droplet in creeping flow. *AIChE, J*, **26**, 536–544.
6. Acharya, S., Baliga, B.R., Karki, K. *et al.* (2007) Pressure-based finite-volume methods in computational fluid dynamic. *ASME J. Heat Transfer*, **129**, 407–424.
7. Achenbach, E. (1972) Experiments on the flow past spheres at very high Reynolds numbers. *J. Fluid Mech*, **54** (3), 565–575.
8. Aerotherm corporation, Ed. (1992) *User's manual non proprietary aerotherm charring material response and ablation program, CMA 925*, Mountain View, California.
9. Ahn, J., Choi, H. and Lee, J.S. (2007) Large eddy simulation of flow and heat transfer in a rotating ribbed channel. *Int. J. Heat Mass Transfer*, **50** (25–26), 4937–4947.
10. Akbar, N.S. and Nadeem, S. (2011) Combined effects of heat and chemical reaction on the peristaltic flow of Carreau fluid model in divergent tube. *Int. J. Num. Meth. Fluids*, **67**, 1818–1832.
11. Al-Amiri, A., Khanafer, K. and Vafai, K. (2014) Fluid-structures interactions in tissue during Hyperthermia. *Num. Heat Transfer Part A*, **66** (1), 1–16.
12. Al-Bakhit, H. and Fakheri, A. (2006) Numerical simulation of heat transfer in simultaneously developing flows in parallel rectangular duct. *Appl. Therm. Engin.*, **26**, 596–603.
13. Ali, N., Sajid, M., Jsved, T. and Abbas, Z. (2010) Heat transfer analysis of peristaltic flow in a curved channel. *Int. J. Heat Mass Transfer*, **53**, 3319–3325.
14. Ambrok, G.S. (1957) The effect of surface temperature variability on heat transfer exchange in laminar flow in a boundary layer. *Soviet Phys.-Tech. Phy*, **2**, 738–748.
15. Amin, M.R. and Gawas, N.L. (2003) Conjugate heat transfer and effects of interfacial heat flux during the solidification process of continuous castings. *J. Heat Transfer*, **125**, 339–349.

16. Antonia, R.A. and Kim, J. (1991) Turbulent Prandtl number in the near-wall region of turbulent channel flow. *Int. J. Heat Mass Trans.*, **34**, 1905–1908.

17. Argyropoulus, C.D. and Markatos, N.C. (2015) Recent advances on the numerical modeling of turbulent flows. *Appl. Math. Model.*, **39** (2), 693–732.

18. Asghar, S., Nussain, Q., Hayat, T. and Alsaadi, F. (2014) Hall and ion slip effects on peristaltic flow and heat transfer analysis with Ohmic heating. *Appl. Math. Mech-English Ed.*, **35** (12), 1509–1524.

19. Ashrafian, A., Andersson, H.I. and Manhart, M. (2004) DNS of turbulent flow in a rod-roughened cannel. *Int. J. Heat Flow*, **25**, 373–383.

20. Atabek, H.B. and Lew, H.S. (1966) Wave propagation through a viscous incompressible fluid contained in an initially stressed elastic tube. *Biophys. J.*, **6**, 481–503.

21. Atabek, H.B. (1968) Wave propagation through a viscous incompressible fluid contained in a tethered, initially stressed, orthotropic elastic tube. *Biophys. J.*, **8**, 626–649.

22. Aydin, O., Avci, M., Bali, T. and Arici, M.E. (2014) Conjugate heat transfer in a duct with an axially varying heat flux. *Int. J. Heat Mass Transfer*, **76**, 385–392.

23. Ayukawa, K., Kawai, T. and Kimura, M. (1981) Streamlines and path lines in peristaltic flows at high Reynolds number. *Bull. Japan Soc. Mech. Engin.*, **24**, 948–955.

24. Baiocco, P. and Bellomi, P. (1996) A coupled thermo-ablative and fluid dynamic analysis for numerical application to solid propellant rockets, *AIAA 96-1811, 31 st. AIAA Thermophysics Conference*, June 17–20, New Orleans, LA.

25. Baldwin, B.S. and Lomax, H. (1978) Thin-later approximation and algebraic model for separated turbulent flows. *AIAA paper*, 78–257.

26. Baliga, B.R. and Patankar, S.V. (1983) A control-volume finite-element method for two-dimensional incompressible fluid flow and heat transfer. *Num. Heat Transfer*, **6**, 245–261.

27. Bamhardt, M. and Candler, G.V. (2012) Detached eddy simulation of the Reentry-F flight experiment. *J. Spacecraft and Rockets*, **49** (4), 691–699.

28. Bang, B.W., Jeong, S., Lee, D.H. *et al.* (2012) The biodurability of covering materials for metallic stents in a bite flow phantom. *Digestive Diseases Sci.*, **57** (4), 1056–1963.

29. Banks, J. and Brssloff, N.W. (2007) Turbulence modeling in three-dimensional stenosed arterial bifurcations. *J. Biomech. Engin. ASME*, **129** (1), 40–50.

30. Barmpas, F., Bouris, D. and Moussiopoulos, N. (2009) 3D Numerical simulation of the transient thermal behavior of a simplified building envelope under external flow. *Appl. Thermal. Engin.*, **19**, 3716–3720.

31. Barozzi, G.S. and Pagliarini, G. (1985) A method to solve conjugate heat transfer problems-the case of fully-developed laminar flow in a pipe. *ASME J. Heat Transfer*, **107** (1), 77–83.

32. Barton, C. and Raynor, S. (1968) Peristaltic flow in tubes. *Bull. Math. Biophys.*, **30**, 663–680.

33. Batchelor, G.K. (1967) *An Introduction to Fluid Dynamics*, Cambridge University Press, New York.

34. Bathe, K.J., Zhang, H. and Ji, S. (1999) Finite element analysis in fluid flows fully coupled with structural interactions. *Comput. Struct.*, **72**, 1–16.

35. Baxter, L.T., (1990) Transport of fluid and macromolecules in normal and neoplastic tissue, *Ph. D. Thesis. Carnegie Mellon University*, Pittsburg, PA.

36. Baxter, L.T. and Jain, R.K. (1991) Transport of fluid and macromolecules in tumors 1V.A microscopic model of the perivascular distribution. *Microvas. Resear.*, **41**, 252–272.

37. Beckers, G. and Dehez, B., (2013) Design and modeling of a quasi-static peristaltic piezoelectric micropamp, *Int. Conf. Elekt. Mash. Syst.*, Busan, South Korea, Okt. 26–29, 2013.

38. Beckers, G. and Dehez, B., (2014) Design and modeling of an electromagnetic peristaltic micropump, *IEEE/ASME Int. Conf. Adven. Intellig. Mechanotron.* Besacon, France, Jul. 08–11.2014.

39. Bejan, A. (1982) Second law analysis in heat transfer and thermal design, *In Advances Heat Transfer*, vol. **15** (eds T.F. Irvine and J.P. Hartnett), pp. 1–58.

40. Bensow, R.E., Persson, T., Fureby, C. *et al.* (2004) *Large Eddy Simulation of the Viscous Flow Around Submarine Hulls*, St. John's Newfoundland and Labrador, Canada, pp. 1–16.

41. Bertuzzi, A., Salinary, S., Mancinelli, R. and Pescatori, M. (1983) Peristaltic transport of a solid bolus. *J. Biomech.*, **16** (7), 459–464.

42. Besharatian, A., Kumar, K., Peterson, R.L., Bernal, L.P. and Najafi, K. (2012) A scalable, modular, multi-stage, peristaltic, electrostatic gas micro-pump, *25 IEEE Int. Conf. Micro Elect. Mech. Syst.*, Paris, France, Jan. 29–feb. 02. 2012.

43. Bhatt, B.S. and Sacheti, N.C. (1979) On the analogy in slip flows. *Indian J. Pure Appl. Math.*, **10**, 303–306.

44. Bird, R.B., Stewart, W.E. and Lightfoot, E.N. (2005) *Transport Phenomena*, 2 edn, Wiley & Sons Inc, New York.

45. Blasé, T.A., Guo, Z.X., Shi, Z. *et al.* (2004) A 3D conjugate heat transfer model for continuous wire casting. *Mater. Sci. Engin. A*, **365**, 318–324.

46. Boyd, R.D. and Zhang, H. (2006) Conjugate heat transfer measurement with single-phase and water flow boiling in a single-side heated monoblock flow channel. *Int. J. Heat Mass Transfer*, **49**, 1320–1328.

47. BPD, Ed. (1994) EBM motor, low density liner characterization, RE-EBM-7101, Ed.3.

48. Bradshow, P., Ferriss, D.H. and Atwell, N.P. (1967) Calculation of boundary layer development using the turbulent energy equation. *J. Fluid Mech.*, **28** (3), 593–616.

49. Brebbia, C.A. and Walker, S. (1980) *Boundary Element Technique in Engineering*, Butterworths, London-Boston.

50. Brebbia, C.A. and Dominguez, J. (2001) *Boundary Elements. An Introductory Course*, Second edn, WIT Press, Boston.

51. Brosh, T., Patel, D., Wacks, D. and Chakraborty, N. (2015) Numerical investigation of localized forced ignition of pulverized coal particle-laden mixtures: A direct numerical simulation (DNS) analysis. *Fuel*, **145**, 50–62.

52. Bukhvostova, A., Russo, E., Kuerten, J.G.M. and Geurts, B.J. (2014) DNS of turbulent droplet-laden heated channel flow with phase transition at different initial relative humidities. *Int. J. Heat Fluid Flow*, **50**, 445–455.

53. Burdo, O.G., Milinchuk, S.L. and Kovalenko, E.A. (2003) Conjugate heat and mass transfer in crystallization from food solutions, *Heat Transfer Res.* **34** (i 5 - 6): 170, 14 pages.

54. Burn, J.C. and Parkes, T. (1967) Peristaltic motion. *J. Fluid Mech.*, **29** (4), 731–743.

55. Bykova, A.A. and Regirer, S.A. (2005) Mathematical models in urinary system mechanics (review). *Fluid Dynamics*, **40** (1), 1–19.

56. Carew, E.O. and Pedley, T.J. (1997) An active membrane model for peristaltic pumping: part 1-Periodic activation waves in an infinite tube. *J. Biomech. Eng. Transact. ASME*, **119**, 66–76.

57. Carlson, G.A., Love, J.T., Urenda, R.S. *et al.* (1980) A portable insulin infusion system with rotary solenoid-driven peristaltic pump. *Med. Prog. Tech.*, **8** (1), 49–56.

58. Carslaw, H.S. and Jaeger, J.C. (1986) *Conduction of Heat in Solids*, 2nd edn, Clarendon Press, Oxford.

59. Cebeci, T. and Smith, A.M.O. (1974) Analysis of turbulent boundary layer, *Ser. In Appl. Math. Mech*, vol. XV, Academic Press.

60. Cebeci, T. and Bradshaw, P. (1984) *Physical and Computational Aspects of Convective Heat Transfer*, Springer, New York.

61. Chakravarty, S. and Chowdhury, A.G. (1988) Response of blood flow through an artery under stenotic conditions. *Rheol. Acta*, **27**, 418–427.

62. Chang, C.M., Cheng, W.T., Huang, C.E. and Du, S.W. (2009) Numerical prediction on the erosion in the hearth of a blast furnace during tapping process. *Int. Commun. Heat Mass Transfe*, **36**, 480–490.

63. Chang, K. (1970) *Separation of Flow*, Pergamon Press, New York.

64. Chang, S.L., Chiu, K.C., Hsu, F.Y. and Chen, J.K. (2012) Design of a novel pump for bio-applications, Ed. Thienpont, H., Mohr, J., Zappe, H. and Nakajima, H., *Conf. Micro-Optics*, Brussels, Belgium, Apr. 16–19. 2012.

65. Chapman, D. and Rubesin, M. (1949) Temperature and velocity profiles in the compressible laminar boundary layer with arbitrary distribution of surface temperature. *J. Aeronaut. Sci.*, **16**, 547–565.

66. Chatterjee, D. and Chakraborty, S. (2005) Large eddy simulation of laser-induced surface-tension-driven flow. *Metallurg. Mater. Transact. B-Process Metallurg. Mater. Sci.*, **36** (6), 743–754.

67. Chen and Pei, D. (1989) A mathematical model of drying processes. *Int. J. Heat Mass Transfer*, **32**, 297–310.

68. Chen, Q., Zhang, Y., Elad, D. *et al.* (2013) Navigating the site for embryo implantation: biomedical and molecular regulation of intrauterine embryo distribution. *Mol. Aspects Med.*, **34** (5), 1024–1042.

69. Chida, K. and Katto, Y. (1976) Conjugate heat transfer of continuously moving surfaces, *Int. J. Heat Mass Transfer* **19** (5) : 461–470.

70. Chiruvella, R.V., Jaluria, Y., Esseghir, M. and Sernas, V. (1996) Extrusion of non-Newtonian fluids in a single-screw extruder with pressure back flow. *Polymer Eng. Sci.*, **36**, 358–367.

71. Chiu, W.K.S., Richards, C.J. and Jaluria, Y. (2001) Experimental and numerical study of conjugate heat transfer in an horizontal channel heated from below. *ASME J. Heat Transfer*, **123** (4), 688–697.

72. Choi, C.Y. and Hsieh, C.K. (1992) Solution of Stefan problems imposed with cyclic temperature and flux boundary conditions. *Int. J. Heat Mass Transfer*, **35**, 1181–1195.

73. Choi, C.Y. (2006) A boundary element solution approach for conjugate heat transfer problem in thermally developing region of a thick walled pipe. *J. Mech. Sci. Techn.*, **20**, 2230–2241.

74. Chrispell, J. and Fauci, L. (2011) Peristaltic pumping of solid particles immersed in viscoelastic fluid. *Math. Model. Natur. Phenom.*, **6** (5), 67–83.

75. Chu, W.K. and Fang, J. (2000) Peristaltic transport in a slip flow. *European Phis. J. B*, **16**, 543–547.

76. Clauser, F.H. (1956) The turbulent boundary layer, *Advances in Applied Mechanics*, vol. IV, Academic Press, New York, pp. 1–51.

77. Coccarelli, A. and Nithiarasu, P. (2015) A robust finite element modeling approach to conjugate heat transfer in flexible elastic tubes and tube network. *Num. Heat Transfer, part A*, **67** (5), 513–530.

78. Cole, K.D. (1997) Conjugate heat transfer from a small heated strip. *Int. J. Heat Mass Transfer*, **40** (11), 2709–2719.

79. Coles, D.E. and Hirst, E.A. (1969) *Computation of turbulent boundary layer, 1968 AFOSR-IFP_Stanford Conference*, vol. II, Stanford University, CA.

80. Comini, G., Nonino, C. and Savino, S. (2008) Modeling of conjugate conduction and heat and mass convection in tube-fin exchangers. *Int. J. Num. Methods Heat Fluid Flow*, **18**, 954–968.

81. Constantinescu, G.S., Pacheco, R. and Squires, K.D. (2002) Detached-eddy simulation of flow over a sphere. *AIAA*, 2002–0425.

82. Croce, G. (2001) A conjugate heat transfer procedure for gas turbine blades. *Annals New York Academy Science*, **934**, 273–280.

83. Cummings, R.M. and Schutte, A. (2013) Detached eddy simulation of the flow field about VFF-2 delta wing. *Aerospace Sci. Techn.*, **24**, 66–76.

84. Dan, C. and Wachs, A. (2010) Direct numerical simulation of particulate flow with heat transfer. *Int. J. Heat Flow*, **31**, 1050–1057.

85. Davie, C.T., Piarce, C.J. and Bicanic, N. (2006) Coupled heat and moisture transport in concrete at evaluated temperatures-effects of capillary pressure and adsorbed water. *Num. Heat Transfer, part A*, **49** (8), 733–763.

86. Davis, E.J. and Gill, W.N. (1970) The effects of axial conduction in the wall on heat transfer with laminar flow. *Int. J. Heat Mass Transfer*, **13**, 459–470.

87. Davis, E.J. (1973) Exact solution for a class of heat and mass transfer problems. *Can. J. Chem. Engin.*, **51**, 562–572.

88. Davis, E.J. and Venkatesh, S. (1979) Solution of conjugated multiphase heat and mass transfer problems. *Chem. Engin. Sci.*, **34** (6), 775–787.

89. De Bonis, M.V. and Ruocco, G. (2014) Conjugate heat and mass transfer by jet impingement over a mist protrusion. *Int. J. Heat Mass Transfer*, **70**, 192–201.

90. Deck, S. (2005) Zonal-detached eddy simulation of the flow around a high-lift configuration. *AIAA J.*, **43** (11), 2372–2384.

91. Deck, S. (2012) Recent improvements in the zonal detached eddy simulation (ZDES) formulation. *Theor. Comput. Fluid Dyn.*, **26** (6), 523–550.

92. Deck, S., Renard, N., Laraufie, R. and Sagaut, P. (2014) Zonal detached eddy simulation (ZDES) of a spatially developing flat plate turbulent boundary layer over the Reynolds number range 3150 <= Re-theta <=14000. *Phys. Fluids*, **26** (2), 025116.

93. Defraeye, T., Blocken, B. and Carmeliet, J. (2012) Analysis of convective heat and mass transfer coefficients for convective drying of a porous flat plate by conjugate modeling. *Int. J. Heat Mass Transfer*, **55**, 112–124.

94. Dirven, S., Xu, W., Cheng, L.K. and Allen, J. (2015) Biomimetic investigation of intrabolus pressure signatures by a peristaltic swallowing robot. *IEEE Transac. Instrum. Measur.*, **64** (4), 967–974.

95. Divo, E., Steinthorsson, E., Kassab, A.J. and Bialecki, R. (2002) An iterative BEM/ FVM protocol for steady-state multi-dimensional conjugate heat transfer in compressible flows. *Engin. Analys. Bound. Elem.*, **26** (5), 447–454.

96. Divo, E. and Kassab, A.J. (2007) An efficient localized radial basis function meshless method for fluid flow and conjugate heat transfer. *ASME J. Heat Transfer*, **129**, 124–136.

97. Dodoulas, I.A. and Navarro-Martinez, S. (2015) Analysis of extinction in a non-premixed turbulent flame using large eddy simulation and the chemical explosion mode analysis. *Comb.Theory Model.*, **19** (1), 107–129.

98. Dogan, T., Sadat-Hosseini, H. and Stern, F. (2015) DES of Delft catamaran at static drift condition. *Naval Engin. J.*, **126** (4), 97–101.

99. Dolinsky, A.A., Dorfman, A.S. and Davydenko, B.V. (1991) Conjugate heat and mass transfer in continuous processes of convective drying. *Int. J. Heat Mass Transfer*, **34** (11), 2883–2889.

100. Dorfman, A.S. (1970) Heat transfer from liquid to liquid in a flow past two sides of a plate. *High Temperature*, **8** (3), 515–520.

101. Dorfman, A.S. (1971) Exact solution of the thermal boundary layer equation with arbitrary temperature distribution on streamlined surface. *High Temperature*, **9**, 870–878.

102. Dorfman, A.S. (1971) Solution heat transfer equation for equilibrium turbulent boundary layer when the temperature distribution on the streamlined surface is arbitrary. *Fluid Dynamics*, **6**, 778–785.

103. Dorfman, A.S. (1972) Calculation of the thermal fluxes and the temperatures of the surface of a plate with heat transfer between fluids flowing around the plate. *High Temperature*, **10**, 293–298.

104. Dorfman, A..S. and Vishnevskii, V.K. (1972) Approximate solution of dynamic and thermal boundary layer equations for non-Newtonian fluids with arbitrary pressure gradients and surface temperature. *Int. Chem. Engin.*, **12**, 288–294.

105. Dorfman, A.S. (1975) Temperature-distribution singularities on the separation surface during heat transfer between a plate and the liquid flowing around it. *High Temperature*, **13** (1), 97–100.

106. Dorfman, A.S. and Lipovetskaya, O.D. (1976) Heat transfer of arbitrary nonisothermic surface with gradient turbulent flow of an incompressible liquid within a wide range of Prandtl and Reynolds numbers. *High Temperature*, **14** (1), 86–92.

107. Dorfman, A.S. and Lipovetskaya, O.D. (1976) Heat transfer to an isothermal flat plate in turbulent flow of a liquid over a wide range of Prandtl and Reynolds numbers. *J. Applied Mechan.Tecnic. Physics*, **17** (4), 530–535.

108. Dorfman, A.S. and Davydenko, B.V. (1980) Conjugate heat exchange for flows past elliptical cylinders. *High Temperature*, **18** (2), 275–280.

109. Dorfman, A.S. and Novikov, V.G. (1980) Heat transfer from a continuously moving surface to surroundings. *High Temperature*, **18**, 898–901.

110. Dorfman, A.S., Grechannyy, O.A. and Novikov, V.G. (1981) Conjugate heat transfer problem for a moving continuous plate in fluid flow. *High Temperature*, **19** (5), 706–714.

111. Dorfman, A.S. (1982) *Heat Transfer in Flow around Nonisothermal Bodies (in Russian)*, Mashinostroenie, Moscow.

112. Dorfman, A.S. (1982) Exact solution of the thermal boundary layer equation for an arbitrary heat flux distribution on a surface. *High Temperature*, **20** (4), 567–574.

113. Dorfman, A.S. (1983) Methods of estimation of coefficients of heat transfer from nonisotheramal walls. *Heat Transfer- Soviet Research*, **15** (6), 35–57.

114. Dorfman, A.S. (1984) Influence of turbulent Prandtl number on heat transfer of a flat plate. *J. Applied Mechan. Tech. Physics*, **25** (4), 572–575.

115. Dorfman, A.S. (1986) Combined heat transfer over the initial segment of a plate in a flow. *Heat Transfer-Soviet Research*, **18**, 52–74.

116. Dorfman, A.S. (1988) Solution of certain problems of optimizing the heat transfer in flow over bodies. *Appl. Therm. Sci.*, **1** (2), 25–34.

117. Dorfman, A.S. (1995) Exact solution of nonsteady thermal boundary layer equation. *ASME J. Heat Transfer*, **117**, 770–772.

118. Dorfman, A.S. (2004) Transient heat transfer between a semi-infinite hot plate and a flowing cooling liquid film. *ASME J. Heat Transfer*, **126** (2), 149–154.

119. Dorfman, A.S. (2009) *Conjugate Problems in Convective Heat Transfer*, CRC Press Taylor & Francis, Boca Raton, Fl.

120. Dorfman, A.S. (2011) Universal functions in boundary layer theory (Review). *Fundam. J. Therm. Sci. Engin.*, **1**, 35–72.

121. Dorfman, A.S. (2013) *Classical and Modern Engineering Methods in Fluid Flow and Heat Transfer*, Momentum Press.

122. Dunin, I.L. and Ivanov, V.V. (1974) Conjugate heat transfer problem with surface radiation taken into account. *Fluid Dynamic*, **9** (4), 667–670.

123. Eckert, E.R.G. and Drake, R.M. (1959) *Heat and Mass Transfer*, McGrew-Hill.

124. Eckstein, E.C. (1970) Experimental and theoretical pressure studies of peristaltic pumping, in *S, M. Thesis. Dep. of Mech. Eng. M. I. T*, Cambridge, Mass.

125. Ede, A.J. (1967) Advances in free convection, in *Advances in Heat Transfer*, vol. **4**, Acad. Press, pp. 1–64.

126. Ellahi, R., Riaz, A. and Nadeem, S. (2014) Three-dimensional peristaltic flow of a Williamson fluid in a rectangular channel having compliant walls. *J. Mech, Med. Biology*, **14** (1), 1450002.

127. El Qarnia, H. (2004) Theoretical study of transient response of a rectangular latent heat thermal energy storage system with conjugate forced convection. *Energy Conver. Maneg.*, **45**, 1537–1551.

128. El-Sayed, M.F., Fldabe, N.T.M., Ghaly, A.Y. and Sayed, H.M. (2011) Effects of chemical reaction, heat, and mass transfer on non-Newtonian fluid flow through porous medium in vertical peristaltic tube. *Transp. Porous Med.*, **89**, 185–212.

129. Erdelyi, A. (ed.) (1954) *Tables of Integral Transforms*, vol. **1**, McGraw-Hill, New York.

130. Eytan, O. and Elad, D. (1999) Analysis of intra-uterine fluid motion induced by uterine contractions. *Bull. Math. Biol.*, **61**, 221–238.

131. Eytan, O., Jaffa, A.J. and Elad, D. (2001) Peristaltic flow in a tapered channel: application to embryo transport within the uterine cavity. *Med. Engin. Physics*, **23** (7), 473–482.

132. Eytan, O., Zaretsky, U., Jaffa, A.J. and Elad, D. (2007) In vitro simulations of embryo transfer in a laboratory model of the uterus. *J. Biomech.*, **40** (5), 1073–1080.

133. Favre, A. (1965) Equations des gaz turbulents compressibles. *J. de Mecanique*, **4** (N 3), 361–390.

134. Favre, T. and Efraimsson, G. (2010) Detached eddy simulation of the effects of different wind gust models on the unsteady aerodynamic of road vehicles. *Proceedings ASME Fluid Engin. Divis. Summer Conf. PTS A-C*, **1**, 2605–2614.

135. Fedorov, A.G. and Viskanta, R. (2000) Three-dimensional conjugate heat transfer in the microchannel heat sink for electronic packaging. *Int. J. Heat Mass Transfer*, **43** (3), 409–415.

136. Fedorovich, E.D. (1959) Heat transfer to a flat plate streamlined by a turbulent boundary layer of incompressible fluid with Pr ≪ 1. *J. Engin. Phys. Thermod*, **2**, 3–11.

137. Fogelson, A.L. and Neevels, K.B. (2015) Fluid mechanics of blood clot formation, Ed. by Davis, S.H. and Moin, P., *Ann. Rev. Fluid Mech.* **47**: 377–403.

138. Forsythe, J.R. and Woodson, S.H. (2005) Unsteady computations of abrupt wing stall using detached eddy simulation. *J. Aircraft*, **42** (3), 606–616.

139. Fung, Y.C. and Yih, C.S. (1968) Peristaltic transport. *Trns. ASME, J. Appl. Mech.*, **35**, 669–675.

140. Fung, Y.C. (1971) *Peristaltic pumping: Bioengineering model, in* Urodynamics: Hydrodynamics of The Ureter And Renal Pelvis, vol. **15** (eds S. Boyarsky, G.W. Gottshalk, E.A. Tanagho and P.D. Zimsjind), pp. 177–198.

141. Gao, T., Wang, Z. and Vanden-Broeck, J.M. (2016) New hydroelastic solitary waves in deep water and their dynamics. *J. Fluid Mech.*, **788**, 469–491.

142. Garzo, V., Fillmer, W.D., Hrenya, C.M. and Xiaolong, Y. (2016) Transport coefficients of solid particles immersed in a viscous gas. *Phys. Rev.*, **93**, 012905.

143. Germano, M., Piomelli, U., Moin, P. and Cabot, W.H. (1991) A dynamic subgrid-scale eddy viscosity model. *Physics Fluid*, **3** (7), 1760–1765.

144. Ghalichi, F., Deng, X., De Champlain, A. *et al.* (1998) Low Reynolds number turbulence modeling of blood flow in arterial stenoses. *Biorheology*, **35**, 281–294.

145. Gibbings, J.C. (2003) Diffusion of the intermittency across the boundary layer in transition. *Proceeding of the institution of Mechanical Engineering, part C: J. Mech. Engin. Sci.* 217/12/1339.

146. Ginevskii, A.S. (1969) *Theory of Turbulent Jets and Wakes, (in Russian)*, Mashinostroenie.

147. http://web2.clarkson.edu/proects/cred/me537/download/02 Pastsphere,pdf

148. Grechannyy, O.A., Nagolkina, Z.I. and Senatos, V.A. (1984) Heat transfer in jet flow over an arbitrary nonisotermal wall. *Heat Transfer –Soviet Research*, **16**, 12–22.

149. Grechannyy, O.A., Dorfman, A.S. and Gorobets, V.G. (1986) Coupled heat transfer and effectiveness of flat finned surfaces in a transverse flow. *High Temperature*, **24** (5), 678–683.

150. Grechannyy, O.A., Dolinsky, A.A. and Dorfman, A.S. (1987) Conjugate heat and mass transfer in continuous processes of the convective drying of thin bodies (in Russian). *Prom. Teplotekhn.*, **9** (4), 27–37.

151. Grechannyy, O.A., Dolinsky, A.A. and Dorfman, A.S. (1988) Flow, heat and mass transfer in the boundary layer on a continuously moving porous sheet. *Heat Transfer- Soviet Research*, **20** (1), 52–64.

152. Grechannyy, O.A., Dolinsky, A.A. and Dorfman, A.S. (1988) Effect of nonuniform distribution of temperature and concentration differences on heat and mass transfer from and to a continuously moving porous plate. *Heat Transfer- Soviet Research*, **20** (3), 355–368.

153. Grober, H., Erk, S. and Grigull, U. (1955) *Die Grengesetze der Wärmeübetragung*, 3 edn, Springer, Berlin.

154. Grotberg, J.B. and Jensen, O.E. (2004) Biofluid mechanics in flexible tubes. *Annu. Rev. Fluid Mech.*, **36**, 121–147.

155. Guedes, R.O.C., Ozisik, M.N. and Cotta, R.M. (1994) Conjugated periodic turbulent forced convection in a parallel plate channel. *ASME J. Heat Transfer*, **116** (1), 40–46.

156. Hajmohammadi, M.R. and Nourazar, S.S. (2014) Conjugate forced convection heat transfer from a heated flat plate of finite thickness and temperature-dependent thermal conductivity. *J. Heat Transfer Engin.*, **35** (9), 863–874.

157. Hakeem, A.E. and Naby, A.E. (2009) Creeping flow of Phan-Thien-Tanner fluids in peristaltic tube with an infinite long wavelength. *J. Appl. Mech. Trans. ASME*, **76**, 064504.

158. Hanin, M. (1968) The flow through a channel due to transversely oscillating walls. *Israel J. Thechnol.*, **6**, 67–71.

159. Haroun, M.H. (2006) On non-linear magnetohydrodynamic flow due to peristaltic transport of an Oldroyd 3-constant fluid. *Zeitschrift für Naturforschung, Section A-A J. Physical Science*, **61**, 263–274.

160. Hassani-Ardekani, H., Ghalichi, F., Niroomand-Oscuii, H. *et al.* (2012) Comparison of blood flow velocity through the internal carotid artery based on Doppler ultrasound and numerical simulation. *Australasian Phys. Engin. Sci. Med.*, **35**, 413–422.

161. Hattori, H., Houra, T. and Nagano, Y. (2007) Direct numerical simulation of stable and unstable turbulent thermal boundary layer. *Int. J. Heat Flow*, **28**, 1262–1271.

162. Hayat, T., Qureshi, M.U. and Hussain, Q. (2009) Effect of heat transfer on the peristaltic flow of an electrical conducting fluid in porous space. *Appl. Math. Model*, **33**, 1862–1873.

163. Hayday, A.A., Bowlus, D.A. and Mcgraw, R.A. (1967) Free convection from a vertical plate with step discontinuities in surface temperature. *Int. J. Heat Mass Transfer*, **89**, 244–250.

164. He, M., Bishop, P.J., Kassab, A.J. and Minardi, A. (1995) A coupled FDM/BEM solution for the conjugate heat transfer problem, *Num. J. Heat Transfer part B* **28** (2): 139–154.

165. Heidmann, J., Rigby, D. and Ameri, A. (2002) A three-dimensional coupled external/internal simulation of a film-cooled turbine vane. *ASME J. Turbomach.*, **122**, 348–359.

166. Hermeth, S., Staffelbach, G., Gicquel, L.Y.M. and Poinsot, T. (2013) LES evaluation of the effects of equivalence ratio fluctuations on the dynamic flame response in a real gas turbine combustion chamber. *Proceeding of the Combustion Institute*, **34**, 3165–3173.

167. Hikman, H.J. (1974) An asymptotic study of the Nusselt-Graetz problem, part I: large x behavior. *J. Heat Transfer*, **96**, 354–358.

168. Horvat, A., Mavko, B. and Catton, I. (2004) The Galerkin method solution of the conjugate heat transfer, *Proceeding of the ASME-ZSIS, Int. Thermal Science Seminar II, 3-Thermal Science*, Bled, Slovenia, June 13–16.

169. aHutch, J.A. (1967) Visco-uretral reflux, *The Ureter*, Ed. by Bergman pp. 465–507.

170. Ikram, Z., Avital, E.J. and Williams, J.J.R. (2012) Detached eddy simulation of free-surface flow around a submerged submarine fairwater. *J. Fluid Engin. Trtransact. ASME*, **134**, 061103–11.

171. Imtiaz, H. and Mahfouz, F.M. (2014) Conjugate heat transfer within a concentric annulus filled with micropolar fluid. *J. Heat Mass Transfer*, **50** (4), 457–468.

172. Incropera, F.P. and Dewitt, D.P. (1996) *Fundamental Heat and Mass Transfer*, Fourth edn, Wiley& Sons, New York.

173. Ishibashi, Y. and Miyaji, K. (2015) Detached eddy simulation of synthetic jets for high-angles-of-attack airfoils. *J. Aircraft*, **52** (1), 168–175.

174. Ishimoto, K. and Gaffney, E.A. (2014) A study of spermatozoa swimming stability near a surface. *J. Theor. Biology*, **360**, 187–199.

175. Issa, R.I., Gosman, A.D. and Watkinc, A.P. (1986) Computation of compressible and incompressible recirculating flows by a non-iterative implicit scheme, *J. Comput. Phys.* **62**: 66–82.

176. Ito, Y., Inokura, N. and Nagasaki, T. (2014) Conjugate heat transfer in air-to-refrigerant airfoil heat exchangers, *J. Heat Transfer-Transactions ASME* **136** (8): article 081703.

177. Jadidi, M., Bazdidi- Tehrani, F. and Kiamansouri, M. (2016) Dynamic sub-grid turbulent Schmidt number approach in large eddy simulation of dispersion around an isolated cubical building. *Building Simul.*, **9**, 183–200.

178. Jaffrin, M.Y. and Shapiro, A.N. (1971) Peristaltic pumping. *Annu. Rev. Fluid Mech.*, **3**, 13–37.

179. Jaluria, Y. (1992) Transport from continuously moving materials undergoing thermal processing, *Annu. Rev. Heat Mass Transfer* **4**, chapter 4, ed. by C.L. Tien, Hemisphere Corp., Taylor and Francis Group, Washington.

180. Javed, N., Hayat, T. and Alsaedi, A. (2014) Peristaltic flow of Burgers' fluid with compliant wall and heat transfer. *Applied Math. Comput.*, **244**, 654–671.

181. Jeong, O.C. and Konishi, S. (2008) The self-generated peristaltic motion of cascaded pneumatic actuators for micro pumps. *J. Micromech. Microeng.*, **18**, 085017.

182. Ji, B., Luo, X.W., Arndt, R.E.A. *et al.* (2015) Large eddy simulation and theoretical investigations of the transient cavitating vortical flow structure around a NACA 66 hydrofoil. *Int. J. Multiphase Flow*, **68**, 121–134.

183. Jians, Y., Zheng, Q., Yue, G. *et al.* (2014) Conjugate heat transfer simulation of turbine blade high efficiency cooling method with mist injection. *Proccedings of the Institution of Mechanical Engineering, part C- J. Mech. Engin. Sci.*, **228** (15), 2738–2749.
184. Jimenez-Lozano, J., Sen, M. and Dunn, P.F. (2009) Particle motion in unsteady two-dimensional peristaltic flow with application to the ureter. *Phys. Rev. E*, **79**, 041901.
185. Johnson, D.A. and King, L.S. (1985) A mathematically simple turbulence closure model for attached and separated turbulent boundary layers. *AIAA J.*, **23** (11), 1684–1692.
186. Jones, T. (1969) Blood flow. *Annu. Rev. Fluid Mech.*, **1**, 223–244.
187. Joukar, A., Nammakie, E. and Niroomand-Osccuii, H. (2015) A comparative study of thermal effects of 3 types of laser in eye: 3D simulation with bioheat equation. *J. Thermal Biology*, **49–50**, 74–81.
188. Joyce, G. and Soliman, H.M. (2009) Analysis of the transient single-phase thermal performance of micro-channel heat sinks. *Heat Transfer Engin.*, **30**, 1058–1067.
189. Kang, B.H., Jaluria, Y. and Karve, M.V. (1991) Numerical simulation of conjugate transport from a continuous moving plate in materials processing. *Num. Heat Transfer*, **19** (2), 151–176.
190. Kanna, P.R. and Das, M.K. (2007) Conjugate heat transfer steady of a two-dimensional laminar incompressible wall jet over a back-ward-facing step. *ASME. J. Heat Transfer*, **129**, 220–231.
191. Kanna, P.R. and Das, M.K., (2009) Effect of geometry on the conjugate heat transfer of wall jet flow over a backward-facing step, *J. Heat Transfer* **131**: 114501-1-7.
192. Kanna, P.R., Taler, J., Anbumalar, V. *et al.* (2015) Conjugate heat transfer sudden expansion using nanofluid. *Num. Heat Transfer, part A*, **67** (1), 75–99.
193. Kant, R., Singh, H., Nayak, M. and Bhattacharya, S. (2013) Optimization of design and characterization of a novel micro-pumping system with peristaltic motion. *Microsyst. Technol.*, **19**, 563–575.
194. Karadimou, D.P. and Markatos, N.C. (2016) Modeling of two-phase, transient airflow and particles distribution in the indoor environment by large eddy simulation. *J. Turbul.*, **17** (2), 216–236.
195. Karniadakis, G.E. and Orszag, S.A. (1993) Nodes, modes and flow codes. *Physics Today*, **46** (3), 34–42.
196. Karwe, M.V. and Jaluria, Y. (1990) Numerical simulation of fluid flow and heat transfer in a single screw extruder for non-Newtonian fluids. *Num. Heat Transfer, part A*, **17**, 167–190.
197. Kassab, A., Divo, E., Heidmann, J. *et al.* (2003) BEM/FVM Conjugate heat transfer analysis of a three-dimensional film cooled turbine blade. *Int. J. Num. Methods Heat Fluid Flow*, **13** (5), 581–610.
198. Kawamura, F., Seki, Y., Iwamoto, K. and Kawamura, H. (2007) DNS of heat transfer in turbulent and transitional channel flow obstructed by rectangular prisms. *Int. J. Heat Flow*, **28**, 1291–1301.
199. Kawamura, H., Abe, H. and Matsuo, Y. (1999) DNS of turbulent heat transfer in channel flow with respect to Reynolds and Prandtl numbers effects. *Int. J. Heat Fluid Flow*, **20**, 196–207.
200. Kawano, K., Minakami, K., Iwasaki, H. and Ishizuka, M. (1998) Development of microchannels heat exchanging, In *Application of Heat Transfer in Equipment, Systems and Education*, Ed. by R.A. Nelson, Jr., L., W. Swanson, M.V.A., Bianchi and C. Camci, HTD-Vol. 361-3/PID-Vol. 3, pp. 173–180, ASME, New York.
201. Kays, W.M. (1980) *Convective Heat and Mass Transfer*, McGraw-Hill, New York.
202. Kestin, J. and Richardson, P.D. (1963) Heat transfer across turbulent incompressible boundary layer. *Int. J. Heat Mass Transfer*, **6** (6), 147–189.
203. Khan, A.A., Ellahi, R., Gulzar, M.M. and Sheikhleslami, M. (2014) Effects of heat transfer on peristaltic motion of Oldroyd fluid in presence of inclined magnetic field. *J. Magnetism, Magnetism Materials*, **372**, 97–106.
204. Kim, K., Baek, S.J. and Sung, H.J. (2002) An implicit velocity decoupling procedure for the incompressible Navier-Stokes equations. *Int. J. Numer. Methods Fluids*, **38**, 125–138.
205. Kim, S., Wilson, P.A. and Chen, Z.M. (2015) Large eddy simulation of the turbulent near wake behind circular cylinder: Reynolds number effect. *Appl. Ocean Research*, **49**, 1–8.

206. Klebanoff, P.S. (1956) *Characteristics of turbulence in a boundary layer with zero pressure gradient*, NACA TN, p. 3178.
207. Kuchumov, A.G., Gilev, V., Popov, V. *et al.* (2014) Non-Newtonian flow of pathological bile in the binary system: experimental investigation and CFD simulation. *Korea-Australia Rheol. J.*, **26** (1), 81–90.
208. Kutateladze, S.S. and Leontev, A.I. (1972) *Heat and Mass Transfer and Friction in Turbulent Boundary Layer (in Russion)*, Energiya Press, Mockow.
209. Kutateladze, S.S. (1973) *Near-Wall Turbulence, (in Russian)*, Nauka, Novosibirsk.
210. Kuznetsov, G.V. and Sheremet, M.A. (2009) Numerical modeling of temperature fields in the elements and units of electronic system. *Microelectronica*, **38**, 344–352.
211. Lai, H.X., Zhang, H.B. and Yan, Y.Y. (2004) Numerical steady of heat and mass transfer in rising inert bubbles using a conjugate flow model. *Num. Heat Transfer part A*, **46** (1), 79–98.
212. Lamnatou, C., Papanicolaou, E., Belessiotis, V. and Kyriakis, N. (2009) Conjugate heat and mass transfer from a drying rectangular cylinder in confined flow. *Num. Heat Mass Transfer, part A*, **56**, 379–405.
213. Laraufie, R. and Deck, S. (2013) Assessment of Reynolds stresses tensor reconstruction methods for synthetic turbulent inflow conditions., Application to hybrid RANS/LES methods. *Int. J. Heat Fluid Flow*, **42**, 68–78.
214. Lathman, T.W. (1966) *Fluid motion in a peristaltic pump*, S. M. Thesis, M. I. T, Mass, Cambridge.
215. Laufer, J. (1951) *Investigation of Turbulent Flow in a Two Dimensional Channel*, NACA, Rep. No. 1053.
216. Laufer, J. (1952) *The Structure of Turbulence in Fully Developed Pipe Flow*, NACA, Rep. No.1174.
217. Lawal, A., Kalyon, D.M. and Yilmazer, U. (1993) Extrusion and lubrication flows of viscoplstic fluids with wall slip. *Chem. Engin. Com.*, **122**, 127–150.
218. Lee, J.H. and Sung, H.J. (2011) Direct numerical simulation of a turbulent boundary layer up to $Re_\theta = 2560$. *Int. J. Heat Fluid Flow*, **32**, 1–10.
219. Lee, K.T. and Yan, W.M. (1993) Transient conjugated forced convection heat transfer with fully developed laminar flow in pipes. *Num. Heat Transfer*, **23** (3), 341–359.
220. Lee, T.S., Liao, W. and Low, H.T. (2001) Development of an artificial compressibility methodology with implicit LU-SGS method, *Int. J. Comp. Fluid Dyn.* **15**: 197–208.
221. Lee, T.S., Liao, W. and Low, H.T. (2003) Numerical simulation of turbulent flow through series stenoses. *Int. J. Num. Meth. Fluid*, **42**, 717–740.
222. Lee, W.C. and Ju, Y.H. (1986) Conjugate Leveque solution for Newtonian fluid in a parallel plate channel. *Int. J. Heat Mass Transfer*, **29**, 941–947.
223. Leonard, A. (1974) Energy cascade in large eddy simulation of turbulent fluid flow. *Advan. Geophys. A*, **18**, 237–248.
224. Leontev, A.I., Mikhin, V.A., Mironov, B.P. and Ivakin, V.P. (1968) *Effect of boundary condition on development of turbulent thermal boundary layer (in Russian), in* Teplo i Massoperenos, vol. **1**, Energiya, Moscow, pp. 125–132.
225. Leontev, A.I., Shishov, E.V., Belov, V.M. and Afanas'ev, V.N. (1977) Mean and fluctuating characteristics of thermal turbulent boundary layer and heat transfer in a diffuser, In: *Teplomassoobmen-V* (Heat and Mass Transfer-V [Proceedings of the 5$^\text{th}$ All-Union Conference on Heat and Mass Transfer, 1, part. 1, [engl. trans., *Heat Transfer-Soviet Research* 9: 48–56], Minsk.
226. Le Pape, A., Richez, F. and Deck, S. (2013) Zonal detached eddy simulation of airfoil in poststall condition. *AIAA J.*, **51** (8), 1919–1931.
227. Levich, V.G. (1962) *Physicochemical Hydrodynamics*, Prentice-Hall.
228. Li, C.H. (1970) Peristaltic transport in circular cylindrical tubes. *J. Biomech.*, **3**, 513–523.
229. Li, M.J. and Brasseur, J.G. (1993) Nonsteady peristaltic transport in finite-length tubes. *J. Fluid Mech.*, **248**, 129–151.
230. Lighthill, M.J. (1950) Contribution to the Theory of Heat Transfer Through a Laminar Boundary Layer, *Proc. Roy. Soc. A*, vol. **202**, London, pp. 359–377.

231. Lilly, D.K. (1966) On the application of the eddy viscosity concept in the inertial subrange of turbulence, NCAR Manuscript **123**.

232. Lin, P. and Jaluria, Y. (1997) Conjugate transport in polymer melt flow through extrusion dies. *Polymer Engin. Sci.*, **37** (9), 1582–1596.

233. Lin, P. and Jaluria, Y. (1998) Conjugate thermal transport in the channel of an extruder for non-Newtonian fluids. *Int. J. Heat Mass Transfer*, **41** (21), 3239–3253.

234. Lin, T.F. and Kuo, J.C. (1988) Transient conjugated heat transfer in fully developed laminar pipe flows. *Int. J. Heat Mass Transfer*, **31** (5), 1093–1102.

235. Lindstedt, M. and Karvinen, R. (2013) Conjugate heat transfer in a plate-one surface at constant temperature and the other cooled by forced or natural convection. *Int. J. Heat Mass Transfer*, **66**, 489–495.

236. Linge, F., Hye, M.A. and Paul, M.C. (2014) Pulsatile spiral plod flow through arterial stenosis. *Comput. Metod Biomech. Biomed. Engin.*, **17**, 1727–1737.

237. Liu, J.Y., Minkowycz, W.J. and Cheng, P. (1986) Conjugated mixed convection-conduction heat transfer along a cylindrical fin in a porous medium. *Int. J. Heat Mass Transfer*, **29**, 769–775.

238. Liu, J., Sun, H.S., Liu, Z.T. and Xiao, Z.X. (2014) Numerical investigation of unsteady vortex breakdown past 80 degrees/65 degrees double-delta wing. *Chinese J. Aeronautics*, **27** (3), 521–530.

239. Lock, G.S.H. and Ko, R.S. (1973) Coupling through a wall between two free convective systems. *Int. J. Heat Mass Transfer*, **16** (11), 2087–2096.

240. Londhe, S.D. and Rao, C.G. (2014) Interaction of surface radiation with mixed convection from a vertical channel with multiple discrete heat sources. *J. Heat Mass Transfer*, **50** (9), 1275–1290.

241. Love, G. (1957) An approximate solution of the laminar heat transfer along a plate with arbitrary distribution of the surface temperature. *J. Aeronaut. Sci.*, **24**, 920–921.

242. Loytsyanskiy, L.G. (1962) *The Laminar Boundary Layer (in Russian)*, Fismatgiz Press, Moskow [English translation, 1966].

243. Ludwieg, H. (1956) Bestimmung des verhaltnisses der austauschkoeffizienten fur warme and impuls bei turbulenten grenzschichten. *ZFW*, **4**, 73–81.

244. Luikov, A.V. (1966) *Heat and Mass Transfer in Capillary-Porous Bodies*, Pergamon Press, Oxford.

245. Luikov, A.V. (1968) *Analytical Heat Diffusion Theory*, Academ. Press, New York.

246. Luikov, A.V., Perelman, T.L., Levitin, R.S. and Gdalevich, L.B. (1970) Heat transfer from a plate in a compressible gas flow. *Int. J. Heat Mass Transfer*, **13** (8), 1261–1270.

247. Luikov, A.V., Aleksashenko, V.A. and Aleksashenko, A.A. (1971) Analytical methods of solution of conjugated problems in convective heat transfer. *Int. J. Heat Mass Transfer*, **14**, 1047–1056.

248. Lund, T., Wu, X. and Squires, D. (1998) Generation of turbulent inflow data for spatially developing boundary layer simulations. *J. Comput. Phys.*, **140**, 233–258.

249. Luong, M.B., Luo, Z., Lu, T. *et al.* (2013) Direct numerical simulations of the ignition of lean primary reference fuel/air mixtures with temperature inhomogeneities. *Comb. Flame*, **160**, 2038–2047.

250. Lykoudis, P.S. (1969) The ureter as a peristaltic pump, *Presented at the Workshop on Hydrodynamics of the Upper Urinary Tract*, Univ. of Chicago, Chicago, Illinois, pp. 24–66, Oktober.

251. Maiti, S. and Misra, J.C. (2011) Peristaltic flow of a fluid in a porous channel: A study having relevance to flow of bile within duct in pathological state. *Int. J. Energ. Sci.*, **49**, 950–966.

252. Mansour, M.K. (2014) Effect of natural convection on conjugate heat transfer characteristics in liquid minichannel during phase change material melting. *Proceedings Institution Mech. Engin., part C-J. Mech. Engn. Sci.*, **228** (3), 491–513.

253. Martyushev, S.G. and Sheremet, M.A. (2013) Surface radiation influences on the regimes of conjugate natural convection in an enclosure with local energy source. *J. Thermophys. Aeromech.*, **20** (4), 417–428.

254. Masmoudi, W. and Prat, M. (1991) Heat and mass transfer between a porous medium and parallel external flow. *Int. J. Heat Mass Transfer*, **34**, 1975–1989.

255. Matheou, G. and Chung, D. (2014) Large eddy simulation of stratified turbulence, par. 11: Application of the stratched-vortex model to the atmospheric boundary layer. *J. Atmosph. Sci.*, **71**, 45–66.

256. McGee, H.A. (1991) *Molecular Engineering*, Mc Graw Hill Press, New York.

257. Meginniss, J.R. (1970) An analytic investigation of flow and hemolysis in peristaltic-type blood pumps, in *S. M. Thesis, M. I. T*, Mass, Cambridge.

258. Mekheimer, K.S., Haroun, M.H. and Elkot, M.A. (2011) Effects of magnetic field, porosity and wall properties for anisotropically elastic multi-stenosis arteries on blood flow characteristics. *Appl. Math. Mech.-Engl. Ed.*, **32**, 1047–1064.

259. Mekheimer, K.S. and Abd Elmabond, Y. (2014) Simultaneous effects of variable viscocity and thermal conductivity on peristaltic flow in a vertical asymmetric channel. *Canadien J. Phys.*, **92** (12), 1541–1555.

260. Mellor, G.L. (1966) Effects of pressure gradients on turbulent flow near a smooth wall. *J. Fluid Mech.*, **24** (2), 255–274.

261. Mellor, G.L. and Gibson, D.M. (1966) Equilibrium turbulent boundary layer. *J. Fluid Mech.*, **24** (2), 225–253.

262. Meneveau, C., Lund, T.S. and Cabot, W.H. (1966) A Langrangian dynamic subgrid-scale model of turbulence. *J. Fluid Mech.*, **319** (1), 353–385.

263. Mhaisekar, A., Kazmierczak, M.J. and Banerjee, R.K. (2005) Steady conjugate heat transfer from X-ray or laser-heated sphere in external flow at low Reynolds number. *Num. Heat Transfer part A*, **47**, 849–874.

264. Miftakhof, R. and Ahkmadeev, N. (2007) Dynamics of intestinal propulsion. *J. Theor. Biology*, **246** (2), 377–393.

265. Miftakhov, R.N. and Wingate, D.L. (1994) Biomechanics of a small bowel motility. *Med. Eng. Phys.*, **16**, 406–416.

266. Mingalev, S.V., Lubimov, D.V. and Lubimova, T.P. (2010) Pressure-driven peristaltic flow, *4th France-Russia Conf. on New Achievement Materials and Environmental,* Okt. 26–29, also in Book Series: *J. Phys. Conf. Ser.* vol. **416**, 012029, 2013.

267. Mironov, B.P., Vasechkin, V.N. and Yarugina, N.I. (1977) Effect of an upstream adiabatic zone on heat transfer in a subsonic and supersonic downstream boundary layer at different flow histories, In: *Teplomassoobmen –V (Heat and Mass Transfer-V [Proceedings of the 5th All-Union Conference on Heat and Mass Transfer])* **Pt.1**: 67–97. Minsk, [Engl. Transl.1977. *Heat Transfer-Soviet Research* 9: 57–65].

268. Misra, J.C., Sinha, A. and Shit, G.C. (2010) Flow of a biomagnetic viscoelastic fluid: application to estimation of blood flow in arteries during electromagnetic hyperthermia, a therapeutic procedure for cancer treatment. *Appl. Math. Mech. Engl. Ed.*, **31**, 1405–1420.

269. Misra, J.C. and Maiti, S. (2012) Peristaltic pumping of blood through small vessels of varying cross-section. *ASME J. Appl. Mech.*, **79** (6), 061003.

270. Mittra, T.K. and Prasad, S.N. (1973) On the influence of wall properties and Poiseuille flow in peristaltic. *Biomech.*, **6**, 681–693.

271. Mohammad, K. (1987) *Conjugated heat transfer from a radiating fluid in a rectangular channel,* Ph. D. Theses Akron Univ, OH.

272. Moin, P. and Mahesh, K. (1998) Direct numerical simulation: A tool in turbulence research. *Annu. Rev. Fluid Mech.*, **30**, 539–578.

273. Mokhtar, A.A.E. and Haroun, M.H. (2008) A new model for study the effect of wall properties on peristaltic transport of a viscous fluid. *Commun. Nonlin Sci. Num. Simul.*, **13**, 752–762.

274. Monin, A.S. and Yaglom, A.M. (1971) *Statistical Fluid Mechanics*, vol. I (ed. J. Lumley).

275. Moretti, P.M. and Kays, W.M. (1965) Heat transfer to a turbulent boundary layer with varying free stream velocity and varying surface temperature-an experimental study. *Int. J. Heat and Mass*, **8**, 1187–1202.

276. Moukalled, F. and Acharya, S. (1989) Improvements to incompressible flow calculation on a non-staggered curvilinear grids. *Numer. Heat Transfer part B*, **15**, 131–152.

277. Mousazadeh, F., van Den Akker, H.E.A. and Mudder, F. (2013) Direct numerical simulation of an exothermic gas-phase reaction in a packed bed with random particle distribution. *Chem. Engin. Sci.*, **100**, 259–265.

278. Mustafa, M., Nina, S., Hayat, T. and Alsaedi, A. (2012) Influence of wall properties on the peristaltic flow of a nanofluid: analytic and numerical solutions. *Int. J. Heat Mass Transfer*, **55** (17–18), 4871–4877.

279. Muthu, P., Kumar, B.V.R. and Chandra, P. (2008) Peristaltic motion of micropolar fluid in circular cylindrical tube: effect of wall properties. *Appl. Math. Model*, **32** (10), 2019–2033.

280. Nam, J.H. and Song, C.S. (2007) Numerical simulation of conjugate heat and mass transfer during multi-dimensional freeze drying of slab-shaped food products. *Int. J. Heat Mass Transfer*, **50**, 4891–4900.

281. Nguyen, H.D. and Chung, J.N. (1992) Conjugate heat transfer from a translating drop in an electric field at low Peclet number. *Int. J. Heat Mass Transfer*, **35** (2), 443–456.

282. Niceno, B. and Sharabi, M. (2013) Large eddy simulation of turbulent heat transfer at supercritical pressures. *Nucl. Engin. Design*, **261**, 44–55.

283. Nicoud, F. and Ducros, F. (1999) Subgrid-scale modeling based on the square of the velocity gradient tensor. *Flow Turbul. Combust.*, **62**, 183–200.

284. Nompelis, I., Drayna, T. and Candler, G.V. (2005) A parallel unstructured implicit solver for hypersonic reacting flow simulation, *AIAA Paper 2005-4867*, June 2005.

285. Nowak, A.J., Biaecki, R.A., Fic, A. *et al.* (2002) Coupling of conductive, convective, and radiative heat transfer in Czochralski crystal growith process. *Comput. Material. Sci.*, **25** (4), 570–576.

286. Nowak, A.J., Biaecki, R.A., Fic, A. and Wecel, G. (2003) Analysis of fluid and energy transport in Czochralski process. *Comput. Fluid*, **32**, 85–95.

287. Nuxoll, E. (2013) BoiMEMSin drug delivery. *Advan. Drug Delivery Rev.*, **65** (11–12), 1611–1625.

288. Olek, S., Elias, E., Wacholder, E. and Kaizerman, S. (1991) Unsteady conjugated heat transfer in laminar pipe flow. *Int. J. Heat Mass Transfer*, **34** (6), 1443–1450.

289. Oliver, D.L.R. and Chung, J.N. (1987) Flow about a fluid sphere at low to moderate Reynolds number. *J. Fluid Mech.*, **177**, 1–18.

290. Oliver, D.L.R. and Chung, J.N. (1990) Unsteady conjugate heat transfer from a translating fluid sphere at moderate Reynolds numbers. *Int. J. Heat Mass Transfer*, **33**, 401–408.

291. Oliveira, L.S. and Haghighi, K. (1998) Conjugate heat and mass transfer in convective drying of porous media. *Num. Heat Transfer, part A*, **34** (2), 105–117.

292. Opheim, L.N. and Lund, W. (1977) Use of peristaltic mini-pumps automatic-analysis. *Analytica Chem. Acta*, **90**, 245–247.

293. Osada, H., Tsunoda, I., Matsuurra, M. *et al.* (1999) Investigation of ovum transport in the oviduct: the dynamics of oviductal fluids in domestic rabbits. *J. Int. Med. Research*, **27** (4), 176–180.

294. Osman, F., Romics, I., Nyirady, P. *et al.*, (2009) Ureteral motility, *Acta Physiologica Hungarica* **96** (4): 407–426.

295. Ozisik, M.N. (1958) *Boundary Value Problem of Heat Conduction*, Int. Textbook Company, Seranton, Pennsylvania.

296. Pagliarini, G. and Borozzi, G.S. (1984) Thermal coupling in laminar double stream heat exchangers, in *Proc. Second National Conf. Heat Transfer*, Bologna, Italy, pp. 103–113.

297. Pagliarini, G. and Barozzi, G.S. (1991) Thermal coupling in laminar flow double-pipe heat exchangers. *ASME J. Heat Transfer*, **113** (3), 526–534.

298. Pan, J.P. and Loth, E. (2005) Detached eddy simulations for iced airfoils. *J. Aircraft*, **42** (6), 1452–1461.

299. Pandey, S.K. and Chaube, M.K. (2011) Peristaltic transport of a Maxwell fluid in a channel of varying cross section induced by asymmetric waves: application to embryo transport within uterine cavity. *J. Mech. Med. Biology*, **11** (3), 675–690.

300. Pandey, S.K. and Chaube, M.K. (2011) Study of wall properties on peristaltic transport of a couple stress fluid. *Meccanica*, **46**, 1319–1330.

301. Panov, D. (1951) *Handbook of Numerical Treatment of Partial Differential Equations* 5th ed.(in Russian) Isd. ANCCR, Moscow.

302. Papoutsakis, E. and Ramkrishna, D. (1981) Conjugated Graetz problems. *Chem. Engng. Sci.*, **36** (8), 1381–1391.

303. Park, S.C., Park, N.S., Kim, D.G. *et al.* (2014) Physical properties of covered stent in gastric acid environment: in vitro study. *Polymer-Korea*, **38** (3), 351–357.

304. Patankar, S.V. and Spalding, D.B. (1970) *Heat and Mass Transfer in Boundary Layers*, 2d edn, Intertex, London.

305. Patankar, S.V. and Spalding, D.B. (1972) A calculation procedure for heat, mass and momentum transfer in three-dimensional parabolic flows. *Int. J. Heat Mass Transfer*, **15**, 1787–1806.

306. Patankar, S.V. (1980) *Numerical Heat Transfer and Fluid Flow*, Taylor & Francis Boca Raton Fl.

307. Perelman, T.L., Levitin, R.S., Gdalevich, L.B. and Khusid, B.M. (1972) Unsteady- state conjugated heat transfer between a semi-infinite surface and incoming flow of a compressible fluid- II. Determination of a temperature field and analysis of result. *Int. J. Heat Mass Transfer*, **15**, 2563–2573.

308. Petukhov, B.S., Detlaf, A.A. and Kirilov, V.V. (1954) Experimental investigation of local heat transfer from a plate to subsonic turbulent air flow. *J. Engin. Phys. Thermod.*, **24**, 1761–1772.

309. Piro, M.H. and Leitch, B.W. (2014) Conjugate heat transfer simulations of advanced research reactor fuels. *Nuclear Engin. Design*, **274**, 30–43.

310. Polyanin, A.D. and Manzhirov, A.V. (1998) *Handbook of Integral Equations*, CRC Press, Boca Raton.

311. Pomeranzev, A.A. (1960) Heating a wall of the plate by a supersonic flow (in Russian). *J. Engin. Physics Thermodyn.*, **3** (8), 39–46.

312. Popovac, M. and Hanjalic, K. (2009) Vortices and heat flux around a wall-mounted cube cooled simultaneously by a jet and a crossflow. *Int. J. Heat Mass Transfer*, **52**, 4047–4062.

313. Pozzi, A. and Lupo, M. (1989) The coupling of conduction with forced convection in a plane duct. *Int. J. Heat Mass Transfer*, **32** (7), 1215–1221.

314. Premachandran, B. and Balaji, C. (2006) Conjugate mixed convection with surface radiation from a horizontal channel with protruding heat sources. *Int. J. Heat Mass Transfer*, **49**, 3568–3582.

315. Qin, W.J., Xie, M.Z., Jia, M. *et al.* (2014) Large eddy simulation of in-cylinder turbulent flows in a DISI gasoline engine. *Appl. Math. Model*, **38** (24), 5967–5985.

316. Raithby, G.D. and Hollands, K.G.T. (1975) A general method of obtaining approximate solution to laminar and turbulent natural convection problems, In *Advances in Heat and Mass Transfer*, ed. by T.F. Irvine, Jr., and J.P. Hartnett **11**: 265–315, Academic Press, New York.

317. Ramis, M.K. and Jilani, G. (2009) Numerical study of a nuclear fuel element dissipating fission heat into its surrounding fluid medium. *Int. J. Heat Mass Transfer*, **52**, 5005–5012.

318. Rao, R. and Venkateshan, S.P. (1996) Experimental study on free convection and radiation in horizontal fin arrays. *Int. J. Heat Mass Transfer*, **39**, 779–789.

319. Rao, V.D., Naidu, S.V., Rao, B.G. and Sharma, K.V. (2006) Heat transfer from a horizontal fin array by natural convection and radiation –a conjugate analysis. *Int. J. Heat Mass Transfer*, **49**, 3379–3391.

320. Reviznikov, D.L. (1995) Coefficients of nonisothermicity in the problem of unsteady - state conjugate heat transfer on the surface on the blunt bodies. *High Temperature*, **33**, 259–264.

321. Reynolds, W.C., Kays, W.M. and Kline, S.T. (1960) A summery of experiments on turbulent heat transfer from nonisothermal flat plate. *Tras. ASME, J. Heat Transfer ser. C*, **4**, 341–348.

322. Richtmyer, R.D. and Morton, K.W. (1967) *Difference Methods for Initial Value Problems*, Interscience Publ, New York.

323. Riera, W., Castillon, L., Marty, J. and Leboeuf, F. (2014) Inlet conditions effects on the tip clearance flow with zonal detached eddy simulation. *J. Turbomach. Transac. ASME*, **136** (4), 041018.

324. Rizzetta, D.P. and Visbal, M.R. (2009) Large eddy simulation of plasma-based turbulent boundary layer separation control, *AIAA 39 Fluid Dyn. Conf. Exib. Locat.* San Antonio, TX, Jun 22–25, 2009.

325. Robertson, G.E., Seinfeld, J.H. and Leal, L.G. (1973) Combined forced and free convection flow past a horizontal flat plate. *AIChE J.*, **19**, 998–1008.
326. Rosenbluth, K.H., Luz, M., Mohr, E. *et al.* (2011) Design an in-dwelling cannula for convection-enhanced delivery. *J. Neuroscien. Methods*, **196** (1), 118–123.
327. Rotta, J.C. (1962) Turbulent boundary layer in incompressible flow. *Progress in Aerospace Sci.*, **2**, 1.
328. Roy, R., Rios, F. and Riahi, D.N. (2011) Mathematical models for flow of chime during gastrointestinal endoscopy. *Appl. Math.*, **2**, 600–607.
329. Rubtsov, N.A., Timofeev, A.M. and Ponomarev, N.N. (1987) On behavior of transfer coefficients in direct differential methods of theory of radiative heat transfer in scattering media. *Izv. SO SSSR Ser. Tekhn. Nauka*, **18**, 3–8.
330. Rubtsov, N.A. and Timofeev, A.M. (1990) Unsteady conjugate problem of radiative-convective heat transfer in a laminar boundary layer on a thin plate. *Num. Heat Transfer*, **17**, 127–142.
331. Rubtsov, N.A. and Sinitsyn, V.A. (2004) Unsteady radiative-convective heat transfer in a flow emitting, absorbing and scattering medium around an ablating plate. *J. Appl. Mech. Technik. Phys.*, **45**, 415–419.
332. Rumsey, C.B., Carter, H.,.S., Hastings, E.C. *et al.* (1969) Initial result from flight measurements of turbulent heat transfer and boundary-layer transition at local Mach numbers near 15 (Reentry F). *NASATM X-1856*.
333. Sakiadis, B.C. (1961) Boundary layer behavior on a continuous solid surface, *AJChE J.* **7**: part 1: 26–28, part 2: 221–225.
334. Sastrohartono, T., Jaluria, Y. and Karve, M.V. (1994) Numerical coupling of multiple-region simulations to study transport in a twin-screw extruder. *Num. Heat Transfer part A*, **25** (5), 541–557.
335. Schlatter, P., Orlu, R., Brethouwer, G. *et al.* (2009) Turbulent boundary layer up to $Re_\theta = 2500$ studied through simulation and experiment. *Phys. Fluids*, **21**, 051702.
336. Schlatter, P., Brethouer, G., Li, Q. *et al.* (2010) Simulations of spatially evolving turbulent boundary layer up to $Re_\theta = 4300$. *Int. J. Heat Fluid Flow*, **31** (3), 251–261.
337. Schley, D., Carare-Nnadi, R., Please, C.P. *et al.* (2006) Mechanism to explain the reverse perivascular transport of solutes out of the brain. *J. Theor. Biol.*, **238**, 962–974.
338. Schlichting, H. (1951, 1968, 1979, 2000) *Boundary layer Theory*, McGraw-Hill, New York. Citations without year pertains to 1979 edition.
339. Schonfeld, T. and Rudgyard, M. (1999) Steady and unsteady flow simulation using hybrid solver AVBP. *AIAA, J.*, **37** (11), 1378–1385.
340. Schwertfirm, F. and Manhart, M. (2007) DNS of passive scalar transport in turbulent channel flow at high Schmidt numbers. *Int. J. Heat Flow*, **28**, 1204–1214.
341. Selverov, A.M. and Stone, H.A. (2001) Perestaltically driven channel flows with applications toward micromuxing. *Phisics Fluids*, **13**, 1837–1860.
342. Semba, T., Fujll, K. and Fujll, Y. (1970) Influence of peristaltic contraction of stomach on blood flow through gastrosplenic vein. *Hiroshima J. Med. Sci.*, **19** (20), 87–97.
343. Shah, R.K. and London, A.L. (1978) *Laminar Flow Forced Convection in Ducts*, Academic Press, New York.
344. Shams, A., Roelofs, F., Komen, E.M.J. and Baglietto, E. (2013) Large eddy simulation of a nuclear pebble bed configuration. *Nucl. Engin.Design*, **261**, 10–19.
345. Shapiro, A.H., Jaffrin, M.Y. and Weinberg, S.L. (1969) Peristaltic pumping with long wavelengths at low Reynolds number. *J. Fluid Mech.*, **37**, 799–825.
346. Shapiro, A.H. and Jaffrin, M.Y. (1971) Reflux in peristaltic pumping: is it determined by the Eulerian or Langrangian mean velocity. *J. Appl. Mech.*, **38** (4), 1060–1062.
347. Sharma, A.K., Velusamy, K., Balaji, C. and Venkateshan, S.P. (2007) Conjugate turbulent natural convection with surface radiation in air filled rectangular enclosure. *Int. J. Heat Mass Transfer*, **50**, 625–639.

348. Shcwing, A.M. and Candler, G.V. (2015) Detached eddy simulation of capsule wake flows and comparison to wind-tunnel test data. *J. Spacecraft Rockets*, **52** (2), 439–449.

349. Sheremet, M.A. and Trifonova, T.A. (2013) Unsteady conjugate natural convection in a vertical cylinder partially filled with a porous medium. *Num. Heat Transfer, part A*, **64** (12), 994–1015.

350. Shugan, I.V., Smirnov, N.N. and Legros, J.C. (2002) Streaming flows in a channel with elastic walls. *Phys. Fluids*, **14**, 3502–3511.

351. Shulman, Z.P. and Berkovskii, B.M. (1966) *Boundary Layer of non-Newtonian Fluids (in Russian)*, Nauka i Technika, Minsk.

352. Shur, M.L., Spalart, P.R., Strelets, M.K. and Travin, A.K. (2008) A hybrid RANS-LES approach with delayed-DES and wall-modeled LES capabilities. *Int. J. Heat Fluid Flow*, **29**, 1638–1649.

353. Shur, M.L., Spalart, P.R., Strelets, M.K. and Travin, A.K. (2011) A rapid and accurate switch from RANS to LES in boundary layers using an overlap region. *Flow Turbl. Comb.*, **86**, 179–206.

354. Shvets, Y.I., Dorfman, A.S. and Didenko, O.I. (1975) Heat transfer between two countercurrently flowing fluids separated by a thin wall. *Heat Transfer- Soviet Research*, **7** (4), 32–39.

355. Shvets, Y.I., Dorfman, A.S. and Didenko, O.I. (1975) Some characteristics of heat transfer between two moving fluids separated by a wall containing heat sources. *J. Heat Transfer-Soviet Research*, **7**, 25–31.

356. Simon, Y.H.O. (2004) A 3D turbulent conjugate heat- transfer model for freezing of food products. *J. Food Sci.*, **69**, 224–231.

357. Sivasubramanian, J. and Fasel, H.F. (2015) Direct numerical simulation of transition in a sharp cone boundary layer at Mach 6: fundamental breakdown. *J. Fluid Mech.*, **768**, 175–218.

358. Smagorinsky, J. (1963) General circulation experiments with the primitive equations. *Mon. Weather Rev.*, **91** (3), 99–164.

359. Smith, J.H. and Humphrey, J.A. (2007) Interstitial transport and transvascular fluid exhange during infusion into brain and tumor tissue. *Microvas. Resear.*, **73** (1), 58–73.

360. Sohal, M.S. and Howell, I.R. (1973) Determination of plate temperature in case of combined conduction, convection and radiation heat exchange. *Int. J. Heat Mass Transfer*, **16** (11), 2055–2066.

361. Sokolova, I.N. (1957) Plate temperature streamlined by supersonic flow (in Russian), in *Theoretical Aerodynamic Investigations*, ZAGI, Oborongis, Moscow, pp. 206–221.

362. Solopov, V.A. (1972) Heat transfer in turbulent boundary layer incompressible fluid with varying pressure gradient and varying surface temperature. *Izv. AN SSSR, Mech. Zhidcosti i Gasa*, **5**, 166–167.

363. Song, W.H. and Li, B.Q. (2002) Finite element solution of conjugate heat transfer problems with and without the use of gap elements, *Int. J. Num. Methods Heat Fluid Flow* **12** : 81–99.

364. Song, W.H. and Lichtenberg, J. (2005) Thermo-pneumatic, single-stroke miropump. *J. Micromech. Microeng.*, **15**, 1425–1432.

365. Spalart, P.R., Jou, W.H., Strelets, M. and Allmaras, S.R. (1997) *Comments on the feasibility of LES for wings, and on a hybrid RANS/LES approach* First AFOSR Int. Conf. on DNS/LES, *Ruston, LA*, Advenc. DNS/LES Greyden Press, Columbus. OH.

366. Spalart, P.R. (2000) Strategies for turbulence modeling and simulations. *Int. J. Heat Fluid Flow*, **21**, 252–263.

367. Spalart, P.R., Deck, S., Shur, M. *et al.* (2006) A new version of detached-eddy simulation, resistant to ambiguous drid densities. *Theor. Comput. Fluid Dyn.*, **20**, 181–195.

368. Spalart, P.R. (2009) Detached-eddy simulation. *Annual Review Fluid Mech.*, **41**, 181–202.

369. Spalding, D.B. and Pun, W.M. (1962) A review of methods for predicting heat transfer coefficients for laminar uniform-property boundary layer flows. *Int. J. Heat Mass Transfer*, **5**, 239–244.

370. Sparrow, E.M. (1958) Combined effects of unsteady flight velocity and surface temperature on heat transfer. *Jet Propulsion*, **28**, 403–405.

371. Sparrow, E.M. and Faghri, M. (1980) Fluid-to- fluid conjugate heat transfer for a vertical pipe-internal forced convection and external natural convection. *ASME J. Heat Transfer*, **102** (3), 402–407.

372. Squires, K.D. (2004) Detached-eddy simulation: current status and perspectives, in *Direct and Large-Eddy Simulation V* (eds R. Friedrich and B. Geurts), Oliver Metais, pp. 465–481.

373. Srinivasulu, C. and Radhakrihnamacharya, G. (2002) Peristaltic transport in non-uniform channel with elastic effects. *Proc. Nat. Sci. India*, **72**, 279–288.

374. Srivastava, V.P. and Srivastava, L.M. (1997) Influence of wall elasticity and Poiseuille flow on peristaltic induced flow of a particle-fluid mixture. *Int. J. Engin. Sci.*, **35** (15), 1359–1386.

375. Srivastava, V.P. and Srivastava, R. (2009) Particulate suspension blood flow through a narrow catheterized artery. *Comput. Math. Applic.*, **58** (2), 227–238.

376. Starner, K.E. and McManus, H.N. Jr. (1963) An experimental investigation of free convection heat transfer from rectangular fin arrays. *ASME J. Heat Transfer*, **85**, 273–278.

377. Stein, C.F., Johansson, P.B., Bergh, J. *et al.* (2002) An analytical asymptotic solution to a conjugate heat transfer problem. *Int. J. Heat Mass Transfer*, **45** (12), 2485–2500.

378. Sucec, J. (1987) Unsteady conjugated forced convective heat transfer in a duct with convection from the ambient. *Int. J. Heat Mass Transfer*, **30** (9), 1963–1970.

379. Sugioka, K. and Tsukada, T. (2015) Direct numerical simulation of drag and lift forces acting on a spherical bubble near a plane wall. *Int. J. Multiphase Flow*, **71**, 32–37.

380. Sugita, R., Sugimura, E., Itoh, M. *et al.* (2003) Pseudolesion of the bile duct caused by flow effect: A diagnostic pitfall of MR cholangiopancreatography. *American J. Roentgenology*, **180**, 467–471.

381. Sunden, B. (1979) A coupled conduction-convection problem at low Reynolds number flow, *Num. Meth, Thermal Problems, Proceeding of the First International Conference*, pp. 412–422, Swansea, Wales, July 2–6.

382. Takabatake, S. and Ayukawa, K. (1982) Numerical study of two-dimensional peristaltic flows. *J. Fluid Mech*, **122**, 439–465.

383. Takagi, D. and Balmforth, N.J. (2011) Peristaltic pumping of rigid object in an elastic tube. *J. Fluid Mech.*, **672**, 219–244.

384. Tang, D. and Rankin, S. (1993) Numerical and asymptotic solutions for peristaltic motion of non-linear viscous flows with elastic free boundaries, *SIAM J. Sci. Comput.* **14** (6): 1300–1319.

385. Tao, Z., Wu, H.W., Chen, G.H. and Deng, H.Y. (2005) Numerical simulation of conjugate heat and mass transfer processes within cylindrical porous media with cylindrical dielectric cores in microware freeze-drying. *Int. J. Heat Mass Transfer*, **48**, 561–572.

386. Tejeda-Martinez, A.E. and Jansen, K.E. (2004) A dynamic Smagorinsky model with dynamic determination of the filter width ratio, https:/www.scorec.rpi.edu/REPORTS/ 2004-4.pdf.

387. Tenchev, R.T., Li, L.Y. and Purkiss, J.A. (2001) Finite element analysis of coupled heat and moisture transfer in concrete subjected to fire. *Num. Heat Transfer, part A*, **39** (7), 685–710.

388. Thomas, T.G. and Williams, J.J.R. (1999) Simulation of skewed turbulent flow past a surface mounted cube. *J. Wind Eng. Ind. Aerodyn.*, **81** (1–4), 347–360.

389. Thornber, B. and Drikakis, D. (2007) Large eddy simulation of shock-wave-induced turbulent mixing. *J. Fluid Engin. Transact. ASME*, **129** (12), 1504–1513.

390. Timoshenko, S. (1974) *Vibration Problems in Engineering*, 4th edn, Wiley, New York.

391. Tong, P. and Vawter, D. (1972) An analysis of peristaltic pumping. *ASME J. Appl. Mech.*, **39**, 857–862.

392. Townsend, A.A. (1976) *The Structure of Turbulent Shear Flow*, Second edn, Cambridge University Press, Cambridge.

393. Trp, A. (2005) An experimental and numerical investigation of heat transfer during technical grade paraffin melting and solidification in a shell-and-tube latent thermal energy storage unit. *Solar Energy*, **79**, 648–660.

394. Tseng, Y.H. and Ferziger, J.H. (2003) A ghost-cell immersed boundary method for flow in complex geometry. *J. Comput. Phys.*, **192** (2), 593–623.

395. Tsou, F.K., Sparrow, E.M. and Goldstein, R.J. (1967) Flow and heat transfer in the boundary layer on a continuous moving surface. *Int. J. Heat mass transfer*, **10**, 219–235.

396. Uehara, Y. and Burnstock, G. (1970) Demonstration of "Gap Junctions" between smooth muscle cells. *J. Cell Biology*, **44** (1), 215–217.

397. Vahidi, B., Fatouraee, N., Imanparast, A. and Moghadam, A.N. (2011) A mathematical simulation of the ureter: effects of the model parameters on ureteral pressure/flow relations. *J. Biomech. Eng. Transact. ASME*, **133** (031004), 9.

398. Vahidi, B. and Fatourace, N. (2012) A biomechanical simulation of uretral flow during peristalsis using intraluminal morphometric data. *J. Theor. Biology*, **298**, 42–50.

399. Vajravelu, K., Sreenadh, S. and Saravana, R. (2013) Combined influence of velocity slip, temperature and concentration jump conditions on MHD peristaltic transport of a Carreau fluid in a nonuniform channel. *Appl. Math. Comput.*, **225**, 656–676.

400. Vajravelu, K., Sreenadh, S., Sucharitha, G. and Lakshminarayana, P. (2014) Peristaltic transport of a conducting Jeffrey fluid in an inclined asymmetric channel. *Int. J. Biomath.*, **7** (6), 1450064.

401. Van Belleghem, M., De Backer, L., Janssens, A. and De Paepe, M (2012) Conjugate modeling of convective drying phenomena in porous building materials, 6th *European Thermal Sci. Conf.*, France, Sept. 4–7, Book Series: *J. of Phys. Conf. Series*, **395**, article 012142.

402. Van Doormaal, J.P. and Raithby, G.D. (1984) Enhancements of the SIMPLE method for predicting incompressible fluid flows. *Numer. Heat Transfer*, **7**, 147–163.

403. Van Driest, E.R. (1956) On turbulent flow near a wall. *J. Aeron. Sci.*, **23**, 1007–1011.

404. Van Dyke, M.D. (1964) *Perturbation Methods in Fluid Mechanics*, Academic Press, New York.

405. Van Dyke, M.D. (1965) *A Method of Series Truncation Applied to Some Problem in Fluid Mechanics*, Stanford University, report SUDAER 247

406. Van Nimwegen, A.T., Schutte, K.C.J. and Portela, L.M. (2015) Direct numerical simulation of turbulent flow in pipes with an arbitrary roughness topography using a combined momentum-mass source immersed boundary method. *Comput. Fluid*, **108**, 92–111.

407. Vaszi, A.Z., Elliott, L.., Ingham, D.B. and Pop, I. (2004) Conjugate free convection from a vertical plate fin with a rounded tip embedded in a porous medium. *Int. J. Heat Mass Transfer*, **47**, 2785–2794.

408. Viskanta, R.V. and Abrams, M. (1971) Thermal interaction of two streams in boundary layer flow separated by a plate. *Int. J. Heat Mass Transfer*, **14** (9), 1311–1321.

409. Viskanta, R. and Lankford, D.W. (1981) Coupling of heat transfer between two natural convection systems separated by a vertical wall. *Int. J. Heat Mass Transfer*, **24** (7), 1171–1177.

410. Viswanath, R. and Jaluria, Y. (1993) Comparison of different solution methodologies for melting and solidification problems in enclosures. *Num. Heat Transfer part A*, **24**, 77–105.

411. Viswanath, R. and Jaluria, Y. (1995) Numerical study of conjugate transient solidification in an enclosed region. *Num. Heat Transfer part A*, **27** (5), 519–536.

412. Vynnycky, M., Kimura, S., Kaneva, K. and Pop, I. (1998) Forced convection heat transfer from a flat plate: the conjugate problem. *Int. J. Heat Mass Transfer*, **41** (1), 45–59.

413. Wang, H.S., Snan, F.L., Piao, Y. *et al.* (2015) IDDES simulation of hydrogen-fueled supersonic combustion using flamelet modeling. *Int. J. Hydrogen Energy*, **40** (1), 683–691.

414. Wang, P. and Olbricht, W.L. (2011) Fluid mechanics in the perivascular space. *J. Theor. Biol.*, **274**, 52–57.

415. Wang, Q.H. and Jaluria, Y. (2004) Three-dimensional conjugate heat transfer in a horizontal channel with discrete heating. *ASME J. Heat Transfer*, **126** (4), 642–647.

416. Wang, X.S., Dagan, Z. and Jlji, L.M. (1989) Conjugate heat transfer between a laminar impinging liquid jet and solid disk. *Int. J. Heat Mass Transfer*, **32** (11), 2189–2197.

417. Wansophark, N., Malatip, A. and Dechauphai, P. (2005) Streamline upwind finite element method of the conjugate heat transfer problems. *Acta Mechanica Sinica*, **21** (5), 436–443.

418. Wecel, G. (2006) BEM/FVM solution of the conjugate radioactive and convective heat transfer problems. *Arch. Comput. Mech. Engn.*, **13**, 171–248.

419. Weinberg, S.L. (1970) A theoretical and experimental treatment of peristaltic pumping and its relation to uretral function, *Ph. D. Thesis, M.I. T.*, Cambridge, Mass.

420. Weinberg, S.L., Eckstein, E.C. and Shapiro, A.H. (1971) An experimentally study of peristaltic pumping. *J. Fluid Mech.*, **49**, 461–479.

421. Wikipedia en. wikipedia.org/wiki/Direct_numerical_simulation

422. Wilcox, D.C. (1994, 2006) *Turbulence Modeling for CFD*, DCW Industries, Inc. La Canada, California.

423. Wu, X. and Moin, P. (2009) Direct numerical simulation of turbulence in a nominally zero-pressure-gradient flat-plate boundary layer. *J. Fluid Mech.*, **630**, 5–41.

424. Xiao, L.H., Xiao, Z.X., Duan, Z.W. and Fu, S. (2015) Improved delayed detached eddy simulation of cavity-induced transition in hypersonic boundary layer. *Int. J. Heat Fluid Flow*, **51**, 138–150.

425. Xiao, Z.X., Liu, J., Luo, K.Y. *et al.* (2013) Investigation of flows around a rudimentary landing gear with advanced detached eddy simulation approach. *AIAA J.*, **51** (1), 107–125.

426. Yamane, T. and Bilgen, E. (2004) Conjugate heat transfer in enclosures with openings for ventilation. *Int. J. Heat Mass Transfer*, **40** (5), 401–411.

427. Yan, P., Cui, Y., Shi, L. and Zhu, J. (2014) Application of conjugate heat transfer and fluid network analysis to improvement design of turbine blades with integrated cooling structures. *Proceeding of the Institution of Mechanical Engineering, part G-J. Aerospace Engin.*, **228** (12), 2286–2299.

428. Yan, W.M., Tsay, Y.L. and Lin, T.F. (1989) Transient conjugated heat transfer in laminar pipe flows. *Int. J. Heat Mass Transfer*, **32** (4), 775–777.

429. Yan, W.M. (1993) Transient conjugated heat transfer in channel flows with convection from the ambient. *Int. J. Heat Mass Transfer*, **36** (5), 1295–1301.

430. Yaniv, S., Jaffa, A.J., Eytan, O. and Elad, D. (2009) Simulation of embryo transport in a closed uterine cavity model. *Europ. J. Obst. Gynec. Rep. Biol.*, **144**, s50–s 60.

431. Yi, M., Bau, H.H. and Hu, H. (2002) Peristaltically induced motion in a closed cavity with two vibrating walls. *Phys. Fluids*, **14**, 184–197.

432. Yin, F. and Fung, Y.C., (1969) Peristaltic waves in circular cylindrical tubes, *J. appl. Mech.* **36** (3): 579–587.

433. Yoo, S.Y. and Jaluria, Y. (2007) Conjugate heat transfer in an optical fiber coating process. *Num. Heat Transfer, part A*, **51** (2), 109–127.

434. Yoo, S.Y. and Jaluria, Y. (2008) Numerical simulation of the meniscus in nonisothermal free surface flow at the exit of a coating die. *Num. Heat Transfer, part A*, **53**, 111–131.

435. Yoshimoto, S., Yamamoto, M. and Toda, K. (2007) Numerical calculations of pressure distribution in the bearing clearance of circular aerostatic thrust bearings with a single air supply inlet. *J. Tribology-Transact. ASME*, **129** (2), 384–390.

436. Yoshinoa, H., Fujii, M., Zhang, X., Takeuchia, T., Toyomasua, T., Conjugate heat transfer from an electronic module package cooled by air in a rectangular duct, http:/www.google.com/#hl=en&source=hp&q(on+&btnG=Google+Search&aq=f&aqi=&oq=&fp=aa7ac5834e645580).

437. Younci, R. and Kocaefe, D., (2007) Numerical and experimental validation of heat and mass transfer during heat treatment of wood, *Proceedings of the 18th IASTED International Conference: modeling and simulation*, pp. 477–482, Montreal, Canada,.

438. Yu, W.S. and Lin, H.T. (1993) Conjugate problems of conduction and free convection on vertical and horizontal flat plates. *Int. J. Heat Mass Transfer*, **36** (5), 1303–1313.

439. Yusoff, S., Mohamed, M., Ahmad, K.A. *et al.* (2009) 3-D conjugate heat transfer analysis of PLCC packages mounted in-line on a printed circuit board. *Int. Comm. Heat Mass Transfer*, **36**, 813–819.

440. Zhang, C., He, J., Zhu, Y. *et al.* (2015) Interface effects on the Kelvin wake of monohull ship represented via a continuous distribution of sources. *European Mech. J. B-Fluids*, **51**, 27–36.

441. Zhang, W. and Samtaney, R. (2016) Assessment of spanwise domain size effect on the transitional flow past an airfoil. *Comput. Fluid*, **124**, 39–53.

442. Zhang, X., Chen, Z. and Huang, Y. (2015) A wave-less microfluidic peristaltic method, *Biomicrofluidics* **9**: 014118. pressure fluctuation in aerostatic bearings, J. Fluid and Structures 40: 42–51.

443. Zhu, J., Chn, H. and Chen, X. (2013) Large eddy simulation of vortex shedding and pressure fluctuation in aerostatic bearing. *J. Fluid Structures*, **40**, 42–51.

444. Zhukauskas, A.A. and Shlanchyauskas, A.A. (1973) *Heat Transfer in Turbulent Flow of Liquids (in Russian)*, Mintas, Vilnyus.

445. Zhuzhgda, I.I. and Zhukauskas, A.A. (1962) Experimental investigation of heat transfer of a plate in a laminar flow of liquid. *Trudy Akad. Nauk Lit. SSR*, **ser. B 4**, 117–126.

446. Zien, T.F. and Ostrach, S. (1970) A long wave approximation to peristaltic motion. *J. Biomech.*, **3**, 63–75.

447. Zinchenko, V.I., Efimov, K.N. and Yakimov, A.S. (2007) Investigation of the characteristics of conjugate heat and mass transfer in spatial flow past a sphere-blunted cone and blowing-in of a gas from the surface of bluntness. *J. Engin. Phys. Thermophys.*, **80**, 751–759.

448. Zuriz, C., Singh, R.P., Moini, S.M. and Henderson, S.M. (1979) Desorption isotherms of rough rice from 10 to $40\,C°$. *Trans. ASAE, St. Joseph, Michigan*, **22** (2), 433–436.

Author Index

Applications of Mathematical Heat Transfer and Fluid Flow Models in Engineering and Medicine,
First Edition. Abram S. Dorfman.
© 2017 John Wiley & Sons Ltd. Published 2017 by John Wiley & Sons Ltd.

Subject Index

Applications of Mathematical Heat Transfer and Fluid Flow Models in Engineering and Medicine,
First Edition. Abram S. Dorfman.
© 2017 John Wiley & Sons Ltd. Published 2017 by John Wiley & Sons Ltd.